LEARN

LEARN PHYSICS WITH FUNCTIONAL PROGRAMMING

A Hands-on Guide to Exploring Physics with Haskell

by Scott N. Walck

San Francisco

Printed in the United States of America

First printing

26 25 24 23 22   1 2 3 4 5

ISBN-13: 978-1-7185-0166-9 (print)
ISBN-13: 978-1-7185-0167-6 (ebook)

Publisher: William Pollock
Managing Editor: Jill Franklin
Production Manager: Sabrina Plomitallo-González
Production Editor: Miles Bond
Developmental Editor: Alex Freed
Cover Illustrator: Gina Redman
Interior Design: Octopod Studios
Technical Reviewer: Gregory Wright
Copyeditor: George Hale
Proofreader: Bart Reed
Indexer: Sanjiv Kumar Sinha

For information on distribution, bulk sales, corporate sales, or translations, please contact No Starch Press, Inc. directly at info@nostarch.com or:

No Starch Press, Inc.
245 8th Street, San Francisco, CA 94103
phone: 1.415.863.9900
www.nostarch.com

Library of Congress Cataloging-in-Publication Data

```
Names: Walck, Scott N., author.
Title: Learn physics with functional programming : a hands-on guide to
   exploring physics with Haskell / by Scott N. Walck.
Description: San Francisco : No Starch Press, [2023] | Includes
   bibliographical references and index.
Identifiers: LCCN 2022018706 (print) | LCCN 2022018707 (ebook) | ISBN
   9781718501669 (print) | ISBN 9781718501676 (ebook)
Subjects: LCSH: Physics--Data processing. | Functional programming
   (Computer science)
Classification: LCC QC52 .W34 2023  (print) | LCC QC52  (ebook) | DDC
   530.0285/5133--dc23/eng20220722
LC record available at https://lccn.loc.gov/2022018706
LC ebook record available at https://lccn.loc.gov/2022018707
```

To Peggy, Carl, Dan, and Jodi

About the Author

Scott N. Walck has a PhD in physics from Lehigh University. He has taught physics, including computational physics, to undergraduates (physics majors and non-majors) for 20 years at Lebanon Valley College, where he has been recognized with a Distinguished Teaching Award. Walck is a three-time NSF grant recipient for research in quantum information and is the author of more than 30 peer-reviewed research articles in physics.

About the Technical Reviewer

Gregory Wright received his PhD in physics from Princeton University, where he built receivers for studying cosmic microwave background radiation. He did postdoctoral studies at Bell Laboratories and was later hired as a Member of Technical Staff there. He was introduced to Haskell when he was a Visiting Industrial Fellow at the University of California's Berkeley Wireless Research Center, where he used it to model wireless communication systems. For many years he maintained MacPort's version of the Glasgow Haskell Compiler (GHC) and contributed to porting GHC to the AMD64 architecture on both macOS and FreeBSD. He was (probably) the first person to build GHC from source on the continent of Antarctica and encourages others to build their tools from source.

BRIEF CONTENTS

PART III: EXPRESSING ELECTROMAGNETIC THEORY AND SOLVING PROBLEMS

CONTENTS IN DETAIL

3 TYPES AND ENTITIES 25

4 DESCRIBING MOTION 35

5 WORKING WITH LISTS 51

9
TUPLES AND TYPE CONSTRUCTORS 113

10
DESCRIBING MOTION IN THREE DIMENSIONS 129

11
CREATING GRAPHS 165

12
CREATING STAND-ALONE PROGRAMS 175

13
CREATING 2D AND 3D ANIMATIONS 187

PART II
EXPRESSING NEWTONIAN MECHANICS AND SOLVING PROBLEMS

14
NEWTON'S SECOND LAW AND DIFFERENTIAL EQUATIONS 205

15
MECHANICS IN ONE DIMENSION 243

18
A VERY SHORT PRIMER ON RELATIVITY 329

19
INTERACTING PARTICLES 339

20
SPRINGS, BILLIARD BALLS, AND A GUITAR STRING 363

PART III
EXPRESSING ELECTROMAGNETIC THEORY AND SOLVING PROBLEMS

23
CURVES, SURFACES, AND VOLUMES 449

24
ELECTRIC CHARGE 461

25
ELECTRIC FIELD 473

29 THE MAXWELL EQUATIONS 551

APPENDIX: INSTALLING HASKELL 581

BIBLIOGRAPHY 595

INDEX 597

ACKNOWLEDGMENTS

A book is never a solo endeavor, and I am grateful to have the opportunity to thank some of the many people who helped me write this book and helped me become the person who could write this book.

I thank Beall Fowler for teaching me physics, Lester Rubenfeld for showing me what mathematics could really be, and Gerald Sussman for conversations about physics and functional programming. I thank Scott Pilzer for being my friend and conversation partner about the most interesting parts of physics.

Lebanon Valley College supported me by granting a sabbatical leave to work on this book.

I thank my colleagues David Lyons, Mike Day, Barry Hurst, Keith Veenhuizen, and Dan Pitonyak for interesting discussions on a number of issues addressed in the book. I thank Brent Yorgey for helping me learn Haskell, and David Thrane Christiansen for conversations about functional programming and publishing.

I thank Alex Heilman, Ben Lippmeier, Greg Horn, and John Kearney for reading sections of the manuscript and providing valuable comments. Nick Durofchalk and Michel Malda helped with Windows installation issues. Ben Gordon gave helpful feedback. I thank Gregory Wright for being the technical reviewer. He looked carefully at the exercises, ran the code, and came up with a number of ideas for making the book better.

I thank Bill Pollock for believing in this project enough to put real resources behind it. I am grateful for the work of Alex Freed, my developmental editor at No Starch. Alex saw the possibility of coolness in using a functional programming language to write about physics and was the principal shepherd of this project through publication, at least from the perspective of the author. She quickly pinpointed the essential weaknesses in my writing, and made substantial edits that improved the exposition without changing the sense of what I was trying to convey.

I am grateful to my medical caregivers, especially Witold Rybka, Seema Naik, Giampaolo Talamo, Mitzi Lowe, and all the nurses at the Penn State Cancer Institute, who kept me alive and well enough to keep working until the project was finished.

My family of origin was instrumental in producing the person who would write this book. My father, B. Roger Walck, sparked my interest in mathematics and computers. My mother, Judy Walck, gave me, and still gives me, unconditional love. My sister, Janice Stout, attempted to teach me to dress with style and appropriateness. The mathematics and unconditional love stuck; if I have had difficulty internalizing the lessons of fashion, the shortcoming is mine alone.

The family that eats, sleeps, lives, and works with me has been essential to my well-being. My wife, Peggy Bright, has steadfastly supported and encouraged me in the work of this book and my endeavors more broadly. My son Carl is a model for me in stepping out of his comfort zone. My son Dan inspires me with his commitment to physical exercise. My daughter Jodi has given me many gifts, including time programming together. All have shown patience, compassion, and good humor over the long course of time that I have been thinking and writing on this topic. I am grateful for their presence in my life.

INTRODUCTION

One of the best ways to learn something is to teach it. It is invaluable to have a person who is willing to listen to what we say, to read what we write, and to respond. Knowing that someone is listening or reading encourages us to spend time and effort creating something of quality. And if our writing incites a response, so much the better, for we have started a conversation that might challenge us to sharpen our understanding.

A pleasant and productive way to learn physics is to teach a computer how to do it. We admit up front that the computer is not as rich a listener as a person and cannot provide the depth or breadth of response to our writing that a person can. On the other hand, the computer is very attentive, willing to listen incessantly, and unwilling to accept statements unless they are expressed in clear language and make sense. The computer can provide us with a useful response because it will happily calculate what we ask it to calculate, and it will quickly tell us if what we just said makes no sense (and hopefully give us a clue about why it makes no sense).

This book is about learning basic theoretical physics by teaching a computer how to do it. We will spend a substantial amount of time with Newton's second law. We will focus on the concept of the *state* of a physical system and see that Newton's second law is the core rule for describing how the state changes in time. We will study basic electromagnetic theory, asking the computer to calculate electric and magnetic fields produced by charge and current distributions. The point is to deepen our understanding of physics by approaching it from a new angle, with a new language. The language we will use is precise and will help to clarify our thinking.

Who This Book Is For

This book arises from a course in computational physics I teach to second-year students of physics at Lebanon Valley College. I expect that you have had one year of introductory physics and at least one semester of calculus. No previous programming experience is required. The purpose of the book is to deepen your understanding of basic physics by exploring it in a new language. By using a formal language to express the ideas of physics, we will stretch our ability to formulate and communicate the ideas of physics as we also calculate quantities we are interested in and make graphs and animations.

Because the book begins with a self-contained introduction to the Haskell programming language for people who have not programmed before, it can be used as a supplement for introductory and intermediate courses in physics in which the instructor or student

- wishes to include a computational component, or
- desires a deeper understanding of the structure of basic physical theories.

The book is also appropriate for self-study by any student who wishes to deepen their understanding of physics by programming.

Why Functional Programming, and Why Haskell?

Many scientists, after learning their second programming language, develop the idea that all programming languages are more or less the same and that the difference between languages is mainly one of syntax. Scientists are busy people, and they have their work to do, so perhaps they can be excused for choosing not to dive into the sea of available programming languages to learn the more complex truth that languages can differ on a semantic level and can have profound effects on the way a person thinks about the problem they are writing code to solve. The style of programming called *functional programming* grows from a different branch of the programming language tree than object-oriented programming, and the two do not mix well together. Neither is clearly better for all applications.

Physics can be encoded in any programming language. Why use a functional language instead of a more mainstream object-oriented language?

Beauty and power are to be found more in verbs than in nouns. Newton found beauty and power not in the world per se, but in the description of how the world changes. Functional programming found beauty and power not in objects but in the functions that take objects as input and produce objects as output, and in the notion that such objects might themselves be functions. Haskell is a good programming language for learning physics for two reasons. First, Haskell is a functional programming language. This means that functions play a central role in the language, including functions that take other functions as arguments and return functions as results. Many physical ideas are naturally expressed in the language of higher-order functions. Second, Haskell's type system provides a clean way to organize our thinking about the physical quantities and procedures of interest in physics. I know of no better way to clarify my thinking than expressing my ideas in functional language.

About This Book

This book is composed of three parts. The first part is an introduction to functional programming in general and Haskell in particular, aimed at people who have never programmed before. The second part shows how to use a functional language to express Newton's second law, and consequently to solve mechanics problems. The third part aims at electromagnetic theory, showing how Faraday and Maxwell's ideas can be expressed in functional language, and how problems that involve electric and magnetic fields can be solved. Throughout, we'll see how functional language is close to mathematics; how it's really a form of mathematics that computers can understand. Many of the deep ideas of physics that are so eloquently and succinctly expressed in mathematical language find beautiful expression in functional language as well.

The book includes the following elements:

Part I: The Haskell Language

Chapter 1: Calculating with Haskell This chapter is all about how to use Haskell as a calculator. Basic mathematical operations are built into Haskell, and these are immediately available to do calculations.

Chapter 2: Writing Basic Functions Here we begin writing functions. Haskell functions are very much like mathematical functions. The simplest Haskell functions take a number as input and produce a number as output. As you might guess, functions play a central role in functional programming languages.

Chapter 3: Types and Entities This chapter introduces the idea of types. The entities, such as numbers and functions, with which Haskell deals are classified into types; every entity has a type. Types guide our thinking about what can be done with an entity. Real numbers can be squared, for example, but it doesn't always make sense to square a function.

Chapter 4: Describing Motion Here we look at how the motion of a particle in one dimension is described in Haskell. We introduce position, velocity, and acceleration, and we notice how these quantities are related by the notion of a derivative from calculus.

Chapter 5: Working with Lists This chapter discusses lists in Haskell. Lists can be lists of numbers, lists of functions, or lists of more complicated things. After functions, lists are probably the most important structures in functional programming because they are used in the process of iteration (doing something over and over again).

Chapter 6: Higher-Order Functions This chapter introduces higher-order functions, which are functions that take other functions as input or produce functions as output. Higher-order functions are central to the power and concision of functional languages. We give examples of how higher-order functions naturally appear in physics.

Chapter 7: Graphing Functions This chapter shows how to graph a function such as the cosine function or a function that you define that takes numbers as input and produces numbers as output.

Chapter 8: Type Classes Here we introduce type classes in Haskell. Type classes own functions that need to be able to work with some, but not all, types. Equality checking is such a function. We want to be able to check equality of numbers, equality of lists, and equality of other things. The equality checking function is owned by a type class.

Chapter 9: Tuples and Type Constructors This chapter introduces tuples, a structure that holds two or more objects. This chapter also discusses type constructors, which are functions at the type level (in other words, functions that take a type as input and produce a type as output).

Chapter 10: Describing Motion in Three Dimensions This chapter is similar in outlook to Chapter 4 in that it focuses on a particular need that physics has (in this case, a need for vectors) and shows how that need is satisfied in the Haskell language.

Chapter 11: Creating Graphs Here we return to the topic of making graphs, first broached in Chapter 7, and include more detail about how to make pleasing and informative graphs.

Chapter 12: Creating Stand-Alone Programs In the beginning of the book, we interact with Haskell primarily through the GHCi interactive compiler. Later in the book, when we start doing animation, we make stand-alone programs. This chapter shows several ways to produce stand-alone programs.

Chapter 13: Creating 2D and 3D Animations This chapter introduces animation, showing how to make simple 2D and 3D animations.

Part II: Newtonian Mechanics

Chapter 14: Newton's Second Law and Differential Equations Here we introduce Newton's first and second laws. We learn how to solve a limited class of mechanics problems in one spatial dimension. We also learn why some mechanics problems are easy to solve and others are difficult. It comes down to what the forces depend on. This chapter covers situations of increasing complexity, starting with constant forces and ending with forces that depend on time and the particle's velocity. This chapter introduces the concept of differential equations, and we write code capable of solving a first-order differential equation.

Chapter 15: Mechanics in One Dimension This chapter continues the path of increasing complexity, looking at forces that depend on time, position, and velocity. Such situations lead to a second-order differential equation, which we solve by introducing state variables.

Chapter 16: Mechanics in Three Dimensions Here we return to the vector setting first seen in Chapter 10, completing the theory of the mechanics of a single object. We show how to express and solve Newton's second law for a single particle in three dimensions.

Chapter 17: Satellite, Projectile, and Proton Motion This chapter gives three extended examples, applying the ideas and tools developed in the previous chapter.

Chapter 18: A Very Short Primer on Relativity This chapter shows what mechanics looks like if we embrace the ideas of special relativity in favor of those of Newton. We see that many of our tools survive the transition, enabling us to solve problems in relativity.

Chapter 19: Interacting Particles This chapter introduces Newton's third law, which is needed when we care about more than one object. We develop a theory of interacting particles and express the key ideas in Haskell.

Chapter 20: Springs, Billiard Balls, and a Guitar String This chapter gives three extended examples of interacting particles in which we use the ideas and tools of Chapter 19. Having dealt with the mechanics of arbitrarily many interacting particles in three dimensions, our treatment of Newtonian mechanics is complete.

Part III: Electromagnetic Theory

Chapter 21: Electricity This chapter looks at the old electric theory of Coulomb, in which electricity is simply a force on charged particles produced by other charged particles, similar in spirit to Newtonian gravity. Coulomb's electricity does not use the electric field.

Chapter 22: Coordinate Systems and Fields Here we introduce the key idea of a field, which is a function of space—a quantity that can have a different value at each position in space. This chapter also introduces Cartesian, cylindrical, and spherical coordinates for three-dimensional space.

Chapter 23: Curves, Surfaces, and Volumes This chapter discusses how we can describe curves, surfaces, and volumes in the Haskell language.

Chapter 24: Electric Charge This chapter covers electric charge, the quantity responsible for electrical phenomena, and the different kinds of charge distributions.

Chapter 25: Electric Field This chapter describes how electric charge produces an electric field, beginning our study of modern Faraday-Maxwell electromagnetic theory in which electric and magnetic fields play such a crucial role.

Chapter 26: Electric Current This chapter discusses electric current and current distributions, paralleling the discussion of charge in Chapter 24.

Chapter 27: Magnetic Field This chapter describes how current produces a magnetic field, paralleling Chapter 25 in that charge is to electric field as current is to magnetic field, at least in static situations.

Chapter 28: The Lorentz Force Law While Chapters 24 through 27 deal with the aspect of electromagnetic theory in which charge creates fields, this chapter discusses the second aspect of electromagnetic theory, in which fields exert forces on charge. The Lorentz force law describes this second aspect.

Chapter 29: The Maxwell Equations This chapter presents the Maxwell equations, in which the first aspect of electromagnetic theory reaches its full sophistication, and we see how electric and magnetic fields are dynamic quantities, interacting and changing in time. Although there are many situations and applications we won't discuss, the Maxwell equations and the Lorentz force law give a complete description of modern electromagnetic theory—a theory important not just for explaining electricity, magnetism, and light, but also for serving as the prototype for present-day theories of elementary particle physics.

Appendix: Installing Haskell This appendix shows how to install the Haskell compiler and software libraries we will use.

This book has been a labor of love, meaning that my motivation for writing it comes from a love of the ideas presented and a desire to share them. I hope that I have created a beautiful book, but even more than that, I hope that this book helps you to express beautiful ideas with beautiful code. Enjoy!

PART I

A HASKELL PRIMER FOR PHYSICISTS

1

CALCULATING WITH HASKELL

In this chapter, we'll see how Haskell can be used as a scientific calculator. Multiple scientific functions are available by default in this calculator. In Chapter 2, we'll write functions of our own that we can load and use. This chapter introduces some features and details of the language that will be useful later. Let's start with a kinematics problem.

A Kinematics Problem

Suppose we have a car on an air track. The car accelerates at a rate of 0.4 $\mathrm{m/s^2}$. At time $t = 0$, the car is stationary. How much time will it take for this car to travel 2 meters?

The acceleration is constant, so we can use the position-time equation.

$$x(t) = \frac{1}{2}a_0t^2 + v(0)t + x(0)$$

Because the car starts at rest, we know $v(0) = 0$. Let's assume that the car starts at position zero, the origin, so that $x(0) = 0$. We're looking for the time t at which $x(t) = 2$ m. A little algebra tells us that

$$t = \sqrt{\frac{2(2\text{ m})}{0.4\text{m}/\text{s}^2}}$$

We could solve this using pen and paper, or we could use a calculator. In this chapter, we'll be using Haskell as our calculator.

In the next section, we'll start a step-by-step explanation of how to use Haskell as a calculator. To close out this section, we'll just show what you'd enter to calculate the time.

```
Prelude> sqrt (2 * 2 / 0.4)
3.1622776601683795
```

As you can see, it takes about 3.2 seconds for the car to travel 2 meters. Now, let's look at how to start up the interactive Haskell "calculator."

The Interactive Compiler

The kinematics example gives us an excuse to introduce the interactive version of the Glasgow Haskell Compiler (GHC). The interactive version of GHC is called GHCi, and we can use it as a calculator. GHCi is included with the Glasgow Haskell Compiler, which is freely available at *https://www.haskell.org*. The method of starting GHCi may depend on the operating system your computer uses. Typically, you can click an icon, choose GHCi from a menu, or enter `ghci` in a command line.

When GHCi starts, we get a prompt at which we can enter expressions. The first prompt we get from GHCi is `Prelude>`. The Prelude is a collection of constants, functions, and operators available by default, and we can immediately use it to construct expressions. GHCi indicates that the Prelude has been loaded for us by including the name `Prelude` in the prompt. GHCi is now waiting for us to enter an expression. If you type `2/3`, followed by ENTER, GHCi will evaluate this expression and print a result.

```
Prelude> 2/3
0.6666666666666666
```

In the expression `2/3`, Haskell interprets the `2` and `3` as numbers and the `/` as a binary operator for division. GHCi performs the requested division and returns the result.

Numeric Functions

Haskell provides functions in the Prelude to perform many of the tasks you expect from calculators. Here's an example:

```
Prelude> log(2)
0.6931471805599453
```

This is the natural logarithm function applied to the number 2. The Haskell language does not need parentheses to apply a function. *Function application* (also known as *function use* or *function evaluation*) is such a basic idea in Haskell that the juxtaposition of two expressions is taken to mean that the first expression is a function and the second is an argument, and that the function is applied to the argument. Therefore, we can type the following:

```
Prelude> log 2
0.6931471805599453
```

Table 1-1 gives a list of numeric functions available in the Prelude.

Table 1-1: Some Common Numeric Functions

Function	Description
exp	exp $x = e^x$
sqrt	Square root
abs	Absolute value
log	Natural logarithm (log base e)
sin	Argument in radians
cos	Argument in radians
tan	Argument in radians
asin	Arcsine (inverse sine)
acos	Arccosine
atan	Arctangent
sinh	sinh $x = (e^x - e^{-x})/2$
cosh	cosh $x = (e^x + e^{-x})/2$
tanh	tanh $x = (e^x - e^{-x})/(e^x + e^{-x})$
asinh	Inverse hyperbolic sine
acosh	Inverse hyperbolic cosine
atanh	Inverse hyperbolic tangent

Haskell also provides the constant π in the Prelude.

```
Prelude> pi
3.141592653589793
```

Here is a trigonometric function:

```
Prelude> cos pi
-1.0
```

Notice that trigonometric functions in Haskell expect an argument in radians.

Now, let's calculate $\cos \frac{\pi}{2}$.

```
Prelude> cos pi/2
-0.5
```

The computer did not give us what we expect here; $\cos \frac{\pi}{2} = 0$, not -0.5. The reason is that function application in Haskell has higher *precedence* than division, so Haskell interprets what we typed as

```
Prelude> (cos pi)/2
-0.5
```

rather than dividing π by 2 first and then taking the cosine. We can get what we want by supplying parentheses.

```
Prelude> cos (pi/2)
6.123233995736766e-17
```

Is the result the computer gave $\cos \frac{\pi}{2}$? Not exactly. Here we see an example of an approximately computed result. My computer gave something times 10^{-17}, which is as close to zero as the computer can get here. It's good to remember that when doing numerical work, the computer (like your calculator) is not giving exact results most of the time. It is giving us approximate results. We must be vigilant in making sure that the results it gives are valuable to us by interpreting them correctly. After we discuss Haskell's system of types in Chapter 3; we'll say more about when we can and when we cannot expect the computer to produce exact results.

Operators

The Haskell Prelude provides several *binary operators*, shown in Table 1-2. Binary operators act on two inputs, or *arguments*, to produce a result. For example, the addition operator (+) is a binary operator because it takes two inputs and adds them. In computer science generally, an operator placed before its arguments is called a prefix operator, an operator placed after its arguments is called a postfix operator, and an operator placed between its arguments is called an infix operator. In Haskell, the term *operator* implies infix operator, although in Chapter 6 we'll see how to turn an infix operator into a prefix operator.

Table 1-2 shows common Haskell operators, along with their precedence and associativity, which are explained next. The operators for addition, subtraction, multiplication, and division work pretty much as you would expect.

Table 1-2: Precedence and Associativity for Common Operators

Operation	Operator(s)	Precedence	Associativity
Composition	.	9	Right
Exponentiation	^, ^^, **	8	Right
Multiplication, division	*, /	7	Left
Addition, subtraction	+, -	6	Left
List operators	:, ++	5	Right
Equality, inequality	==, /=	4	
Comparison	<, >, <=, >=	4	
Logical AND	&&	3	Right
Logical OR	||	2	Right
Application	$	0	Right

The table also shows three different operators for exponentiation. This proliferation is related to Haskell's type system, which we'll say more about in Chapters 3 and 8.

The caret operator (^) can only handle nonnegative integer exponents. The expression x^n means the product of n factors of x. The double caret (^^) can handle any integer exponent. The ** operator can handle any real exponent. For now, I recommend using ** for exponentiation.

The equality, inequality, and comparison operators can be used between numeric expressions.

```
Prelude> pi > 3
True
```

The result of a comparison is a Boolean expression, either True or False. Chapter 5 covers the list operators : and ++ from Table 1-2.

Precedence and Associativity

As we saw earlier when we tried to take the cosine of $\pi/2$, function application takes precedence over infix operators. In addition, some operators take precedence over other operators. In the expression

```
Prelude> 1 + 2 * 3
7
```

the multiplication of 2 and 3 will occur before the addition with 1. This is consistent with usual mathematical convention. To carry this out, binary operators in Haskell have a precedence associated with them that describes which operations should be carried out first. Binary operators have a precedence from 0 to 9. An operation with a higher precedence number means that operation will be carried out first. For example, addition and subtraction have a precedence of 6 in Haskell, while the precedence of multiplication and division is 7 and the precedence of exponentiation is 8. The OR operation `||` between Boolean values has a precedence of 2, and the AND operation `&&` has a precedence of 3.

The far-right column of Table 1-2 lists the *associativity* of some operators. Consider the expression `8 - 3 - 2`. There are two ways in which this expression might be interpreted. The standard mathematical convention is that the expression is shorthand for (8 − 3) − 2, which evaluates to 3. But another interpretation is that the expression is shorthand for 8 − (3 − 2), which evaluates to 7. Clearly, it's important for us to understand which of these two interpretations is correct for the original expression, and that is where associativity comes in. Looking at Table 1-2, we see that subtraction is left associative. This means that the leftmost subtraction is carried out first, resulting in the first interpretation given earlier (resulting in 3, not 7). Precedence and associativity allow us to unambiguously determine which operators act first.

The purpose in learning the precedence and associativity rules is so we can avoid using parentheses as much as possible. Multiple levels of nested expressions make things hard to read. My advice is to never try to use more than two levels of nested parentheses. In addition to knowing the precedence and associativity rules, there are other ways to avoid the use of parentheses, such as defining a local variable. We'll discuss these later.

Let's add parentheses to the following expression to indicate the order in which Haskell's precedence and associativity rules would evaluate the expression.

```
8 / 7 / 4 ** 2 ** 3 > sin pi/4
```

Function application takes precedence over all operators, so `sin pi` is the first thing calculated.

```
8 / 7 / 4 ** 2 ** 3 > (sin pi)/4
```

Next, exponentiation is the operator in the expression with the highest precedence in Table 1-2. Exponentiation occurs twice, and because it is right associative, the rightmost exponentiation occurs next.

```
8 / 7 / 4 ** (2 ** 3) > (sin pi)/4
```

Next is the left exponentiation.

```
8 / 7 / (4 ** (2 ** 3)) > (sin pi)/4
```

Next is division. There are three divisions. The rightmost division is unproblematic, but we need to resolve the two divisions on the left of the expression by associativity rules. Division is left associative.

```
(8 / 7) / (4 ** (2 ** 3)) > ((sin pi)/4)
```

Note we have inserted two sets of parentheses in the last step. One is for the rightmost division and one is for the leftmost division. Now we can put parentheses in for the final division, which occurs before the comparison.

```
((8 / 7) / (4 ** (2 ** 3))) > ((sin pi)/4)
```

The last operator to act is the comparison operator >. There is no need to put parentheses around the entire expression, so we are done. The fully parenthesized expression is

```
((8 / 7) / (4 ** (2 ** 3))) > ((sin pi)/4)
```

The Application Operator

The operator `$` in Table 1-2 is called the function application operator. No operator is required to apply a function. Juxtaposition of two expressions means that the first is a function and the second is an argument, and that the function is to be applied to the argument.

The function application operator does nothing but apply the function on its left to the expression on its right.

```
Prelude> cos pi
-1.0
Prelude> cos $ pi
-1.0
Prelude> cos $ pi / 2
6.123233995736766e-17
```

The key to its use lies in the fact that it has a precedence of 0. This means that the operator `$` changes application from the *first* thing to be done to the *last* thing to be done. In this way, the `$` serves as a kind of one-symbol parentheses. Instead of putting parentheses around `pi / 2` in the example above, we can use the single-symbol function application operator. Because it has right associativity, the function application operator is a frequent Haskell idiom that makes nested applications (`h $ g $ f x`) easier to read.

Functions with Two Arguments

All of the functions in Table 1-1 take one real number as input and give a real number as output (assuming the input is in the domain of the function). There are also a couple of useful numeric functions that take two real numbers as input. These are listed in Table 1-3.

Table 1-3: Numeric Functions with Two Arguments

Function	Example
`logBase`	`logBase 10 100` = 2
`atan2`	`atan2 1 0` = $\pi/2$

Let's look at these functions in action:

```
Prelude> logBase 10 100
2.0
Prelude> atan2 1 0
1.5707963267948966
```

The `logBase` function takes two arguments: the first is the base of the logarithm and the second is the number we wish to take the log of.

The `atan2` function solves a problem you may have run into if you've tried to use the inverse tangent function to convert from Cartesian to polar coordinates. Consider the following equations for polar coordinates (r, θ) in terms of Cartesian coordinates (x, y):

$$r = \sqrt{x^2 + y^2}$$
$$\theta = \tan^{-1}(y/x)$$

Suppose that we're trying to find the polar coordinates associated with the point $(x, y) = (-1, -\sqrt{3})$. The answer needs to be a point in the third quadrant because x and y are both negative. This means θ should be in the range $\pi < \theta < 3\pi/2$ (or $-\pi < \theta < -\pi/2$). But if we mechanically apply the formula above for θ, we'll calculate $\tan^{-1}(\sqrt{3})$, which our calculator or computer will tell us is $\pi/3$. The problem is the domain of the inverse tangent function, and a solution is to use the `atan2` function instead of the `atan` function. The result of `atan2` y x will give the angle, in radians, in the correct quadrant.

Note how the two arguments are given to the functions `logBase` and `atan2`. In particular, there is no comma between the two argument values, as would be required in traditional mathematical notation.

Numbers in Haskell

A few details about numbers in Haskell are not intuitive for someone coming to the language for the first time. In this section, we point out some issues related to negative numbers, decimals, and exponential notation.

Negative Numbers in Haskell

If you try the code

```
Prelude> 5 * -1

<interactive>:15:1: error:
    Precedence parsing error
        cannot mix '*' [infixl 7] and prefix `-' [infixl 6] in the same infix
            expression
```

you will get an error, although the meaning of the expression seems clear enough: we want to multiply 5 by −1. The trouble here is that the minus sign acts as both a binary operator (as in the expression 3 − 2) and a unary operator (as in the expression −2). Binary operators play an important role in Haskell, and the syntax of the language supports their use in a consistent, unified way. Unary operators in Haskell are much more of a special case; in fact, the minus sign is the only one. Because of decisions the Haskell designers made (see *https://wiki.haskell.org/Unary_operator*), negative numbers are sometimes not recognized as readily as you might expect.

The solution is simply to enclose negative numbers in parentheses. For example,

```
Prelude> 5 * (-1)
-5
```

evaluates to −5.

Decimal Numbers in Haskell

A number containing a decimal point must have digits (0 through 9) both before and after the decimal point. Thus, we must write `0.1` and not `.1`; instead of `5.`, we must write either `5.0` or `5` without a decimal point. The reason is that the dot character serves another role in the language (namely function composition, mentioned in the top line of Table 1-2, which we'll study further in Chapter 2). The rule requiring digits before and after a decimal point helps the compiler distinguish the meaning of the dot character.

Exponential Notation

You can use exponential notation to describe numbers in Haskell that are very big or very small. Here are a few examples:

Mathematical notation	Haskell notation
3.00×10^{8}	`3.00e8`
6.63×10^{-34}	`6.63e-34`

Haskell will also use exponential notation to show you numbers that turn out to be very big or very small.

Expression		Evaluates to
`8**8`	⇝	`1.6777216e7`
`8**(-8)`	⇝	`5.960464477539063e-8`

Approximate Calculation

Most of the calculations we will do are not exact calculations. When we ask the computer to find the square root of 5,

```
Prelude> sqrt 5
2.23606797749979
```

it gives a very accurate result, but it is not an exact result. This is because the computer uses a finite number of bits to represent this number and cannot represent every conceivable number.

If you evaluate `sqrt 5 ^ 2` in GHCi, you may not get exactly 5.0 as a result.

```
Prelude> sqrt 5 ^ 2
5.000000000000001
```

The computer does not represent $\sqrt{5}$ exactly. We can even ask the following in GHCi:

```
Prelude> sqrt 5 ^ 2 == 5
False
```

My computer gives `False` because of the approximate calculation.

Another source of non-exactness in calculation comes from the computer's use of a binary (base 2) internal representation of numbers. When I multiply 3×0.2, I don't get exactly 0.6. Why? The reason is that 0.2, which has a nice finite decimal (base 10) representation, has a repeating binary (base 2) representation. Just like the fraction 1/3 has an infinite repeating representation in base 10 (0.333333. . .), the fraction 1/5 has an infinite repeating representation in base 2. Table 1-4 shows representations of some simple fractions in decimal (base 10) and binary (base 2).

Table 1-4: Representations of Numbers in Base 10 and Base 2

Number	Decimal	Binary
1/2	0.5	0.1
1/3	0.333333...	0.01010101...
1/4	0.25	0.01
1/5	0.2	0.001100110011...

The computer converts every number that we supply in decimal into its internal binary form and only keeps a finite number of digits (bits, really). Most of the time, we don't need to be concerned with this, but it explains why some calculations that seem like they should be exactly calculable are not. One moral of this story is never to do equality checking of numbers when either number has been approximately calculated.

Errors

People make mistakes. This is as it should be. When you enter something the computer does not understand, it will give you an error message. These messages can appear intimidating, but they are a great opportunity for learning, and it's worthwhile to learn how to read them.

We saw a "Precedence parsing error" in "Numbers in Haskell" when we tried to multiply 5 by −1 without enclosing the −1 in parentheses. Another error you're bound to run into sooner or later is "No instance for Show."

There are some completely legitimate, well-defined expressions in Haskell that have no good way of being shown on the screen. Functions are the most common example. Because a function can accept a wide variety of inputs and produce a wide variety of outputs, there is in general no good way of displaying the "value" of a function. If you ask GHCi to tell you what the square root function "is," it will complain that it knows no way to show it to you by saying that is has "No instance for Show."

```
Prelude> sqrt

<interactive>:25:1: error:
    • No instance for (Show (Double -> Double))
        arising from a use of 'print'
        (maybe you haven't applied a function to enough arguments?)
    • In a stmt of an interactive GHCi command: print it
```

This message is not due to an error at all. The `sqrt` function is a completely legitimate Haskell expression. GHCi is merely saying that it knows no way to display it.

Getting Help and Quitting

To ask GHCi for help, enter `:help` (or `:h`). To leave GHCi, enter `:quit` (or `:q`).

Commands that start with a colon do not belong to the Haskell programming language proper but rather to the GHCi interactive compiler, and they control its operation. We will see more of these commands that start with colons later.

More Information

To learn more about the Haskell programming language (of which GHCi is a popular implementation), you can visit the website *https://www.haskell.org*.

The *haskell.org* website has links to many online and paper sources for learning the language. Some particularly good ones are *Learn You a Haskell for Great Good!* [1] (*http://learnyouahaskell.com*) and *Real World Haskell* [2] (*http://book.realworldhaskell.org*).

Summary

In this chapter, we saw how to use GHCi as a calculator. GHCi comes equipped with an array of scientific functions and operators. The operators have precedence and associativity rules that determine the order in which they are carried out. These rules are useful to know and use so we can reduce the number of parentheses in our expressions. Function application has higher precedence than any operator. Negative numbers sometimes need parentheses around them to be parsed correctly. Numbers with a decimal point require a digit both before and after it. Many calculations the computer makes are approximate. Errors should be thought of as helpful hints; it is a useful attitude to be curious about errors. Patience and persistence in the face of errors are part of the path to understanding.

In the next chapter, we'll show how to define our own functions and load them into GHCi.

Exercises

Exercise 1.1. Evaluate `sin 30` in GHCi. Why is it not equal to 0.5?

Exercise 1.2. Add parentheses to the following expressions to indicate the order in which Haskell's precedence and associativity rules would evaluate the expressions:

(a) `2 ^ 3 ^ 4`

(b) `2 / 3 / 4`

(c) `7 - 5 / 4`

(d) `log 49/7`

Exercise 1.3. Use GHCi to find $\log_2 32$.

Exercise 1.4. Use the `atan2` function in GHCi to find the polar coordinates (r, θ) associated with Cartesian coordinates $(x, y) = (-3, 4)$.

Exercise 1.5. Find a new example of a calculation in which the computer produces a result that is just a little bit different from the exact result.

Exercise 1.6. Why is there no associativity listed for the equality, inequality, and comparison operators in Table 1-2? (Hint: write down the simplest expression you can think of that would require the associativity rules to resolve the precedence of comparison operators and try to make sense of it.)

2

WRITING BASIC FUNCTIONS

The function is the central idea of functional programming. In this chapter, we'll learn how to define functions and constants and how to use those functions and constants in GHCi. We'll discuss the language we use to talk about functions, and we'll see how communicating with computers often requires more precision than communicating with people. We'll then introduce Haskell's system of *anonymous functions*, which are functions without a name. After a brief glimpse of Haskell's type system (which we'll describe more in Chapter 3), we'll show how the function composition operator can be used to compose functions. Finally, we'll show the kind of error you get if you use a name that has not been defined.

Constants, Functions, and Types

Programming in Haskell is a process of defining functions. Functions express to the computer how to calculate something we want. Haskell functions are much like mathematical functions: they take inputs and produce an output that depends on the inputs. Like mathematical functions, Haskell functions have a domain, describing the kind of entities that can be used as input, and a codomain (sometimes called range), describing the kind of entities that will be produced as output.

Unlike mathematical functions, Haskell functions must be *constructive*. They must give a clear, well-defined recipe for constructing the output from the inputs. Abelson and Sussman, in their wonderful book *Structure and Interpretation of Computer Programs* [3], note that the square root function, defined as a number that is nonnegative and squares to equal the input, is a perfectly legitimate mathematical function. But this definition does not give a recipe for how to construct the square root from the input, so it cannot be made into a Haskell function. Fortunately, there are other definitions of square root that are constructive and can be made into Haskell functions.

There is a way to define functions inside GHCi, but since we'll want to use most functions we define more than once, it's better to define our functions in a *source code file*, also called a *program file*, and then load that file into GHCi.

We'll need a text editor to create such a file. Examples of popular text editors are GNU Emacs, Vim, and gedit.

Word processing programs you might use to type a letter or a document are not appropriate for this purpose because they store the text you type with additional information (such as font type and size) that will make no sense to the Haskell compiler.

Using a text editor, let's create a file named *first.hs* for our first program. (The *.hs* extension indicates a Haskell program.) Put the following lines in the file:

```
-- First Haskell program

-- Here we define a constant
e :: Double
e = exp 1

-- Here we define a function
square :: Double -> Double
square x = x**2
```

This program file defines a constant and a function. The lines that begin with a double hyphen are *comments*. The Haskell compiler ignores any line that begins with a double hyphen; in fact, it ignores whatever is written after a double hyphen until the end of the line, unless the double hyphen is part of a string or certain other special environments. Comments are meant to help humans read the code.

The first two non-comment lines of the file define the constant *e*, the base of natural logarithms. Unlike π, *e* is not included in the Haskell Prelude. The line

```
e :: Double
```

declares the *type* of e to be Double. A *type* is a description of commonality in how entities can be used. Every expression in Haskell has a type, which tells the compiler about the situations in which the expression can be used and the situations in which it cannot. For example, the Double type tells the compiler that e is an approximation of a real number, sometimes called a floating-point number. The name Double is used for historical reasons to mean a *double-precision* floating-point number. This type of number is capable of about 15 decimal digits of precision, compared with a single-precision number that is capable of about seven decimal digits of precision. Haskell has a type Float for single-precision numbers. Unless there is a compelling reason to do otherwise, we'll always use type Double for our (approximations of) real numbers.

In addition to Double, there are several other types we might want to use. Haskell has a type Int for small integers (up to at least a few billion) and a type Integer for arbitrary-size integers. Chapter 3 is all about types.

Let's get back to our *first.hs* program file. As we said earlier, the first non-comment line of the file declares the type of the name e to be Double. This kind of line, with a name followed by a double colon followed by a type, is called a *type signature*. We may also call such a line a *declaration*, because it declares the name e to have type Double.

The second non-comment line of the file actually *defines* e. Here, we use the built-in function exp applied to the number 1 to produce the constant e. Remember that we don't need parentheses to apply a function to an argument.

Next, we have a type signature for the function square. The type of square is declared to be Double -> Double. A type containing an arrow is called a *function type*. (Function types will be explored in more detail in the next chapter.) It says that square is a function that takes a Double as input and produces a Double as output. The last line defines the function square. Note the ** operator used for exponentiation.

To load this program file into GHCi, use GHCi's :load command (:l for short).

```
Prelude> :l first.hs
[1 of 1] Compiling Main             ( first.hs, interpreted )
Ok, one module loaded.
*Main> square 7
49.0
*Main> square e
7.3890560989306495
```

After we load *first.hs*, the GHCi prompt changes from Prelude> to *Main>. This indicates that our program file has been successfully loaded and given the default name Main. We now have access to the constant and function defined in the file.

The names e and square defined in the file *first.hs* are examples of *variable identifiers* in Haskell. Variable identifiers must begin with a lowercase letter, followed by zero or more uppercase letters, lowercase letters, digits, underscores, and single quotes. Names that begin with an uppercase letter are reserved for types, type classes (which we'll discuss in Chapter 8), and module names.

If you forget or don't know the type of something, you can ask GHCi for the type with the :type command (:t for short).

```
*Main> :t square
square :: Double -> Double
```

The notation used for defining a function in Haskell is similar to mathematical notation in some ways and different in a few others. Let's comment on the differences. Table 2-1 shows a few examples.

Table 2-1: Comparison of Function Definitions in Traditional Mathematical Notation with Function Definitions in Haskell

Mathematical definition	Haskell definition
$f(x) = x^3$	`f x = x**3`
$f(x) = 3x^2 - 4x + 5$	`f x = 3 * x**2 - 4 * x + 5`
$g(x) = \cos 2x$	`g x = cos (2 * x)`
$v(t) = 10t + 20$	`v t = 10 * t + 20`
$h(x) = e^{-x}$	`h x = exp (-x)`

First, notice that traditional mathematical notation (and some computer algebra systems) use juxtaposition to represent multiplication. For example, $2x$ means 2 multiplied by x, just because the symbols are next to each other. Haskell requires use of the multiplication operator *. In Haskell, juxtaposition means function application.

Next, notice that traditional mathematical notation requires that function arguments be put in parentheses after the function name. This is true for function definitions (compare $f(x) = x^3$ with Haskell's f x = x**3) as well as function applications (compare $f(2)$ with Haskell's f 2). Haskell does not require parentheses in function definition or application. Haskell uses parentheses to indicate the order of operations.

Lastly, traditional mathematical notation tries to get away with single-letter function names, such as f. Haskell allows single-letter function names, but it is much more common to use a multi-letter word for a function name (such as square above), especially when the word can serve as a good description of what the function does.

How We Talk About Functions

Suppose we define a function f as $f(x) = x^2 - 3x + 2$. It's common in mathematics and physics to speak of "the function $f(x)$." Haskell invites us to think a bit more carefully and precisely about this bad habit. (Well, it really *requires* us to think more carefully about it, but it's always nicer to be invited than required, no?) Instead of saying "the function $f(x)$," we should say one of the following, depending on what we mean:

- The function f
- The value $f(x)$
- Given a number x, the value of the function f at x

The second and third bullet points are two ways of saying the same thing. The first bullet point is saying something different from the second and third.

What's wrong with saying "the function $f(x)$?" It's common in mathematics and physics to use "the function f" and "the function $f(x)$" interchangeably, with the second expression merely indicating explicitly that f depends on x. We think of mathematical notation as being a precise representation of an idea, but this is a case where the commonly used notation is not precise.

One reason for shunning the language "the function $f(x)$" is that if $f(x) = x^2 - 3x + 2$, then $f(y) = y^2 - 3y + 2$. The letter x really has nothing to do with the function f. Granted, we need *some* letter to use to make the definition, but it doesn't matter which one. We call x a *dummy variable* when it's used to define something else.

In Haskell, we say $f(x)$ when we want to evaluate the function f using the input x. We say f when we want to speak about the function itself, not evaluate it (that is, not give it any input). What else is there to do with a function except give it an input? Well, you could *integrate* the function between given limits. You could *differentiate* the function to obtain another function. You could, in some cases, apply the function twice. In short, there are many things we might want to do with a function other than simply evaluating it.

Haskell's type system helps us understand the key distinction between f and $f(x)$. The variable x is a number, so it has a type like `Double`. Now f is a function, so it has a type like `Double -> Double`. Finally, $f(x)$ means the function f evaluated at the number x, so $f(x)$ has type `Double`. Things that have type `Double -> Double` are functions. Things that have type `Double` are numbers. The table below summarizes these distinctions.

Math notation	Haskell notation	Haskell type
f	`f`	`Double -> Double`
$f(3)$	`f 3`	`Double`
$f(x)$	`f x`	`Double`

Computers are notorious for being inflexible in understanding what a person means. Computers look at exactly what you say, and they give warnings and errors if your input doesn't meet their requirements for format and interpretation. Most of the time, this is a pain in the neck. We would like to have an assistant that understands what we mean and does what we want.

However, in the case of types and functions, Haskell's inflexibility is a great teaching aid. Haskell is helping us to organize our thinking so we will be prepared to do more complex things in a structured and organized way. In Chapter 6, which covers higher-order functions, we'll see examples of how careful thinking about types and functions allows us to encode more complex ideas simply and easily.

When we use Haskell, we make a trade-off. We agree to use language in a precise and careful way (the compiler is going to check us on this), and in exchange, we'll be able to say things in the language that are rather complex and difficult to say in a language that accommodates imprecision. Thus, we'll be able expose the essential structure of a physical theory like Newtonian mechanics.

Anonymous Functions

Haskell provides a way to specify a function without naming it. For example, the function that squares its argument can be written as `\x -> x**2`.

A function specified in this way is called an *anonymous function* or a *lambda function* after the lambda calculus developed by Alonzo Church in the 1930s. (Church was Alan Turing's PhD advisor.) The backslash character (\) was thought by Haskell's creators to look a bit like the lowercase Greek letter lambda (λ).

Table 2-2 shows examples of mathematical functions written as lambda functions. This is an alternative way to define the functions in Table 2-1.

Table 2-2: Comparison of Function Definitions in Traditional Mathematical Notation with Lambda Functions Defined in Haskell

Mathematical function	Haskell lambda function
$f(x) = x^3$	`f = \x -> x**3`
$f(x) = 3x^2 - 4x + 5$	`f = \x -> 3 * x**2 - 4 * x + 5`
$g(x) = \cos 2x$	`g = \x -> cos (2 * x)`
$v(t) = 10t + 20$	`v = \t -> 10 * t + 20`
$h(x) = e^{-x}$	`h = \x -> exp (-x)`

The real power of lambda functions comes from using them in places where we need a function but don't want to spend the effort (that is, a declaration and a definition) to name a new function. We'll see examples of how this is useful in Chapter 6, where we discuss higher-order functions that take other functions as input. These other functions are sometimes conveniently expressed as anonymous functions.

We can apply the anonymous squaring function `\x -> x**2` to the argument 3 by writing `(\x -> x**2) 3` at the GHCi prompt.

```
*Main> (\x -> x**2) 3
9.0
```

Notice that when we write `\x -> x**2`, we are *not* defining what `x` is. Instead we are saying that if we temporarily allow `x` to stand for the argument of the function (such as `3` above), we have a rule for determining the value of the function applied to the argument. The same remark is true of (named) mathematical functions; when we define $f(x) = x^2$, this is a definition for f, not a definition for x. The function `\x -> x**2` is the same as the function `\y -> y**2`; the variable we use to name the argument is not important. Both are the function that squares its argument.

Table 2-3 shows examples of the application of anonymous functions to an argument.

Table 2-3: Examples of Applying Anonymous Functions to an Argument

Expression		Evaluates to
`(\x -> x**2) 3`	⇝	`9.0`
`(\y -> y**2) 3`	⇝	`9.0`
`(\x -> x**3) 3`	⇝	`27.0`
`(\x -> 3 * x**2 - 4 * x + 5) 3`	⇝	`20.0`
`(\x -> cos (2 * x)) pi`	⇝	`1.0`
`(\t -> 10 * t + 20) 3`	⇝	`50`
`(\x -> exp (-x)) (log 2)`	⇝	`0.5`

These examples can be evaluated at the GHCi prompt.

Composing Functions

Writing $\cos^2 x$ is shorthand for $(\cos x)^2$, which means "take the cosine of x and then square the result." When we use the output of one function f as the input to another function g, we are *composing* the two functions to produce a new function. We write $g \circ f$, called g *after* f, for the function that applies f to its input and then applies g to the result.

$$(g \circ f)(x) = g(f(x)) \tag{2.1}$$

The *function composition operator* (`.`) from Table 1-2 serves the role that $\circ$ serves in mathematical notation. The following four functions are equivalent ways of defining the cosine squared function:

```
cosSq :: Double -> Double
cosSq x = square (cos x)

cosSq' :: Double -> Double
cosSq' x = square $ cos x
```

```
cosSq'' :: Double -> Double
cosSq'' x = (square . cos) x

cosSq''' :: Double -> Double
cosSq''' = square . cos
```

The first function, cosSq, defines the square of the cosine of a number in the most straightforward way. It's clear from the parentheses that the cosine acts on x first, and then the function square gets applied. The second function, cosSq', does the same thing, but it uses the function application operator $ instead of parentheses (see "The Application Operator" in Chapter 1). The third function, cosSq'', shows how the composition operator can be used to compose the functions square and cos. The expression square . cos is like the $g \circ f$ on the left side of Equation 2.1, where square plays the role of g and cos plays the role of f. The fourth function, cosSq''', shows how Haskell lets us define a function without mentioning the argument to which it will be applied. Such a definition is called *point-free style*. If h is the function defined by $h(x) = g(f(x))$, mathematical notation allows us to alternatively define h by $h = g \circ f$. The function cosSq'' expresses the former definition, and the function cosSq''' expresses the latter. If you feel the need to define a cosine squared function, any of the four functions would be perfectly acceptable. The choice is a matter of style. The last definition is my favorite because of its concision.

The definitions just shown are examples of Haskell's delightful feature of allowing tick marks (single quotes) in identifiers. This is nice because it supports our mathematical usage of the concise "x prime" for something that is related to x.

The function composition operator can be used between any two functions in which the output type of the first function applied matches the input type of the second. In practice, the function composition operator often serves as a way to avoid naming a new function. If the functions square and cos are available, there is really no need to make any of the four definitions because square . cos is a perfectly good function that can be used anywhere cosSq can be.

Variable Not in Scope Error

One of the simplest types of error comes from using a name that has not been defined. If we ask GHCi for the value of x without having defined x, we'll get a "Variable not in scope" error.

```
*Main> x

<interactive>:6:1: error: Variable not in scope: x
```

The *scope* of a name is the set of situations in which the name can be used and properly understood by the compiler. The "Variable not in scope" error might be better called "name not recognized." Any name the compiler expects to recognize but doesn't will produce this error. The error comes from the use of a name we haven't defined or haven't told the compiler where to find. This applies to functions, constants, and local variables (which we will introduce later)—essentially any entity that can have a name. Common identifiers, such as `x`, can be reused, and there are ways to unambiguously control which `x` we are referring to at a particular point in our program.

Summary

In this chapter, we saw how to define functions in a source code file and load them into GHCi to use them. We showed how anonymous functions can be used in places where we need a function but don't want to name it. The need for and usefulness of anonymous functions will become clearer in Chapter 6. The function composition operator can be used to compose any two functions in which the output type of the first matches the input type of the second. We saw how a "Variable not in scope" error can arise when the computer thinks it should know the meaning of a name but doesn't. In the next chapter, we'll look more deeply into Haskell's system of types, which provides a powerful tool to organize our thinking and reflect that organization in our writing.

Exercises

Exercise 2.1. In a Haskell program file (a new file with a new name that ends in *.hs*), define the function $f(x) = \sqrt{1+x}$. As we did for the function `square`, give both a type signature and a function definition. Then load this file into GHCi and check that $f(0)$ gives 1, $f(1)$ gives about 1.414, and $f(3)$ gives 2.

Exercise 2.2. Consider throwing a rock straight upward from the ground at 30 m/s. Ignoring air resistance, find an expression $y(t)$ for the height of the rock as a function of time.

Add on to your program file *first.hs* by writing a function

```
yRock30 :: Double -> Double
```

that accepts as input the time (after the rock was thrown) in seconds and gives as output the height of the rock in meters.

Exercise 2.3. Continuing with the rock example, write a function

```
vRock30 :: Double -> Double
```

that accepts as input the time (after the rock was thrown) in seconds and gives as output the upward velocity of the rock in meters per second. (A downward velocity should be returned as a negative number.)

Exercise 2.4. Define a function `sinDeg` that computes the sine of an angle given in degrees. Test your function by evaluating `sinDeg 30`.

Exercise 2.5. Write Haskell function definitions for the following mathematical functions. In each case, write a type signature (the type should be `Double -> Double` for each function) and a function definition. You will need to pick alternative names for some of these functions because Haskell functions must begin with a lowercase letter. Do not use more than two levels of nested parentheses.

(a) $f(x) = \sqrt[3]{x}$

(b) $g(y) = e^y + 8^y$

(c) $h(x) = \dfrac{1}{\sqrt{(x-5)^2 + 16}}$

(d) $\gamma(\beta) = \dfrac{1}{\sqrt{1-\beta^2}}$

(e) $U(x) = \dfrac{1}{10+x} + \dfrac{1}{10-x}$

(f) $L(l) = \sqrt{l(l+1)}$

(g) $E(x) = \dfrac{1}{|x|^3}$

(h) $E(z) = \dfrac{1}{(z^2+4)^{3/2}}$

Exercise 2.6.

(a) Express $\gamma(\beta) = \dfrac{1}{\sqrt{1-\beta^2}}$ as an anonymous function.

(b) Write an expression that applies the anonymous function from part (a) to the argument 0.8. What result do you get from GHCi?

3

TYPES AND ENTITIES

The idea that every expression has a type is central to Haskell. Haskell has several built-in types available to us in the Prelude and a system for making our own types. In this chapter, we'll discuss some of the built-in types, and in Chapter 10, we'll see how to make types of our own.

Basic Types

Types reflect the nature of information. For example, in physics, we want to know whether something is a scalar or a vector. These are different types. It makes no sense to add a scalar to a vector, and the computer can prevent us from making this mistake if we use a good system of types.

Table 3-1 shows Haskell's most important basic types.

Table 3-1: Haskell's Basic Types

Type	Description	Examples
`Bool`	Boolean	`False, True`
`Char`	Character	`'h', '7'`
`String`	String	`"101 N. College Ave."`
`Int`	Small (machine-precision) integer	`42`
`Integer`	Arbitrarily large integer	`18446744073709551616`
`Float`	Single-precision floating point	`0.33333334`
`Double`	Double-precision floating point	`0.3333333333333333`

The `Bool` type is for values that are either true or false, like the result of a comparison. For example, `3 > 4` evaluates to `False`.

```
Prelude> 3 > 4
False
```

The `Char` type is for single characters. The `String` type is for a sequence of characters. The types `Int`, `Integer`, `Float`, and `Double` are for numbers.

Let's take a closer look at each of these types.

The Boolean Type

The `Bool` type has only two possible values: `False` and `True`. The type is used for expressions that are meant to represent claims that might be true or false.

Haskell has an `if-then-else` expression whose value depends on a Boolean. The expression has the form `if` *b* `then` *c* `else` *a*. Here *b* is an expression of type `Bool` called the *condition* is called the *consequent* is called the *alternative*. Haskell's type system demands not only that *b* have type `Bool` but also that the consequent *c* and the alternative *a* have the same type (this can be any type, `Bool` or something else). If the condition *b* evaluates to `True`, the entire `if-then-else` expression evaluates to *c*; if the condition *b* evaluates to `False`, the entire `if-then-else` expression evaluates to *a*.

If you are familiar with an imperative language like Python or C, it may help to realize that Haskell's `if-then-else` construction is an *expression*, not a statement. An expression evaluates to a value. In imperative languages, `if-then` constructions are typically statements that are executed if the condition is true and ignored otherwise. In an imperative language, the `else` clause is optional; that is, it's only used when there are statements to be executed if the condition is false. Because the `if-then-else` construction in a functional language is an expression, the `else` clause is mandatory, not optional. *Some* value must be returned whether the condition is true or false.

As an example of the `if-then-else` expression, consider the following function (which is sometimes called the *Heaviside step function* or the *unit step function*):

$$H(x) = \begin{cases} 0 & , \quad x \leq 0 \\ 1 & , \quad x > 0 \end{cases} \tag{3.1}$$

We can write a definition for this function in Haskell using the if-then-else construction. In Haskell, we are not allowed to begin the names of constants or functions with a capital letter (recall the discussion of variable identifiers in the last chapter), so we'll call this function stepFunction.

```
stepFunction :: Double -> Double
stepFunction x = if x <= 0
                 then 0
                 else 1
```

The function stepFunction accepts a Double as input (called x in the definition) and returns a Double as output. The expression x <= 0 is the condition, the expression 0 is the consequent, and the expression 1 is the alternative.

The Prelude provides a few functions that work with Booleans. The first is not, which has type Bool -> Bool, meaning it accepts a Boolean as input and gives another Boolean as output. The function not returns True if its input is False and returns False if its input is True. You can see this for yourself in GHCi if you type

```
Prelude> not False
True
```

or

```
Prelude> not True
False
```

at the GHCi prompt.

As you saw in Chapter 2, GHCi has a command :type (:t for short) that asks about the type of something. You can ask GHCi for the type of not by entering

```
Prelude> :t not
not :: Bool -> Bool
```

at the GHCi prompt. GHCi commands that start with a colon are not part of the Haskell language itself. You cannot use the colon commands in a Haskell program file.

The Boolean AND operator && takes two Booleans as input and gives one Boolean as output. The output is True only when both inputs are True, and it's False otherwise. Table 3-2 describes the behavior of the && operator.

Table 3-2: Definition of the AND Operator

x	*y*	*x* && *y*
False	False	False
False	True	False
True	False	False
True	True	True

The Boolean OR operator || takes two Booleans as input and gives one Boolean as output. The output is False only when both inputs are False, and it's True otherwise. Table 3-3 describes the behavior of the || operator.

Table 3-3: Definition of the OR Operator

x	*y*	*x* || *y*
False	False	False
False	True	True
True	False	True
True	True	True

These operators are listed in Table 1-2 with their precedence and associativity. You can play with them in GHCi, asking for evaluations of expressions such as

```
Prelude> True || False && True
True
```

at the GHCi prompt.

The Character Type

The Char type is for single characters, including uppercase and lowercase letters, digits, and some special characters (like the newline character, which produces a new line of text). Here are some examples of character definitions:

```
ticTacToeMarker :: Char
ticTacToeMarker = 'X'

newLine :: Char
newLine = '\n'
```

There is very little reason to make these definitions because any place we could use newLine, for example, we could just as easily use '\n', which takes up less space. We do it here only to show the relationship between the term 'X' and the type Char. As shown in the examples above, a character can be formed by enclosing a single letter or digit in single quotes.

The String Type

A string is a sequence of characters. (In Chapter 5, we'll learn that a string is a *list* of characters, where list has a precise meaning.) Here are some examples:

```
hello :: String
hello = "Hello, world!"
```

```
errorMessage :: String
errorMessage = "Can't take the square root of a Boolean!"
```

These definitions are not as useless as the ones shown earlier for characters, because although `"Hello, world!"` is entirely equivalent to `hello`, the name `hello` is at least shorter and easier to type than the string it represents. If such a string was needed at several different places in a program, that would justify the definition of a name such as `hello`. To form a string from a sequence of characters, we enclose the character sequence in double quotes.

Numeric Types

The basic numeric types are `Int`, `Integer`, `Float`, and `Double`. The `Int` type is for small integers. A 32-bit machine will use 32 bits to represent an `Int`, which gives numbers up to a few billion. A 64-bit machine will use 64 bits to represent an `Int`, which gives numbers up to about 10^{18}. The `Integer` type is for arbitrary integers. The computer will use whatever number of bits it needs to represent an `Integer` exactly. On my 64-bit machine, I get the following results:

```
Prelude> 10^18 :: Int
1000000000000000000
Prelude> 10^18 :: Integer
1000000000000000000
Prelude> 10^19 :: Int
-8446744073709551616
Prelude> 10^19 :: Integer
10000000000000000000
```

Notice that I get no error message about going too high with `Int`; I just get the wrong answer. The `Int` type is good for almost any kind of counting you might ask the computer to do. The computer can't count up to 10^{18} because it takes too long.

The `Float` type is for approximations to real numbers and has a precision of about 7 decimal digits. The `Double` type is for approximations to real numbers and has a precision of about 15 decimal digits. I always choose `Double` for my real numbers unless I am using a library written by someone else that uses `Float`.

The numeric examples in the rightmost column of Table 3-1 *can be* expressions of the type indicated, but an expression by itself, such as `42`, does not *necessarily* have type `Int`. To be specific, `False` and `True` must have type `Bool`, `'h'` and `'7'` must have type `Char`, and `"101 N. College Ave."` must have type `String`. On the other hand, `42` could have type `Int`, `Integer`, `Float`, or `Double`. Clarifying this ambiguity is one reason to give a type signature with each name you define in a Haskell program. Without a type signature, the compiler cannot tell which of the four numeric types you might want for a number like 18446744073709551616. Any of the four numeric types would try to hold the number, but only Integer would represent the number exactly. The complexity of numeric types in Haskell is related to a more advanced language feature called *type classes*, which we'll discuss in Chapter 8.

The four numeric types in Table 3-1 are not the only numeric types in the Prelude. The Prelude includes a Rational type for rational numbers that we won't use in this book but that you can explore on your own if you are interested. Complex numbers are provided by a library module called Data.Complex. We won't use complex numbers in this book.

Function Types

Haskell provides several ways to form new types from existing types. Given any two types a and b, there is a type a -> b for functions that take an expression of type a as input and produce an expression of type b as output. Here is an example:

```
isX :: Char -> Bool
isX c = c == 'X'
```

The function isX takes a character as input and gives a Boolean as output. The function returns True if the input character is 'X' and returns False otherwise. Adding parentheses may help in reading the function definition. The definition is equivalent to

```
isX c = (c == 'X')
```

In general in a definition, the name on the left of the single equal sign (=) is being defined (isX in this case), and the expression on the right of the single equal sign is the body of the definition. The expression c == 'X' uses the equality operator == from Table 1-2 to ask if the input character c is the same as 'X'.

If we put this function definition into a Haskell program file (for example, *FunctionType.hs*) and load it into GHCi,

```
Prelude> :l FunctionType.hs
[1 of 1] Compiling Main             ( FunctionType.hs, interpreted )
Ok, one module loaded.
```

we can ask about the types of things. If we ask about the type of isX,

```
*Main> :t isX
isX :: Char -> Bool
```

we see what we wrote in our type signature. In GHCi, we can also ask for the type of isX 't':

```
*Main> :t isX 't'
isX 't' :: Bool
```

This makes sense because the expression `isX 't'` represents the function `isX` applied to the character argument `'t'`. Therefore, the type represents the type of the output of `isX`, namely `Bool`.

We can also ask GHCi for the *value* of `isX 't'` (as opposed to the type of the expression). If we enter `isX 't'` at the GHCi prompt,

```
*Main> isX 't'
False
```

we see that the value of `isX 't'` is `False` because `'t'` is not equal to `'X'`.

Here is an example of a function with type `Bool -> String`:

```
bagFeeMessage :: Bool -> String
bagFeeMessage checkingBags = if checkingBags
                             then "There is a $100 fee."
                             else "There is no fee."
```

The function `bagFeeMessage` takes a Boolean as input and gives a string as output. The input Boolean (called `checkingBags`) is intended to represent an answer (`True` or `False`) to the question of whether a passenger is checking bags. The style of naming a variable by sticking words together without spaces and using a capital letter at the beginning of the second and subsequent words is common in Haskell programming.

There is an alternative way to write the function `bagFeeMessage` that uses a facility in Haskell called *pattern matching*. Some data types have one or more patterns that values of that type fall into. The idea behind pattern matching for `Bool` is that the only possible values are `False` and `True`, so why not just give the output for each possible input? The fundamental way of achieving pattern matching is with the `case-of` construction. Here is what the function looks like using pattern matching:

```
bagFeeMessage2 :: Bool -> String
bagFeeMessage2 checkingBags = case checkingBags of
                                False -> "There is no fee."
                                True  -> "There is a $100 fee."
```

This doesn't look so different from the `if-then-else` construction, but the `case-of` construction is more general because it can be used with other data types, not just `Bool`. In Chapter 5, for example, we will see that every list falls into one of two patterns that can be distinguished using the `case-of` construction.

Although the `case-of` construction is the basic way to do pattern matching, Haskell provides some syntactic sugar for the special case in which we want to pattern match on the input to a function.

```
bagFeeMessage3 :: Bool -> String
bagFeeMessage3 False = "There is no fee."
bagFeeMessage3 True  = "There is a $100 fee."
```

By using pattern matching on the input, we have avoided using the `if-then-else` construction. Also, we no longer need the variable `checkingBags`, which held the input value.

Summary

Haskell has built-in types and facilities for making our own types. Types are intended to describe the meaning of data. This chapter looks at seven of the most common built-in types: `Bool`, `Char`, `String`, `Int`, `Integer`, `Float`, and `Double`. It also considers function types, which are very important to the language because functions play such a central role. We got a first glimpse of pattern matching, both with the `case-of` construction and by pattern matching on the input. In the next chapter, we begin our physics work, starting with motion in one dimension.

Exercises

Exercise 3.1. Add parentheses to the following expressions to indicate the order in which Haskell's precedence and associativity rules (Table 1-2) would evaluate the expressions. Some of the expressions are well-formed and have a clear type. In those cases, give the type of the (entire) expression. Also identify expressions that are not correctly formed (and consequently do not have a clear type) and say what is wrong with them.

(a) `False || True && False || True`

(b) `2 / 3 / 4 == 4 / 3 / 2`

(c) `7 - 5 / 4 > 6 || 2 ^ 5 - 1 == 31`

(d) `2 < 3 < 4`

(e) `2 < 3 && 3 < 4`

(f) `2 && 3 < 4`

Exercise 3.2. Write Haskell function definitions for the following mathematical functions. In each case, write a type signature (the type should be `Double -> Double` for each function) and a function definition.

(a) $f(x) = \begin{cases} 0 & , \quad x \leq 0 \\ x & , \quad x > 0 \end{cases}$

(b) $E(r) = \begin{cases} r & , \quad r \leq 1 \\ \frac{1}{r^2} & , \quad r > 1 \end{cases}$

Exercise 3.3. Define a function `isXorY` with type signature

```
isXorY :: Char -> Bool
```

that will return `True` if the input character is `'X'` or `'Y'` (capital X or Y) and return `False` otherwise. Test your function by loading it into GHCi and giving it inputs of `'X'`, `'Y'`, `'Z'`, and so on.

Exercise 3.4. Define a function bagFee with type signature

```
bagFee :: Bool -> Int
```

that will return the integer 100 if the person is checking bags and the integer 0 if not. Use an if-then-else construction for this function. Then define a second function, bagFee2, with the same type signature that uses pattern matching on the input instead of the if-then-else construction.

Exercise 3.5. Define a function greaterThan50 with type signature

```
greaterThan50 :: Integer -> Bool
```

that will return True if the given integer is greater than 50 and return False otherwise.

Exercise 3.6. Define a function amazingCurve with type signature

```
amazingCurve :: Int -> Int
```

that will double a student's score on an exam. However, if the new score after doubling is greater than 100, the function should output 100.

Exercise 3.7. What is the *type* of the expression bagFee False using the definition of bagFee you wrote in Exercise 3.4? What is the *value* of the expression bagFee False using that definition of bagFee?

Exercise 3.8. "Give every function a type signature." In Haskell, it is good practice to give every function you define in your program file a type signature. We have been doing this all along. Type signatures serve as a form of documentation to readers of your program (including yourself).

Add type signatures for each of the definitions in the code below:

```
circleRadius = 3.5

cot x = 1 / tan x

fe epsilon = epsilon * tan (epsilon * pi / 2)

fo epsilon = -epsilon * cot (epsilon * pi / 2)

g nu epsilon = sqrt (nu**2 - epsilon**2)
```

Exercise 3.9. There are only a finite number of functions with type Bool -> Bool. How many are there? What would be good names for them? How many functions have type Bool -> Bool -> Bool?

Exercise 3.10. Devise an expression using True, False, &&, and || that would come out differently if the precedence of || was higher than the precedence of &&.

4

DESCRIBING MOTION

The description of the motion of an object in terms of its position, velocity, and acceleration is called *kinematics*. In this chapter, we'll give a succinct review of one-dimensional kinematics while showing how the Haskell language naturally encodes its ideas and equations. We'll use the Haskell functions introduced in Chapter 2 and the types we introduced in Chapter 3. The equations of kinematics, being largely definitional, have an almost one-to-one correspondence with Haskell functions.

Position and Velocity on an Air Track

Have you ever seen an air track? An air track is a fun toy, or, if you're feeling more serious, a piece of experimental equipment. It consists of a long horizontal rail (maybe 2 or 3 meters long) with little holes that allow air to shoot upward out of the rail. A small car (maybe 5 cm wide and 10 cm long) with no wheels rides atop this air track. The air eliminates most of the friction that would exist between the car and the rail as the car slides along the rail, so the car can slide freely along the air track. The cross section of the rail is shaped so that the car can only slide back and forth along the length of the rail; the car cannot slide sideways or move up or down.

We can make marks on the air track to allow us to talk about the *position* of the car. Let's imagine that we have an air track that has already been carefully marked in meters. For a particular motion of the car on the air track, we define x to be the function that associates each time t with the position of the car at that time. We say that $x(t)$ is the position of the car at time t.

Velocity is defined to be the rate at which position changes. The *average velocity* over the time interval that starts at time t_0 and ends at time t_1 is

$$\bar{v}_{[t_0,t_1]} = \frac{x(t_1) - x(t_0)}{t_1 - t_0} \tag{4.1}$$

Average velocity is change in position divided by change in time.

One of the advantages of using the Haskell programming language is that there is almost a one-to-one correspondence between the equations of physics and the code we write to describe them. In Haskell, the following lines, which could appear in a source code file, have the same meaning as Equation 4.1:

```
averageVelocity :: Time -> Time -> PositionFunction -> Velocity
averageVelocity t0 t1 x = (x t1 - x t0) / (t1 - t0)
```

The first line of Haskell code is a type signature saying that `averageVelocity` is a function that takes a time, a second time, and a position function as input and gives a velocity as output. We can use the arrow (`->`) to chain together inputs. The last term is the output type, and all of the other terms are input types. There is a deeper reason for this notation that we will explore in Chapters 6 and 9.

The second line of Haskell code above is the definition for the function `averageVelocity`. The definition says that if we call `t0` the first time, `t1` the second time, and `x` the position function, the velocity will be given by the expression on the right of the equal sign. The inputs `t0` and `t1` are numbers, but the input `x` is a function. The practice of using functions as inputs to other functions is common in Haskell; we will discuss this in much more detail in Chapter 6.

Table 4-1 shows a comparison of mathematical notation and Haskell notation.

Table 4-1: Comparison of Mathematical Notation and Haskell Notation

Mathematical notation	Haskell notation
t_0	`t0`
t_1	`t1`
x	`x`
$x(t_0)$	`x t0`
$x(t_1)$	`x t1`
$\bar{v}$	`averageVelocity`

As we saw in Chapter 1, parentheses are not required to apply a function to an argument. The `x` in the code is a function, just like the x in Equation 4.1. When we write $x(t_0)$, we mean the function x applied to (or evaluated at) the time t_0. Similarly, when we write `x t0`, we mean the function `x` applied to (or evaluated at) the time `t0`. Functions play such a central role in Haskell that juxtaposition of names implies that the first is a function and the second is an argument, and that the function is to be applied to the argument.

The notation $\overline{v}_{[t_0,t_1]}$ explicitly shows that average velocity depends on the initial time t_0 and the final time t_1 of the time interval, but the dependence on the position function is implicit and not shown in Equation 4.1. Haskell code shows all dependencies explicitly.

Types for Physical Quantities

It makes our thinking simpler if we can talk about velocity at a single time rather than over a time interval. We can take a step in this direction by labeling the average velocity with the time at the center of the interval and the length of the interval, rather than with the beginning and ending times.

$$\overline{v}(t, \Delta t) = \frac{x(t + \Delta t/2) - x(t - \Delta t/2)}{\Delta t} \tag{4.2}$$

In Haskell, Equation 4.2 looks like the following:

```
averageVelocity2 :: Time -> TimeInterval -> PositionFunction
                 -> Velocity
averageVelocity2 t dt x = (x (t + dt/2) - x (t - dt/2)) / dt
```

Here, we do need to enclose `t + dt/2` in parentheses so that adding `t` to `dt/2` takes place before the function `x` is applied.

Notice that in the code above, the types from the type signature match up with the arguments in the definition in the same order, as emphasized in Table 4-2.

Table 4-2: Matching of Arguments and Types for the Function `averageVelocity2`

Argument	Type
`t`	`Time`
`dt`	`TimeInterval`
`x`	`PositionFunction`

At this point, we're only dealing with motion in one dimension, so time, position, and velocity are all represented by numbers. We tell the Haskell compiler this with the following lines, which are called *type synonyms*.

```
type R = Double

type Time         = R
type TimeInterval = R
type Position     = R
type Velocity     = R

type PositionFunction = Time -> Position
type VelocityFunction = Time -> Velocity
```

The only type in the lines above that Haskell understands by default is `Double`. I prefer to think of this as the type of real numbers (not every real number can be represented by a `Double`, but we are willing to do approximate calculation), so I like the name `R` better than `Double`.

The first line says that whenever I use the type `R`, it means the same thing as `Double`. The next four lines say that time, time interval, position, and velocity are all just real numbers at this point. Finally, the last two lines define function types. The type `PositionFunction` is the type of a function that takes time as input and gives position as output. Recall that the argument `x` above was a function with this type. Since `Time` is the same as `R` and `Position` is the same as `R`, the type `PositionFunction` is the same as `R -> R`, which takes a real number as input and produces a real number as output. For a similar reason, the type `VelocityFunction` is also the same as the function type `R -> R`.

A type synonym merely gives an additional name to an existing type. The compiler sees `Double`, `R`, `Position`, and `Velocity` as identical and is unable to warn us if we attempt to use a `Velocity` where a `Position` is called for. In Chapter 10, we will introduce a way to define a new type that is distinct from all existing types and allows the compiler to check that we haven't confused one type with another. There is also a Haskell package called `units` (*https://hackage.haskell.org/package/units*) specifically designed to allow physical units such as meters per second to be attached to numerical quantities.

Introducing Derivatives

If in making the time interval Δt shorter and shorter we find that the average velocity converges to a particular value, we call this value the *instantaneous velocity*.

$$v(t) = \lim_{\Delta t \to 0} \frac{x(t + \Delta t/2) - x(t - \Delta t/2)}{\Delta t} \tag{4.3}$$

A limit of a ratio of differences occurs often enough that it gets awarded the name *derivative*. Given a function x of one variable, the derivative of x, denoted Dx, x', or $\dot{x}$, is the function of one variable defined as follows:

$$Dx(t) = \lim_{\epsilon \to 0} \frac{x(t + \epsilon/2) - x(t - \epsilon/2)}{\epsilon} \tag{4.4}$$

HANDLING CONTINUOUS TIME

The mathematical definitions of limit and derivative require a continuity of real numbers that may not be present in the physical world. There's a bit of idealization when we imagine time to be continuous. Time certainly seems to be continuous, and no one has made any direct measurements that suggest a discreteness to time, but there are quantum theoretical reasons to believe that at some very short scale, time may not behave as a perfect continuum.

To make a guess at a time scale when quantum effects are likely to disturb the continuity of time, we can use *dimensional analysis*, which means combining parameters that are relevant to the situation to construct a quantity with dimensions of time. We haven't said anything at all about the situation, so all we have to work with are fundamental constants of physics, namely the Newtonian constant of gravitation G, Planck's constant $\hbar$, and the speed of light in a vacuum c. There is only one way to multiply powers of these fundamental constants to produce a result with dimensions of time. The resulting time is called the *Planck time*, and it looks like this:

$$t_P = \sqrt{\frac{\hbar G}{c^5}} \approx 5.4 \times 10^{-44} \text{ s}$$

The Planck time is many, many orders of magnitude smaller than the smallest time scales that can currently be probed. For the physics that is experimentally accessible today, time appears continuous, and the derivative is in no danger of becoming obsolete.

We can say that instantaneous velocity is the derivative of position.

$$v = Dx \tag{4.5}$$

Note that Equation 4.5 is an equality of functions: the instantaneous velocity function is on the left of the equality and the derivative of the position function is on the right. When two functions are equal, they give equal results for equal inputs, so we can also write

$$v(t) = Dx(t) \tag{4.6}$$

for any time t. The right-hand side is the function Dx evaluated at the time t. We can think of the derivative operator as taking the entire position function as input and returning the velocity function as output.

It's more common to see the notation

$$v(t) = \frac{dx(t)}{dt} \tag{4.7}$$

to define velocity. Equation 4.5 is more succinct, but Equations 4.5, 4.6, and 4.7 all mean the same thing. We say that $v(t)$ is the velocity of the car at time t. Note that the velocity of the car on the air track can be negative (meaning the position is decreasing) or positive (meaning the position is increasing).

Derivatives in Haskell

A derivative takes a function as input and gives a function as output. In other words, a derivative is a function from functions to functions. A function that takes another function as input or returns a function as output is called a *higher-order function*. If the idea of functions as inputs and outputs of other functions is new to you, it will take some practice and examples to get used to, but I assure you that it is worth the effort. There are many ideas in physics, the derivative being just one, that are naturally expressed as higher-order functions. Chapter 6 is entirely devoted to such higher-order functions.

One possible type synonym for a derivative looks like this:

```
type Derivative = (R -> R) -> R -> R
```

Since arrows are right associative, with the rightmost arrow having highest precedence, the type `(R -> R) -> R -> R` is the same as `(R -> R) -> (R -> R)`.

We can write a numerical derivative in Haskell like this:

```
derivative :: R -> Derivative
derivative dt x t = (x (t + dt/2) - x (t - dt/2)) / dt
```

This numerical derivative does not take a limit but instead uses a small interval, `dt`, supplied by the user. If the interval is small enough, the result should be a good approximation to the derivative.

Let's play the matching game with arguments and types for the function `derivative` like we did for the function `averageVelocity2`. At first glance, it seems that `dt` has type `R`, `x` has type `Derivative`, and `t` is left with no type at all. This does not make any sense; the trouble is that we need to expand the `Derivative` type. When fully expanded, `derivative` has the following type:

```
derivative :: R -> (R -> R) -> R -> R
```

Now, `dt` has type `R`, `x` has type `R -> R`, `t` has type `R`, and the final `R` is the return type.

In playing the matching game, we are thinking of `derivative` as a function with three inputs and one output. The arrow notation may seem a strange way to specify a function with three inputs. There is a deeper meaning to the notation that we will discuss briefly now and treat more fully in Chapter 6.

Since arrows associate to the right, the following three types are the same to the compiler:

- `R -> (R -> R) -> R -> R`
- `R -> (R -> R) -> (R -> R)`
- `R -> ((R -> R) -> (R -> R))`

Using the thinking from the matching game, it appears that the first type takes three inputs, the second type takes two inputs, and the third type takes one input.

There are three ways to think about the function `derivative`:

- `derivative` takes three inputs, with types `R`, `R -> R`, and `R`, and produces one output with type `R`. In this way of thinking, `derivative` takes a time interval, a position function, and a time and returns the numeric velocity at that time. This was our thinking in the matching game.
- `derivative` takes two inputs, with types `R` and `R -> R`, and produces one output with type `R -> R`. In this way of thinking, `derivative` takes a time interval and a position function and returns a velocity function.
- `derivative` takes one input with type `R` (namely `dt`) and produces one output with type `(R -> R) -> R -> R` (or type `Derivative`). This is the way the compiler thinks of it.

These three ways of thinking are mathematically equivalent, but they seem to sit differently with the brain. The second way of thinking is my favorite because I like to think of a derivative as something that takes a function as input and gives a function as output.

Transforming a function with two inputs into a function with one input whose output is a function is called *currying*, named after the logician Haskell Curry (who donates his first name to the programming language we are using). Currying is discussed more fully in Chapters 6 and 9. Currying allows the compiler to treat all functions as having a single input and a single output, with the understanding that the input and/or the output might be a function.

Modeling the Car's Position and Velocity

Suppose we have a car position function

$$x_C = \cos t \tag{4.8}$$

in which t is in seconds and x_C is in meters. The corresponding Haskell code is as follows:

```
carPosition :: Time -> Position
carPosition t = cos t
```

Using Equation 4.5, we can find the velocity function for the car.

$$v_C = D x_C \tag{4.9}$$

The corresponding Haskell code could be something like the following:

```
carVelocity :: Time -> Velocity
carVelocity = derivative 0.01 carPosition
```

In the Haskell code, `derivative 0.01` is playing the role of the derivative operator D in the mathematical expression. These are not exactly the same thing because D is a true mathematical derivative and `derivative 0.01` is only a numerical derivative, but we can get decent approximate results by using it, and we can improve our results to the degree that we like by using a number smaller than `0.01`. Moreover, `derivative 0.01 sin` is a perfectly valid function with type `R -> R` in the Haskell language and is every bit as legitimate as the function `cos` (also type `R -> R`) that it approximates. It can be evaluated, graphed, differentiated, integrated, or used anywhere that a function with type `R -> R` can be used.

Equation 4.9 is an equality of functions, and the corresponding Haskell code defines the function `carVelocity` without using any function arguments. This is the point-free style introduced in Chapter 2.

It's just as valid to write Haskell code based on Equation 4.6. The mathematical equation would then look like the following:

$$v_C(t) = Dx_C(t) \tag{4.10}$$

And the Haskell code would look like this:

```
carVelocity' :: Time -> Velocity
carVelocity' t = derivative 0.01 carPosition t
```

We use a prime to denote an alternate way of writing the function. As far as the computer is concerned, `carVelocity` and `carVelocity'` mean the same thing. The difference is one of notational preference. We'll often use primes in code to indicate an alternate way of writing something. This prime has nothing to do with the derivative.

The position function for the car is given analytically, so we can take the derivative analytically and write an explicit equation for the velocity of the car.

$$v_C(t) = -\sin t \tag{4.11}$$

We can also write this in Haskell as

```
carVelocityAnalytic :: Time -> Velocity
carVelocityAnalytic t = -sin t
```

but in this book we're not asking the computer to do symbolic algebra or to take derivatives analytically. The function `carVelocityAnalytic` is *not* the same function as `carVelocity` or `carVelocity'`. The numeric value of `carVelocity 2` is close to, but not exactly the same as, the numeric value of `carVelocityAnalytic 2`. In this book, we'll only ask the computer to do things that a scientific calculator can do. Nevertheless, we'll find that Haskell's notation will aid our thinking through attention to the types of expressions, through the concision allowed by higher-order functions and through the simplicity of writing language that avoids mutable state.

There is an even better way to express Equation 4.5 in Haskell. The function `velFromPos` accepts any position function as input and supplies the corresponding velocity function as output.

```
velFromPos :: R                      -- dt
           -> (Time -> Position)  -- position function
```

```
         -> (Time -> Velocity)  -- velocity function
velFromPos dt x = derivative dt x
```

We see that the function to find velocity from position is none other than the derivative function we defined earlier. Note also that we can split a type signature across multiple lines. This is often good practice, and it gives the code writer the opportunity to give a short comment about the meaning of each type in the signature.

If the velocity happens to be constant, say v_0, we can integrate both sides of Equation 4.5 or 4.7 to obtain

$$v_0 t = x(t) - x(0)$$

If velocity is constant, position is a linear function of time.

$$x(t) = v_0 t + x(0)$$

Here is the corresponding Haskell code:

```
positionCV :: Position -> Velocity -> Time -> Position
positionCV x0 v0 t = v0 * t + x0
```

The `CV` at the end of the name is an abbreviation for constant velocity. Notice again the different ways to read the type: we can think of `positionCV` as a function that takes three arguments and returns a `Position`, as a function that takes two arguments and returns a function `Time -> Position`, or as a function that takes one argument and returns a function `Velocity -> Time -> Position`. The expression `positionCV 5 10 2` is the `Position` of an object at time 2 s, if it moves with a constant velocity of 10 m/s and started at a position of 5 m when time was 0. The expression `positionCV 5 10` is the `PositionFunction` that describes an object moving with a constant velocity of 10 m/s that started at a position of 5 m when time was 0.

In everyday speech we often use the terms *velocity* and *speed* interchangeably. The language of physics makes a technical distinction between these two terms. *Speed* is the magnitude (absolute value) of velocity. Speed is never negative. Although speed (that is, how fast something is going) is the simpler idea to understand, velocity is much more important to the theory of motion. Velocity contains more information than speed because it tells the direction of travel as well as the speed. The velocity of a rock thrown straight upward makes a continuous regular change, both while the rock travels upward and while it travels downward. Speed, on the other hand, decreases on the way up and increases on the way down, unnecessarily making it seem like the situation needs to be thought of as two processes. There will be times when it is convenient to speak about speed, and it is certainly worth having a concept and a word for it. When we discuss motion in more than one dimension, velocity will require a vector for its description, while speed will remain a number.

Modeling Acceleration

Acceleration is defined to be the rate at which velocity changes. We define a to be the function that associates with each time t the rate at which the velocity is changing at time t. In the language of calculus, we can write

$$a = Dv \tag{4.12}$$

or

$$a(t) = \frac{dv(t)}{dt} \tag{4.13}$$

to define acceleration. Equation 4.12 is more succinct, but the two equations mean the same thing. We say that $a(t)$ is the acceleration of the car at time t.

We're only dealing with one-dimensional motion in this chapter, so we represent acceleration with a number.

```
type Acceleration = R
```

Equation 4.12 can be encoded in a function `accFromVel` that produces an acceleration function from a velocity function.

```
accFromVel :: R                        -- dt
           -> (Time -> Velocity)       -- velocity function
           -> (Time -> Acceleration)   -- acceleration function
accFromVel = derivative
```

Again, this function is just the derivative. Here, we use point-free style to emphasize the equality of the two functions.

If the acceleration happens to be constant, say a_0, we can integrate both sides of Equation 4.12 or 4.13 to obtain

$$a_0 t = v(t) - v(0)$$

If acceleration is constant, velocity is a linear function of time.

$$v(t) = a_0 t + v(0) \tag{4.14}$$

Here is the Haskell code for Equation 4.14:

```
velocityCA :: Velocity -> Acceleration -> Time -> Velocity
velocityCA v0 a0 t = a0 * t + v0
```

The `CA` at the end of the name is an abbreviation for constant acceleration.

To believe that we really know what is happening with an object in motion, we want an expression that gives the position of the object as a function of time. Since position is the antiderivative or integral of velocity, we can obtain such a relationship by integrating both sides of Equation 4.14 to obtain

$$x(t) - x(0) = \frac{1}{2} a_0 t^2 + v(0) t$$

If acceleration is constant, position is a quadratic function of time.

$$x(t) = \frac{1}{2}a_0 t^2 + v(0)t + x(0) \tag{4.15}$$

Here is the Haskell code for Equation 4.15:

```
positionCA :: Position -> Velocity -> Acceleration
           -> Time -> Position
positionCA x0 v0 a0 t = a0 * t**2 / 2 + v0 * t + x0
```

Equations 4.14 and 4.15 are known as *constant acceleration equations.* They are used over and over again in a typical introductory physics course. Later we will learn some techniques to deal with situations in which acceleration is not constant.

The relationships between time, position, velocity, and acceleration are known as *kinematics*, or the description of motion. These are the quantities required to give a description of how a car on an air track is moving. We will need to introduce additional ideas, like force and mass, before we have a theory that can account for the causes of motion.

Approximate Algorithms and Finite Precision

In mathematics, the derivative is defined by the limit in Equation 4.4. In many cases, a mathematical derivative of an explicitly specified function can be calculated exactly. The numerical `derivative` we defined in this chapter does not take a limit, relying instead on a small but finite ϵ. For this reason, the numerical derivative calculates an approximation to the derivative of a function. We call the rule for computing the derivative from a finite value of ϵ an *approximate algorithm.*

The use of approximate algorithms is the second source of inexactness in our computations. In Chapter 1, we saw that numbers with type `R`, or `Double`, are generally not represented exactly by the computer. Some numbers can be represented exactly, but even a number as seemingly innocent as 0.1 is not represented exactly as an `R`. This is because, like the number 0.2 in Table 1-4, 0.1 requires an infinite binary expansion (0.0001100110011. . .), which the computer truncates at some point. This is usually not a problem because an `R` provides about 15 significant figures of precision, which is enough for our purposes but still *finite precision.* If we add a very small number to a very big number,

```
Prelude> 1e9 + 1e-9
1.0e9
```

the computer just throws away the very small number.

Even if the computer doesn't go to the extreme of throwing away a small number, the relative precision with which the small number is represented gets worse when it is added to a big number. For example, each of the numbers from $1/3$ down to $1/3 \times 10^{18}$ in Table 4-3 has about 15 decimal digits of precision, as indicated by the number of threes in its expression.

Table 4-3: Fractions Evaluated to About 15 Decimal Digits of Precision

Expression		Evaluates to
`1/3`	⇝	`0.3333333333333333`
`1/3000`	⇝	`3.333333333333333e-4`
`1/3e6`	⇝	`3.3333333333333335e-7`
`1/3e9`	⇝	`3.333333333333333e-10`
`1/3e12`	⇝	`3.3333333333333334e-13`
`1/3e15`	⇝	`3.333333333333333e-16`
`1/3e18`	⇝	`3.3333333333333334e-19`

However, when each of these numbers is added to the relatively large number 1, a different number of threes is kept, as shown in Table 4-4, depending on the relative size of the two numbers being added. When adding $1/3 \times 10^9$, for example, only 6 of its 15 threes are retained.

Table 4-4: How Adding a Small Number to a Relatively Big Number Reduces the Relative Precision of the Small Number

Expression		Evaluates to
`1 + 1/3`	⇝	`1.3333333333333333`
`1 + 1/3000`	⇝	`1.0003333333333333`
`1 + 1/3e6`	⇝	`1.0000003333333334`
`1 + 1/3e9`	⇝	`1.0000000003333334`
`1 + 1/3e12`	⇝	`1.0000000000003333`
`1 + 1/3e15`	⇝	`1.0000000000000004`
`1 + 1/3e18`	⇝	`1.0`

The process of adding a small number to a big number is central to the idea of a derivative. We would like ϵ to be small, but the fact that an `R` is only an approximation to a real number means we don't want it to be too small.

Table 4-5 shows the relative error in a numerical derivative of the function f, where $f(x) = x^4/4$. The exact derivative is $Df(x) = x^3$. The derivative is evaluated at $x = 1$, so the exact result is 1. Each row of the table shows the relative error for values of ϵ that range from 1 down to 10^{-18}.

Table 4-5: Relative Error of a Numerical Derivative That Decreases as ϵ Gets Smaller and Increases as ϵ Gets Smaller Still

Expression		Evaluates to
`derivative 1     (\x -> x**4 / 4) 1 - 1`	⇝	`0.25`
`derivative 1e-3  (\x -> x**4 / 4) 1 - 1`	⇝	`2.499998827953931e-7`
`derivative 1e-6  (\x -> x**4 / 4) 1 - 1`	⇝	`1.000088900582341e-12`
`derivative 1e-9  (\x -> x**4 / 4) 1 - 1`	⇝	`8.274037099909037e-8`
`derivative 1e-12 (\x -> x**4 / 4) 1 - 1`	⇝	`8.890058234101161e-5`
`derivative 1e-15 (\x -> x**4 / 4) 1 - 1`	⇝	`-7.992778373592246e-4`
`derivative 1e-18 (\x -> x**4 / 4) 1 - 1`	⇝	`-1.0`

As ϵ decreases from 1 down to 10^{-6}, the error gets smaller. For these values of ϵ, the approximate nature of the derivative algorithm is a larger contribution to the error than the finite precision used by the computer. But as ϵ continues to decrease, the error in the table gets larger. For these values of ϵ, the finite precision used in the calculation and representation of numbers is a larger contribution to the error than the approximate algorithm for computing the derivative.

In the case of the numerical derivative, finite precision wants ϵ to be large so its relative precision is maintained, but the algorithm wants ϵ to be small to approach the true derivative. The best results are obtained somewhere in the middle, around $\epsilon = 10^{-6}$ for the situation in Table 4-5.

These two sources of inaccuracy, finite precision and approximate algorithms, are going to be with us throughout our journey in computational physics. The algorithms for solving differential equations that we introduce later are also approximate algorithms, relying on small but finite steps to solve differential equations that are continuous. We will introduce rules of thumb for choosing such small finite parameters. Our attitude in this book is not to deeply study the interesting subject of numerical analysis, or to take an alarmist view toward inaccuracy, but rather simply to be aware of the nature of approximate calculation so we can produce meaningful results.

Summary

This chapter introduced the notions of position, velocity, acceleration, and time, as well as their relationships, which are articulated by the mathematical idea of a derivative. We saw how various kinematic equations can be encoded in Haskell. In the next chapter, we take a look at lists, which play a role almost as central as functions in functional programming because they are the basis of most of our iteration.

Exercises

Exercise 4.1. Consider the following function:

$$f(x) = \frac{1}{2}x^2$$

The derivative of this function is $Df(x) = x$. In this case, Df is the identity function on real numbers. Because

$$\frac{f(x + \epsilon/2) - f(x - \epsilon/2)}{\epsilon} = \frac{(x + \epsilon/2)^2 - (x - \epsilon/2)^2}{2\epsilon} = \frac{2x\epsilon}{2\epsilon} = x \tag{4.16}$$

even before we take the limit, our numerical `derivative` should give exact results for any ϵ that we use. Write Haskell code to take the derivative of f using `derivative 10`, `derivative 1`, and `derivative 0.1`. You should find that `derivative 10` and `derivative 1` produce the identity function exactly and that `derivative 0.1` comes very close but is not exact. Why does `derivative 0.1` not produce exactly the identity function on real numbers?

Exercise 4.2. Consider the following function:

$$f(x) = x^3$$

The derivative of this function is $Df(x) = 3x^2$. The error introduced by the numerical derivative at a particular value of x is the absolute value of the difference between the numerical derivative evaluated at x and the exact derivative evaluated at x. Write Haskell code to take the derivative of f using `derivative 1`. By evaluating the derivative at different values of x, see if you can find a pattern for the error introduced by the numerical derivative. After you find the pattern for this error, extend your exploration to `derivative a` for different values of `a`. Can you give an expression for the error in terms of `a`?

When $x = 4$, $Df(4) = 48$. What value of `a` produces an error of 1 percent at $x = 4$? When $x = 0.1$, $Df(0.1) = 0.03$. What value of `a` produces an error of 1 percent at $x = 0.1$?

Exercise 4.3. Find a function and a value of its independent variable so that `derivative 0.01` produces at least a 10 percent error compared to the exact derivative.

Exercise 4.4. Consider the cosine function, `cos`, and its numerical derivative `derivative a cos`. For what values of the independent variable (let's call it `t`) is the numerical derivative most sensitive to the value of `a`? For what values is it least sensitive? You should be able to find some values of `t` where `a` can be made very large and the numerical derivative is still a good approximation.

Exercise 4.5. Consider the following position function:

```
pos1 :: Time -> Position
pos1 t = if t < 0
         then 0
         else 5 * t**2
```

Write functions

```
vel1Analytic :: Time -> Velocity
vel1Analytic t = undefined
```

and

```
acc1Analytic :: Time -> Acceleration
acc1Analytic t = undefined
```

for the corresponding velocity and acceleration functions by taking an analytic derivative of the position function.

The `undefined` function can be used as a placeholder for code not yet written. The compiler will accept `undefined` and happily compile the code, but if you try to use a function based on `undefined`, you will get a runtime error.

Write functions

```
vel1Numerical :: Time -> Velocity
vel1Numerical t = undefined
```

and

```
acc1Numerical :: Time -> Acceleration
acc1Numerical t = undefined
```

for the corresponding velocity and acceleration functions by taking a numerical derivative of the position function using `derivative 0.01`. Can you find any values of `t` where `vel1Analytic t` and `vel1Numerical t` differ substantially? Can you find any values of `t` where `acc1Analytic t` and `acc1Numerical t` differ substantially?

5

WORKING WITH LISTS

People naturally make lists. Whether it's a bucket list, shopping list, or list of top ten favorite books, an ordered sequence of items that share something in common sits easy on the brain. The history of functional programming is entwined with that of lists. The early functional language Lisp even has a name that is short for "list processor." In Haskell, lists are just as important because the way we think about iteration in functional programming is often in terms of constructing a list and then using it to produce the result we want.

In this chapter, we'll learn about lists and the functions that work with them. We'll start off with list basics, such as how to construct a list, how to select a particular element of a list, and how to concatenate lists. We'll then see how to give a type to a list. Lists of numbers have a special role to play. There is special syntax for arithmetic sequences, and there are multiple Prelude functions for working with lists of numbers. After that, we'll introduce the idea of type variables. We'll take a short diversion to talk about type conversion before introducing list comprehension, a very useful way to form new lists from old. We'll end the chapter with pattern matching, identifying the data constructors for the list type.

List Basics

A *list* in Haskell is an ordered sequence of data, all with the same type. Here is an example of a list:

```
physicists :: [String]
physicists = ["Einstein","Newton","Maxwell"]
```

The type [String] indicates that physicists is a list of Strings.

Square brackets around the type indicate a list. A list with type [String] can have any number of items (including zero), but each item must have type String. In the second line, we define physicists by enclosing its elements in square brackets and separating the elements by commas. The empty list is denoted as [].

Using the type synonym

```
type R = Double
```

here is a list of real numbers:

```
velocities :: [R]
velocities = [0,-9.8,-19.6,-29.4]
```

Selecting an Element from a List

The list element operator !! can be used to learn the value of an individual element of a list. We use the operator between the list and the place, or *index*, of the element we want. The first element of a list is considered to be element number 0.

```
Prelude> :l Lists.hs
[1 of 1] Compiling Main             ( Lists.hs, interpreted )
Ok, one module loaded.
*Main> velocities !! 0
0.0
*Main> velocities !! 1
-9.8
*Main> velocities !! 3
-29.4
```

The first command loads the file *Lists.hs*, which contains the code in this chapter. This file, along with code files for other chapters, is available at *https://lpfp.io*. After the file is loaded, we can make reference to velocities, a name that would be unknown to GHCi before we loaded the file.

Concatenating Lists

Lists of the same type can be concatenated with the ++ operator shown in Table 1-2. For example, if we have another list that has type [R],

```
moreVelocities :: [R]
moreVelocities = [-39.2,-49.0]
```

we can concatenate velocities with this list:

```
*Main> velocities ++ moreVelocities
[0.0,-9.8,-19.6,-29.4,-39.2,-49.0]
*Main> :t velocities ++ moreVelocities
velocities ++ moreVelocities :: [R]
```

Notice that the concatenation has the same type that each component list has, in this case [R].

Attempting to concatenate lists with different underlying types produces an error. For example, physicists ++ velocities gives an error.

```
*Main> physicists ++ velocities

❶ <interactive>:7:15: error:
  ❷ • Couldn't match type Double with [Char]
    ❸ Expected type: [String]
      ❹ Actual type: [R]
  ❺ • In the second argument of (++), namely velocities
      In the expression: physicists ++ velocities
      In an equation for it: it = physicists ++ velocities
```

This sort of error is called a *type error*. While physicists has a well-defined type, namely [String], and velocities has a well-defined type, namely [R], the expression physicists ++ velocities cannot be given a well-defined type. A type error results when we attempt to apply a function (in this case, the function is the concatenation operator ++) that expects input of one type to a value that actually has a different type. The concatenation operator expects its second argument to have type [String] because physicists has type [String]. However, we gave a second argument of velocities, which has type [R] and thus does not match [String]. In Chapter 6, we will discuss the type of the concatenation operator.

Let's try to understand the error message. The word "interactive" ❶ indicates that the error occurred at the GHCi prompt, rather than in a source code file. The numbers ❶ are the line number and column number where the error occurred. This is useful information for an error in a source code file, but it is not really needed for an error at the GHCi prompt. The text "Couldn't match type" ❷ indicates a type error. The types that don't match are Double and [Char] ❷. Since R is a type synonym for Double and String is a type synonym for [Char], as we will see later in the chapter, the compiler tells us that R and String do not match.

Next, the compiler tells us that a function expected an expression with type `[String]` (the "Expected type") ❸ but was actually given an expression with type `[R]` (the "Actual type") ❹. The error message then tells us that the location of this discrepancy is the second argument of the `++` operator ❺. It's useful to be able to read type errors like this. There is no shame in making such an error. The compiler is helping us by checking that what we say makes sense. In physics, we can use Haskell's type system to great advantage by assigning different types to different conceptual entities. We know, for example, that it makes no sense to add a number to a vector. By giving numbers and vectors different types, we engage Haskell's type system to help us ensure that the code we write does not attempt to add a number to a vector.

Any number of lists of the same type can be concatenated with the `concat` function. If we define a list of strings,

```
shortWords :: [String]
shortWords = ["am","I","to"]
```

then we can make the following concatenation:

```
*Main> concat [shortWords,physicists,shortWords]
["am","I","to","Einstein","Newton","Maxwell","am","I","to"]
```

Arithmetic Sequences

An *arithmetic sequence* is a list formed with two dots (`..`), like so:

```
ns :: [Int]
ns = [0..10]
```

The list `ns` contains the integers from 0 to 10. I chose the name `ns` because it looks like the plural of the name `n`, which seems like a good name for an integer. It is a common style in Haskell programs to use names that end in `s` for lists, but it is by no means necessary.

If we enter a list into GHCi, GHCi will evaluate each element and return the list of evaluated elements:

```
*Main> [0,2,5+3]
[0,2,8]
```

If we give GHCi an arithmetic sequence, GHCi will expand the sequence for us:

```
*Main> [0..10]
[0,1,2,3,4,5,6,7,8,9,10]
```

A second form of arithmetic sequence allows us to increment from one term to the next with a value that is different from `1`:

```
*Main> [-2,-1.5..1]
[-2.0,-1.5,-1.0,-0.5,0.0,0.5,1.0]
```

In this second form, we specify the first, second, and last entries of the desired list. We can even do a decreasing list:

```
*Main> [10,9.5..8]
[10.0,9.5,9.0,8.5,8.0]
```

List Types

List types are a second way to form a new type from an existing type (the first way being function types, as we saw in "Function Types" in Chapter 3). Given any type `a` (`Int`, `Integer`, `Double`, and so on), there is a type `[a]` for lists with elements of type `a`.

You can, for example, make a list of functions. Recall the `square` function we defined in Chapter 2.

```
square :: R -> R
square x = x**2
```

We can define the following list, where `cos` and `sin` are functions defined in the Haskell Prelude:

```
funcs :: [R -> R]
funcs = [cos,square,sin]
```

Why would we want a list of functions? One reason will show up in Chapter 11 when we meet a function that takes a list of functions and plots them all on the same set of axes. A second reason appears in Chapter 14 when the forces that act on an object are functions of time or velocity. Our function for solving Newton's second law will take a list of these force functions to describe the forces that act on the object.

Functions for Lists of Numbers

Haskell has a few Prelude functions that work with lists of numbers. The first two are `sum` and `product`, shown in Table 5-1. As you might expect from its name, `sum` returns the sum of the items in a list, returning `0` for the empty list. The function `product` returns the product of the items in a list, returning `1` for the empty list. The functions `maximum` and `minimum` return the largest and smallest items in a list, respectively, producing errors if you give them the empty list.

Table 5-1: Functions for Lists of Numbers

Expression		Evaluates to
`sum [3,4,5]`	⇝	`12`
`sum []`	⇝	`0`
`product [3,4,5]`	⇝	`60`
`product []`	⇝	`1`
`maximum [4,5,-2,1]`	⇝	`5`
`minimum [4,5,-2,1]`	⇝	`-2`

When Not to Use a List

There will be times when you want to "bundle together" expressions of different types. For example, we may wish to form pairs composed of a person's name (a `String`) and age (an `Int`). A list is not the right structure to use for this job. All elements of a list must have the same type. In Chapter 9, we'll learn about *tuples*, which are a good way to bundle together items of different types.

Type Variables

In the previous section, we saw how the list element operator `!!` returns a specified element of a list. The list element operator doesn't care what type of elements the list contains. We would write `physicists !! 2` to get the number-two element of `physicists` in the same way that we would write `velocities !! 2` to get the number-two element of velocities, even though the former list has type `[String]` while the latter list has type `[R]`.

There are other functions that also don't care what element type a list has. Table 5-2 shows several such functions from the Prelude. The types of these functions are expressed in terms of a *type variable* (`a` in this case). A type variable must start with a lowercase letter and can stand for any type.

Table 5-2: Some Prelude Functions for Working with Lists

Function		Type	Description
`head`	`::`	`[a] -> a`	Returns first item of list
`tail`	`::`	`[a] -> [a]`	Returns all but first item of list
`last`	`::`	`[a] -> a`	Returns last item of list
`init`	`::`	`[a] -> [a]`	Returns all but last item of list
`reverse`	`::`	`[a] -> [a]`	Reverses order of list
`repeat`	`::`	`a -> [a]`	Infinite list of a single item
`cycle`	`::`	`[a] -> [a]`	Infinite list repeating given list

The function `head` returns the first element of a list. You can see some uses of `head`, as well as the other list functions, in Table 5-3.

Table 5-3: Use of List Functions

Expression		Evaluates to
`head ["Gal","Jo","Isaac","Mike"]`	⇝	`"Gal"`
`head [1, 2, 4, 8, 16]`	⇝	`1`
`tail ["Gal","Jo","Isaac","Mike"]`	⇝	`["Jo","Isaac","Mike"]`
`tail [1, 2, 4, 8, 16]`	⇝	`[2,4,8,16]`
`last ["Gal","Jo","Isaac","Mike"]`	⇝	`"Mike"`
`last [1, 2, 4, 8, 16]`	⇝	`16`
`init ["Gal","Jo","Isaac","Mike"]`	⇝	`["Gal","Jo","Isaac"]`
`init [1, 2, 4, 8, 16]`	⇝	`[1,2,4,8]`
`length ["Gal","Jo","Isaac","Mike"]`	⇝	`4`
`length [1, 2, 4, 8, 16]`	⇝	`5`

The function head can accept a list of type [Double], a list of type [Char], or a list of type [Int]. Because head doesn't care about the type of the payload, the best way of expressing the type of input head takes is by using a type variable a to say that head accepts an input of type [a]. The same type variable a appears also in the output; the return type of head is a.

You can see a type variable if you ask GHCi for the type of the empty list.

```
*Main> :t []
[] :: [a]
```

Let's look at a few more of the functions from Table 5-2. The tail function returns everything but the first element of a list. The function last returns the last element of a list. The function init returns everything except the last element. The book *Learn You a Haskell for Great Good!* has a cute picture of a caterpillar (*http://learnyouahaskell.com/starting-out#an-intro-to-lists*) that visually explains these list functions. Table 5-2 gives the types of these functions, and Table 5-3 shows some examples of how to use them.

Having introduced type variables, we are in a good position to take a short diversion into type conversion.

Type Conversion

GHCi appears to allow a Double to be divided by an Int:

```
*Main> 0.4 / 4
0.1
```

However, that is not what is happening here. The number 0.4 can be a Float or a Double. The number 4 can be an Int, Integer, Float, or Double. The division operator demands that the types of the two numbers being divided are the same. In this case, both must be interpreted as Float or both as Double. Addition, subtraction, multiplication, and division require that the two expressions being combined have the same type. In terms of the type variables we introduced earlier, addition, subtraction, multiplication, and division all have type a -> a -> a, meaning that each of the two numbers being combined must have the same type a, and then the result of the operation will also have type a. (The full story of the types of arithmetic operations like addition is more complex. You can't add two Strings, but a function with type a -> a -> a must be able to accept two Strings as input and produce a String as output. The missing piece involves the idea of type classes, which we'll discuss in Chapter 8.)

The Haskell compiler will refuse to divide a Double by an Int. If we give explicit types to some numbers,

```
oneDouble :: Double
oneDouble = 1

twoInt :: Int
twoInt = 2
```

we can see the error the compiler produces:

```
*Main> oneDouble / twoInt

<interactive>:42:13: error:
    • Couldn't match expected type Double with actual type Int
    • In the second argument of (/), namely twoInt
      In the expression: oneDouble / twoInt
      In an equation for it: it = oneDouble / twoInt
```

This is another example of a type error. The first input to the division is a `Double`, so the "expected type" for the second argument of division is also `Double`. The "actual type" we supplied is `Int`, which doesn't match. The compiler will similarly refuse to add a `Float` to a `Double`.

The fact that division can occur only between numbers with the same type can be irritating, especially if we expect the compiler to automatically convert one type into another. The solution is to use a type-conversion function to convert, say, an `Int` to a `Double`.

There are two important type-conversion functions you may need to use from time to time. The first is `fromIntegral`, which converts an `Int` or `Integer` to some other kind of number. The compiler can usually figure out which type to convert to, but it demands your explicit permission through the use of this function. The second conversion function is `realToFrac`, which converts a `Float` to a `Double` or a `Double` to a `Float`. Again, you do not usually need to explicitly specify the type to convert to; you just need to give permission for the conversion to be done. Here is an example:

```
*Main> oneDouble / fromIntegral twoInt
0.5
```

The rationale behind the requirement for conversion is that, in Haskell, most errors are type errors. Type errors often indicate that we haven't completely thought through the code we've written. It could be that dividing a `Double` by an `Int` is not what we intended and we are grateful to the type checker for producing an error rather than quietly converting the `Int` to a `Double`.

This concludes our diversion into type conversion. We can now return to our regularly scheduled program, namely lists.

The Length of Lists

The Prelude provides a function `length` that returns the number of items in a list.

```
*Main> length velocities
4
*Main> length ns
11
```

```
*Main> length funcs
3
```

In the early days of Haskell, `length` was a simple function with a simple type. The type of `length` was `[a] -> Int`, meaning you could give `length` a list of anything, and it would give you back an integer. That was nice and simple. If we ask GHCi today for the type of `length`,

```
*Main> :t length
length :: Foldable t => t a -> Int
```

we see a more complex type. This type involves the idea of a *type class*, which we'll explore in Chapter 8. But for now, we can make our own length function with the simple type `[a] -> Int`:

```
len :: [a] -> Int
len = length
```

If we look at the type of `len` in GHCi,

```
*Main> :t len
len :: [a] -> Int
```

we see the simple type that we want.

There is no great advantage to defining our own length function because we are free to use `length` even though it has a complicated type, but understanding the types of the functions we use gives us real insight into what we are doing. We want to understand the types of the functions we write and use, and we want them to be as simple as they can be. There will, of course, be trade-offs between simplicity and power. The decision of the Haskell designers to give the `length` function a more complicated type means that it can be used in a wider variety of situations. In this case, the designers made a decision favoring power over simplicity. We will often favor simplicity over power.

A String Is a List of Characters

Now that I have introduced lists, I can tell you that a string in Haskell is nothing but a list of characters. In other words, the type `String` is exactly the same as the type `[Char]`; in fact, `String` is defined in the Haskell Prelude to be a type synonym for `[Char]` in exactly the same way that we defined `R` to be a type synonym for `Double`. Haskell provides some special syntax for strings, namely the ability to enclose a sequence of characters in double quotes to form a `String`. This is obviously more pleasant than requiring an explicit list of characters, such as `['W','h','y','?']`. You can ask GHCi whether this is the same as `"Why?"`:

```
*Main> ['W','h','y','?'] == "Why?"
True
```

GHCi responds with `True`, indicating that it regards these two expressions as identical.

The identity of the types `String` and `[Char]` also means that a string can be used in any function that expects a list of something. For example, we can use the function `length` on a string to tell us how many characters it has.

Readers who have some experience with programming may worry about the efficiency implications of representing strings as lists of characters. Rest assured that Haskell has some other options that are more efficient for programmers who need to process a lot of strings. However, for our purposes, we will not need to process a lot of strings, so the basic `String` type is just fine for us.

List Comprehensions

Haskell offers a powerful way to make new lists out of old lists. Suppose you have a list of times (in seconds),

```
ts :: [R]
ts = [0,0.1..6]
```

and you want to have a list of positions for a rock that you threw up in the air at 30 m/s, with each position corresponding to one of the times in the time list. In Exercise 2.2, you wrote a function `yRock30` to produce the position of the rock when given the time. Perhaps your function looked something like the following:

```
yRock30 :: R -> R
yRock30 t = 30 * t - 0.5 * 9.8 * t**2
```

The code below produces the desired list of positions:

```
xs :: [R]
xs = [yRock30 t | t <- ts]
```

The definition of `xs` is an example of a *list comprehension*. The syntax for a list comprehension consists of square brackets, a vertical bar, and a left arrow, as follows:

[function of *item* | *item* `<-` *list*]

This means that, given a function and a list, Haskell will compute that function for each item in the list and then form a list of the resulting values. In our example above, for each `t` in `ts`, Haskell will compute `yRock30 t` and form a list of these values. The list `xs` of positions will be the same length as the original list `ts` of times.

List comprehensions, in conjunction with the `sum` and `product` functions, allow us to write elegant Haskell expressions that mimic the sigma and pi notation in mathematics for sums and products. Table 5-4 shows the correspondence between mathematical and Haskell notations.

Table 5-4: Sum and Product Notation in Haskell

Mathematical notation	Haskell notation
$\sum_{i=m}^{n} f(i)$	`sum [f(i) | i <- [m..n]]`
$\prod_{i=m}^{n} f(i)$	`product [f(i) | i <- [m..n]]`

Infinite Lists

Haskell is a lazy language, meaning that it does not always evaluate everything in the order you might expect. Instead, it waits to see if values are needed before doing any actual work. Haskell's laziness allows for the possibility of infinite lists. Of course, Haskell never actually creates an infinite list, but you can think of the list as infinite because Haskell is willing to continue down the list as far as it needs to. The list `[1..]` is an example of an infinite list. If you ask GHCi to show you this list, it will go on indefinitely. You can enter CTRL-C or something similar to stop the endless printing of numbers.

An infinite list can be convenient when you don't know in advance exactly how much of a list you will want or need. For example, we might want to compute a list of positions of a particle at 0.01 s time increments. We may not know in advance the length of time over which we want this information. If we write our function so that it returns an infinite list of positions, the function will be simpler because it doesn't need to know the total number of positions to calculate.

A good way to view the first several elements of an infinite list is with the `take` function. Try the following in GHCi:

```
*Main> take 10 [3..]
[3,4,5,6,7,8,9,10,11,12]
```

GHCi shows you the first 10 elements of the infinite list `[3..]`.

Two Prelude functions from Table 5-2 create infinite lists. The function `repeat` takes a single expression and returns an infinite list with the expression repeated an infinite number of times. By itself, this function doesn't seem very useful, but in combination with other functions and techniques we'll learn about later, it can be useful.

The Prelude function `cycle` takes a (finite) list and returns the infinite list formed by cycling through the elements of the finite list over and over again. You can get an idea of what `cycle` does by asking GHCi to show you the first several elements of such a list, like the following:

```
*Main> take 10 (cycle [4,7,8])
[4,7,8,4,7,8,4,7,8,4]
```

List Constructors and Pattern Matching

The colon (`:`) operator (called *cons* for historical reasons having to do with the early functional programming language Lisp) from Table 1-2 can be used to attach a single item of type `a` to a list with type `[a]`. For example, `3:[4,5]` is the same as `[3,4,5]`, and `3:[]` is the same as `[3]`.

In Chapter 3, we saw how we could use pattern matching on the `Bool` type. The `Bool` type has two patterns, `False` and `True`. The list type also has two patterns. A list is either the empty list `[]` or the cons `x:xs` of an item `x` with a list `xs`. Every list is exactly one of these two mutually exclusive and exhaustive possibilities. In fact, Haskell internally regards lists as being formed out of the two *constructors* (also called *data constructors*,) `[]` and `:`. Each type in Haskell has one or more data constructors that are used to form expressions of that type. Therefore, a data constructor is a way of making an expression of a particular type. When we define our own types in Chapter 10, we will see that data constructors are an essential part of the definition for the type.

The list we think of as `[13,6,4]` is represented internally as `13:6:4:[]`, which means `13:(6:(4:[]))` when we allow for the right associativity of `:`. Table 5-5 shows the data constructors for the Boolean and list types.

Table 5-5: Data Constructors for the Boolean and List Types

Type	Data constructors
`Bool`	`False, True`
`[a]`	`[], :`

The fundamental mechanism for pattern matching in Haskell is the `case-of` construction. If we are pattern matching on a `Bool`, there will be two cases corresponding to the two data constructors `False` and `True` from which all values of type `Bool` are constructed. If we are pattern matching on a list, there will be two cases corresponding to the two data constructors `[]` and `:` from which all lists are constructed.

Let's look at an example of defining a function on lists using pattern matching. Let's define a function `sndItem` that returns the second element of a list, or gives an error if the list has fewer than two elements. The idea is that `sndItem [8,6,7,5]` should return `6`. Our first definition uses the `case-of` construction:

```
sndItem :: [a] -> a
sndItem ys = case ys of
               []     -> error "Empty list has no second element."
               (x:xs) -> if null xs
                         then error "1-item list has no 2nd item."
                         else head xs
```

In the case where `ys` is the empty list, we use the `error` function, which has type `[Char] -> a`, meaning that it takes a string as input and can serve as any type. The error function halts execution and returns the given string as a message.

In the case where the input is the cons of an item and a list, the notation (x:xs) indicates that the item will be assigned the name x and the list will be assigned the name xs for use in the body of the definition (the body is the expression to the right of the arrow). For example, if ys is the list [1879,3,14], then x will be assigned the value 1879 and xs will be assigned the value [3,14]. This assignment of the names x and xs is *local*, meaning that it only holds in the body of the definition. Outside of the definition body, the names x and xs may have another meaning or no meaning at all.

The expression null xs returns True if xs is empty and False otherwise. If xs is empty, the original list x:xs had only one element, so we give the one-element error. If xs is not empty, its first element (its head) is the second element of the original list, and hence the value we should return.

If the value we are pattern matching (ys from earlier) is also an input to a function, we can do pattern matching directly on the input rather than using the case-of construction.

```
sndItem2 :: [a] -> a
sndItem2 []     = error "Empty list has no second element."
sndItem2 (x:xs) = if null xs
                  then error "1-item list has no 2nd item."
                  else head xs
```

Notice that we no longer need the local variable ys in sndItem2.

In pattern matching on the input, the definition is expressed in parts, with one part for each data constructor of the input type. In this case, the input type is a list, and the data constructors for a list are the empty list and cons, so part 1 of the definition defines sndItem2 for the empty list and part 2 of the definition defines the function for the cons of an item with a list.

We can make an even nicer function definition by going one step further and using additional pattern matching on the xs list in sndItem. Let's define a function sndItem3 that does the same thing as sndItem but uses even more pattern matching:

```
sndItem3 :: [a] -> a
sndItem3 ys = case ys of
                []       -> error "Empty list has no second element."
                (x:[])   -> error "1-item list has no 2nd item."
                (x:z:_)  -> z
```

Case 1 of sndItem remains unchanged. Case 2 of sndItem splits into two subcases, depending on whether the xs in sndItem is the empty list or the cons of an item with a list.

Notice the underscore character (_) in the last line. Names that begin with an underscore in Haskell are names that we never intend to refer to. We could have written (x:z:zs) in place of (x:z:_). The underscore means that we can't be bothered to give the list a real name because we have no intention of using it or referring to it again. We make no reference to this list in the definition body, so there is no motivation to give it a proper name. Sometimes it is helpful to the code reader (who might be you) to give a name to something that never gets used. If you want to give something a name *and* signal that it never gets used, you can use a name that begins with an underscore, such as `_zs`. Note finally that the cons operator (`:`) is right associative, so the expression `x:z:_` is read as `x:(z:_)`.

Because the value we are pattern matching is the input to our function, we can do pattern matching on the input rather than use the `case-of` construction.

```
sndItem4 :: [a] -> a
sndItem4 []       = error "Empty list has no second element."
sndItem4 (x:[])  = error "1-item list has no 2nd item."
sndItem4 (x:z:_) = z
```

Summary

This chapter introduced lists. Each member of a list must have the same type. Square brackets serve two roles for lists. A type enclosed in square brackets is a list type, and a sequence of items enclosed in square brackets and separated by commas forms a list. A type variable serves as a placeholder for any type. Several list functions have types with type variables because they don't care about the underlying type of a list. Since addition, subtraction, multiplication, and division only work between numbers with the same type in Haskell, we introduced two type-conversion functions for situations in which conversion is necessary. A list comprehension is a method to form a new list from an existing list. Haskell allows for infinite lists because it is a lazy language. Lists are formed from two constructors: the empty list and the cons operator. Pattern matching can be used to define a function that takes a list as input. There are two patterns that a list may have: either it is the empty list or it is the cons of an element and another list.

Exercises

Exercise 5.1. Give an abbreviation for the following list using the double dot (`..`) notation. Use GHCi to check that your expression does the right thing.

```
numbers :: [R]
numbers = [-2.0,-1.2,-0.4,0.4,1.2,2.0]
```

Exercise 5.2. Write a function `sndItem0 :: [a] -> a` that does the same thing as `sndItem` but does not use any pattern matching.

Exercise 5.3. What is the type of the following expression?

```
length "Hello, world!"
```

What is the value of the expression?

Exercise 5.4. Write a function with type `Int -> [Int]` and describe in words what it does.

Exercise 5.5. Write a function `null'` that does the same thing as the Prelude function `null`. Use the Prelude function `length` in your definition of `null'` but do not use the function `null`.

Exercise 5.6. Write a function `last'` that does the same thing as the Prelude function `last`. Use the Prelude functions `head` and `reverse` in your definition of `last'` but do not use the function `last`.

Exercise 5.7. Write a function `palindrome :: String -> Bool` that returns `True` if the input string is a palindrome (a word like *radar* that is spelled the same backward as it is forward) and `False` otherwise.

Exercise 5.8. What are the first five elements of the infinite list `[9,1..]`?

Exercise 5.9. Write a function `cycle'` that does the same thing as the Prelude function `cycle`. Use the Prelude functions `repeat` and `concat` in your definition of `cycle'` but do not use the function `cycle`.

Exercise 5.10. Which of the following are valid Haskell expressions? If an expression is valid, give its type. If an expression is not valid, say what is wrong with it.

(a) `["hello",42]`

(b) `['h',"ello"]`

(c) `['a','b','c']`

(d) `length ['w','h','o']`

(e) `length "hello"`

(f) `reverse` (Hint: this is a valid Haskell expression, and it has a well-defined type, even though GHCi cannot print the expression.)

Exercise 5.11. In an arithmetic sequence, if the specified last element does not occur in the sequence,

```
*Main> [0,3..8]
[0,3,6]
*Main> [0,3..8.0]
[0.0,3.0,6.0,9.0]
```

the result seems to depend on whether you are using whole numbers. Explore this and try to find a general rule for where an arithmetic sequence will end.

Exercise 5.12. In the 1730s, Leonhard Euler showed that

$$\sum_{n=1}^{\infty} \frac{1}{n^2} = \frac{\pi^2}{6}$$

Write a Haskell expression to evaluate

$$\sum_{n=1}^{100} \frac{1}{n^2}$$

Exercise 5.13. The number $n!$, called "n factorial," is the product of the positive integers less than or equal to n:

$$n! = n(n-1)\ldots 1$$

Here's an example:

$$5! = 5 \times 4 \times 3 \times 2 \times 1 = 120$$

Using the `product` function, write a factorial function.

```
fact :: Integer -> Integer
fact n = undefined
```

Exercise 5.14. The exponential function is equal to the following limit:

$$\exp(x) = \lim_{n\to\infty} \left(1 + \frac{x}{n}\right)^n$$

Write a function

```
expList :: R -> [R]
expList x = undefined
```

that takes a real number x as input and produces an infinite list of successive approximations to $\exp(x)$:

$$\left[\left(1+\frac{x}{1}\right)^1, \left(1+\frac{x}{2}\right)^2, \left(1+\frac{x}{3}\right)^3, \ldots\right]$$

How big does n need to be to get within 1 percent of the correct value for $x = 1$? How big does n need to be to get within 1 percent of the correct value for $x = 10$?

Exercise 5.15. The exponential function is equal to the following infinite series:

$$\exp(x) = \sum_{m=0}^{\infty} \frac{x^m}{m!}$$

Write a function

```
expSeries :: R -> [R]
expSeries x = undefined
```

that takes a real number x as input and produces an infinite list of successive approximations to $\exp(x)$:

$$\left[1, 1 + x, 1 + x + \frac{x^2}{2}, \ldots\right]$$

How big does n need to be for

$$\sum_{m=0}^{n} \frac{x^m}{m!}$$

to be within 1 percent of the correct value for $x = 1$? How big does n need to be to get within 1 percent of the correct value for $x = 10$? You may want to use the function `fromIntegral` here.

6

HIGHER-ORDER FUNCTIONS

Higher-order functions are central to functional programming and flow naturally from the idea that functions should be "first-class objects" in a language, with all of the rights and privileges that numbers or lists have. A *higher-order function* is one that accepts a function as input and/or returns a function as output. Many things we want the computer to do for us find a natural expression as a higher-order function.

In this chapter, we'll first look at higher-order functions that produce functions as output. We'll see that these higher-order functions can be viewed as having one input, or alternatively, as having multiple inputs. From there, we'll consider mapping, the idea of applying a function to every element of a list and producing a list of the results. We'll then show how we can use the higher-order function `iterate`, another function that takes a function as input, to do our iteration. After looking at anonymous higher-order functions and operators, we'll look at predicate-based higher-order functions. Finally, we'll look in detail at numerical integration, a central tool in physics that has a natural expression as a higher-order function.

How to Think About Functions with Parameters

Consider the force of a linear spring with a spring constant k. We usually write this as

$$F_{\text{spring}} = -kx$$

where the minus sign indicates that the force acts in the direction opposite the displacement.

Suppose we wish to write a Haskell function to give the force in Newtons produced by a spring with a spring constant of 5,500 N/m. We could write the following:

```
springForce5500 :: R -> R
springForce5500 x = -5500 * x
```

This is a fine function, but it handles only the force produced by a spring with a spring constant of 5500 N/m. It would be nicer to have a function that could handle a spring with any spring constant.

Note that as usual, we are using the type synonym

```
type R = Double
```

because we like to think of these numbers as real numbers.

Now consider the following function:

```
springForce :: R -> R -> R
springForce k x = -k * x
```

Because arrows between types associate to the right, `springForce`'s type, `R -> R -> R`, is equivalent to `R -> (R -> R)`, meaning that if we send the `spring Force` function an `R` (the spring constant), it will return to us a *function* with type `R -> R`. This latter function wants an `R` as input (the displacement) and will give an `R` as output (the force).

We can look at the types of these functions using GHCi's `:type` command (abbreviated `:t`):

```
Prelude> :l HigherOrder.lhs
[1 of 1] Compiling Main             ( HigherOrder.lhs, interpreted )
Ok, one module loaded.
*Main> :t springForce
springForce :: R -> R -> R
```

Next, let's look at the function `springForce 2200`:

```
*Main> :t springForce 2200
springForce 2200 :: R -> R
```

The function `springForce 2200` represents the force function (input: displacement, output: force) for a spring with a spring constant of 2200 N/m. It has the same type and plays the same role as the `springForce5500` function above. However, it looks funny because it is a function made up of two parts: the `springForce` part and the `2200` part.

Finally, look at the type of springForce 2200 0.4:

```
*Main> :t springForce 2200 0.4
springForce 2200 0.4 :: R
```

This is not a function but just a number representing the force exerted by a spring with a spring constant of 2200 N/m when extended by a distance of 0.4 m.

A function that takes another function as input or returns another function as a result is called a higher-order function. The function springForce is a higher-order function because it returns a function as its result. Figure 6-1 shows that the springForce function takes a number as input (the spring constant k :: R) and returns a function as output (springForce k :: R -> R). The function springForce k then takes a number as input (the displacement x :: R) and returns a number as output (the force springForce k x :: R).

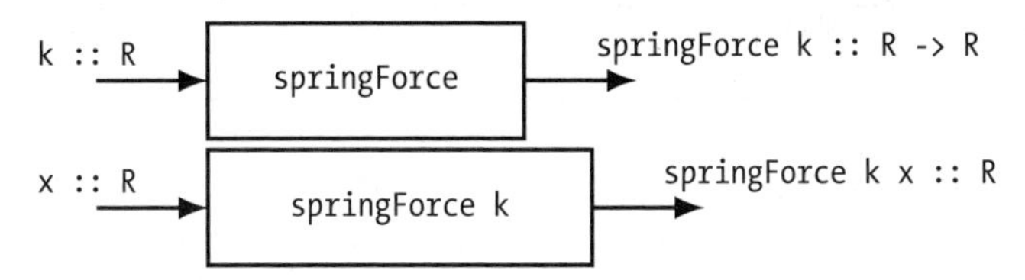

Figure 6-1: The higher-order function springForce takes a number as input and returns the function springForce k as output. The function springForce k then takes a number as input and returns a number as output.

Higher-order functions give us a convenient way to define a function that takes one or more parameters (like the spring constant) in addition to its "actual" input (like the displacement). Table 6-1 shows some higher-order functions from the Prelude that return a function as output.

Table 6-1: Some Higher-Order Functions from the Prelude That Produce a Function as Output

Function		Type
take	::	Int -> [a] -> [a]
drop	::	Int -> [a] -> [a]
replicate	::	Int -> a -> [a]

Consider the higher-order function take. The take function produces a list by taking a given number of elements from a given list. Table 6-2 shows some examples of its use.

Table 6-2: Examples of the Use of take

Expression		Evaluates to
take 3 [9,7,5,3,17]	⇝	[9,7,5]
take 3 [3,2]	⇝	[3,2]
take 4 [1..]	⇝	[1,2,3,4]
take 4 [-10.0,-9.5..10]	⇝	[-10.0,-9.5,-9.0,-8.5]

Let's look at the type of take:

```
*Main> :t take
take :: Int -> [a] -> [a]
```

According to the type of take, when given an Int, it should return a function with type [a] -> [a]. What function should take return? If we give the integer *n* to take, the returned function will accept a list as input and return a list of the first *n* elements of the input list.

There are two ways to think about the higher-order function take (and others like it that return a function as output), as shown in Table 6-3.

Table 6-3: Two Ways of Thinking About the Higher-Order Function take

Way of thinking	Input to take	Output from take
One-input thinking	Int	[a] -> [a]
Two-input thinking	Int and then [a]	[a]

We have already described, with springForce, the "one-input thinking" in which we read the type signature of take as expecting a single Int as input and producing a [a] -> [a] (read "list of a to list of a") as output. Figure 6-2 shows the one-input picture of take.

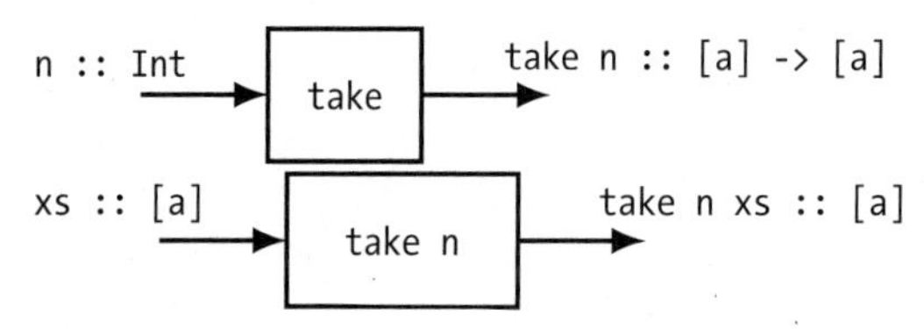

Figure 6-2: One-input thinking about the higher-order function take

An alternative way to think about the type signature Int -> [a] -> [a] is that the function expects two inputs, the first of type Int and the second of type [a], and produces an output of type [a]. Figure 6-3 shows the two-input picture of take.

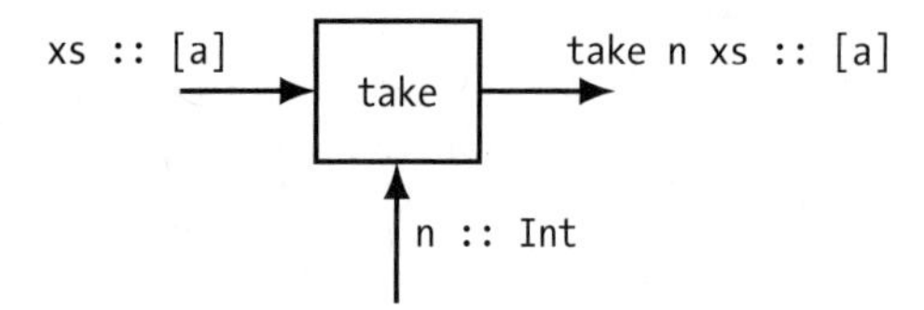

Figure 6-3: Two-input thinking about the higher-order function take

As readers and writers of higher-order functions, we have a choice in how to think about them. Sometimes it is convenient to think of a higher-order function as accepting multiple inputs, but it can also be very useful to think of every function, higher-order functions included, as accepting exactly one input. The Haskell compiler regards every function as having exactly one input.

Like `take`, the higher-order function `drop` is another everyday tool for working with lists. The function `drop` produces a list by discarding a given number of elements from a given list. Table 6-4 shows some examples of its use.

Table 6-4: Examples of the Use of `drop`

Expression		Evaluates to
`drop 3 [9,7,5,10,17]`	⇝	`[10,17]`
`drop 3 [4,2]`	⇝	`[]`
`drop 37 [-10.0,-9.5..10]`	⇝	`[8.5,9.0,9.5,10.0]`

The function `replicate` produces a list by repeating one item a given number of times. Table 6-5 shows some examples of its use.

Table 6-5: Examples of the Use of `replicate`

Expression		Evaluates to
`replicate 2 False`	⇝	`[False,False]`
`replicate 3 "ho"`	⇝	`["ho","ho","ho"]`
`replicate 4 5`	⇝	`[5,5,5,5]`
`replicate 3 'x'`	⇝	`"xxx"`

In this section, we've focused on higher-order functions that return functions as output. Now let's take a look at a higher-order function that takes a function as input.

Mapping a Function Over a List

Table 6-6 shows some higher-order Prelude functions that take other functions as input.

Table 6-6: Some Higher-Order Functions from the Prelude That Accept a Function as Input

Function		Type
`map`	`::`	`(a -> b) -> [a] -> [b]`
`iterate`	`::`	`(a -> a) -> a -> [a]`
`flip`	`::`	`(a -> b -> c) -> b -> a -> c`

The Prelude function `map` is a nice example of a higher-order function that takes another function as input. The function `map` will apply the function you give to every element of the list you give. Table 6-7 shows some examples of the use of `map`.

Table 6-7: Examples of the Use of `map`

Expression		Evaluates to
`map sqrt [1,4,9]`	⇝	`[1.0,2.0,3.0]`
`map length ["Four","score","and"]`	⇝	`[4,5,3]`
`map (logBase 2) [1,64,1024]`	⇝	`[0.0,6.0,10.0]`
`map reverse ["Four","score"]`	⇝	`["ruoF","erocs"]`

In the first example listed in Table 6-7, we say that the function `sqrt` gets "mapped" over the list, meaning that it gets applied to each element of the list.

Note that the parentheses around the function type `a -> b` in the type signature for `map` are essential. The type `a -> b -> [a] -> [b]`, in which no parentheses appear, is an entirely different type. This latter type would take as input a value of type `a`, a value of type `b`, and a list of type `[a]` and produce a list of type `[b]` as output. This latter type is short for `a -> (b -> ([a] -> [b]))` because arrows associate to the right.

A list comprehension can do the work of `map`. Whether you choose `map` or a list comprehension is a matter of style.

```
*Main> map sqrt [1,4,9]
[1.0,2.0,3.0]
*Main> [sqrt x | x <- [1,4,9]]
[1.0,2.0,3.0]
```

The idea of mapping a function over a structure actually extends past the structure of a list to other structures like trees. Haskell has a function `fmap` for this, although we will not have occasion to use it in this book.

Iteration and Recursion

Iteration is an essential feature in any programming language. The ability to do things over and over again is one of the chief sources of power that computers have. How do people express the idea of iteration in a programming language?

Imperative programming languages provide ways to write *loops*, which are instructions to do something over and over, either a fixed number of times or until some condition is reached.

Functional programming languages have a different way to express iteration. The most popular way to achieve iteration among functional programmers is to write *recursive functions*, that is, functions that call themselves. Recursive functions are very powerful, but it takes a person approaching recursive functions for the first time some time and effort to comprehend how to write them.

The Haskell Prelude has many built-in recursive functions that we can use without needing to write our own explicitly. In this book, we'll avoid writing *explicitly* recursive functions, meaning that you can see from the function definition that the function calls itself. We will write functions using the built-in recursive functions from the Prelude, and these functions that we write may legitimately be called recursive because their behavior is recursive in that something is calling itself under the hood. However, the functions we write are not *explicitly* recursive because they will not call themselves.

We can understand and explain most of the recursive functions in the Prelude by their behavior through examples without needing to understand how they achieve this behavior through the power of recursion. Recursion is certainly very interesting, and I encourage you to look into it if you have the time. I learned recursion from *The Little Schemer* [4], a book using the Scheme language that I enthusiastically recommend. The book *Learn You a Haskell for Great Good!* [1], deals with recursion in its Chapter 4, showing how to write Prelude functions like `take`, `reverse`, `replicate`, and `repeat`.

How will we achieve iteration in Haskell without writing explicitly recursive functions? We'll use the Prelude function `iterate`. Rather than thinking imperatively about what we want the computer to *do*, we will think functionally about what we want to *have* and arrange to have these things in a list. "Lists instead of loops" is the functional programming slogan we might apply here.

The Prelude function `iterate`, whose type is given in Table 6-6, is a higher-order function that takes a function as input. Iteration and the `iterate` function will be important when we solve Newton's second law in Part II. The function `iterate` produces an infinite list as follows: if `f :: a -> a` and `x :: a` (read "`f` has type `a` to `a`" and "`x` has type `a`"), then `iterate f x` produces this infinite list:

```
[x, f x, f (f x), f (f (f x)), ...]
```

In other words, the result is a list with `f` applied zero times, once, twice, three times, and so on. Table 6-8 shows some examples of the use of `iterate`.

Table 6-8: Examples of the Use of `iterate`

Expression		Evaluates to
`iterate (\n -> 2*n) 1`	⇝	`[1,2,4,8,...]`
`iterate (\n -> n*n) 1`	⇝	`[1,1,1,1,...]`
`iterate (\n -> n*n) 2`	⇝	`[2,4,16,256,...]`
`iterate (\v -> v - 9.8*0.1) 4`	⇝	`[4.0,3.02,2.04,1.06,...]`

Figure 6-4 shows how `iterate f` applies the function `f` to its input zero times, one time, two times, and so on, and collects the results in a list.

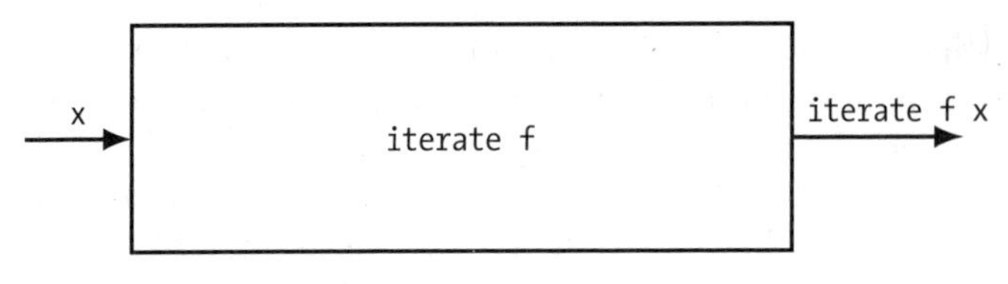

is equivalent to

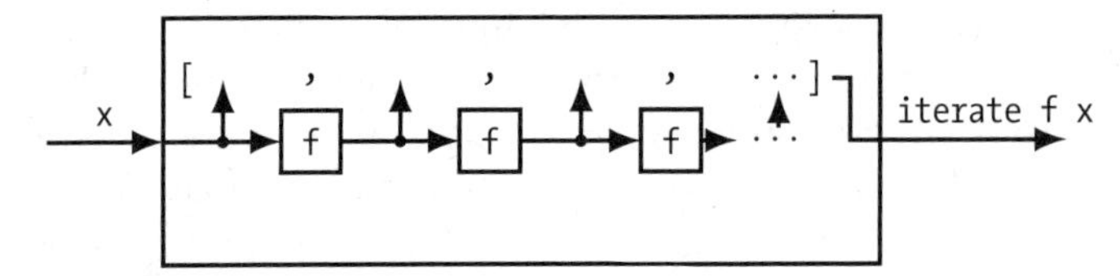

Figure 6-4: The function `iterate f` applies the function f to the input zero, one, two, three, and more times and collects the results in an infinite list.

Anonymous Higher-Order Functions

In Chapter 2, we discussed anonymous functions as a way to describe a function without giving it a name. We can do the same thing for higher-order functions, describing them without giving them a name.

Let's return to the function `springForce` discussed earlier. How could we write `springForce` without naming it? There are actually two ways to write this function as an anonymous function, corresponding to the one-input thinking and two-input thinking we described in "How to Think About Functions with Parameters." In one-input thinking, we regard the input to `springForce` as being a number (`R`) and the output as being a function `R -> R`. The anonymous function for one-input thinking is shown in the first row of Table 6-9. We can see from the form of that function that it returns a function, namely `\x -> -k*x`.

Table 6-9: Two Ways of Writing the `springForce` Function as an Anonymous Function

Way of thinking	Anonymous function
One-input thinking	`\k -> \x -> -k*x`
Two-input thinking	`\k x -> -k*x`

In two-input thinking, we regard the inputs to `springForce` as an `R` for the spring constant and a second `R` for the position and the output as being simply an `R`. The anonymous function for two-input thinking is shown in the second row of Table 6-9. The form of the anonymous function shows us that it returns a number. Either form is completely legitimate. In fact, the two forms describe the same function.

Operators as Higher-Order Functions

In Chapter 1, we introduced several infix operators in Table 1-2. Any infix operator can be converted into a higher-order function by enclosing it in parentheses. Table 6-10 shows examples of how infix operators may be written as higher-order functions.

Table 6-10: Infix Operators Transformed into Prefix Functions by Enclosing in Parentheses

Infix expression	Equivalent prefix expression
`f . g`	`(.) f g`
`'A':"moral"`	`(:) 'A' "moral"`
`[3,9] ++ [6,7]`	`(++) [3,9] [6,7]`
`True && False`	`(&&) True False`
`log . sqrt $ 10`	`($) (log . sqrt) 10`

Table 6-11 shows the types of some higher-order functions obtained from operators. If you want to ask GHCi for the type of an operator, you must enclose the operator in parentheses when using the GHCi `:t` command.

Table 6-11: Infix Operators Viewed as Higher-Order Functions

Function		Type
`(.)`	`::`	`(b -> c) -> (a -> b) -> a -> c`
`(:)`	`::`	`a -> [a] -> [a]`
`(++)`	`::`	`[a] -> [a] -> [a]`
`(&&)`	`::`	`Bool -> Bool -> Bool`
`(||)`	`::`	`Bool -> Bool -> Bool`
`($)`	`::`	`(a -> b) -> a -> b`

In Chapter 5, we discussed a type error that occurred when trying to concatenate lists with different types. Now that we know the type of `(++)`, we are in a better position to understand type errors more generally. A type error comes from trying to apply a function that expects input of one type to an expression that actually has a different type. When we apply the concatenation function `(++)` to the expression `physicists`,

```
physicists :: [String]
physicists = ["Einstein","Newton","Maxwell"]
```

we obtain a function `(++) physicists`, which has type `[String] -> [String]`:

```
*Main> :t (++)
(++) :: [a] -> [a] -> [a]
*Main> :t physicists
```

```
physicists :: [String]
*Main> :t (++) physicists
(++) physicists :: [String] -> [String]
```

The concatenation function accepts a list of any type a. When given a list of strings, the concrete type String is substituted for all occurrences of the type variable a so that the function (++) physicists has type [String] -> [String]. This function (++) physicists expects an input with type [String], so if we give it something with a different type instead, we get a type error.

```
velocities :: [R]
velocities = [0,-9.8,-19.6,-29.4]
```

In GHCi, we get:

```
*Main> :t (++) physicists velocities

<interactive>:1:17: error:
    • Couldn't match type Double with [Char]
      Expected type: [String]
        Actual type: [R]
    • In the second argument of (++), namely velocities
      In the expression: (++) physicists velocities
```

The "Expected type" [String] is the type expected as input by the function (++) physicists, while the "Actual type" [R] is the type of the list velocities.

A class of functions and operators known as *combinators* that can be thought of as standard connectors can make it easier to use higher-order functions. We turn to these next.

Combinators

Combinators, in the broad sense of the term, are functions that combine things. They are often functions that have very general applicability, and their types show this by being full of type variables. They tend not to be specific to numbers, lists, Booleans, or any particular basic type. They are standard plumbing fixtures that make it easier to work with and connect higher-order functions. Table 6-12 gives a short list of Haskell functions that are considered to be combinators.

Table 6-12: Some Haskell Functions Regarded as Combinators

Function		Type
id	::	a -> a
const	::	a -> b -> a
flip	::	(a -> b -> c) -> b -> a -> c
(.)	::	(b -> c) -> (a -> b) -> a -> c
($)	::	(a -> b) -> a -> b

The identity function `id` seems at first to be a silly function, because when I apply it to some value, I get that same value back. So what purpose does the identity function serve? The relevance of the identity function, like the anonymous functions we studied in Chapter 2, lies in its use as an input to a higher-order function. In Chapter 22, we will write a higher-order function to display vector fields. One of the inputs to this function will be a scaling function that accepts one number and returns another. The most trivial scaling is achieved by using the identity function, which leaves every input unchanged. The higher-order function requires *some* scaling function to be given, and the identity function `id` can be given when no scaling is desired.

The combinator `const` turns a value into a constant function that returns that value. For example, `const 3` is the function that returns `3` regardless of its input. So, `const 3 7` evaluates to `3`, `const 3 5` evaluates to `3`, and `const 3 "Hi"` evaluates to `3`. In Chapter 14, we will write a function to solve mechanics problems that takes a list of forces as input. The forces are functions that depend on inputs like velocity, so if we want to specify a constant force, we can use the `const` combinator. In Chapter 23, we will write surfaces like hemispheres and spheres. Our surfaces will be specified with two parameters that take values over some range. The limits for the parameters are allowed to be functions, so we can write, for example, a triangular surface. If we want a constant function, again we can use `const`.

The `flip` combinator takes a higher-order function and exchanges the positions of its inputs. It is often used in conjunction with point-free style. For example, the exponentiation function (**) takes two inputs: the base and the exponent. If we provide the first input but not the second, as in `(**) 2`, we get the "two to the power of" function. What if we want the function that cubes its input? We would need to fix the second input of (**) at `3` while leaving the first open to be the input to the cubing function. We could do this with an anonymous function, as `\x -> x ** 3`. We could also do it with flip, as `flip (**) 3`. Either of these is an expression for the cubing function.

The composition combinator was listed in Table 1-2 and discussed in Chapter 2. We will use it in Chapter 16 when we solve mechanics problems by passing information through a sequence of functions. The full solution is given as a composition of the functions in the sequence.

The function application combinator was listed in Table 1-2 and discussed in Chapter 1. It seems like a silly and useless operator until we realize that its precedence allows it to act like parentheses.

The next section discusses a class of higher-order functions used to classify data.

Predicate-Based Higher-Order Functions

A *predicate* is a function with type `a -> Bool`, where `a` is any valid Haskell type. (For example, `a` could be a concrete type like `Int` or `R -> R`, a type variable like a, or a type that contains type variables like [a] or even a -> [b].) A predicate expresses a property that an element of type a may or may not have. For example, the property of an integer being greater than or equal to 7 is a predicate. We can define such a predicate in Haskell.

```
greaterThanOrEq7 :: Int -> Bool
greaterThanOrEq7 n = if n >= 7 then True else False
```

Table 6-13 shows a few higher-order functions that take a predicate as the first argument.

Table 6-13: Some Predicate-Based Higher-Order Functions from the Prelude

Function		Type
`filter`	`::`	`(a -> Bool) -> [a] -> [a]`
`takeWhile`	`::`	`(a -> Bool) -> [a] -> [a]`
`dropWhile`	`::`	`(a -> Bool) -> [a] -> [a]`

Let's examine the use of these functions. Suppose we define the following "less than 10" predicate:

```
lt10 :: Int -> Bool
lt10 n = n < 10
```

Table 6-14 shows examples of how to use the higher-order functions in Table 6-13.

Table 6-14: Examples of the Use of Some Predicate-Based Higher-Order Functions

Expression		Evaluates to
`filter    lt10 [6,4,8,13,7]`	⇝	`[6,4,8,7]`
`takeWhile lt10 [6,4,8,13,7]`	⇝	`[6,4,8]`
`dropWhile lt10 [6,4,8,13,7]`	⇝	`[13,7]`
`any       lt10 [6,4,8,13,7]`	⇝	`True`
`all       lt10 [6,4,8,13,7]`	⇝	`False`

Let's go through the functions in this table. The `filter` function returns all elements in a list that satisfy the predicate, regardless of where they occur in the list. The `takeWhile` function returns elements of a list that satisfy the predicate *until* it finds one that fails to satisfy the predicate, and it returns the initial list of satisfying elements. Elements in the input list that occur after the first non-satisfying element are not even considered for inclusion in the result list. The `dropWhile` function returns a list that begins with the first non-satisfying element of the input list and includes every element from that point on, regardless of whether it satisfies the predicate. The `any` function returns `True` if one or more elements of the input list satisfy the predicate and returns `False` otherwise. The `all` function returns `True` if all elements of the input list satisfy the predicate, and it returns `False` otherwise.

A list comprehension can also do the work of `filter`.

```
*Main> filter lt10 [6,4,8,13,7]
[6,4,8,7]
*Main> [x | x <- [6,4,8,13,7], x < 10]
[6,4,8,7]
```

To filter a list using a list comprehension, include a Boolean expression (`x < 10` in the example above) after a comma on the right side of the list comprehension. Such an expression is called a *guard*. Only terms that satisfy the Boolean guard will be included in the resulting list.

Numerical Integration

Acceleration is the rate at which velocity changes. If we know how velocity depends on time, we can use a derivative to find acceleration, as we did in Chapter 4. What about the converse problem? What if we know how acceleration depends on time and we want to know velocity? This is the purpose of integration in calculus. Integrating is the opposite of differentiating (or taking the derivative); this claim is the content of the fundamental theorem of calculus.

Introducing Integrators

If $a(t)$ is the acceleration of an object at time t, the velocity $v(t)$ of the object can be found by integrating

$$v(t) = v(0) + \int_0^t a(t')\, dt' \tag{6.1}$$

where $v(0)$ is the velocity of the object at time 0.

If we are integrating and the variable of integration is time, we can imagine a device that takes as input the value of acceleration at time t and produces as output the value of velocity at time t. Let's call such a device an *integrator*, and we can picture it as in Figure 6-5.

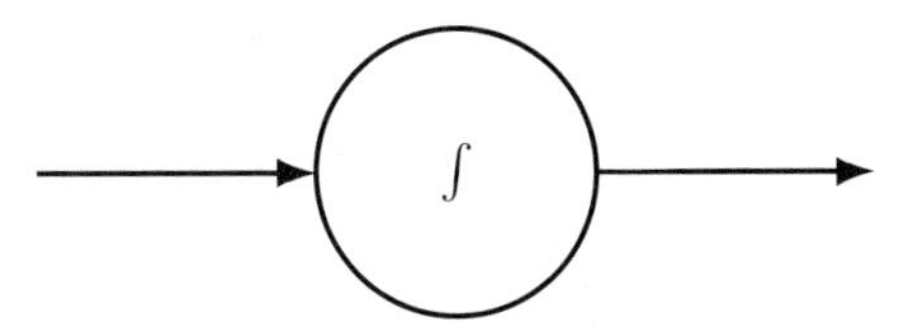

Figure 6-5: An integrator is continuous and stateful. The continuously changing output at the right depends on the continuously changing input at the left in addition to some stored state.

The input to the integrator describes the rate at which the output is to change. For example, the input could be the volume flow rate of water from a faucet into a tub in gallons per minute. The output is the volume of water in the tub in gallons. Or the input could be the current flowing into a capacitor, and the output the charge on a capacitor plate. The current flowing into the capacitor describes the rate at which charge is deposited on the capacitor plate. Table 6-15 shows some physical quantities related by integration.

Table 6-15: Physical Quantities Related by Integration

Input to integrator	Output of integrator
Acceleration	Velocity
Velocity	Position
Flow rate	Volume
Capacitor current	Capacitor charge

The integrator is continuous and stateful. By *continuous*, I mean that the input is ever-present and the output is continuously changing. *Stateful* means that the integrator must maintain some internal information; that is, the output is not a pure function of the input. For the integrator to output the volume of water in the tub from the flow rate, it needs to maintain a value of the volume that will be updated by the flow rate. Similarly, for the integrator to output the charge on the capacitor from the current, it needs to maintain a value of the charge that will be updated by the input current.

Digital Integration

We'd like to be able to teach Haskell how to integrate. Unfortunately, while nature supplies us with many good candidates for integrators, digital computation does not. The integrator is an analog continuous device. To model the integrator, we switch from continuous to discrete and work in time steps that are smaller than any time scales of interest in our problem.

Figure 6-6 shows how we can model an integrator.

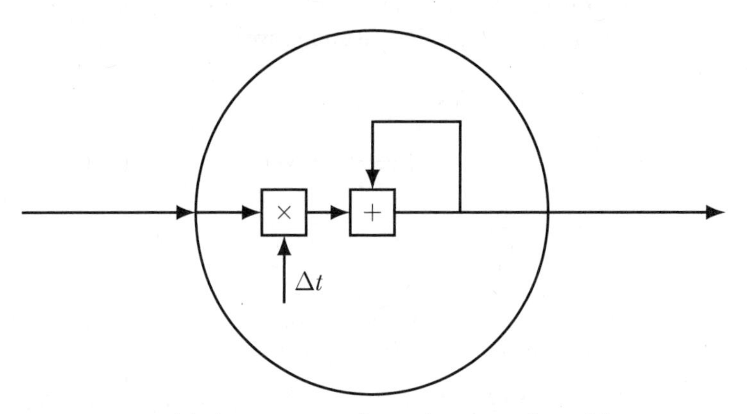

Figure 6-6: Model of an integrator that is discrete and stateful

The model is intended to be used in a discrete way. At instants of time separated by Δt, we sample the input, multiply it by Δt, and then add it to the current value of the output to produce a new output. If Δt is small compared with time scales on which the input changes, this discrete model will emulate the continuous integrator very well.

The model in Figure 6-6 forms the basis for a method of numerical integration. To integrate a function, choose a small time step Δt, sample the values the function produces at discrete times, multiply each function output by Δt, and sum the results.

If we unwrap the stateful integrator in Figure 6-6, we obtain the state-free integrator in Figure 6-7, which is shown for the case in which we are integrating acceleration to obtain velocity.

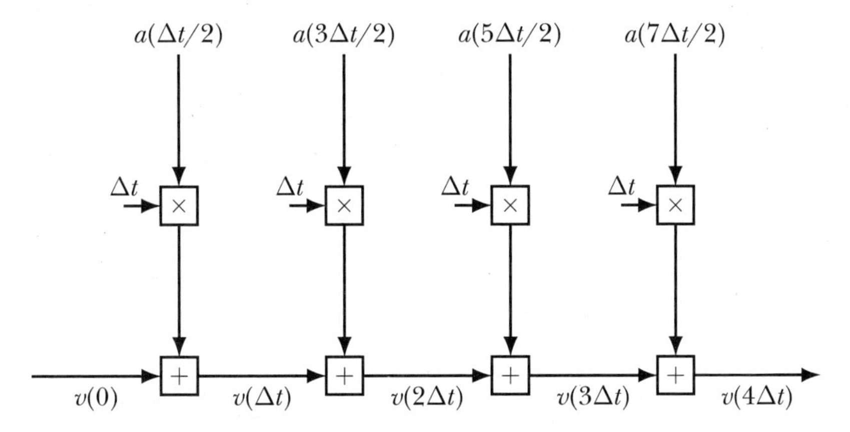

Figure 6-7: A functional model of an integrator that is discrete and state free. This integrator uses the midpoint rule.

All of the rectangles are purely functional operations (addition and multiplication); the state contained in the circular integrator now exists only in the wires between pure functions. To approximate the velocity at time Δt, we sample the acceleration at time $\Delta t/2$, multiply it by Δt, and add that to the velocity at time 0. Sampling at $\Delta t/2$ is called the *midpoint rule* for numerical integration. If Δt is small compared with the important time scales for the situation, we will get a good approximation from this method. Figure 6-7 shows four samples of the acceleration function; usually we will ask the computer to do many more.

What we would like to be able to do is to give the computer a function f, give the computer limits a and b, and ask it to compute the number:

$$\int_a^b f(t)\, dt \tag{6.2}$$

Viewed in this way, `Integration` is a function that takes a function `R -> R` as input, along with two limits, and gives a number as output.

```
type Integration = (R -> R)  -- function
                 -> R        -- lower limit
```

```
    -> R          -- upper limit
    -> R          -- result
```

Integration is often thought of as finding the area under the curve. The midpoint rule for numerical integration samples the function at the midpoint of each interval of the independent variable, as shown in Figure 6-8.

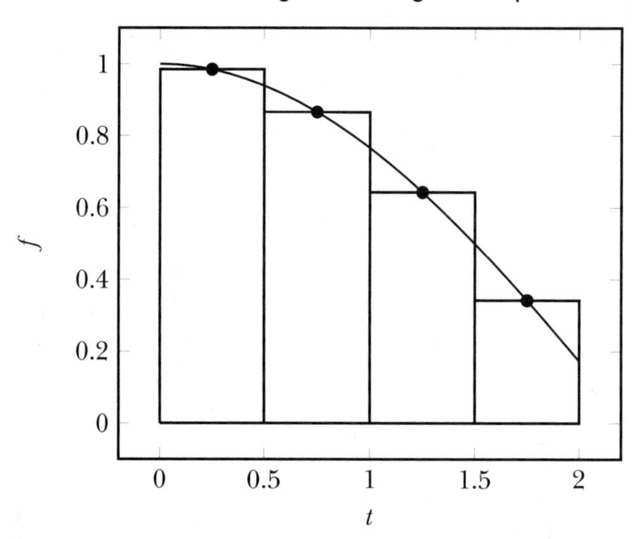

Figure 6-8: Numerical integration using the midpoint rule, Δt = 0.5

Here is the Haskell code for a numerical integrator that uses the midpoint rule:

```
integral :: R -> Integration
integral dt f a b
    = sum [f t * dt | t <- [a+dt/2, a+3*dt/2 .. b - dt/2]]
```

The first argument to the function is a step size to use for the numerical integration. We have been speaking as though the independent variable is time, but in mathematics it could be anything. The second argument to `integral` is the function to be integrated. Note how we use a single identifier (`f`) to name the function that the user of `integral` passes in. Also, we don't need to define the function `f`; what we are doing here is *naming* the function that the user of `integral` is sending in.

We use an arithmetic sequence to specify the times at which we want to sample the function. We use a list comprehension to give back a list of the same size that contains the products of the values the function returns with the step size. These products are the areas of the rectangles in Figure 6-8. All that remains is to add up these areas with the `sum` function.

Let's test out our integral function.

```
*Main> integral 0.01 (\x -> x**2) 0 1
0.33332499999999987
```

Here we use an anonymous function to specify the function that squares its input because it's easier than writing a function definition to name that function. In Chapter 2, we mentioned that anonymous functions would be useful as inputs to higher-order functions, and now we see an example of that. The exact value of this definite integral is 1/3, as shown here:

$$\int_0^1 x^2 \, dx = \frac{1}{3} x^3 \bigg|_0^1 = \frac{1}{3}$$

Implementing Antiderivatives

The type synonym `Integration` corresponds to the idea of a definite integral, where one has a function and two limits and expects to get a number as output, as in Expression 6.2. However, there is a second way of thinking about integration in which one integrates a function to obtain another function. In Equation 6.1, for example, we integrate an acceleration function to obtain a velocity function.

In calculus, a distinction is often made between a definite integral on one hand and an indefinite integral, or *antiderivative*, on the other. The antiderivative wants to be the inverse function to the derivative, but there is a catch, because functions like `sin` and `\x -> sin x + 7` have the same derivative, namely `cos`. Thus, the derivative, as a higher-order function from functions to functions, does not have a well-defined inverse function. An antiderivative of a function f is any function F whose derivative is f. In other words, F is an antiderivative of f exactly when $DF = f$. If F is an antiderivative of f, we write

$$\int f(x) \, dx = F(x) + C \tag{6.3}$$

using the indefinite integral symbol without limits and the constant of integration C. For example, we write

$$\int \cos(x) \, dx = \sin(x) + C$$

where C is an undetermined constant of integration. There is a relationship between a definite integral and an indefinite integral. The fundamental theorem of calculus claims that if F is any antiderivative of f, then

$$\int_a^b f(x) \, dx = F(b) - F(a)$$

Renaming variables and rearranging terms, we obtain an expression that allows us to relate the constant of integration to an initial value of F.

$$F(x) = F(a) + \int_a^x f(x') \, dx' \tag{6.4}$$

For any real number a, an antiderivative F is a function whose value $F(x)$ at x is the sum of its "initial" value $F(a)$ at a and the definite integral of f from a to x. If we associate the indefinite integral $\int f(x)\,dx$ of Equation 6.3 with the definite integral $\int_a^x f(x')\,dx'$ of Equation 6.4, we can associate the initial value $F(a)$ in Equation 6.4 with $-C$, where C is the constant of integration in Equation 6.3. In this sense, constants of integration are related to initial values.

A typical function has many functions that can serve as its antiderivative. How can we select one function in particular? There are two ways: we can specify a lower limit or we can specify an initial value. Specifying a lower limit corresponds to the definite integral `Integration` that we explored earlier. Specifying an initial value leads to the `AntiDerivative` we explore below.

Let's call an antiderivative a function that takes an initial value (such as $v(0)$ in Equation 6.1) and a function (such as a) and returns a function (such as v).

```
type AntiDerivative =  R             -- initial value
                    -> (R -> R)      -- function
                    -> (R -> R)      -- antiderivative of function
```

The idea of the antiderivative is closely related to the idea of the integral. We can define a function `antiDerivative` in terms of the `integral` we have already defined.

```
antiDerivative :: R -> AntiDerivative
antiDerivative dt v0 a t = v0 + integral dt a 0 t
```

Chapter 4 showed how to implement Equations 4.5 and 4.12 as the functions `velFromPos` and `accFromVel`. Now let's implement Equation 6.1 in Haskell.

```
velFromAcc :: R                              -- dt
           -> Velocity                       -- initial velocity
           -> (Time -> Acceleration)         -- acceleration function
           -> (Time -> Velocity)             -- velocity function
velFromAcc dt v0 a t = antiDerivative dt v0 a t
```

We see that finding a velocity function from an acceleration function is nothing other than the antiderivative.

How about finding a position function from a velocity function?

$$x(t) = x(0) + \int_0^t v(t')\,dt' \tag{6.5}$$

In Haskell, this is:

```
posFromVel :: R                          -- dt
           -> Position                   -- initial position
           -> (Time -> Velocity)         -- velocity function
           -> (Time -> Position)         -- position function
posFromVel = antiDerivative
```

Again, nothing but the antiderivative. Here we use point-free style to show an alternative way of writing the function and to emphasize the equality of the two functions. The functions `velFromAcc` and `posFromVel` from earlier are the same, and each is the same as `antiDerivative`. Both integrate a given function from 0 to t and add on an initial value.

Type synonyms show that time, velocity, and acceleration are all treated as numbers.

```
type Time         = R
type Position     = R
type Velocity     = R
type Acceleration = R
```

Perhaps you'd rather provide your numerical integrator with the number of steps to take from the lower limit to the upper limit rather than the step size. This is not hard to do.

```
integralN :: Int -> Integration
integralN n f a b
    = let dt = (b - a) / fromIntegral n
      in integral dt f a b
```

The `let` keyword introduces local variables and/or functions that can be used in the body after the `in` keyword. The variable `dt` is a local variable defined inside the function `integralN`. This `dt` is not visible outside of the definition of `integralN`. Any `dt` we use outside of this function has an independent meaning separate from this. Local variables are particularly useful to define if they are used more than once in the remainder of the definition. In this case, we use `dt` only once, so we could have inserted the definition of `dt` directly in the final line.

```
integralN' :: Int -> Integration
integralN' n f a b
    = integral ((b - a) / fromIntegral n) f a b
```

Using the local variable `dt` saves us a set of parentheses and makes the code easier to read because the name `dt` has meaning to us as a step size. I encourage you to define local variables with `let` whenever you can think of a name that has meaning. It will help readers of the code, including the writer.

The division operator `(/)` can be used only between numbers of the same type. Since `b - a` has type `R` and `n` has type `Int`, we can't directly divide one by the other. The solution is to transform `n` into something with type `R`, and `fromIntegral` does the trick.

Summary

A higher-order function takes another function as input and/or produces a function as output. Higher-order functions that produce functions as output can be thought of as taking multiple inputs. Numerical integration is a key example of a higher-order function that takes a function as input. The lambda notation for writing anonymous functions can be used for higher-order functions as well. Mapping a function over a list is an example of a higher-order function that takes another function as input and is similar to a list comprehension. We can make binary infix operators into higher-order functions by enclosing them in parentheses. Haskell can achieve iteration via the higher-order function `iterate`, which takes a function and a starting value and repeatedly applies the function to form an infinite list. Some higher-order functions, such as `filter`, take a predicate as input. We can also achieve filtering through list comprehensions. In the next chapter, we introduce a library that allows us to plot functions.

Exercises

Exercise 6.1. Let us return to the example of throwing a rock straight upward. Perhaps we don't want to throw it upward at 30 m/s but would like to be able to throw it upward with whatever initial velocity we choose. Write a function

```
yRock :: R -> R -> R
```

that takes as input an initial velocity and returns as output a function that takes as input a time and returns as output a height. Also, write a function

```
vRock :: R -> R -> R
```

that takes as input an initial velocity and returns as output a function that takes as input a time and returns as output a velocity.

Exercise 6.2. Give the type of `take 4`.

Exercise 6.3. The function `map` has type `(a -> b) -> [a] -> [b]`. This means that `map` is expecting a function with type `a -> b` as its first argument. The function `not` has type `Bool -> Bool`. Can `not` be the first argument to `map`? If so, what is the type of `map not`? Show how, starting from the types of `map` and `not`, you can figure out the type of `map not`.

Exercise 6.4. Write a function

```
greaterThanOrEq7' :: Int -> Bool
greaterThanOrEq7' n = undefined
```

that does the same thing as `greaterThanOrEq7` but doesn't use an `if-then-else` construction. (Hint: look at the function `lt10`.)

Exercise 6.5. Write a function with type `Int -> String -> Bool` and describe in words what it does.

Exercise 6.6. Write a predicate expressing the property "has more than six elements" that takes a list as input. Include a type signature with your predicate definition.

Exercise 6.7. Table 6-5 gives examples of the use of the `replicate` function. In the first three examples, a list is created with the requested length of the requested item. In the last case, a string is created. This seems different. Explain what is going on here.

Exercise 6.8. Make a list of the first 1,000 squares. Don't print the list; just print your definition. You could print the first 10 squares to see if your method is working.

Exercise 6.9. Use `iterate` to define a function `repeat'` that does the same thing as the Prelude function `repeat`.

Exercise 6.10. Use `take` and `repeat` to define a function `replicate'` that does the same thing as the Prelude function `replicate`.

Exercise 6.11. A car starts from rest and accelerates at 5 m/s^2 on a straight, level highway. Use `iterate` to make an infinite list of velocities for this car, with one velocity every second. (The list should look like `[0,5,10,15,...]`. Use the `take` function to see the first several elements of your infinite list.)

Exercise 6.12. List comprehensions can be used as an alternative to the `map` function. To prove this, write a function

```
map' :: (a -> b) -> [a] -> [b]
```

that does the same thing as `map`. Use a list comprehension to write your definition.

Exercise 6.13. List comprehensions can be used as an alternative to the `filter` function. To prove this, write a function

```
filter' :: (a -> Bool) -> [a] -> [a]
```

that does the same thing as `filter`. Use a list comprehension to write your definition.

Exercise 6.14. Write a function

```
average :: [R] -> R
average xs = undefined
```

that finds the average of a list of numbers. You can assume the list has at least one number. You may want to use the `fromIntegral` function.

Exercise 6.15. Produce one-input and two-input pictures, similar to Figures 6-2 and 6-3, for the higher-order function `drop`. Do the same thing for the higher-order function `replicate`.

Exercise 6.16. An alternative rule to the midpoint rule for numerical integration is the trapezoidal rule. In the trapezoidal rule, we approximate the area under a curve by the sum of the areas of a bunch of trapezoids, as shown in Figure 6-9.

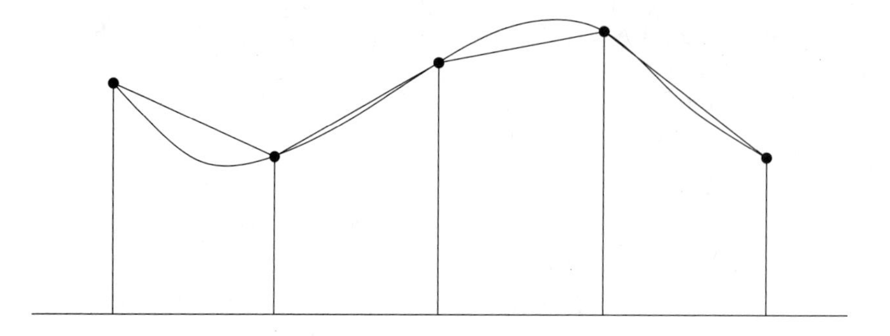

Figure 6-9: The trapezoidal rule

For the example, in Figure 6-9, the area of the first trapezoid is

$$\frac{1}{2}[f(x) + f(x + \Delta x)]\Delta x$$

and the area of all four trapezoids in the figure is

$$\left(\frac{1}{2}f(x) + f(x + \Delta x) + f(x + 2\Delta x) + f(x + 3\Delta x) + \frac{1}{2}f(x + 4\Delta x)\right) \Delta x$$

Write a definition for the function

```
trapIntegrate :: Int           -- # of trapezoids n
              -> (R -> R)      -- function f
              -> R             -- lower limit a
              -> R             -- upper limit b
              -> R             -- result
trapIntegrate n f a b = undefined
```

that takes a number of trapezoids, a function, and two limits as its arguments and gives back (an approximation to) the definite integral, using the trapezoidal rule. Test your integrator on the following integrals and see how close you can get to the correct values:

$$\int_0^1 x^3 \, dx = 0.25$$

$$\int_0^{10^{-6}} x^3 \, dx = 2.5 \times 10^{-25}$$

$$\int_0^1 e^{-x^2} \, dx \approx 0.7468$$

7

GRAPHING FUNCTIONS

Functions with the type `R -> R` are functions that can be plotted on a graph. This chapter shows how to plot such functions. Tools for making graphs are not part of the Prelude, so we'll start by discussing how to install and use library modules.

Using Library Modules

There are functions other people have written that we want to use but that are not included in the Prelude. However, such functions exist in library modules that can be imported into our source code file or loaded directly into GHCi. A standard set of library modules comes with GHC (the Glasgow Haskell Compiler we have been using), but other modules require installation.

Standard Library Modules

The library module `Data.List` is one of the standard library modules. It includes functions for working with lists. To load it into GHCi, use the `:module` command (`:m` for short).

```
Prelude> :m Data.List
```

Now, we can use functions from this module, such as sort.

```
Prelude Data.List> :t sort
sort :: Ord a => [a] -> [a]
Prelude Data.List> sort [7,5,6]
[5,6,7]
```

Note that the GHCi prompt that normally says `Prelude>` has been expanded to include the name of the module we just loaded.

To use the sort function in a source code file, include the line

```
import Data.List
```

at the top of your source code file.

Documentation for standard libraries is available online at *https://www.haskell.org* under Documentation and then Library Documentation, or you can access it directly at *https://downloads.haskell.org/~ghc/latest/docs/html/libraries/index.html*.

Other Library Modules

Library modules outside of the standard libraries are organized into *packages*. The appendix describes how to install Haskell library packages. Each package contains one or more modules. For the plotting we'll do in this chapter, we want the `Graphics.Gnuplot.Simple` module, which is supplied by the `gnuplot` package.

Follow the instructions in the appendix to install gnuplot. Several steps are required. The installation ends with commands such as the following:

```
$ cabal install gnuplot
```

or

```
$ stack install gnuplot
```

After installing the `gnuplot` package, you can restart GHCi and load the `Graphics.Gnuplot.Simple` module into GHCi, like so:

```
Prelude Data.List> :m Graphics.Gnuplot.Simple
```

Before we start the next section, let's unload the `Graphics.Gnuplot.Simple` module so we're starting with a clean slate:

```
Prelude Graphics.Gnuplot.Simple> :m
```

Issuing the `:m` command without any module name will clear any loaded modules.

Plotting

There are times when you want to make a quick plot to see what a function looks like. Here is an example of how to do this using GHCi:

```
Prelude> :m Graphics.Gnuplot.Simple
Prelude Graphics.Gnuplot.Simple> plotFunc [] [0,0.1..10] cos
```

The first command loads a graphics module that can make graphs. The second command plots the function `cos` from 0 to 10 in increments of 0.1. This is carried out by the `plotFunc` function, which is one of the functions provided by the `Graphics.Gnuplot.Simple` module. The `plotFunc` function takes a list of attributes (in this case, the empty list, `[]`), a list of values at which to compute the function (in this case, `[0,0.1..10]`, which is a list of 101 numbers from 0 to 10 in increments of 0.1), and a function to plot (in this case, `cos`).

100 points is usually enough to get a nice smooth graph. If it's not smooth enough for you, you could use 500 points or more. If you use only 4 points, you won't get a smooth graph (try it and see what happens). In Chapter 11, we'll learn how to make a nice plot with a title and axis labels for a presentation or an assignment.

If you wish to plot a function that is defined in a program file, you have a few options:

- Put only the function you want to plot in the program file.
- Use the program file to import the graphing module and define the function you want to plot.
- Use the program file to import the graphing module, define the function you want to plot, and define the plot.

We'll explore each of these options in turn.

Function Only

Suppose we want to plot the `square` function we defined in Chapter 2 from $x = -3$ to $x = 3$. Let's unload the `Graphics.Gnuplot.Simple` module so that we're starting with a clean slate:

```
Prelude Graphics.Gnuplot.Simple> :m
```

Now, we issue the following sequence of commands:

```
Prelude> :m Graphics.Gnuplot.Simple
Prelude Graphics.Gnuplot.Simple> :l first.hs
[1 of 1] Compiling Main             ( first.hs, interpreted )
Ok, one module loaded.
*Main Graphics.Gnuplot.Simple> plotFunc [] [-3,-2.99..3] square
```

The first command loads the graphing module, the second loads the file with the function definition, and the third makes the graph. Using the

`:module` command clears any source code file previously loaded with the `:load` command, so you must load the module before the source code file.

Function and Module

If we know that a program file contains a function or functions we will want to plot, we can import the `Graphics.Gnuplot.Simple` module in the program file so we don't have to do it at the GHCi command line. Instead of entering the `:m Graphics.Gnuplot.Simple` line into GHCi, we can put the following line at the top of our program file:

```
import Graphics.Gnuplot.Simple
```

Let's suppose this augmented program file is called *firstWithImport.hs*. Let's start with a clean slate by unloading the file and the module:

```
*Main Graphics.Gnuplot.Simple> :l
Ok, no modules loaded.
Prelude Graphics.Gnuplot.Simple> :m
```

Issuing the `:l` command without any filename will clear a loaded program file, leaving any loaded modules intact.

Now in GHCi we do the following:

```
Prelude> :l firstWithImport.hs
[1 of 1] Compiling Main             ( firstWithImport.hs, interpreted )
Ok, one module loaded.
*Main> plotFunc [] [-3,-2.99..3] square
```

You should see the same plot you saw in the last section.

Function, Module, and Plot Definition

If we know in advance what plot we want, we can include the plotting commands in the program file itself. In our source code file, we'll include the `import` command,

```
import Graphics.Gnuplot.Simple
```

the type synonym that defines the type `R`,

```
type R = Double
```

the function we'll plot,

```
square :: R -> R
square x = x**2
```

and the plot we want,

```
plot1 :: IO ()
plot1 = plotFunc [] [-3,-2.99..3] square
```

Notice the type `IO ()` (pronounced "eye oh unit") of `plot1`. The `IO` stands for input/output, and it signals the type of an impure function that has a side effect. In this case, the side effect is the graph popping up on the screen. Anything with type `IO ()` is something that is done only for its effect and not because we expect a value to be returned.

Let's make a clean slate in GHCi.

```
*Main> :l
Ok, no modules loaded.
Prelude> :m
```

If the source code file is called *QuickPlotting.hs*, we just load our file and give the name of our plot.

```
Prelude> :l QuickPlotting.hs
[1 of 1] Compiling Main             ( QuickPlotting.hs, interpreted )
Ok, one module loaded.
*Main> plot1
```

You should again see the plot.

Summary

This chapter introduced library modules, including standard library modules as well as those that require installation. We installed the `gnuplot` package, which provides the `Graphics.Gnuplot.Simple` module, and showed how to use the function `plotFunc` to make basic plots. The chapter also showed different ways to use the functions provided by a module, either by loading the module into GHCi with the `:module` command or by importing the module into a source code file using the `import` keyword.

The next chapter introduces *type classes*, a mechanism to take advantage of commonality among types.

Exercises

Exercise 7.1. Make a plot of $\sin(x)$ from $x = -10$ to $x = 10$.

Exercise 7.2. Make a plot of your `yRock30` function from $t = 0$ to $t = 6$ s.

Exercise 7.3. Make a plot of your `yRock 20` function from $t = 0$ to $t = 4$ s. You will need to enclose `yRock 20` in parentheses when you use it as an argument to `plotFunc`.

8

TYPE CLASSES

We have seen functions that have a concrete type, such as `not`, which takes a `Bool` as input and returns a `Bool` as output. We have also seen functions with types that use a type variable to express that they work with all types, such as `head`, which takes a list of any type and returns the first element. Use of a type variable expresses commonality over all types.

Between the extremes of functions that work with a single type and functions that work with all types is a need to express a more limited commonality among types that does not extend to all types. For example, we would like addition to be available for numeric types like `Int`, `Integer`, and `Double`, without the need to define addition for *all* types. The term *parametric polymorphism* is used to express commonality among all types. The `head` function mentioned earlier is parametric polymorphic on the underlying type of the input list. The term *ad-hoc polymorphism* is used to express more limited commonality. Type classes are Haskell's way of providing a mechanism for ad-hoc polymorphism. They express commonality among types that doesn't extend to all types.

In this chapter, we'll introduce the idea of a type class, along with a number of type classes from the Prelude. We'll describe which of the basic types are members of these type classes, and why. The explanation for why Haskell has three different exponentiation operators is based on type classes. A *section* is a function based on an operator and one of its arguments, and many sections have types that involve type classes. While type classes provide a nice way to express commonality, they also allow for the possibility that the compiler cannot figure out concrete types for values when it needs to, and we give an example where the code writer must provide extra type information to the compiler.

Type Classes and Numbers

Let us begin by asking GHCi about the type of the number `4`.

```
Prelude> :t 4
4 :: Num p => p
```

It would be entirely reasonable to expect the type of the number `4` to be `Int` or `Integer`. But the designers of the Haskell language wanted a number like `4` to be able to be an `Int`, an `Integer`, a `Double`, or even a few other types, depending on the programmer's needs. For this reason (and other, more compelling reasons), they invented the idea of type classes.

A type class is like a club to which a type can belong, and it makes certain functions available to that type. The types `Int`, `Integer`, and `Double` all belong to the type class `Num` (short for number). Each of these types has addition, subtraction, and multiplication available to it because these functions are owned by the `Num` type class. When a type belongs to a type class, we say that it is an *instance* of that type class.

The type signature `4 :: Num p => p` can be read as "`4` has type `p` as long as `p` is in type class `Num`." The letter `p` is a type variable in this type signature. It can stand for any type. The conditions to the left of the double arrow (`=>`) are *type class constraints*. In the type signature above, there is one type class constraint, `Num p`, which says that `p` must belong to type class `Num`.

Type classes are a way to express commonality among types. The types `Int`, `Integer`, `Float`, and `Double` have quite a bit in common; namely, we want to do the same sorts of things with them. We want to be able to add, subtract, and multiply numbers with these types. By having the type class `Num` own addition, subtraction, and multiplication, we allow the same addition operator that works with `Int`s to also work with `Double`s.

In the type signature `4 :: Num p => p`, GHCi hasn't committed to a concrete type for the number `4` yet. But this noncommittal attitude about the type of `4` can't go on forever. At some point, the Haskell compiler will demand that every value have a concrete type. The inability of the compiler to assign a concrete type to a value can be a source of trouble. However, GHCi has some type-defaulting rules to make our lives easier. For example, if you put the line

```
x = 4
```

into a program file (say *typetest.hs*), giving x no type signature, load it into GHCi, and then ask for the type of x,

```
Prelude> :l typetest.hs
[1 of 1] Compiling Main             ( typetest.hs, interpreted )
Ok, one module loaded.
*Main> :t x
x :: Integer
```

GHCi will tell you that x has type Integer. Here, GHCi has committed to a concrete type without our specifying the type.

There are other situations, like that in "Example of Type Classes and Plotting" at the end of this chapter, where GHCi feels unable to assign a concrete type, and you will need to help it out by adding type signatures to your code.

Type Classes from the Prelude

Table 8-1 shows several type classes provided by the Prelude. The table also shows which of the basic types are instances of each of the type classes.

Table 8-1: Basic Types That Are Instances of Various Type Classes

Type class	Bool	Char	Int	Integer	Float	Double
Eq	X	X	X	X	X	X
Ord	X	X	X	X	X	X
Show	X	X	X	X	X	X
Num			X	X	X	X
Integral			X	X		
Fractional					X	X
Floating					X	X

The following sections discuss the purpose and use of the type classes listed in Table 8-1.

The Eq Type Class

We want to be able to ask the computer whether two things are equal. That's what the == (equality) operator (first introduced in Table 1-2) does. But if we are serious about functions having types, what should the type of (==) be?

A function that takes two strings as input and gives as output a Boolean value (true or false) that indicates whether the two strings are equal should have the type String -> String -> Bool. A function that takes two integers as input and gives as output a Boolean value indicating whether the two integers are equal should have the type Integer -> Integer -> Bool. It would be an unfortunate state of affairs if we needed a different function for each type (String, Integer, and so on) that we wanted to check for equality. Perhaps a type variable could solve the problem, and the type of (==) could be a -> a -> Bool. That is almost correct, but the type a -> a -> Bool implies that every type a can be checked for equality, while there are some types that can't be checked for equality (such as function types).

The type class Eq is for types that have a notion of equality. In other words, types for which equality checking makes sense will be instances of Eq. These are the types for which the operators == and /= are defined. You can see in Table 8-1 that all of the six basic types (Bool, Char, Int, Integer, Float, and Double) are instances of Eq.

The type of the function (==) is

```
*Main> :t (==)
(==) :: Eq a => a -> a -> Bool
```

which means that we can use the == operator between any two expressions of the same type a, as long as a is an instance of Eq. What sort of type would not be an instance of Eq? Generally function types are not instances of Eq. For example, the type R -> R is not an instance of Eq.

The reason is that it is usually difficult or impossible to check whether two functions are equal. (There is a rigorous mathematical result called Richardson's theorem that gives sufficient conditions, which are quite mild, for when function equality is undecidable.)

From the perspective of computational physics, it's a bad idea that Float and Double are instances of Eq. Because these two types are used for approximate calculation, you should never test Floats and Doubles for equality. (The one exception to this rule is that you might check whether a Float or Double is zero before attempting to divide by it.) From the perspective of the computer, these types are each represented by a finite number of bits, and the computer will happily check whether each bit of one Double is the same as the corresponding bit of another Double. But as we saw in Chapter 1, the bits of sqrt 5 ^ 2 are not the same as the bits of 5. They are very close, but not the same. If you want to go deeper, a nice introduction to floating-point computation is [5]. The take-home message for computational physics is to avoid using == for approximate types like Double. The issue of equality checking with floating-point numbers is not unique to Haskell or to functional programming. Approximately calculated results should not be tested for equality in any language.

The Show Type Class

The type class Show is for types whose values can be shown using text. As indicated in Table 8-1, all of the basic types (Bool, Char, Int, Integer, Float, and Double) are instances of Show.

Function types are not typically instances of Show. If I enter the name of a function at the GHCi prompt, I get a message that complains that there is no Show instance for sqrt:

```
*Main> sqrt

<interactive>:5:1: error:
    • No instance for (Show (Double -> Double))
        arising from a use of print
        (maybe you haven't applied a function to enough arguments?)
    • In a stmt of an interactive GHCi command: print it
```

GHCi knows how to apply the sqrt function to numbers and show you the result, but it does not know how to show you the sqrt function itself. The reason behind this is the design decision that anything that is a member of the Show type class should also be a member of the Read type class. This means that what can be rendered from its internal form to a String (Show instance) can also be converted from a String back to its internal form (Read instance). If a function's Show instance returns the function's name, or even its source definition, it wouldn't usually be convertible back into its internal representation because the source file context would be missing. Note that sqrt is a perfectly acceptable Haskell expression, with a well-defined type, even though it cannot be shown.

The Num Type Class

As we saw earlier, the type class Num is for numeric types. You can see in Table 8-1 that the types Int, Integer, Float, and Double are instances of Num, while Bool and Char are not. The functions (+) (addition), (-) (subtraction), and (*) (multiplication) are owned by Num. The type of the function (+) is

```
*Main> :t (+)
(+) :: Num a => a -> a -> a
```

meaning that we can use the operator + between any two expressions of the same type a, as long as a is an instance of Num. The result will be an expression of type a. The type class Num allows the addition function (+) to act as though it has the type Int -> Int -> Int or Integer -> Integer -> Integer or Float -> Float -> Float or Double -> Double -> Double.

We can ask for information about the Num type class by using GHCi's :info (or :i) command:

```
*Main> :i Num
class Num a where
  (+) :: a -> a -> a
  (-) :: a -> a -> a
  (*) :: a -> a -> a
  negate :: a -> a
  abs :: a -> a
  signum :: a -> a
  fromInteger :: Integer -> a
  {-# MINIMAL (+), (*), abs, signum, fromInteger, (negate | (-)) #-}
```

```
  -- Defined in 'GHC.Num'
instance Num Word -- Defined in 'GHC.Num'
instance Num Integer -- Defined in 'GHC.Num'
instance Num Int -- Defined in 'GHC.Num'
instance Num Float -- Defined in 'GHC.Float'
instance Num Double -- Defined in 'GHC.Float'
```

Here we see that `Num` owns addition, subtraction, multiplication, and a few other functions. We also see some concrete types that are instances of `Num`.

The Integral Type Class

The `Integral` type class is for types that behave like integers. A type has to be an instance of `Num` before it can be an instance of `Integral`. You can see in Table 8-1 that the types `Int` and `Integer` are instances of `Integral`, while `Float` and `Double` are not. The type of the function `rem`, which finds the remainder of one integer divided by another, is

```
*Main> :t rem
rem :: Integral a => a -> a -> a
```

meaning that we can use the function `rem` between any two expressions of the same type `a`, as long as `a` is an instance of `Integral`, and the result will be an expression of type `a`.

The Ord Type Class

We want to be able to compare things, but most people would agree that

```
"kitchen" > 4
```

doesn't make any sense. The designers of the Haskell language decided that such an expression should not evaluate to `True` or to `False` but instead should be regarded as a type error. To use one of the comparison operators (`<`, `<=`, `>`, or `>=`), two requirements must be met:

- The two things being compared must have the same type. Let us call this type `a`.
- Type `a` must belong to the `Ord` type class.

The `Ord` type class is for types that have a notion of order. A type must first be an instance of `Eq` before it may be an instance of `Ord`. The type of the function `(<)` is

```
*Main> :t (<)
(<) :: Ord a => a -> a -> Bool
```

which means we can use the operator `<` between any two expressions of the same type `a`, as long as `a` is an instance of `Ord`. The type of `(<)` expresses both requirements we listed above.

Some types have no obvious notion of comparison. Three-dimensional vectors, for which we'll define the Vec type in Chapter 10, have no obvious notion of order. This is not to say that there is no way of defining comparison for vectors. We could compare their magnitudes or their x-components, for example. The point is that there is not a single, obvious candidate for what comparison would mean. Because vectors have no obvious notion of comparison, Vec will not belong to Ord.

What about the expression

```
x > y
```

if x has the value 4.2 and y has the value 4? Most people would agree that the expression should evaluate to True. But Haskell will regard this expression as an error if x has type Double and y has type Int, because Double and Int are not the same type. To compare two things, we must explicitly convert the type of one thing into the type of the other. To avoid a rounding error, we'd want to convert y (the Int) into a Double. For this, we can use the Prelude function fromIntegral, replacing our original expression x > y with

```
x > fromIntegral y
```

The type of fromIntegral is

```
*Main> :t fromIntegral
fromIntegral :: (Integral a, Num b) => a -> b
```

indicating that fromIntegral will convert any type in Integral into any type in Num. The type checker will figure out that in this case, since x has type Double, y needs to be converted to a Double.

Some other programming languages have a process of *type coercion* that changes the type of a value so that it can be compared with or used with another value. For example, an integer will automatically be changed into a floating-point number when the two are being compared. Haskell does not have automatic type conversion, and the decision to not include this was intentional. The designers of the language believed that many or most type coercions are really mistakes that the programmer did not intend rather than automatic help that the compiler could provide.

The Fractional Type Class

The Fractional type class is for numeric types that support division. A type has to be an instance of Num before it can be an instance of Fractional. We see in Table 8-1 that the types Float and Double are instances of Fractional, while Int and Integer are not. The type of the function (/) is

```
*Main> :t (/)
(/) :: Fractional a => a -> a -> a
```

meaning that we can use the / operator between any two expressions of the same type a, as long as a is an instance of Fractional. The result will be an expression of type a.

The Floating Type Class

The Floating type class is for numeric types that are stored by the computer as floating-point numbers, that is, as inexact approximations. A type has to be an instance of Fractional before it can be an instance of Floating. You can see in Table 8-1 that the types Float and Double are instances of Floating, while Int and Integer are not. The type of the function cos is

```
*Main> :t cos
cos :: Floating a => a -> a
```

meaning that we can use the cos function on any expression of type a, as long as a is an instance of Floating. The result will be an expression of type a.

Figure 8-1 shows the relationship among the numeric type classes we've just discussed.

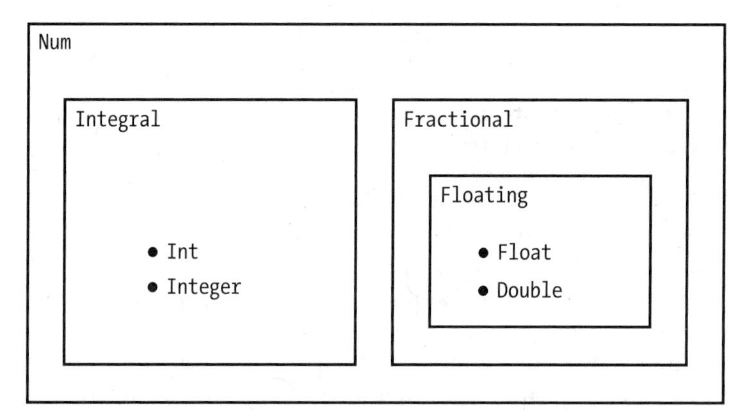

Figure 8-1: Relationship among the numeric type classes Num, Integral, Fractional, and Floating

In Figure 8-1, types are preceded by a bullet and type classes are not. As you can see, the types Int and Integer are instances of type classes Integral and Num. The types Float and Double are instances of type classes Floating, Fractional, and Num.

Exponentiation and Type Classes

Haskell offers three operators for exponentiation, shown in Table 8-2. The difference between these operators lies in the type class constraints that each work under and in the methods of implementation. The single caret operator (^) requires that the exponent be a nonnegative integer. Exponentiation by ^ repeatedly multiplies a Num by itself. The double-caret operator (^^) requires that the exponent be an integer, enforced by the Integral b constraint. The ^^ operator can be implemented by repeated multiplication and taking reciprocals, and it can accept a negative exponent. The double-star operator (**) requires that the base and the exponent have the same type and that that type be an instance of Floating. This operator requires a more complicated implementation, taking a logarithm and using the exponential function.

Table 8-2: Haskell's Three Functions for Exponentiation

Function		Type
(^)	::	(Integral b, Num a) => a -> b -> a
(^^)	::	(Fractional a, Integral b) => a -> b -> a
(**)	::	Floating a => a -> a -> a

Since Float and Double are members of the Floating type class, and thus represent numbers approximately, the ** operator is generally going to do an approximate calculation. This is certainly what you want for a non-integral exponent. Under the right circumstances, the caret and double-caret operators can do exact calculation.

Table 8-3 shows which of types (Int, Integer, Float, and Double) are allowable for the base x and the exponent y in the expression x ^^ y.

Table 8-3: Possible Types for x and y with the Double-Caret Exponentiation Operator

	y :: Int	y :: Integer	y :: Float	y :: Double
x :: Int				
x :: Integer				
x :: Float	^^	^^		
x :: Double	^^	^^		

Since the base must have a type that is an instance of Fractional, only Float and Double can serve as a type for the base. Since the exponent must have a type that is an instance of Integral, only Int and Integer can serve as a type for the exponent.

Sections

An infix operator expects an argument on its left and an argument on its right. If only one of these two arguments is given, the resulting expression can be thought of as a function waiting for the other argument. Haskell allows us to make such functions by enclosing an operator and one of its arguments in parentheses. A function formed by enclosing an operator and one argument in parentheses is called a *section*.

Table 8-4 shows examples of sections, with their types. Many useful sections have types with type-class constraints.

Table 8-4: Examples of Sections

Function		Type
`(+1)`	`::`	`Num a => a -> a`
`(2*)`	`::`	`Num a => a -> a`
`(^2)`	`::`	`Num a => a -> a`
`(2^)`	`::`	`(Integral b, Num a) => b -> a`
`('A':)`	`::`	`[Char] -> [Char]`
`(:"end")`	`::`	`Char -> [Char]`
`("I won't " ++)`	`::`	`[Char] -> [Char]`
`($ True)`	`::`	`(Bool -> b) -> b`

For example, `(+1)`, which could also be written `(1+)`, is a function that adds 1 to its argument, and `(2*)`, which could also be written `(*2)`, is a function that doubles its argument. However, the sections `(^2)` and `(2^)` are not the same function; the former is the squaring function and the latter is the "2 to the power of" function.

Sections can be useful in the same way that anonymous functions are useful: they provide a quick way of specifying a function without naming it, often as an input to a higher-order function.

Let's use a section to integrate the squaring function. Using the `integral` function from Chapter 6,

```
type R = Double

integral :: R -> (R -> R) -> R -> R -> R
integral dt f a b
    = sum [f t * dt | t <- [a+dt/2, a+3*dt/2 .. b - dt/2]]
```

we can use a section for the squaring function as follows:

```
*Main> :l TypeClasses
[1 of 1] Compiling Main             ( TypeClasses.lhs, interpreted )
Ok, one module loaded.
*Main> integral 0.01 (^2) 0 1
0.33332499999999987
```

Like anonymous functions, sections give the programmer tools to create functions "on the fly" without having to name them. And like anonymous functions, using sections requires a bit of care since it is easy to forget why you defined a particular section. This is because the terse syntax provides no clue about what you meant when you wrote it. If the meaning of a section is not immediately clear, a short function definition with an evocative name may be a better choice.

Example of Type Classes and Plotting

At the beginning of this chapter, we mentioned that the Haskell type checker will sometimes complain if it cannot determine concrete types for each expression it needs to deal with. The solution is to add a type signature or type annotation to your code. As an example of this, create a new program file called *typeTrouble.hs* with the following code:

```
import Graphics.Gnuplot.Simple

plot1 = plotFunc [] [0,0.01..10] cos
```

When I try to load this file into GHCi, I get this horrible-looking error message:

```
typeTrouble.hs:3:9: error:
    • Ambiguous type variable 'a0' arising from a use of 'plotFunc'
      prevents the constraint '(Graphics.Gnuplot.Value.Atom.C
                                  a0)' from being solved.
      Probable fix: use a type annotation to specify what 'a0' should be.
      These potential instances exist:
        instance [safe] Graphics.Gnuplot.Value.Atom.C Integer
          -- Defined in 'Graphics.Gnuplot.Value.Atom'
        instance [safe] Graphics.Gnuplot.Value.Atom.C Double
          -- Defined in 'Graphics.Gnuplot.Value.Atom'
        instance [safe] Graphics.Gnuplot.Value.Atom.C Float
          -- Defined in 'Graphics.Gnuplot.Value.Atom'
        ...plus one other
        ...plus 11 instances involving out-of-scope types
        (use -fprint-potential-instances to see them all)
    • In the expression: plotFunc [] [0, 0.01 .. 10] cos
      In an equation for 'plot1': plot1 = plotFunc [] [0, 0.01 .. 10] cos
  |
3 | plot1 = plotFunc [] [0,0.01..10] cos
  |         ^^^^^^^^^^^^^^^^^^^^^^^^^^^^
```

Don't panic. This error message contains much more information than we need to solve the problem. The most useful part of the message is the first line, which tells where the problem is in the code (line 3, column 9). At line 3, column 9 of our code is the function `plotFunc`. Let's look at the type of `plotFunc`.

```
*Main> :t plotFunc

<interactive>:1:1: error: Variable not in scope: plotFunc
```

Hmmm. Life just got worse. But this latter error is an easy one. "Variable not in scope" means that GHCi doesn't know this function. That makes sense, actually, because it's not included in the Prelude (the collection of built-in functions that are loaded automatically when we start up GHCi), and GHCi refused to load our *typeTrouble.hs* file because it had a problem with it. At the moment, it has no knowledge of `plotFunc`. The `plotFunc` function is defined in the Graphics.Gnuplot.Simple module. We can get access to plotFunc by loading the plotting module manually, like we first did to make a quick plot.

```
*Main> :m Graphics.Gnuplot.Simple
```

Now, let's ask again for the type of plotFunc.

```
Prelude Graphics.Gnuplot.Simple> :t plotFunc
plotFunc
  :: (Graphics.Gnuplot.Value.Atom.C a,
      Graphics.Gnuplot.Value.Tuple.C a) =>
     [Attribute] -> [a] -> (a -> a) -> IO ()
```

There are a couple of type class constraints to the left of the =>. I don't know the specifics of those type classes, but as long as a (a type variable) belongs to those two type classes, the type of plotFunc is the following:

```
[Attribute] -> [a] -> (a -> a) -> IO ()
```

In other words, plotFunc needs a list of Attributes (we have given an empty list in our examples so far), a list of as, and a function that takes an a as input and gives back an a as output. If we give plotFunc all this stuff, it will give us back an IO (), which is a way of saying that it will actually *do* something for us (make a plot).

The key to fixing an "ambiguous type variable" error like this lies in the suggestion of the fifth line in the error message shown earlier. Add a type signature to your code. The Haskell type checker would like more help figuring out the types of things. In particular, it can't figure out the types of [0,0.01..10] and cos. Let's ask GHCi about the types of these two.

```
Prelude Graphics.Gnuplot.Simple> :t [0,0.01..10]
[0,0.01..10] :: (Fractional a, Enum a) => [a]
Prelude Graphics.Gnuplot.Simple> :t cos
cos :: Floating a => a -> a
```

Both of these expressions contain type class constraints.

One solution to the problem is to give the list [0,0.01..10] a name and a type signature. Let's make a program file called *typeTrouble2.hs* with the following lines:

```
import Graphics.Gnuplot.Simple

xRange :: [Double]
xRange = [0,0.01..10]

plot2 = plotFunc [] xRange cos
```

This program file should load fine and give you a nice plot when you enter plot2. Try it and see.

A second solution is to specify the type of the list [0,0.01..10] on the line where it's used. We could make a program file called *typeTrouble3.hs* with the following lines:

```
import Graphics.Gnuplot.Simple

plot3 = plotFunc [] ([0,0.01..10] :: [Double]) cos
```

A third solution, which is my favorite because it involves the fewest keystrokes, is to tell the compiler that the final element of the list, 10, has type Double. This implies that all of the elements in the list have type Double.

```
import Graphics.Gnuplot.Simple

plot4 = plotFunc [] [0,0.01..10 :: Double] cos
```

The moral of the story is that you should include type signatures for all of the functions you define, and you should be prepared to add more type signatures if the type checker complains.

Summary

This chapter introduced the idea of a type class, something that contains types and owns certain functions. We discussed several standard type classes from the Prelude, in addition to functions that have type-class constraints. We saw that a section is a function without a name, formed by combining an operator with one of its arguments. Many sections have types with type-class constraints. Finally, we gave an example where it was necessary to add a type annotation to our code to satisfy the Haskell type checker.

With the introduction of type classes, we have described almost all of Haskell's type system. There are basic types, function types, list types, type variables, and type classes. Once we cover tuple types in the next chapter, we'll have described Haskell's type system in full.

Exercises

Exercise 8.1. Is it possible for a type to belong to more than one type class? If so, give an example. If not, why not?

Exercise 8.2. We said in this chapter that function types are typically not instances of Eq because it's too hard to check whether two functions are equal.

(a) What does it mean mathematically for two functions to be equal?

(b) Why is it usually very hard or impossible for the computer to check if two functions are equal?

(c) Give a specific example of a function type that would be easy to check for equality.

Exercise 8.3. The function (*2) is the same as the function (2*). Is the function (/2) the same as the function (2/)? Explain what these functions do.

Exercise 8.4. In Chapter 2, we defined a function `square`. Now that we know that Haskell has sections, we can see that we didn't need to define `square`. Show how to use a section to write the function that squares its argument. How about a section for the function that cubes its argument?

Exercise 8.5. You can get information from GHCi about a type or a type class by using the GHCi command `:info` (`:i` for short), followed by the name of the type or type class you want information about. If you ask for information about a type, GHCi will tell you the type classes of which your type is an instance (the line `instance Num Double`, for example, means that the type `Double` is an instance of the type class `Num`). If you ask for information about a type class, GHCi will tell you the types that are instances of your type class.

(a) We showed in Table 8-1 that the type `Integer` was an instance of type classes `Eq`, `Ord`, `Show`, `Num`, and `Integral`. There are a few more type classes that we did not discuss of which `Integer` is also an instance. Find these.

(b) Type class `Enum` is for types that can be enumerated, or listed. Which Prelude types are instances of `Enum`?

Exercise 8.6. Find the types of the following Prelude Haskell expressions (some are functions and some are not):

(a) `42`

(b) `42.0`

(c) `42.5`

(d) `pi`

(e) `[3,1,4]`

(f) `[3,3.5,4]`

(g) `[3,3.1,pi]`

(h) `(==)`

(i) `(/=)`

(j) `(<)`

(k) `(<=)`

(l) `(+)`

(m) `(-)`

(n) `(*)`

(o) `(/)`

(p) `(^)`

(q) `(**)`

(r) `8/4`

(s) `sqrt`

(t) `cos`

(u) `show`

(v) `(2/)`

Exercise 8.7. If 8/4 = 2, and `2 :: Num a => a` (2 has type `a` for every type `a` in type class `Num`), why does `8/4 :: Fractional a => a`?

Exercise 8.8. The functions `quot`, `rem`, `div`, and `mod` all have to do with integer division and remainders. All of these functions work with types that are instances of type class `Integral`, as shown in the following table:

Function		Type
`quot`	`::`	`Integral a => a -> a -> a`
`rem`	`::`	`Integral a => a -> a -> a`
`div`	`::`	`Integral a => a -> a -> a`
`mod`	`::`	`Integral a => a -> a -> a`

By playing around with these functions, try to explain in words what each does. What is the difference between `quot` and `div`? What is the difference between `rem` and `mod`?

Exercise 8.9. Make a table like Table 8-3, showing which of types `Int`, `Integer`, `Float`, and `Double` can be used for the base `x` and the exponent `y` in the expression `x ^ y`. Do the same for the expression `x ** y`. Finally, find a pair of types `a` and `b` for which there is no exponentiation operator that allows the base to have type `a` and the exponent to have type `b`.

9

TUPLES AND TYPE CONSTRUCTORS

A tuple is an ordered collection of values. There are tuples for ordered pairs, ordered triples, ordered quadruples, and so on. The types of the values in a tuple are generally different, but they could be the same.

In this chapter, we'll discuss tuples and their generalization: the type constructor. Function types, list types, and tuple types are all examples of type constructors. We'll present a unified way to think about these type constructors, which on the surface appear different, but all share essential features. Finally, we'll revisit numerical integration to show how tuples, in conjunction with iteration, give a way to perform numerical integration that we will later generalize to solve differential equations.

Pairs

The simplest tuple is the pair. A pair type is specified by giving an ordered pair `(a,b)` of types `a` and `b`, separated by a comma and enclosed in parentheses. If `x :: a` and `y :: b`, then `(x,y) :: (a,b)`. The value `(x,y)` has the type `(a,b)`. The comma and parentheses have two uses: they form the type `(a,b)` and they form the value `(x,y)` of that type.

For example, here is a pair composed of a `String` describing a person's name and an `Int` representing the person's score on an exam:

```
nameScore :: (String,Int)
nameScore = ("Albert Einstein", 79)
```

In this example, the pair type is `(String,Int)` and the pair value is `("Albert Einstein", 79)`.

To get some experience with tuples, let's write a function `pythag` that computes the hypotenuse of a right triangle from the lengths of its two sides. We'll pass the two side lengths to the function using a pair. Here is one way to write this function:

```
pythag :: (R,R) -> R
pythag (x,y) = sqrt (x**2 + y**2)
```

This type signature shows us that `pythag` expects a pair of two `R`s as input and produces an `R` as output. The fact that in the second line we call the input `(x,y)` (rather than a simple variable like `p`) means that this definition uses pattern matching on the input. This is similar to the pattern matching we saw earlier for `Bool` and for lists. Pattern matching for pairs is simple because there is only one pattern: every pair has the form `(x,y)`. Recall that `Bool` has two patterns (`True` and `False`) and that lists have two patterns (the empty list `[]` and the cons of an element and a list `x:xs`).

A couple of Prelude functions deal with pairs. The `fst` function takes a pair as input and returns the first component of the pair as output. The `snd` function takes a pair as input and returns the second component of the pair as output. We can test this behavior in GHCi:

```
Prelude> fst ("Albert Einstein", 79)
"Albert Einstein"
```

The types of functions `fst` and `snd` are given entirely in terms of type variables, indicating that these functions care nothing for the type of the payload.

```
Prelude> :t fst
fst :: (a, b) -> a
Prelude> :t snd
snd :: (a, b) -> b
```

In general, there are two ways to get data out of a type like a pair that contains multiple pieces of data. One way is pattern matching, and the other is using functions like `fst` and `snd`, generally called *eliminators*. While it is theoretically possible to write code that uses only pattern matching and never uses eliminators, eliminators are sometimes simpler, so it is nice to have access to both. The eliminators `fst` and `snd` are particularly useful if you process some data that results in a pair and you just want one part of it. An example is the function `integral'` in the last section of this chapter.

Currying a Function of Two Variables

In Chapter 6, we talked about currying as a way to think of a function as accepting more than one argument. Using a higher-order function, we can write our hypotenuse function as follows:

```
pythagCurried :: R -> R -> R
pythagCurried x y = sqrt (x**2 + y**2)
```

Tuples offer an alternative way to write a function of two variables. Although `pythag` and `pythagCurried` are different Haskell functions with different types, they're doing the same thing mathematically: they are both expressing the mathematical function that finds the hypotenuse. Let's call `pythag` the *tuple form* of the hypotenuse function and `pythagCurried` the *curried form* of the hypotenuse function.

These two ways of encoding a function of two variables are mutually exclusive. You need to pick one or the other for a particular function; you can't use both. Notice that the tuple form `pythag` requires the use of parentheses and a comma around the two arguments. That's because you need to have a tuple as input! Notice that the curried form `pythagCurried` has no parentheses and no comma. It's not that the comma is optional; no comma can be present.

Sometimes you might use one form but realize later on that you wish you had used the other. Haskell provides two functions to let you convert between forms. To convert from tuple form to curried form, Haskell provides the function `curry`. The function `curry pythag` is exactly the same function as `pythagCurried` we defined earlier. The function `uncurry pythagCurried` is the same as the function `pythag`. However, it does not make sense to write `curry pythagCurried` or `uncurry pythag`; these constructions will produce type errors when the compiler tries to read them.

If we load these functions into GHCi (code files are available at *https://lpfp.io*),

```
Prelude> :l Tuples
[1 of 1] Compiling Main             ( Tuples.lhs, interpreted )
Ok, one module loaded.
```

we can see that the types of `pythagCurried` and `curry pythag` are the same:

```
*Main> :t pythagCurried
pythagCurried :: R -> R -> R
*Main> :t curry pythag
curry pythag :: R -> R -> R
```

We can also see that the types of `pythag` and `uncurry pythagCurried` are the same:

```
*Main> :t pythag
pythag :: (R, R) -> R
```

```
*Main> :t uncurry pythagCurried
uncurry pythagCurried :: (R, R) -> R
```

Take a look at the types for curry and uncurry:

```
*Main> :t curry
curry :: ((a, b) -> c) -> a -> b -> c
*Main> :t uncurry
uncurry :: (a -> b -> c) -> (a, b) -> c
```

Figure 9-1 attempts to explain these complex types. The two types shown in the figure are alternative ways to encode a function of two variables. The higher-order functions curry and uncurry transform one two-variable function type into the other.

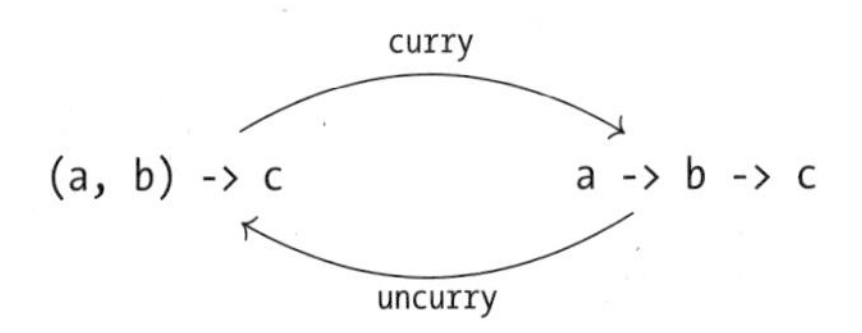

Figure 9-1: Two types for a function of two variables. The higher-order functions curry and uncurry transform a two-variable function of one type into the other type.

Triples

In addition to pairs, you can make triples, or tuples with even more components. However, the functions fst and snd work only with pairs. To access elements of triples and larger tuples, the standard method is to use pattern matching. For example, functions that pick out the components of triples can be defined as follows:

```
fst3 :: (a,b,c) -> a
fst3 (x,y,z) = x

snd3 :: (a,b,c) -> b
snd3 (_,y,_) = y

thd3 :: (a,b,c) -> c
thd3 (_x,_y,z) = z
```

The definitions of fst3, snd3, and thd3 use pattern matching to assign names to the items in the triple. These names can then be used on the right-hand side of the definition to indicate the value we want the function to return. In the function fst3, the values y and z are not used. Because they are not used, it is in some sense superfluous to give them names. In the definition of snd3, the _ (underscore) character is used as a placeholder to represent a quantity that doesn't get used in the expression that follows. In the definition of snd3, we use underscores in the first and third slots of the triple. This signals that it is superfluous to give these items names since the names are not used in the definition. In the definition of thd3, we show an alternate use of underscores. Here we start the variable name with an underscore, indicating it won't be used, but we give it a name for our own use, perhaps to remind us of the purpose of that variable. Using underscores for quantities that aren't used is a best practice, because the compiler will generate warnings about unused variables, which are often errors. To distinguish these genuine errors (frequently misspellings) from items you don't want to use, employ the underscore.

Comparing Lists and Tuples

A tuple is different from a list in that every element of a list must have the same type. On the other hand, unlike a list, the type of a tuple says exactly how many elements the tuple has. If an expression has type `[Int]`, for example, it can be a list of zero, one, two, or more `Int`s. However, if an expression has type `(String,Int)`, it is a pair consisting of exactly one `String` and exactly one `Int`. If you want to combine exactly two things or exactly three things, a tuple is what you want. Beyond three items, tuples rapidly become unwieldy. Lists, on the other hand, are often very long. A list can happily contain thousands of elements.

Before we discuss lists of pairs, we'll take a short detour to look at a class of types that will be helpful for that discussion.

Maybe Types

We saw in Chapter 5 that for any type `a` there is another type `[a]` consisting of lists of elements that each have type `a`. Such a list may have zero, one, two, or more elements of type `a`.

Similarly, for any type `a` there is another type `Maybe a` consisting of zero or one element of type `a`. To motivate this new data type, imagine that we are writing a function `findFirst` that will search through a list for the first element that meets some criterion. We might want such a function to have the following type:

```
findFirst :: (b -> Bool) -> [b] -> b
```

The type indicates our intent to have the function `findFirst` accept a predicate and a list of elements of type `b` as input and provide a single element of type `b` as output. But what if the list contains no element that meets our criterion? In that case, there is a problem because the function `findFirst` has no way to come up with an element of type `b`, but the type *demands* that the function return an element of type `b`. One possibility is for `findFirst` to use the `error` function if no suitable element is found, but this is an extreme measure and will halt the program so that no later recovery is possible.

A better solution is to use a different type signature, like so:

```
findFirstMaybe :: (b -> Bool) -> [b] -> Maybe b
```

If `findFirstMaybe` finds an element `x :: b` that meets the criterion, it will return `Just x`. If it finds no element of type `b` that meets the criterion, `findFirstMaybe` will return `Nothing`. Let's see what the function definition looks like:

```
findFirstMaybe p xs = case dropWhile (not . p) xs of
                        []    -> Nothing
                        (x:_) -> Just x
```

The expression `dropWhile (not . p) xs` is what remains of the list `xs` after the longest possible sequence of elements that do *not* satisfy the predicate are removed from the front of the list. The `case` construction allows us to do pattern matching on the expression `dropWhile (not . p) xs`, asking which of the two list patterns the expression matches and returning an appropriate result in each case.

The type `Maybe a` has two patterns. (Recall that `Bool` has two patterns, lists have two patterns, and tuples have one pattern.) A value of `Maybe a` is either `Nothing` or `Just x` for some `x :: a`. The value `Nothing` is the way of specifying zero elements of type `a`, and the value `Just x` is the way of specifying one element of type `a` (namely `x`). Table 9-1 shows some expressions involving `Maybe` and their types.

Table 9-1: Expressions Involving `Maybe` and Their Types

Expression		Type
`Nothing`	`::`	`Maybe a`
`Just "me"`	`::`	`Maybe [Char]`
`Just 'X'`	`::`	`Maybe Char`
`Just False`	`::`	`Maybe Bool`
`Just 4`	`::`	`Num a => Maybe a`

Table 9-2 shows a comparison of expressions having `Maybe` types with expressions having the underlying type.

Table 9-2: Comparison of `Maybe` Type and Underlying Type Expressions

Type	Expressions with this type
`Bool`	`False, True`
`Maybe Bool`	`Just False, Just True, Nothing`
`Char`	`'h', '7'`
`Maybe Char`	`Just 'h', Just '7', Nothing`
`String`	`"Monday", "Tuesday"`
`Maybe String`	`Just "Monday", Just "Tuesday", Nothing`
`Int`	`3, 7, -13`
`Maybe Int`	`Just 3, Just 7, Just (-13), Nothing`

Table 9-2 shows four things: for each type a there is a type Maybe a; an expression of type Maybe String, unless it is Nothing, holds a value of type String; an expression of type String can be made into an expression of type Maybe String by prefixing the expression with the constructor Just; and these observations about the String type are also valid for Bool, Char, or any other type.

Now that we have the Maybe type under our belt, let's take a look at lists of pairs.

Lists of Pairs

Just as we can form lists of lists, we can make pairs of pairs, lists of pairs, pairs of lists, and more complicated things. The list of pairs is probably the most useful of these (although lists of lists are very useful), for reasons we will see next.

To form a list of pairs, we can use the Prelude function zip:

```
*Main> :t zip
zip :: [a] -> [b] -> [(a, b)]
```

The zip function takes two lists and pairs their first elements, their second elements, and so on until the end of the shorter list. Table 9-3 shows some examples of how to use zip.

Table 9-3: Examples of zip and zipWith

Expression		Evaluates to
zip [1,2,3] [4,5,6]	⤳	[(1,4),(2,5),(3,6)]
zip [1,2] [4,5,6]	⤳	[(1,4),(2,5)]
zip [5..7] "who"	⤳	[(5,'w'),(6,'h'),(7,'o')]
zipWith (+) [1,2,3] [4,5,6]	⤳	[5,7,9]
zipWith (-) [1,2,3] [4,5,6]	⤳	[-3,-3,-3]
zipWith (*) [1,2,3] [4,5,6]	⤳	[4,10,18]

The Prelude function zipWith is a high-power relative of zip that goes one step further and applies a function to each pair of values that zip would have generated.

```
*Main> :t zipWith
zipWith :: (a -> b -> c) -> [a] -> [b] -> [c]
```

The first argument to zipWith is a higher-order function that describes what to do with an element of type a (from the first list) and an element of type b (from the second list). The second argument to zipWith is the first list, and the third argument to zipWith is the second list.

The Prelude function unzip takes a lists of pairs and turns it into a pair of lists.

```
*Main> :t unzip
unzip :: [(a, b)] -> ([a], [b])
```

One use for a list of pairs is a lookup table. In a lookup table, the first item of each pair serves as a *key* and the second item of each pair serves as a *value*. Such a pair is referred to as a *key-value pair*. The following list of pairs is a lookup table containing the final numeric grade for the History of Western Civilization course taken by three famous scientists. The name of each person acts as the key, and the grade is the value.

```
grades :: [(String, Int)]
grades = [ ("Albert Einstein", 89)
         , ("Isaac Newton"   , 95)
         , ("Alan Turing"    , 91)
         ]
```

The Prelude function `lookup` takes a key and a lookup table and returns the corresponding value, if there is one. The function `lookup` returns a `Maybe` type to allow for the possibility that the key is not found in the lookup table.

```
*Main> :t lookup
lookup :: Eq a => a -> [(a, b)] -> Maybe b
*Main> lookup "Isaac Newton" grades
Just 95
*Main> lookup "Richard Feynman" grades
Nothing
```

Tuples and List Comprehensions

Later, we'll want to have a way to use list comprehensions to form lists of pairs (x, y) we want to plot. In Chapter 11, for example, we'll meet a plotting function called `plotPath` that takes a list of pairs of numbers as input, usually `[(R,R)]`, and produces a plot. We can use list comprehensions to transform our data into a form suitable for plotting. If we wanted to plot position as a function of time, we could form a list of time-position pairs as follows:

```
txPairs :: [(R,R)]
txPairs = [(t,yRock30 t) | t <- [0,0.1..6]]

type R = Double
```

The same list of pairs can be formed with `map`:

```
txPairs' :: [(R,R)]
txPairs' = map (\t -> (t,yRock30 t)) [0,0.1..6]

yRock30 :: R -> R
yRock30 t = 30 * t - 0.5 * 9.8 * t**2
```

Besides mapping, a list comprehension can filter data based on a Boolean expression. Let's continue our example of forming a list of time-position pairs with yRock30. Suppose we want to only have pairs in our list while the rock is in the air ($y > 0$).

```
txPairsInAir :: [(R,R)]
txPairsInAir
    = [(t,yRock30 t) | t <- [0,0.1..20], yRock30 t > 0]
```

After we give the list that the values of t come from, we add a comma and then the Boolean expression to use for filtering. The computer will form a list, as before, but now only keep values for which the Boolean expression returns True.

We can achieve the same effect with a combination of map and filter. We can do the filtering first,

```
txPairsInAir' :: [(R,R)]
txPairsInAir'
    = map (\t -> (t,yRock30 t)) $
      filter (\t -> yRock30 t > 0) [0,0.1..20]
```

or we can do the mapping first:

```
txPairsInAir'' :: [(R,R)]
txPairsInAir''
    = filter (\(_t,y) -> y > 0) $
      map (\t -> (t,yRock30 t)) [0,0.1..20]
```

The application operator $ from Table 1-2 has a precedence of 0, so the expressions on each side of it are evaluated before they are combined. In this way, the application operator serves as a kind of one-symbol parentheses. The same effect could have been produced by enclosing the entire map line above in parentheses.

Note the use of the _ (underscore) character in the anonymous function just shown. Since the conditional expression only depends on the second item in the pair, there is no need to give a name to the first item in the pair.

The type of a pair is formed from two existing types. The type of a triple is formed from three existing types. The idea of a *type constructor*, which we explore in the next section, provides a unifying framework for constructing new types from old.

Type Constructors and Kinds

Maybe Int is a type, Maybe Bool is a type, and Maybe R is a type, but Maybe itself is not a type. It's what's called a *type constructor*. A type constructor is an object that takes zero or more types as input and produces a type as output. Maybe is a one-place type constructor, taking the type Int as input and producing the type Maybe Int as output. In other words, Maybe is a function at the type level. A zero-place type constructor is the same as a type. To keep track of this complexity, Haskell assigns a *kind* to each type and type constructor. A type, such as `R`, has the kind `*`. GHCi has the command `:kind` (or `:k` for short) to ask about the kind of something.

```
*Main> :k R
R :: *
```

A one-place type constructor, such as `Maybe`, has the kind `* -> *`.

```
*Main> :k Maybe
Maybe :: * -> *
```

Once we apply `Maybe` to `R`, the resulting `Maybe R` is once again a type, with the kind `*`.

```
*Main> :k Maybe R
Maybe R :: *
```

Types have the kind `*`, one-place type constructors have the kind `* -> *`, and there are objects with more complicated kinds as well. It is interesting to note that you can ask GHCi for the kind of a type class.

```
*Main> :k Num
Num :: * -> Constraint
```

This kind means that, when provided with a type, the type class `Num` produces a constraint.

Function types, list types, and tuple types are all special cases of types constructed with a type constructor. Haskell provides special syntax for function types, list types, and tuple types, so it may aid our understanding to give regular names to the type constructors that produce functions, lists, and tuples. The `type` keyword, which was introduced in Chapter 4 to make `R` a synonym for `Double`, can also be used for parameterized types.

```
type List a = [a]
type Function a b = a -> b
type Pair a b = (a,b)
type Triple a b c = (a,b,c)
```

`List`, like `Maybe`, is a one-place type constructor. It takes one type as input and produces a type as output. `Function` and `Pair` are two-place type constructors. They take two types as input and produce a type as output. `Triple` is a three-place type constructor. It takes three types as input and produces a type as output.

Table 9-4 shows the kinds of some type constructors and type classes.

Table 9-4: Kinds of Several Type Constructors and Type Classes

Type constructor/class		Kind
`Integer`	`::`	`*`
`R -> R`	`::`	`*`
`[String]`	`::`	`*`
`(Int,String)`	`::`	`*`
`Maybe Int`	`::`	`*`
`()`	`::`	`*`
`List`	`::`	`* -> *`
`[]`	`::`	`* -> *`
`Maybe`	`::`	`* -> *`
`IO`	`::`	`* -> *`
`Function`	`::`	`* -> * -> *`
`(->)`	`::`	`* -> * -> *`
`Pair`	`::`	`* -> * -> *`
`(,)`	`::`	`* -> * -> *`
`Either`	`::`	`* -> * -> *`
`Triple`	`::`	`* -> * -> * -> *`
`(,,)`	`::`	`* -> * -> * -> *`
`Num`	`::`	`* -> Constraint`
`Foldable`	`::`	`(* -> *) -> Constraint`

Basic types, function types, list types, tuple types, `Maybe` types, and the unit type all have the kind `*`. One-place type constructors, such as `List`, `Maybe`, and `IO`, have the kind `* -> *`, which indicates that they take a type as input and produce a type as output. Note that the symbol `[]`, which is the empty list and hence is a data constructor for the list type, serves double duty as a type constructor for the list type. In this role it does the same thing as the `List` type constructor we defined above. `IO` is a type constructor that turns a pure type into a type with side effects; we'll discuss it in Chapter 11.

Two-place type constructors, such as `Function`, `Pair`, and `Either` have the kind `* -> * -> *` to indicate that they take two types as input and produce a type as output. The symbol `(->)` is the same as the `Function` type constructor we defined above and `(,)` is the same as `Pair`. A three-place type constructor, such as `Triple`, has the kind `* -> * -> * -> *` to indicate that it takes three types as input and produces a type as output. The symbol `(,,)` is the same as `Triple`.

Table 9-5 shows the meanings of various kinds in Haskell.

Table 9-5: Meanings of Kinds

Kind	Meaning
`*`	Type
`* -> *`	One-place type constructor
`* -> * -> *`	Two-place type constructor
`* -> * -> * -> *`	Three-place type constructor
`* -> Constraint`	Type class for types
`(* -> *) -> Constraint`	Type class for type constructors

Each type class also has a kind. A type class takes a type or type constructor as input and produces a constraint as output. The basic type classes we discussed in Chapter 8 have the kind `* -> Constraint`, meaning they take a type as input and produce a type-class constraint as output. The type class `Foldable` has the kind `(* -> *) -> Constraint`, meaning that it takes a type constructor (such as `List` or `Maybe`) as input and produces a type-class constraint as output. `List` and `Maybe` are instances of `Foldable`, but `IO` is not.

Our final use of tuples in this chapter is in numerical integration. By using a tuple together with `iterate`, we get a method for numerical integration that we can later generalize to a method for solving a differential equation.

Numerical Integration Redux

We looked at numerical integration in Chapter 6, where we used a list comprehension to sum the areas of rectangles under a curve. Now that we have tuples, we can present an alternative method for numerical integration that comes closer to the method we'll later use to solve differential equations. The idea is that if we pair up the current value of the integration variable (let's call it time) with the accumulating value of the integral, we can proceed step by step to get the entire integral. To take one step forward, we increment the time by the time step and increment the running total that will ultimately be our integral by the area of one rectangle under the curve.

The function that advances one step looks like this:

```
oneStep :: R          -- time step
        -> (R -> R)   -- function to integrate
        -> (R,R)      -- current (t,y)
        -> (R,R)      -- updated (t,y)
oneStep dt f (t,y) = let t' = t + dt
                         y' = y + f t * dt
                     in (t',y')
```

The function `oneStep` names the incoming time step `dt`, the function to be integrated `f`, the current value of the integration variable `t`, and the current accumulating value of the integral `y`. It then returns a pair with the integration variable increased by the time step and the current value of the integral increased by the area `f t * dt` of one rectangle under the function `f`.

To compute the integral, we iterate the single step as long as the independent variable is less than the upper limit.

```
integral' :: R -> Integration
integral' dt f a b
    = snd $ head $ dropWhile (\(t,_) -> t < b) $
      iterate (oneStep dt f) (a + dt/2,0)
```

The expression `oneStep dt f :: (R,R) -> (R,R)` is a function that updates the current time-integral pair by one time step. Since this function has the type `a -> a`, it can be iterated with `iterate`. The expression `iterate (oneStep dt f) (a + dt/2,0)` produces an infinite list of time-integral pairs, starting with the time `a + dt/2`, which is in the middle of the first time interval, and an initial value of 0 for the value of the integral that will accumulate as we iterate.

By acting on the infinite list with `dropWhile (\(t,_) -> t < b)`, we drop the initial pairs whose times are less than the upper limit `b` to obtain an infinite list whose first pair has a time very close to the upper limit `b`. Acting on this infinite list with `head` returns the pair whose time is very close to the upper limit. Finally, acting on this pair with `snd` returns the value of the integral.

For convenience, here is the `Integration` type from Chapter 6 that we used earlier in the type signature of `integral'`:

```
type Integration = (R -> R)  -- function
                   -> R      -- lower limit
                   -> R      -- upper limit
                   -> R      -- result
```

Summary

This chapter introduced tuples, a way to combine two or more values into a single value. We then looked at type constructors, functions at the type level that form an output type from an input type. We ended the chapter by using tuples to introduce an alternative way to achieve numerical integration using `iterate`.

In the next chapter, we'll return to physics, look at kinematics in three dimensions, and develop the data type we'll use for vectors.

Exercises

Exercise 9.1. Write a function

```
polarToCart :: (R,R) -> (R,R)
```

that takes as input polar coordinates (r, θ), with θ in radians, and returns as output a pair (x, y) of Cartesian coordinates.

Exercise 9.2. Explain in words the meaning of the types of `curry` and `uncurry`.

Exercise 9.3. The Prelude function

```
head :: [a] -> a
```

is slightly problematic in that it causes a runtime error if it is passed an empty list. Write a function

```
headSafe :: [a] -> Maybe a
headSafe = undefined
```

that returns `Nothing` if passed the empty list and `Just x` otherwise, where `x` is the first element (the head) of the given list. Replace the `undefined` with your own code. (You can use `undefined` in your own functions as a placeholder if you want your code to load before you are finished writing it.)

Exercise 9.4. We mentioned earlier that the type `Maybe a` is a bit like the type `[a]`, except that elements of `Maybe a` are constrained to have zero or one element. To make this analogy precise, write a function

```
maybeToList :: Maybe a -> [a]
maybeToList = undefined
```

that makes a list out of a `Maybe` type. What list should `Nothing` map to? What list should `Just x` map to?

Exercise 9.5. Find out and explain what happens when `zip` is used with two lists that don't have the same length.

Exercise 9.6. Define a function

```
zip' :: ([a], [b]) -> [(a, b)]
zip' = undefined
```

that turns a pair of lists into a list of pairs. (Hint: consider using `curry` or `uncurry`.)

Exercise 9.7. The dot operator `(.)` is for function composition. If we do `unzip` followed by `zip'`, we have a function with the following type signature:

```
    zip' . unzip :: [(a, b)] -> [(a, b)]
```

Is this the identity function? (In other words, does it always return the expression it was given?) If so, how do you know? If not, give a counterexample.

If we do `zip'` followed by `unzip`, we have a function with the following type signature:

```
unzip . zip' :: ([a], [b]) -> ([a], [b])
```

Is this the identity function?

Exercise 9.8. Using the `grades` lookup table from earlier, show how to use the `lookup` function to produce the value `Just 89`. Also show how to use the `lookup` function to produce the value `Nothing`.

Exercise 9.9. Translate the following mathematical function into Haskell:

$$x(r, \theta, \phi) = r \sin\theta \cos\phi$$

Use a triple for the input to the function `x`. Give a type signature as well as a function definition.

Exercise 9.10. A car starts from rest and accelerates at 5 m/s^2 along a straight, level highway. We want to make an infinite list `tvPairs` of time-velocity pairs for this car, one every second. Here is our code for `tvPairs`:

```
tvPairs :: [(R,R)]
tvPairs = iterate tvUpdate (0,0)
```

Write a type signature and function definition for `tvUpdate`.

```
tvUpdate = undefined
```

The list `tvPairs` should look like `[(0,0),(1,5),(2,10),(3,15),...]`. After you write the function `tvUpdate`, use the `take` function to see the first several elements of `tvPairs`.

Exercise 9.11. A Fibonacci sequence is one in which each term is the sum of the previous two terms. The first several terms are 1, 1, 2, 3, 5, 8, 13, 21, 34, 55. Write a sequence

```
fibonacci :: [Int]
fibonacci = undefined
```

for the (infinite) list of Fibonacci numbers.

As a suggestion, write a helping sequence

```
fibHelper :: [(Int,Int)]
fibHelper = undefined
```

using the function `iterate`. The first several terms of `fibHelper` should be `[(0,1),(1,1),(1,2),(2,3),(3,5),...]`. Then write the sequence `fibonacci` using `fibHelper`.

Exercise 9.12. The factorial function takes a non-negative integer and returns the product of all positive integers up to and including the given integer. It is usually denoted with an exclamation point. For example, $5! = 5 \times 4 \times 3 \times 2 \times 1 = 120$. We define $0! = 1$. The purpose of this exercise is to write a factorial function

```
fact :: Int -> Int
```

using `iterate`. The suggestion in Exercise 9.11 is useful here as well (write a sequence `factHelper :: [(Int,Int)]` using `iterate` and then define `fact` to get its values from this sequence using the `!!` operator).

Exercise 9.13. Write the following function using a list comprehension rather than a map:

```
pick13 :: [(R,R,R)] -> [(R,R)]
pick13 triples = map (\(x1,_,x3) -> (x1,x3)) triples
```

Exercise 9.14. Suppose we throw a rock straight up in the air at 15 m/s. Use a list comprehension to make a list of (time, position, velocity) triples (type `[(R,R,R)]`) for an interval of time while the rock is in the air. Your list should have enough triples to let the data make a reasonably smooth graph if plotted.

Exercise 9.15. Tuples can be nested, like `((3,4),5)`. Although this pair contains three numbers, it is not the same as a triple. Write a function

```
toTriple :: ((a,b),c) -> (a,b,c)
toTriple = undefined
```

that converts a pair whose first component is a pair into a triple.

10

DESCRIBING MOTION IN THREE DIMENSIONS

In Chapter 4, we reviewed one-dimensional kinematics, describing quantities like velocity and acceleration with real numbers. In this chapter, we'll look at three-dimensional kinematics, describing velocity and acceleration as vectors. Haskell does not have a built-in type for vectors, but it does have powerful facilities for making your own types, which we'll use to create a `Vec` type for vectors. Before deciding how to implement the `Vec` type, we'll take a careful look at the meaning and use of vectors in physics so we can produce an implementation that aligns well with how we think and write about vectors.

Three-Dimensional Vectors

The notion of a three-dimensional vector is essential in physics. In physics, vectors are geometric objects used to describe quantities that have a magnitude and a direction in space. They are best thought of as arrows in which the length of the arrow represents the magnitude and the arrow points in some direction. Living on the surface of the earth as we do, directions can sometimes be described briefly in words, such as "up," "north," and so on. We can only ever specify the direction of a vector relative to some object, such as the earth; there is no universal or absolute notion of direction.

When we discussed one-dimensional motion in Chapter 4, we did it in the context of an air track that came already marked in meters. The markings on the air track amount to a coordinate system in one dimension. A *coordinate system* is a method for describing position in terms of numbers.

Nature does not typically grant us a coordinate system to use; instead, we choose the coordinate system we want to use. In three dimensions, this amounts to choosing a location and orientation for a set of three mutually perpendicular directions. Along each direction (let's call them x, y, and z), we make (real or imaginary) marks, in meters. The place where $x = y = z = 0$ is called the *origin* of the coordinate system. Once we have chosen a coordinate system, a position can be described by three numbers (x, y, and z), indicating the (positive or negative) distance from the origin in each direction.

To describe motion in three dimensions, we usually need to introduce a coordinate system. But the laws of physics shouldn't depend on any particular coordinate system. They should work with whatever coordinate system we want to use. Vectors are geometric objects; along with the laws of physics, vectors have an existence apart from any coordinate system. Vectors allow various operations we can describe without a coordinate system. We'll give geometric (coordinate-free) descriptions of the important properties and operations of vectors and then show how the same operations appear once we have introduced a coordinate system.

Before we get into the various vector operations, I want to write the code that must exist at the top of the source code file for this chapter:

```
{-# OPTIONS -Wall #-}

module SimpleVec where

infixl 6 ^+^
infixl 6 ^-^
infixr 7 *^
infixl 7 ^*
infixr 7 ^/
infixr 7 <.>
infixl 7 ><
```

The first line turns on compiler warnings, which is a good idea to help avoid some common mistakes that are legal code but may not mean what you think they mean. If there are warnings, you will see them when you load the file.

The next line gives the code in this chapter a module name so that this code can be imported into another source code file later. The module name `SimpleVec` must match the name of the file that contains it, so the filename should be *SimpleVec.hs*. The remaining lines specify precedence levels and associativities for the operators we define later in the chapter. The precedence levels are numbers from 0 to 9, described in Chapter 1, that determine which operators act first in an expression with multiple operators. The keyword `infixl` is for an operator with left associativity, while `infixr` is for right associativity.

Coordinate-Free Vectors

Now that we have this code at the top of our file and have a basic idea of what vectors are, let's look some of their geometric properties. We'll give the geometric definitions for vector addition, scalar multiplication, vector subtraction, the dot product, the cross product, and the derivative of a vector-valued function of a real number. If you're interested, the book *Modern Classical Physics* [6] by Kip Thorne and Roger Blandford gives an elegant motivation for the coordinate-free, geometric view of vectors.

Geometric Definition of Vector Addition

We can combine two vectors using what we'll call *vector addition*. Geometrically, we define the sum of two vectors to be the vector that points from the tail of the first to the tip of the second when the two vectors are placed tip-to-tail. You can see from Figure 10-1 that the order in which they are placed tip-to-tail does not matter; consequently, vector addition is commutative (**A** + **B** = **B** + **A**).

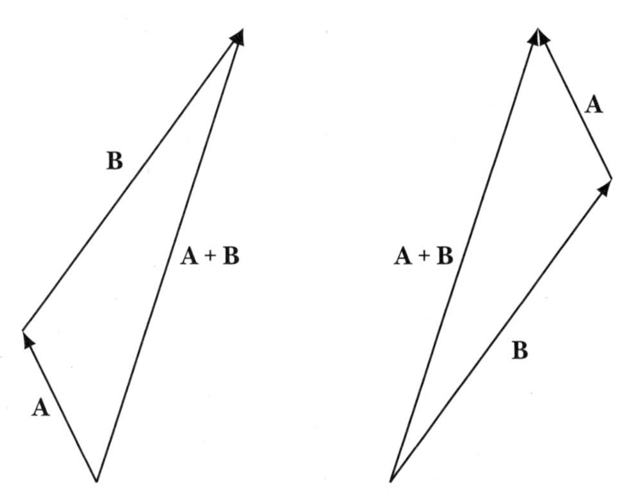

Figure 10-1: Vector addition. The vector ***A + B*** *is the sum of vectors* ***A*** *and* ***B****.*

Physicists need to know whether a symbol stands for a number or a vector; thus, the theory of Newtonian mechanics (and most other theories in physics) invites us to think in terms of types. The mathematical notation that physicists typically use to indicate a vector is syntactic and identifies a vector by a boldface symbol. In Haskell, the distinction between numbers and vectors is not syntactic; names for each are simply identifiers that begin with a lowercase letter. In Haskell, the distinction between numbers and vectors is semantic and is captured by the type of the value: `R` for numbers and `Vec` for vectors.

In mathematical notation, we use the same + sign for vector addition that we use for addition of numbers, even though vectors and numbers are very different things and it makes no sense to add a vector to a number. In Haskell, we'll use a different symbol (`^+^`) for vector addition than we do for number addition. If `a` and `b` are vectors (we write `a :: Vec` to say that `a` has type `Vec`), then `a ^+^ b` will be their vector sum. At the end of the chapter, we'll show how to define the `Vec` type and the `^+^` operator.

NOTE *In this chapter, we introduce new operators for vector addition, subtraction, and scalar multiplication. An alternative path, which we do not follow, would be to expand the definitions of addition (+), subtraction (-), and multiplication (*) so they work for vectors as well as numbers. The Haskell language is certainly powerful enough to do this. The reason we don't follow this course is that we prefer to have simple, concrete types for our vector operations, rather than types involving type classes. The names, such as `^+^`, that we use for the new operators are borrowed from Conal Elliott's `vector-space` package [7], a more sophisticated and general way of dealing with vectors than we present here.*

Geometric Definition of Scaling a Vector

We define scaling a vector by a number (also called *scalar multiplication* or multiplying a number by a vector) as follows: if the number is positive, we multiply the magnitude of the vector by the number and leave the direction of the vector unchanged. If the number is negative, we multiply the magnitude of the vector by the absolute value of the number and flip the direction of the vector. If the number is 0, the result is the zero vector.

We define division of a vector $\mathbf{A}$ by a number m to be scalar multiplication of the vector by the reciprocal of m.

$$\mathbf{A}/m = \frac{1}{m}\mathbf{A}$$

In Figure 10-2, we show the results of scaling a vector by 2, -1, and $-1/2$. Scaling by a positive number multiplies the length of the vector, keeping the direction the same. Scaling by a negative number multiples the vector's length and flips its direction.

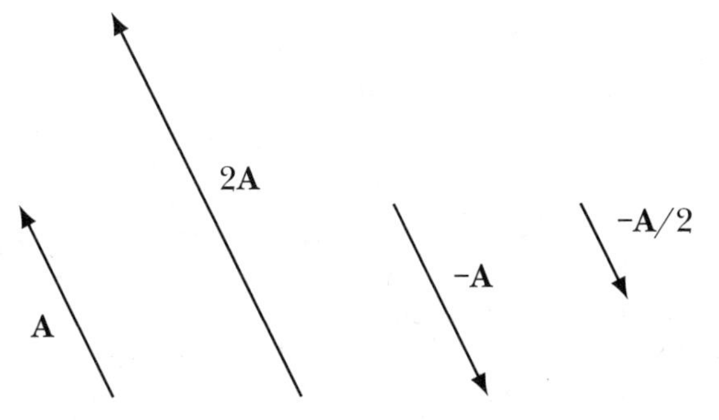

*Figure 10-2: Scalar multiplication. Scaling **A** by 2, –1, and –1/2 results in 2 **A**, –**A**, and –**A**/2, respectively.*

In mathematical notation, we use the same juxtaposition of symbols (placing a number beside a vector) for scalar multiplication that we use for multiplication of numbers, even though the operation is different. Similarly, we use the same sign (/) for dividing a vector by a number, even though the operation is different.

In Haskell, we'll use different symbols for scalar multiplication than we do for number multiplication, and a different symbol for division of a vector by a number than we use for division of a number by a number. If `m` is a number and `a` is a vector, then `m *^ a` and `a ^* m` each mean the scaling of `a` by `m`. Note that the caret sign is closer to the vector in each case. To divide `a` by `m`, we write `a ^/ m`.

Geometric Definition of Vector Subtraction

Another way to combine two vectors is with what we'll call *vector subtraction*. The difference of two vectors is defined to be the vector that points from the tip of the first to the tip of the second when the two vectors are placed tail-to-tail. Figure 10-3 shows that the difference of two vectors is equal to the sum of one vector and the opposite of the other vector. In symbols, **B** – **A** = **B** + (–**A**).

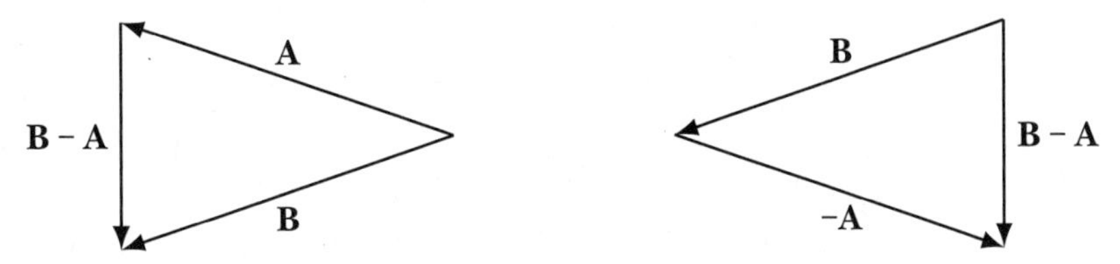

*Figure 10-3: Vector subtraction. The vector **B** – **A** is the difference of vectors **B** and **A**.*

In mathematical notation, we use the same sign (–) for vector subtraction that we use for subtraction of numbers, even though vectors and numbers are very different things. In Haskell, if `a` and `b` are vectors, we'll define `a ^-^ b` to be their vector difference.

Geometric Definition of Dot Product

There are (at least) two important products of vectors in physics. One is the *dot product*, or *inner product*. The dot product of two vectors is a scalar, or number. Here's the geometric definition:

$$\mathbf{A} \cdot \mathbf{B} = AB \cos \theta$$

In this equation, θ is the angle between the two vectors when they are placed tail-to-tail, and we use the standard notation of allowing an italic symbol to stand for the magnitude of the vector whose boldface symbol has the same letter. In other words, $A = |\mathbf{A}|$ and $B = |\mathbf{B}|$.

Figure 10-4 shows that the dot product of two vectors is the product of the magnitude of one vector (B or A) and the projection of the second vector onto the first ($A \cos \theta$ or $B \cos \theta$, respectively). Note that the projection will be negative when $\theta > 90°$.

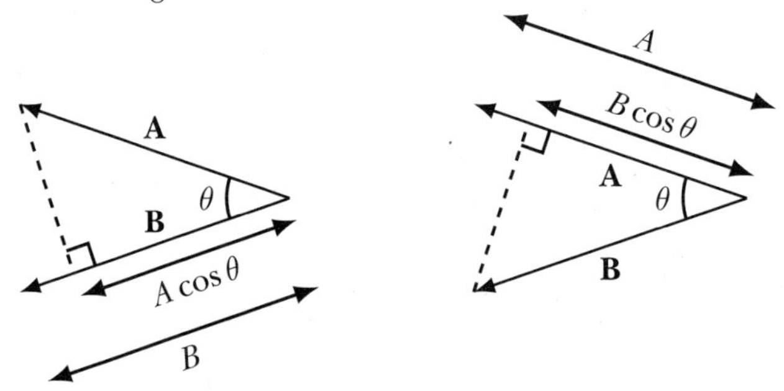

Figure 10-4: The dot product of two vectors is the product of the magnitude of one vector and the projection of the second vector onto the first.

Notice that the dot product is commutative: $\mathbf{A} \cdot \mathbf{B} = \mathbf{B} \cdot \mathbf{A}$. Also, the dot product is related to the magnitude of a vector.

$$\mathbf{A} \cdot \mathbf{A} = |\mathbf{A}|^2 \cos(0) = |\mathbf{A}|^2 = A^2$$

Therefore,

$$A = \sqrt{\mathbf{A} \cdot \mathbf{A}}$$

The dot product distributes over a vector sum.

$$\mathbf{C} \cdot (\mathbf{A} + \mathbf{B}) = \mathbf{C} \cdot \mathbf{A} + \mathbf{C} \cdot \mathbf{B}$$

In Haskell, if `a` and `b` are vectors, then `a <.> b` will be their dot product.

Geometric Definition of Cross Product

The *cross product* of two vectors is the vector whose magnitude is given by

$$|\mathbf{A} \times \mathbf{B}| = AB \sin \theta$$

and whose direction is perpendicular to both **A** and **B**.

Figure 10-5 shows the plane containing vectors **A** and **B**. To find the direction of $\mathbf{A} \times \mathbf{B}$, imagine rotating vector **A** about its tail through an angle less than 180° until it aligns with vector **B**. If counterclockwise rotation is required to carry this out, the direction of $\mathbf{A} \times \mathbf{B}$ is out of the page. If instead clockwise rotation is required, the direction of $\mathbf{A} \times \mathbf{B}$ is into the page.

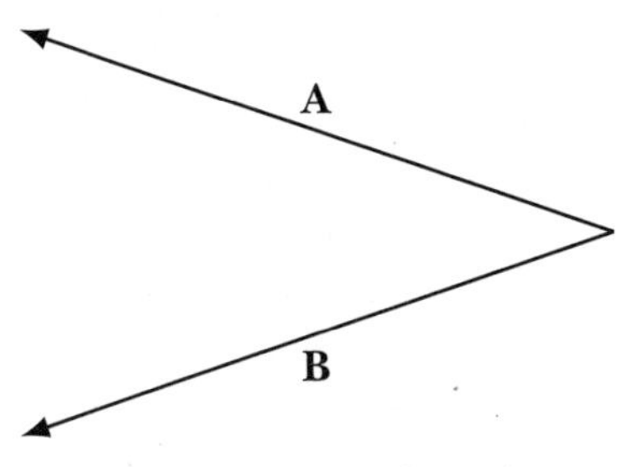

Figure 10-5: Cross product. The vector $\mathbf{A} \times \mathbf{B}$ *points out of the page. The vector* $\mathbf{B} \times \mathbf{A}$ *points into the page.*

For **A** and **B** in Figure 10-5, the vector $\mathbf{A} \times \mathbf{B}$ points out of the page and the vector $\mathbf{B} \times \mathbf{A}$ points into the page. The cross product is anti-commutative: $\mathbf{A} \times \mathbf{B} = -\mathbf{B} \times \mathbf{A}$. Note also that the cross product of any vector with itself is 0. The magnitude of the cross product gives the area of the parallelogram formed by the two vectors when two additional parallel sides are added.

The cross product distributes over a vector sum:

$$\mathbf{C} \times (\mathbf{A} + \mathbf{B}) = \mathbf{C} \times \mathbf{A} + \mathbf{C} \times \mathbf{B}$$

In Haskell, if `a` and `b` are vectors, we'll define `a >< b` to be their cross product. (The operator `><` is supposed to look like a cross product.)

NOTE *If you are interested in mathematical innovations, the geometric product is more sophisticated than the dot and cross products, but it contains the essence of both. The book* Geometric Algebra for Physicists *[8] by Chris Doran and Anthony Lasenby is a nice introduction. The book* Space-Time Algebra *[9] by David Hestenes is another great resource.*

Derivative of a Vector-Valued Function

Suppose **V** is a function that takes one real variable (such as time) as input and gives a vector (such as velocity) as output. Because we can subtract vectors, and because we can divide a vector by a number, we can define the derivative of a vector-valued function of a real number. In Haskell, the type of such a function is `R -> Vec`.

The derivative of **V**, denoted $D\mathbf{V}$, $\mathbf{V}'$, or $\dot{\mathbf{V}}$, is the function of one variable defined as follows:

$$D\mathbf{V}(t) = \lim_{\epsilon \to 0} \frac{\mathbf{V}(t + \epsilon/2) - \mathbf{V}(t - \epsilon/2)}{\epsilon} \tag{10.1}$$

Notice that the leftmost minus sign is a subtraction of vectors. We are using the same D symbol for the derivative of a vector-valued function that we used for the derivative of a number-valued function.

The vector derivative takes a vector-valued function of one variable (a function with type `R -> Vec`) as input and provides a vector-valued function of one variable as output.

```
type VecDerivative = (R -> Vec) -> R -> Vec
```

The type `(R -> Vec) -> R -> Vec` is the same as the type `(R -> Vec) -> (R -> Vec)`. Here is a vector derivative in Haskell:

```
vecDerivative :: R -> VecDerivative
vecDerivative dt v t = (v (t + dt/2) ^-^ v (t - dt/2)) ^/ dt
```

Like the function `derivative` from Chapter 4, this numerical derivative does not take a limit but instead uses a small interval `dt`, supplied by the user of the function.

Table 10-1 shows a comparison of mathematical notation with Haskell notation for the vector operations we've introduced.

Table 10-1: Comparison of Mathematical Notation with Haskell Notation for Vector Operations

Math notation	Haskell notation	Haskell type
t	`t`	`R`
m	`m`	`R`
$\mathbf{A}$	`a`	`Vec`
$\mathbf{B}$	`b`	`Vec`
$\mathbf{V}$	`v`	`R -> Vec`
$\mathbf{V}(t)$	`v t`	`Vec`
$\mathbf{A}+\mathbf{B}$	`a ^+^ b`	`Vec`
$m\mathbf{A}$	`m *^ a`	`Vec`
$\mathbf{A}m$	`a ^* m`	`Vec`
$\mathbf{A}/m$	`a ^/ m`	`Vec`
$\mathbf{A}-\mathbf{B}$	`a ^-^ b`	`Vec`
$\mathbf{A}\cdot\mathbf{B}$	`a <.> b`	`R`
$\mathbf{A}\times\mathbf{B}$	`a >< b`	`Vec`
$D\mathbf{V}$	`vecDerivative 0.01 v`	`R -> Vec`
$D\mathbf{V}(t)$	`vecDerivative 0.01 v t`	`Vec`
$\frac{d\mathbf{V}(t)}{dt}$	`vecDerivative 0.01 v t`	`Vec`

Let's move on and see what happens with vectors when we introduce a coordinate system.

Coordinate Systems

We choose a coordinate system by choosing a location and orientation for a set of three mutually perpendicular directions. We define $\hat{\mathbf{i}}$ to be a vector with magnitude 1 that points in the direction of increasing x. A vector with magnitude 1 is also known as a *unit vector*. Vectors augmented with a hat are unit vectors. Figure 10-6 shows a coordinate system along with coordinate unit vectors in each coordinate direction.

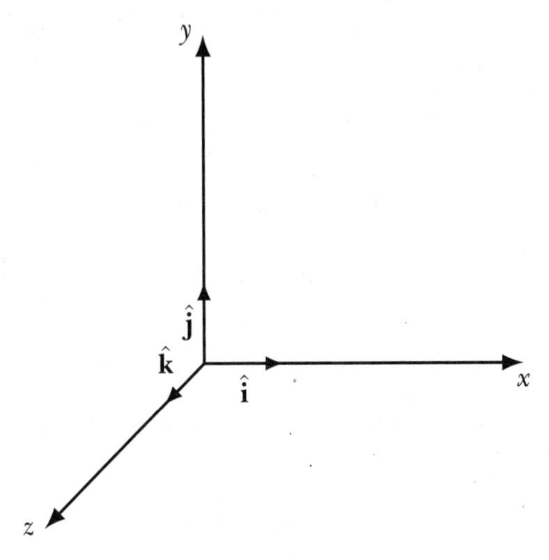

Figure 10-6: A right-handed coordinate system. The z-axis is to be imagined as coming out of the page.

Because $\hat{\mathbf{i}}$ has magnitude 1, we know that $\hat{\mathbf{i}} \cdot \hat{\mathbf{i}} = 1$. Similarly, we define $\hat{\mathbf{j}}$ to be a unit vector that points in the direction of increasing y, and we define $\hat{\mathbf{k}}$ to be a unit vector that points in the direction of increasing z. The reason they are called $\hat{\mathbf{i}}, \hat{\mathbf{j}}$, and $\hat{\mathbf{k}}$ goes back to William Rowan Hamilton and his quaternions. (Search for the William Rowan Hamilton video by A Capella Science for a wonderful musical biography of this mathematical physicist, set to a tune written for a political figure with the same last name.) Because $\hat{\mathbf{i}}$ and $\hat{\mathbf{j}}$ are perpendicular, we know that $\hat{\mathbf{i}} \cdot \hat{\mathbf{j}} = 0$. By similar reasoning, we can find the dot products of all the coordinate unit vectors.

$$\begin{array}{lll} \hat{\mathbf{i}} \cdot \hat{\mathbf{i}} = 1 & \hat{\mathbf{i}} \cdot \hat{\mathbf{j}} = 0 & \hat{\mathbf{i}} \cdot \hat{\mathbf{k}} = 0 \\ \hat{\mathbf{j}} \cdot \hat{\mathbf{i}} = 0 & \hat{\mathbf{j}} \cdot \hat{\mathbf{j}} = 1 & \hat{\mathbf{j}} \cdot \hat{\mathbf{k}} = 0 \\ \hat{\mathbf{k}} \cdot \hat{\mathbf{i}} = 0 & \hat{\mathbf{k}} \cdot \hat{\mathbf{j}} = 0 & \hat{\mathbf{k}} \cdot \hat{\mathbf{k}} = 1 \end{array} \tag{10.2}$$

Because the cross product of any vector with itself is 0, we know that $\hat{\mathbf{i}} \times \hat{\mathbf{i}} = 0$. Because the three directions of our coordinate system are mutually perpendicular, we know that $\hat{\mathbf{i}} \times \hat{\mathbf{j}} = \pm\hat{\mathbf{k}}$. To resolve the sign ambiguity, we usually agree to use a *right-handed coordinate system*, which means that $\hat{\mathbf{i}} \times \hat{\mathbf{j}} = \hat{\mathbf{k}}$. Figure 10-6 shows a right-handed coordinate system. By similar reasoning, we can find the cross products of all the coordinate unit vectors.

$$\begin{array}{lll} \hat{\mathbf{i}} \times \hat{\mathbf{i}} = 0 & \hat{\mathbf{i}} \times \hat{\mathbf{j}} = \hat{\mathbf{k}} & \hat{\mathbf{i}} \times \hat{\mathbf{k}} = -\hat{\mathbf{j}} \\ \hat{\mathbf{j}} \times \hat{\mathbf{i}} = -\hat{\mathbf{k}} & \hat{\mathbf{j}} \times \hat{\mathbf{j}} = 0 & \hat{\mathbf{j}} \times \hat{\mathbf{k}} = \hat{\mathbf{i}} \\ \hat{\mathbf{k}} \times \hat{\mathbf{i}} = \hat{\mathbf{j}} & \hat{\mathbf{k}} \times \hat{\mathbf{j}} = -\hat{\mathbf{i}} & \hat{\mathbf{k}} \times \hat{\mathbf{k}} = 0 \end{array} \tag{10.3}$$

Once we have a coordinate system and the coordinate unit vectors it gives rise to, we can "break a vector into components." Any vector $\mathbf{A}$ can be expressed as a *linear combination* of $\hat{\mathbf{i}}$, $\hat{\mathbf{j}}$, and $\hat{\mathbf{k}}$. A linear combination of vectors means a number times the first, plus a number times the second, and so on.

$$\mathbf{A} = A_x\hat{\mathbf{i}} + A_y\hat{\mathbf{j}} + A_z\hat{\mathbf{k}} \tag{10.4}$$

We call A_x the x component of $\mathbf{A}$, and similarly for y and z. The collection of the three numbers A_x, A_y, and A_z is called the *components* of $\mathbf{A}$ with respect to the coordinate system. By dotting the equation above by $\hat{\mathbf{i}}$, we get an expression for A_x in terms of $\hat{\mathbf{i}}$ and $\mathbf{A}$. We can do the same for A_y and A_z.

$$\begin{aligned} A_x &= \hat{\mathbf{i}} \cdot \mathbf{A} \\ A_y &= \hat{\mathbf{j}} \cdot \mathbf{A} \\ A_z &= \hat{\mathbf{k}} \cdot \mathbf{A} \end{aligned}$$

The Haskell code at the end of the chapter defines a default coordinate system you can use. The default coordinate system provides coordinate unit vectors `iHat`, `jHat`, and `kHat` that play the role of $\hat{\mathbf{i}}$, $\hat{\mathbf{j}}$, and $\hat{\mathbf{k}}$, respectively.

Let's revisit the vector operations introduced above in a geometric way and see how they look with a coordinate system.

Vector Addition with Coordinate Components

The components of a sum are the sums of the components. If $\mathbf{C} = \mathbf{A} + \mathbf{B}$,

$$C_x = \hat{\mathbf{i}} \cdot \mathbf{C} = \hat{\mathbf{i}} \cdot (\mathbf{A} + \mathbf{B}) = \hat{\mathbf{i}} \cdot \mathbf{A} + \hat{\mathbf{i}} \cdot \mathbf{B} = A_x + B_x$$

and similarly for the y- and z-components. If $\mathbf{C} = \mathbf{A} + \mathbf{B}$, then

$$\begin{aligned} C_x &= A_x + B_x \\ C_y &= A_y + B_y \\ C_z &= A_z + B_z \end{aligned} \tag{10.5}$$

In Haskell, this is

```
*SimpleVec> vec 1 2 3 ^+^ vec 4 5 6
vec 5.0 7.0 9.0
```

You can think of the caret on each side of the plus sign as a reminder that there is a vector on the left and a vector on the right.

Vector Scaling with Coordinate Components

If $\mathbf{C} = m\mathbf{A}$, then

$$C_x = \hat{\mathbf{i}} \cdot \mathbf{C} = \hat{\mathbf{i}} \cdot (m\mathbf{A}) = m\hat{\mathbf{i}} \cdot \mathbf{A} = mA_x$$

and similarly for the y- and z-components. If $\mathbf{C} = m\mathbf{A}$, then

$$\begin{aligned} C_x &= mA_x \\ C_y &= mA_y \\ C_z &= mA_z \end{aligned} \tag{10.6}$$

To scale a vector, we can use the `*^` operator.

```
*SimpleVec> 5 *^ vec 1 2 3
vec 5.0 10.0 15.0
```

Notice that the caret goes on the right of the asterisk because the vector is on the right. You can multiply a `Vec` on the left by an `R` on the right with the `^*` operator.

```
*SimpleVec> vec 1 2 3 ^* 5
vec 5.0 10.0 15.0
```

Since the vector is on the left, the caret is on the left. Similarly, we can divide by an `R` with the `^/` operator.

```
*SimpleVec> vec 1 2 3 ^/ 5
vec 0.2 0.4 0.6
```

Vector Subtraction with Coordinate Components

If $\mathbf{C} = \mathbf{A} - \mathbf{B}$, then

$$C_x = \hat{\mathbf{i}} \cdot \mathbf{C} = \hat{\mathbf{i}} \cdot (\mathbf{A} - \mathbf{B}) = \hat{\mathbf{i}} \cdot \mathbf{A} - \hat{\mathbf{i}} \cdot \mathbf{B} = A_x - B_x$$

and similarly for the y and z components. If $\mathbf{C} = \mathbf{A} - \mathbf{B}$, then

$$\begin{aligned} C_x &= A_x - B_x \\ C_y &= A_y - B_y \\ C_z &= A_z - B_z \end{aligned} \tag{10.7}$$

We say "the components of a difference are the differences of the components." The first use of the word *difference* refers to vector difference, while the second use refers to number difference. If `a` and `b` are vectors, then `xComp (a ^-^ b)` and `xComp a - xComp b` evaluate to the same number.

Here is an example of vector subtraction:

```
*SimpleVec> vec 1 2 3 ^-^ vec 4 5 6
vec (-3.0) (-3.0) (-3.0)
```

Dot Product with Coordinate Components

Suppose **A** and **B** are vectors. Given a coordinate system, we can express each vector in components using Equation 10.4 and then use the distributive property of the dot product along with Equation 10.2 to simplify the result.

$$\begin{aligned}\mathbf{A} \cdot \mathbf{B} &= (A_x\hat{\mathbf{i}} + A_y\hat{\mathbf{j}} + A_z\hat{\mathbf{k}}) \cdot (B_x\hat{\mathbf{i}} + B_y\hat{\mathbf{j}} + B_z\hat{\mathbf{k}}) \\ &= A_xB_x + A_yB_y + A_zB_z \end{aligned} \tag{10.8}$$

If we know the components of two vectors **A** and **B**, Equation 10.8 gives a handy way to find their dot product.

You can take the dot product of two `Vec`s with the `<.>` operator.

```
*SimpleVec> vec 1 2 3 <.> vec 4 5 6
32.0
```

Cross Product with Coordinate Components

Suppose **A** and **B** are vectors. Given a coordinate system, we can express each vector in components using Equation 10.4 and then use the distributive property of the cross product along with Equation 10.3 to simplify the result.

$$\begin{aligned}\mathbf{A} \times \mathbf{B} &= (A_x\hat{\mathbf{i}} + A_y\hat{\mathbf{j}} + A_z\hat{\mathbf{k}}) \times (B_x\hat{\mathbf{i}} + B_y\hat{\mathbf{j}} + B_z\hat{\mathbf{k}}) \\ &= (A_yB_z - A_zB_y)\hat{\mathbf{i}} + (A_zB_x - A_xB_z)\hat{\mathbf{j}} + (A_xB_y - A_yB_x)\hat{\mathbf{k}} \end{aligned} \tag{10.9}$$

If we know the components of two vectors **A** and **B**, Equation 10.9 gives a good way to find their cross product.

You can take the cross product of two `Vec`s with the `><` operator.

```
*SimpleVec> vec 1 2 3 >< vec 4 5 6
vec (-3.0) 6.0 (-3.0)
```

If you need the components of a vector, you can get them with the `xComp` function.

```
*SimpleVec> xComp $ vec 1 2 3 >< vec 4 5 6
-3.0
```

There are also functions `yComp` and `zComp`.

The unary minus sign (`-`) will not work to negate a vector, but you can negate a vector with `negateV`.

```
*SimpleVec> negateV $ vec 1 2 3 >< vec 4 5 6
vec 3.0 (-6.0) 3.0
```

Derivative with Coordinate Components

Suppose **V** is a vector-valued function of one real variable. If $\mathbf{W} = D\mathbf{V}$, then

$$W_x(t) = \hat{\mathbf{i}} \cdot \mathbf{W}(t) = \hat{\mathbf{i}} \cdot [D\mathbf{V}(t)] = D[\hat{\mathbf{i}} \cdot \mathbf{V}(t)] = DV_x(t)$$

and similarly for the y- and z-components. If $\mathbf{W} = D\mathbf{V}$, then

$$\begin{aligned} W_x &= DV_x \\ W_y &= DV_y \\ W_z &= DV_z \end{aligned} \tag{10.10}$$

Here is an example of a vector-valued function:

```
v1 :: R -> Vec
v1 t = 2 *^ t**2 *^ iHat ^+^ 3 *^ t**3 *^ jHat ^+^ t**4 *^ kHat
```

Note that we cannot write the x-component of this vector-valued function in the most obvious way, namely `xComp v1`. This would produce a type error because `xComp` takes a `Vec` as input, not a function `R -> Vec`. What we really mean when we talk about the x-component of a vector-valued function is the scalar-valued function that takes an input `t`, applies the vector-valued function, and returns the x-component. In Haskell, the x-component of a vector-valued function can be written as follows:

```
xCompFunc :: (R -> Vec) -> R -> R
xCompFunc v t = xComp (v t)
```

In words, Equation 10.10 says that the x-component of the derivative is the same as the derivative of the x-component. In Haskell, the same equations say that the x-component of the (vector) derivative

```
xCompFunc . vecDerivative dt
```

is the same as the (scalar) derivative of the x-component:

```
derivative dt . xCompFunc
```

We can check this in GHCi for a particular vector-valued function evaluated at a particular value of the independent variable.

```
*SimpleVec> (xCompFunc . vecDerivative 0.01) v1 3
11.999999999999744
*SimpleVec> (derivative 0.01 . xCompFunc) v1 3
11.999999999999744
```

We defined the scalar derivative in Chapter 4 and repeat it here for convenience.

```
type Derivative = (R -> R) -> R -> R

derivative :: R -> Derivative
derivative dt x t = (x (t + dt/2) - x (t - dt/2)) / dt
```

Table 10-2 shows the types of the vector functions and expressions we've been working with.

Table 10-2: Expressions and Functions for Working with Vectors

Expression		Type
zeroV	::	Vec
iHat	::	Vec
(^+^)	::	Vec -> Vec -> Vec
(^-^)	::	Vec -> Vec -> Vec
(*^)	::	R -> Vec -> Vec
(^*)	::	Vec -> R -> Vec
(^/)	::	Vec -> R -> Vec
(<.>)	::	Vec -> Vec -> R
(><)	::	Vec -> Vec -> Vec
negateV	::	Vec -> Vec
magnitude	::	Vec -> R
xComp	::	Vec -> R
vec	::	R -> R -> R -> Vec
sumV	::	[Vec] -> Vec

Now that we've seen some of the key properties of vectors in both geometric and coordinate settings, let's take a look at how vectors get used to describe kinematics in three dimensions.

Kinematics in 3D

The essential quantities of kinematics are time, position, velocity, and acceleration. Time will continue to be a real number, as it was in Chapter 4. Velocity and acceleration we'll now treat as vectors, using the `Vec` type that we'll define at the end of the chapter. What about position?

Position is really not a vector. It doesn't make sense to add positions, nor does it make sense to scale a position by a number. However, subtracting positions does make sense, and it produces a displacement vector from one position to the other. In Chapter 22, we'll create a proper type for position, which allows us to use Cartesian, cylindrical, and spherical coordinates to describe positions. However, our aims are more modest at the moment and simplicity suggests that we use the `Vec` type for position, even though we just gave some reasons not to. Displacement is certainly a vector, so we can think of a vector-valued position as a displacement from the origin of the default coordinate system.

We'll use the following type synonyms in this chapter:

```
type Time         = R
type PosVec       = Vec
type Velocity     = Vec
type Acceleration = Vec
```

We use the type `PosVec` to denote the type of a position when position is being represented by a vector. This will keep us from confusing it with the `Position` type we'll define in Chapter 22, which, again, is not a vector.

Defining Position, Velocity, and Acceleration

For a particular motion of an object, we define $\mathbf{r}$ to be the function that associates with each time t the position at which the object is located at time t. We say that $\mathbf{r}(t)$ is the position of the object at time t.

The velocity function for an object is the derivative of its position function.

$$\mathbf{v} = D\mathbf{r} \tag{10.11}$$

Note that Equation 10.11 is an equality of functions, the instantaneous velocity function on the left of the equality, and the derivative of the position function on the right. Equation 10.11 can be written in Haskell as the function `velFromPos`, which takes a small time step and position function to return a velocity function.

```
velFromPos :: R                        -- dt
           -> (Time -> PosVec  )       -- position function
           -> (Time -> Velocity)       -- velocity function
velFromPos = vecDerivative
```

As you can see from the definition, the function `velFromPos` is just the vector derivative we defined earlier in the chapter.

When two functions are equal, they give equal results for equal inputs, so we can also write

$$\mathbf{v}(t) = D\mathbf{r}(t) \tag{10.12}$$

for any time t. The righthand side is the function $D\mathbf{r}$ evaluated at time t. We can think of the derivative operator as taking the entire position function as input and returning the velocity function as output. It's also common to see the notation

$$\mathbf{v}(t) = \frac{d\mathbf{r}(t)}{dt} \tag{10.13}$$

to define velocity.

Speed is the magnitude of the velocity vector. *Acceleration* is defined to be the rate at which velocity changes. We define $\mathbf{a}$ to be the function that associates with each time t the rate at which the velocity is changing at time t. In the language of calculus, we can write

$$\mathbf{a} = D\mathbf{v} \tag{10.14}$$

or

$$\mathbf{a}(t) = \frac{d\mathbf{v}(t)}{dt} \tag{10.15}$$

to define acceleration. Equation 10.14 can be encoded in a function `accFromVel` that produces an acceleration function from a velocity function. Again, this function is just the vector derivative.

```
accFromVel :: R                            -- dt
           -> (Time -> Velocity)           -- velocity function
           -> (Time -> Acceleration)       -- acceleration function
accFromVel = vecDerivative
```

If the velocity happens to be constant, say $\mathbf{v}_0$, we can integrate both sides of Equation 10.11 to obtain

$$\mathbf{v}_0 t = \mathbf{r}(t) - \mathbf{r}(0)$$

If velocity is constant, position is a linear function of time.

$$\mathbf{r}(t) = \mathbf{v}_0 t + \mathbf{r}(0)$$

Here is the corresponding Haskell code:

```
positionCV :: PosVec -> Velocity -> Time -> PosVec
positionCV r0 v0 t = v0 ^* t ^+^ r0
```

The `CV` at the end of the name is an abbreviation for constant velocity.

If the acceleration happens to be constant, say $\mathbf{a}_0$, we can integrate both sides of Equation 10.14 or 10.15 to obtain

$$\mathbf{a}_0 t = \mathbf{v}(t) - \mathbf{v}(0)$$

If acceleration is constant, velocity is a linear function of time.

$$\mathbf{v}(t) = \mathbf{a}_0 t + \mathbf{v}(0) \tag{10.16}$$

I like to call Equation 10.16 the *velocity-time equation* for constant acceleration because it gives the velocity $\mathbf{v}(t)$ of an object at any time t, provided we know the constant acceleration $\mathbf{a}_0$ and initial velocity $\mathbf{v}(0)$. Here is the Haskell code for Equation 10.16:

```
velocityCA :: Velocity -> Acceleration -> Time -> Velocity
velocityCA v0 a0 t = a0 ^* t ^+^ v0
```

The `CA` at the end of the name is an abbreviation for constant acceleration. We can integrate both sides of Equation 10.16 to obtain

$$\mathbf{r}(t) - \mathbf{r}(0) = \frac{1}{2}\mathbf{a}_0 t^2 + \mathbf{v}(0)t$$

If acceleration is constant, position is a quadratic function of time.

$$\mathbf{r}(t) = \frac{1}{2}\mathbf{a}_0 t^2 + \mathbf{v}(0)t + \mathbf{r}(0) \tag{10.17}$$

I like to call Equation 10.17 the *position-time equation* for constant acceleration because it gives the position $\mathbf{r}(t)$ of an object at any time t, provided we know the constant acceleration $\mathbf{a}_0$, initial position $\mathbf{r}(0)$, and initial velocity $\mathbf{v}(0)$. Here is the Haskell code for Equation 10.17:

```
positionCA :: PosVec -> Velocity -> Acceleration
           -> Time -> PosVec
positionCA r0 v0 a0 t = 0.5 *^ t**2 *^ a0 ^+^ v0 ^* t ^+^ r0
```

Equations 10.16 and 10.17 are known as *constant acceleration equations*. They are used over and over again in a typical introductory physics course. Later we'll learn some techniques to deal with situations in which acceleration is not constant.

Having introduced the definitions of velocity and acceleration in the three-dimensional vector setting, we can now look at how acceleration is composed of two qualitatively different components.

Two Components of Acceleration

If at any moment the velocity of an object is 0, any acceleration the object has serves to give the object some velocity in the direction of the acceleration. If, on the other hand, $\mathbf{v}(t) \neq 0$, the relative directions of velocity and acceleration determine the qualitative motion of the object. In everyday speech, people often use the word *acceleration* to mean increase in speed. However, in physics, acceleration means change in velocity per unit tine, and velocity can change in either magnitude or direction. In physics, acceleration is responsible not only for increases in speed but for decreases in speed and changes in direction.

If $\mathbf{v}(t) \neq 0$, we can decompose the acceleration into a component parallel to the velocity and a component perpendicular to the velocity.

$$\mathbf{a}(t) = \mathbf{a}_{||}(t) + \mathbf{a}_{\perp}(t) \tag{10.18}$$

Since $\mathbf{v}(t) \neq 0$, we can define a unit vector in the direction of the velocity.

$$\hat{\mathbf{v}}(t) = \frac{\mathbf{v}(t)}{v(t)} = \frac{\mathbf{v}(t)}{|\mathbf{v}(t)|}$$

The parallel and perpendicular components of acceleration are given by the following equations:

$$\mathbf{a}_{||}(t) = [\hat{\mathbf{v}}(t) \cdot \mathbf{a}(t)]\, \hat{\mathbf{v}}(t)$$
$$\mathbf{a}_{\perp}(t) = \mathbf{a}(t) - [\hat{\mathbf{v}}(t) \cdot \mathbf{a}(t)]\, \hat{\mathbf{v}}(t)$$

Here are Haskell functions that calculate the parallel and perpendicular components of acceleration:

```
aParallel :: Vec -> Vec -> Vec
aParallel v a = let vHat = v ^/ magnitude v
                in (vHat <.> a) *^ vHat

aPerp :: Vec -> Vec -> Vec
aPerp v a = a ^-^ aParallel v a
```

The parallel component $\mathbf{a}_{||}(t)$ is also called the *tangential component* of acceleration, and it's responsible for the change in speed of the object. The perpendicular component $\mathbf{a}_{\perp}(t)$ is also called the *radial* or *transverse component* of acceleration, and it's responsible for the change in direction of the object.

The dot product of velocity and acceleration depends on the angle between them and thus contains useful information. Let's take the time derivative of the square of the speed $v(t)^2 = \mathbf{v}(t) \cdot \mathbf{v}(t)$:

$$\frac{d}{dt}\left[\mathbf{v}(t) \cdot \mathbf{v}(t)\right] = \mathbf{v}(t) \cdot \frac{d\mathbf{v}(t)}{dt} + \frac{d\mathbf{v}(t)}{dt} \cdot \mathbf{v}(t) = 2\mathbf{v}(t) \cdot \mathbf{a}(t) \tag{10.19}$$

We can see that the dot product of velocity and acceleration controls how the speed changes:

Dot product	Effect on speed
$\mathbf{v}(t) \cdot \mathbf{a}(t) > 0$	Speed increases
$\mathbf{v}(t) \cdot \mathbf{a}(t) = 0$	Speed remains constant
$\mathbf{v}(t) \cdot \mathbf{a}(t) < 0$	Speed decreases

In this chapter, v stands for the speed function, which is a different convention from Chapter 4 where v was the one-dimensional velocity function. One-dimensional velocity can be negative, but speed can't.

Figure 10-7 shows how the relative direction of velocity and acceleration controls the qualitative behavior of an object's motion.

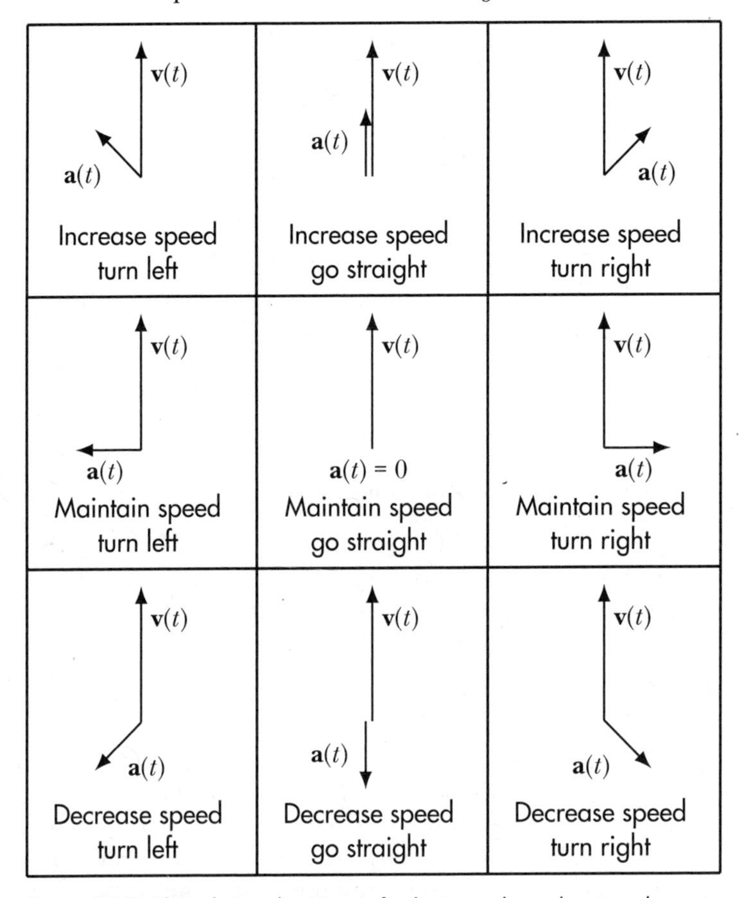

Figure 10-7: The relative directions of velocity and acceleration determine the qualitative motion of an object.

When acceleration has a component in the direction of velocity, an object speeds up. When acceleration has a component opposite the direction of velocity, an object slows down. When acceleration has only a component perpendicular to the velocity, an object maintains its speed. No coordinate system is necessary to make these conclusions; this aspect of the qualitative behavior of motion is purely geometric.

We've seen how the tangential component of acceleration is related to speeding up and slowing down. We can make an even stronger statement: the rate of change of speed is directly related to the tangential component.

$$\begin{aligned}\frac{dv(t)}{dt} &= \frac{d}{dt}\left[\sqrt{\mathbf{v}(t)\cdot\mathbf{v}(t)}\right] \\ &= \frac{1}{2}\left[\mathbf{v}(t)\cdot\mathbf{v}(t)\right]^{-1/2}\frac{d}{dt}\left[\mathbf{v}(t)\cdot\mathbf{v}(t)\right] \\ &= \frac{\mathbf{v}(t)\cdot\mathbf{a}(t)}{v(t)} \\ &= \mathbf{a}(t)\cdot\hat{\mathbf{v}}(t)\end{aligned} \tag{10.20}$$

Here is the rate of change of speed as a Haskell function. Given the velocity and acceleration of an object, this function returns the rate at which speed is increasing, with a negative result meaning that speed is decreasing.

```
speedRateChange :: Vec -> Vec -> R
speedRateChange v a = (v <.> a) / magnitude v
```

The magnitude of the tangential component of acceleration equals the magnitude of the rate of change in speed.

$$a_{||}(t) = \left|\frac{dv(t)}{dt}\right| \tag{10.21}$$

If there is a transverse component of acceleration, it causes the object to turn (in other words, to change direction). We can calculate a radius of curvature for this turning motion. Figure 10-8 shows the trajectory of a particle with speed $v(t)$ and transverse acceleration $a_\perp(t)$. Over a small interval of time Δt, the particle will move forward a distance $v(t)\Delta t$ and in a perpendicular direction a distance $a_\perp(t)\Delta t^2/2$.

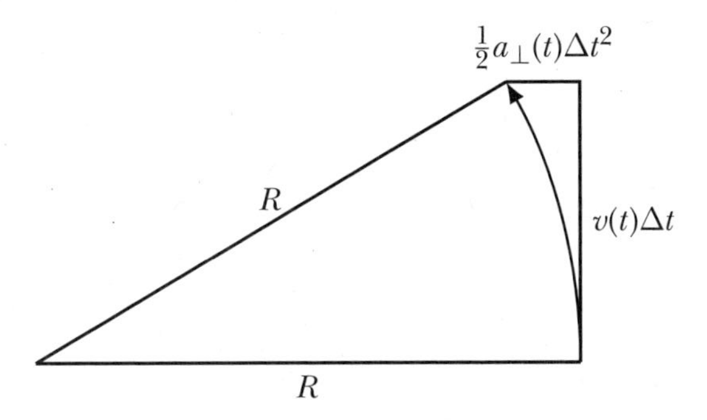

Figure 10-8: Determination of a radius of curvature from transverse acceleration

Figure 10-8 gives a way to find an expression for the radius of curvature in terms of the speed and the transverse acceleration. Here we write down the Pythagorean theorem for the right triangle in the figure:

$$[v(t)\Delta t]^2 + \left[R - \frac{1}{2}a_\perp(t)\Delta t^2\right]^2 = R^2$$

Expanding this equation and taking the limit as $\Delta t \to 0$, which discards the term proportional to Δt^4, we arrive at the following equation for the radius of curvature:

$$R = \frac{[v(t)]^2}{a_\perp(t)} \tag{10.22}$$

Here is a Haskell function that computes the (instantaneous) radius of curvature of an object's motion from the object's velocity and acceleration:

```
radiusOfCurvature :: Vec -> Vec -> R
radiusOfCurvature v a = (v <.> v) / magnitude (aPerp v a)
```

If we wish, we can invert Equation 10.22 to give an expression for the transverse acceleration in terms of the radius of curvature R.

$$a_\perp(t) = \frac{[v(t)]^2}{R} \tag{10.23}$$

As we see from Equations 10.21 and 10.23, the tangential component of acceleration controls the change in speed, and the radial component of acceleration controls the change in direction.

Projectile Motion

One of the very first uses to which vectors are typically put in physics is the study of projectile motion. A projectile is any object thrown, launched, or shot near the surface of the earth. The problem is to predict its motion *after* the throwing, launching, or shooting force is gone; in fact, rather than speak of the launching force, we instead assume that the action of launching simply gives the projectile some initial velocity.

It is the presence of Earth's gravity that makes projectile motion interesting. Physics offers four theories of gravity, three of which we will discuss in this book:

1. Gravity causes objects near Earth's surface to accelerate. An object near Earth's surface that is allowed to move or fall freely will accelerate toward the center of the earth at a rate of about 9.81 m/s^2.
2. Gravity is a force produced by the earth on objects near its surface. We engage with this view of gravity in Chapters 15 and 16 after we've started mechanics and discussed the ideas of force, mass, and Newton's second law. The Haskell function `earthSurfaceGravity` in Chapter 16 describes this force of gravity.

3. Gravity is a force between any two objects with mass. This is called Newton's law of universal gravitation. We discuss it in Chapters 16 and 19 and describe it with the Haskell function `universalGravity` after we have introduced Newton's third law.
4. Gravity is the curvature of spacetime. This is Einstein's general theory of relativity. We will not get to it in this book. The book *Gravitation* [10] by Charles Misner, Kip Thorne, and John Wheeler is an excellent introduction to general relativity. *Functional Differential Geometry* [11] by Gerald Sussman and Jack Wisdom looks at general relativity from a computational point of view, describing it with the functional programming language Scheme. Other introductions to general relativity by Rindler [12], Carroll [13], and Schutz [14] are also recommended.

Each theory in this list is more sophisticated than the previous theory. In that sense, the later theories are "more correct" than the earlier ones, although the earlier ones are often more useful since they are simpler and easier to apply. General relativity, in particular, while beautiful and accurate, is rather nontrivial to apply and to calculate with.

Some physicists may not agree that the first two theories on my list deserve to be called theories, arguing that they are instead approximations to Newton's law of universal gravitation applicable to simple situations. It is not my concern to argue whether the first two deserve to be called theories; the important point is that they are different ways to incorporate gravity into our calculations.

The simplest way to approach projectile motion, and the one we'll follow in this chapter, is based on theory 1. We assume the projectile accelerates only because of Earth's gravitational attraction; therefore, the projectile's acceleration is given by the acceleration of gravity, $\mathbf{g}$, which is a vector pointing toward the center of the earth with magnitude 9.81 $\mathrm{m/s^2}$.

Because the acceleration of gravity is constant, we can use the position-time equation, Equation 10.17, to give an expression for position as a function of time for a projectile.

$$\mathbf{r}(t) = \frac{1}{2}\mathbf{g}t^2 + \mathbf{v}(0)t + \mathbf{r}(0) \qquad (10.24)$$

If the z-direction is up and we use SI units, the following function returns position as a function of time, where `r0` is the initial position of the projectile and `v0` is the initial velocity:

```
projectilePos :: PosVec -> Velocity -> Time -> PosVec
projectilePos r0 v0 = positionCA r0 v0 (9.81 *^ negateV kHat)
```

In Exercise 10.5, you are asked to write a function `projectileVel` that returns the velocity of a projectile as a function of time.

Projectile motion with air resistance requires theory 2 because we view air resistance and gravity on an equal footing by seeing them both as forces. To the extent that projectile motion occurs near Earth's surface, it rarely requires theories 3 or 4, which would produce only the slightest difference in results for a substantial increase in computation complexity. Theory 3 will have other uses, like satellite motion, that we'll see in later chapters.

Having seen projectile motion as one early application of vectors, let's turn to the project of creating the vector data type we have been using.

Making Your Own Data Type

Haskell has a sophisticated and flexible type system. One of the language features that makes the type system so powerful is the ability to create your own types.

In discussing pattern matching in Chapters 3 and 5, we noted that each type has one or more data constructors to construct values of that type. In making our own data type, we must provide one or more brand new data constructors as ways of constructing values of our new data type.

We'll first look at making a new data type with a single data constructor before turning to making a new data type with multiple data constructors.

Single Data Constructor

In Chapter 4, we used the `type` keyword to make type synonyms. In a type synonym, such as

```
type R = Double
```

the compiler regards the types `R` and `Double` as interchangeable. In Chapter 4, the types `Time`, `Velocity`, `R`, and `Double` were all interchangeable. This was convenient, but it does not empower the Haskell type checker to help the code writer avoid confusing a `Time` with a `Velocity` or using a `Velocity` where a `Time` was expected.

We use the `data` keyword to define new types that are not interchangeable with any existing types. Time and mass are each described by a real number in physics, but we should never provide a mass in an equation that calls for a time. Let's define a new data type `Mass` that holds a real number but cannot be confused with `R` or any other existing data type.

```
data Mass = Mass R
            deriving (Eq,Show)
```

We use the `data` keyword to define a new data type. We give the name of the new data type, `Mass` in this case, after the `data` keyword, followed by an equal sign. To the right of the equal sign we give a data constructor, `Mass` in this case, followed by the information our new data type contains, in this case `R`. The name of the new data type and the name of the data constructor can be the same or different. In this case, they have the same name, but they are different things. In defining data types with a single constructor, it's common to use the same name for the data constructor that we use for the type, but it's not necessary.

By default, a new data type is not an instance of any type classes. Since it's common to want a new data type to be an instance of some of the standard type classes, such as `Eq` and `Show`, Haskell provides a `deriving` keyword that attempts to make the new type an instance of the type classes listed.

To construct a value with type `Mass`, we use the data constructor `Mass`.

```
*SimpleVec> Mass 9
Mass 9.0
```

If we ask for the type of this value, GHCi will tell us the following:

```
*SimpleVec> :t Mass 9
Mass 9 :: Mass
```

In GHCi's response, the `Mass` on the left of the double colon is the data constructor, and the `Mass` on the right of the double colon is the data type.

The data constructor itself has a function type. It takes an `R` as input and returns a `Mass` as output.

```
*SimpleVec> :t Mass
Mass :: R -> Mass
```

We are asking GHCi for the type of the data constructor, not the type of the type, which wouldn't make any sense. Again, the `Mass` on the left of the double colon is the data constructor, and the `Mass` on the right of the double colon is the data type.

If we now accidentally supply a `Mass` in a place where an `R` is called for, the type checker will give us a type error, helping us to identify our mistake rather than silently doing the wrong thing.

We can provide more than a single piece of information under the data constructor. In Chapter 9 we used a list of pairs to hold grade information. Let's define a new data type called `Grade` that holds a `String` and an `Int`, representing the name of a person and their grade on some assignment.

```
data Grade = Grade String Int
             deriving (Eq,Show)
```

We give the data constructor the same name as the type and simply list the types of information that are to be contained in the new data type.

Here is a list of grades for a few people:

```
grades :: [Grade]
grades = [Grade "Albert Einstein" 89
         ,Grade "Isaac Newton"    95
         ,Grade "Alan Turing"     91
         ]
```

To construct a value of type `Grade`, we use the data constructor `Grade`, followed by a `String` and an `Int`.

If we look at the type of the data constructor Grade,

```
*SimpleVec> :t Grade
Grade :: String -> Int -> Grade
```

we see that it takes a String and an Int as input and returns a Grade as output. As before, the Grade on the left of the double colon is the data constructor, and the Grade on the right of the double colon is the data type.

There is an alternative syntax for defining a new data type, called *record syntax,* that gives names to each of the pieces of data under a constructor. Let's define a new data type called GradeRecord that is essentially the same as the type Grade but uses record syntax for its definition.

```
data GradeRecord = GradeRecord { name  :: String
                               , grade :: Int
                               } deriving (Eq,Show)
```

To use record syntax, we enclose the names and types of each piece of information in curly braces after the data constructor. Use of record syntax automatically creates a new function for each named piece of information.

```
*SimpleVec> :t name
name :: GradeRecord -> String
*SimpleVec> :t grade
grade :: GradeRecord -> Int
```

The function name takes a GradeRecord as input and returns the String that holds the name in that GradeRecord. The function grade takes a GradeRecord as input and returns the Int that holds the grade in that GradeRecord. By default, the names name and grade are placed in the global namespace so they may not be reused as names for fields in another data type. This default behavior is simple but too constraining in some situations, so the language option DuplicateRecordFields is available to override this default behavior, although we will not explore it in this book.

If we use record syntax to define our new data type, there are two ways to construct a value of that type. First, we can use the same syntax we used for the type Grade above, simply giving the data constructor followed by a String and an Int.

```
gradeRecords1 :: [GradeRecord]
gradeRecords1 = [GradeRecord "Albert Einstein" 89
                ,GradeRecord "Isaac Newton"    95
                ,GradeRecord "Alan Turing"     91
                ]
```

Second, we can use record syntax to construct a value of type GradeRecord.

```
gradeRecords2 :: [GradeRecord]
gradeRecords2 = [GradeRecord {name = "Albert Einstein", grade = 89}
                ,GradeRecord {name = "Isaac Newton"   , grade = 95}
```

```
             ,GradeRecord {name = "Alan Turing"    , grade = 91}
             ]
```

Here we use curly braces and give the pieces of information by name rather than by position.

The decision on whether to use record syntax should be based on the usefulness of having names to describe the pieces of data that the new type holds. If you don't need the names, you should use the basic syntax. If the names seem useful, record syntax is a good choice.

We've seen how to define a new data type with a single data constructor. Now let's look at data types with more than one data constructor.

Multiple Data Constructors

The Prelude type `Bool` has two data constructors, `False` and `True`, as we saw when we did pattern matching on `Bool`. Neither of the data constructors contains any information beyond the name of the constructor itself.

Let's define a new data type called `MyBool` that works the same way `Bool` does. We need a fresh name because `Bool` is already defined in the Prelude.

```
data MyBool = MyFalse | MyTrue
              deriving (Eq,Show)
```

We begin with the `data` keyword as before, followed by the name `MyBool` of our new data type. On the right of the equal sign, we give the first data constructor, which we call `MyFalse`, then a vertical bar, and then the second data constructor `MyTrue`. We need fresh names for the data constructors because the names `False` and `True` are already taken.

The vertical bar in the definition can be read as "or," in the sense that a value of type `MyBool` is either `MyFalse` or `MyTrue`.

Having defined the new data type `MyBool`, we can ask about the type of `MyFalse`,

```
*SimpleVec> :t MyFalse
MyFalse :: MyBool
```

and we are not surprised to find that it has type `MyBool`.

When we have multiple data constructors, they usually have different names from the data type itself.

Let's define our own version of `Maybe`, called `MyMaybe`. Recall from Chapter 9 that `Maybe` is a type constructor, which means that it takes a type as input to produce a new type.

```
data MyMaybe a = MyNothing
               | MyJust a
                deriving (Eq,Show)
```

The type variable a stands for any type. The fact that we are using a type variable after `MyMaybe` in this data type definition makes `MyMaybe` a type constructor rather than a type. Here we have two data constructors, but unlike for the type `MyBool`, the data constructor `MyJust` contains some information, namely a value of type `a`. A value of type `MyMaybe a` is either `MyNothing` or `MyJust x` for some `x :: a`.

Let's look at the types of the data constructors.

```
*SimpleVec> :t MyNothing
MyNothing :: MyMaybe a
*SimpleVec> :t MyJust
MyJust :: a -> MyMaybe a
```

For comparison, we can look at the types of the Prelude data constructors for `Maybe`.

```
*SimpleVec> :t Nothing
Nothing :: Maybe a
*SimpleVec> :t Just
Just :: a -> Maybe a
```

We see that `Nothing` is not even a function but just a value of type `Maybe a`. On the other hand, `Just` is a function that takes a value of type `a` and returns a value of type `Maybe a`.

In Chapter 19, when we're talking about systems of particles, we'll define a new data type called `Force` that has two constructors: one for an external force and one for an internal force.

Now that we've talked about how to define a new data type, let's move on to define the `Vec` type we've been using in this chapter.

Defining a New Data Type for 3D Vectors

Haskell does not come with a built-in type for vectors, so we have to define it ourselves. In the beginning of this chapter, we looked at how vectors are defined and used in physics. With this knowledge, we'll turn to the question of how to implement three-dimensional vectors in Haskell. The new type must hold three real numbers, for the three components of a vector in some coordinate system, or something equivalent. We have several options.

Possible Implementations

Let's consider some possible implementations for the type `Vec` before making a final choice.

Option 1: Use a List

We could use a list of real numbers to hold the three components of a vector. A type synonym for this definition would look like the following:

```
type Vec = [R]  -- not our definition
```

This type of vector can hold all possible triples of real numbers. The trouble with this definition is that the type can also hold lists of real numbers that don't have a length of three. This potential type doesn't match our requirements as nicely as we would like; it's a little too big, given that the empty list or a list of two real numbers would satisfy the type checker as a legitimate value of this type.

Option 2: Use a Tuple

A better option would be to choose a triple of real numbers. A type synonym would look like the following:

```
type Vec = (R,R,R)  -- not our definition
```

This is a better match for our requirements because this type guarantees that there must be three components. The only downside to this option is the possible confusion of a triple representing the three components of a vector with a triple representing three other numbers, such as the spherical coordinates of a position. Since option 2 uses a type synonym, the type checker could not help us catch an accidental use of our new vector type in a place where some other triple of real numbers is called for.

Option 3: Make a New Data Type

A third option is to define a new data type for `Vec` that cannot be confused with any other data type, even if the other data type is essentially a collection of three real numbers like `Vec`. We want a fundamental concept in physics, like the three-dimensional vector, to be reflected in the type system so the type system can help us keep things straight in a way that respects how we think about the subject. This is the option we'll pursue next.

Data Type Definition for Vec

Here is our data type definition:

```
data Vec = Vec { xComp :: R  -- x component
               , yComp :: R  -- y component
               , zComp :: R  -- z component
               } deriving (Eq)
```

We decide to use the same name `Vec` for the data constructor we used for the type. We use record syntax because that automatically produces functions `xComp`, `yComp`, and `zComp` for the three components of a vector. We ask the compiler, through the `deriving` keyword, to create an instance of `Eq` for the new `Vec` data type. However, we do not ask for an automatic instance of `Show` because we want to define that manually.

Next, we'll show how to make the type `Vec` an instance of type class `Show`. The general way to make a type an instance of a type class is with the `instance` keyword.

```
instance Show Vec where
    show (Vec x y z) = "vec " ++ showDouble x ++ " "
                         ++ showDouble y ++ " "
                         ++ showDouble z
```

After the instance keyword, we give the type class, then the type to be an instance of that type class, and then the keyword where, before giving definitions for the functions owned by the type class.

Starting on the second line, we define the functions owned by the type class, and we say how they are supposed to work for the specific case of the type in question, Vec in this case. The only function that needs to be defined in an instance of Show is the function show, which describes how to turn a Vec into a String so it can be shown.

Our way of showing a vector begins with the string "vec ", followed by each of the three components in turn. The function showDouble does the work of turning each real number into a string.

The definition of the show function inside the instance definition must be indented with respect to the instance keyword. Any other functions that get defined inside the instance definition must be indented by the same amount.

There are actually two other functions owned by type class Show, called showsPrec and showList, but these will get default definitions if we don't define them, which we didn't earlier. Using :i Show in GHCi gives a list of the functions owned by type class Show, along with which functions *must* be defined in an instance definition.

Here is the function showDouble, used earlier in the function show:

```
showDouble :: R -> String
showDouble x
    | x < 0      = "(" ++ show x ++ ")"
    | otherwise  = show x
```

The type Double is already an instance of Show, as noted in Chapter 8, so the function show is already available to turn a Double into a String. Our function showDouble uses the function show and simply encloses negative numbers in parentheses. The reason for enclosing negative components in parentheses is so the way a Vec gets shown is a legitimate expression for the Vec, meaning it can be used as input wherever a Vec is called for. To carry out this plan for having an expression like vec 3.1 (-4.2) 5.0 accepted as a legitimate value of type Vec, we need a vec function.

```
-- Form a vector by giving its x, y, and z components.
vec :: R  -- x component
    -> R  -- y component
    -> R  -- z component
    -> Vec
vec = Vec
```

This vec function does the same thing that the data constructor Vec does.

Why not just use the data constructor `Vec` to form and show our vectors, thus eliminating the need to define a `Show` instance for `Vec` and define the function `vec`? This is indeed a possibility, and not a bad one. The main reason I did not do it is that I wanted to use record syntax, and the automatically generated `Show` instance obtained by using the `deriving` keyword would have used record syntax to show the vector. This in itself is unproblematic, but when we get into lists of vectors, or lists of tuples of vectors, we will want a brief way to show our vectors.

Haskell has traditionally favored that what can be shown can also be read. The `Read` type class is for types that can be read from a `String`, and it serves as a sort of inverse to the `Show` type class for types that can be shown as a `String`. This is the reason why the `Show` instance looks just like the application of the `vec` function to three components.

Vec Functions

Here are unit vectors in the x-, y-, and z-directions:

```
iHat :: Vec
iHat = vec 1 0 0

jHat :: Vec
jHat = vec 0 1 0

kHat :: Vec
kHat = vec 0 0 1
```

We give the zero vector the special name `zeroV`.

```
zeroV :: Vec
zeroV = vec 0 0 0
```

The unary minus sign will not work in front of a vector, so we define a function `negateV` that returns the additive inverse of a vector (the negative of the vector).

```
negateV :: Vec -> Vec
negateV (Vec ax ay az) = Vec (-ax) (-ay) (-az)
```

Vector addition and subtraction are just the addition and subtraction of the corresponding Cartesian components.

```
(^+^) :: Vec -> Vec -> Vec
Vec ax ay az ^+^ Vec bx by bz = Vec (ax+bx) (ay+by) (az+bz)

(^-^) :: Vec -> Vec -> Vec
Vec ax ay az ^-^ Vec bx by bz = Vec (ax-bx) (ay-by) (az-bz)
```

It'll be useful to have a function that adds a whole list of vectors. We'll use this function when we do numeric integrals of vector-valued functions.

```
sumV :: [Vec] -> Vec
sumV = foldr (^+^) zeroV
```

The function foldr is defined in the Prelude. The definition of sumV is written in point-free style, which means it is short for sumV vs = foldr (^+^) zeroV vs. Roughly speaking, foldr takes a binary operator ((^+^) in this case), an initial value, and a list of values, and "folds" the initial value and an element from the list into an accumulated value, after which it continues to fold the accumulated value with the next element of the list to form a new accumulated value until the list is gone and the final accumulated value is returned. It's a fairly powerful function, but here it's used just to keep adding the members of the list until there are no more.

There are three ways to multiply three-dimensional vectors. The first is scalar multiplication, in which we multiply a number by a vector or a vector by a number. We use (*^) and (^*) for scalar multiplication. The first takes a number on the left and a vector on the right. The second takes a vector on the left and a number on the right. The vector always goes next to the caret symbol. The second vector multiplication method is the dot product. We use (<.>) for the dot product. The third vector multiplication method is the cross product. We use (><) for the cross product.

Here are the definitions for the three vector multiplications:

```
(*^)  :: R   -> Vec -> Vec
c *^ Vec ax ay az = Vec (c*ax) (c*ay) (c*az)

(^*)  :: Vec -> R   -> Vec
Vec ax ay az ^* c = Vec (c*ax) (c*ay) (c*az)

(<.>) :: Vec -> Vec -> R
Vec ax ay az <.> Vec bx by bz = ax*bx + ay*by + az*bz

(><)  :: Vec -> Vec -> Vec
Vec ax ay az >< Vec bx by bz
    = Vec (ay*bz - az*by) (az*bx - ax*bz) (ax*by - ay*bx)
```

The first two definitions are for scalar multiplication. If the vector is to the right of the number, we use the operator with the caret on the right. If the vector is to the left of the number, we use the operator with the caret on the left. In either case, the definition says that scalar multiplication is carried out by multiplying each Cartesian component by the scaling number. The dot product is defined via Equation 10.8. The cross product is defined via Equation 10.9.

We can also divide a vector by a scalar.

```
(^/) :: Vec -> R -> Vec
Vec ax ay az ^/ c = Vec (ax/c) (ay/c) (az/c)
```

Finally, we define a magnitude function to take the magnitude of a vector.

```
magnitude :: Vec -> R
magnitude v = sqrt(v <.> v)
```

This completes our data type definition for the new data type `Vec`, along with supporting functions that allow us to write about vectors in the way we think about them and use them in physics.

Summary

This chapter discussed kinematics in three spatial dimensions. In three dimensions, time is represented by a real number while velocity and acceleration are represented by vectors. Position is not truly a vector, but in this chapter we keep it simple and regard position as the displacement vector from the origin of some preferred or default coordinate system.

Vectors are fundamentally geometric entities; to use numbers to describe the components of a vector, we must introduce a coordinate system. In any situation in which an object is moving, we can decompose the acceleration into a component parallel to the velocity and a component perpendicular to the velocity. This decomposition is coordinate independent.

With a system for vectors in place, we are now in a position to do all the projectile motion problems we ever wanted to do. We showed Haskell's facility for defining our own data types, and we used that system to implement the `Vec` type for three-dimensional vectors.

Exercises

Exercise 10.1. Translate the following mathematical definitions into Haskell definitions:

(a) $\mathbf{v}_0 = 20\hat{\mathbf{i}}$ (Use `v0` in Haskell for $\mathbf{v}_0$.)

(b) $\mathbf{v}_1 = 20\hat{\mathbf{i}} - 9.8\hat{\mathbf{k}}$ (Use `v1` for $\mathbf{v}_1$.)

(c) $\mathbf{v}(t) = 20\hat{\mathbf{i}} - 9.8t\hat{\mathbf{k}}$ (Use `v` for $\mathbf{v}$.)

(d) $\mathbf{r}(t) = 30\hat{\mathbf{j}} + 20t\hat{\mathbf{i}} - 4.9t^2\hat{\mathbf{k}}$ (Use `r` for $\mathbf{r}$.)

(e) $x(t) = \hat{\mathbf{i}} \cdot \mathbf{r}(t)$ (Use `x` for x.)

What are the Haskell types of `v0`, `v1`, `v`, `r`, and `x`?

Exercise 10.2. Write an integration function

```
vecIntegral :: R              -- step size dt
            -> (R -> Vec)     -- vector-valued function
            -> R              -- lower limit
            -> R              -- upper limit
            -> Vec            -- result
vecIntegral = undefined
```

for vector-valued functions of a real variable that is similar to the function

```
integral :: R -> (R -> R) -> R -> R -> R  -- from Chapter 6
```

that we wrote in Chapter 6.

Exercise 10.3. Write a function

```
maxHeight :: PosVec -> Velocity -> R
maxHeight = undefined
```

that returns the maximum z-component for projectile motion in which the initial position and the initial velocity of an object are given. Assume gravity acts in the negative z-direction.

Exercise 10.4. Write a function

```
speedCA :: Velocity -> Acceleration -> Time -> R
speedCA = undefined
```

that, given an initial velocity and a constant acceleration, returns a function giving speed as a function of time.

Exercise 10.5. In the spirit of the function `projectilePos`, write a type signature and function definition for a function `projectileVel` that computes the velocity of a projectile at a given time.

Exercise 10.6. Define a new type `Vec2D` for two-dimensional vectors. Then define functions

```
magAngleFromVec2D :: Vec2D -> (R,R)
magAngleFromVec2D = undefined

vec2DFromMagAngle :: (R,R) -> Vec2D
vec2DFromMagAngle = undefined
```

that find the magnitude and angle of a two-dimensional vector and construct a two-dimensional vector from a magnitude and angle. You may want to use the `atan` or `atan2` functions, which we discussed in Chapter 1.

Exercise 10.7. Define a function

```
xyProj :: Vec -> Vec
xyProj = undefined
```

that computes the projection of a vector into the xy plane. For example, `xyProj (vec 6 9 7)` should evaluate to `vec 6 9 0`.

Exercise 10.8. Define a function

```
magAngles :: Vec -> (R,R,R)
magAngles = undefined
```

that returns a triple (v, θ, ϕ) for a vector $\mathbf{v}$ in which

$$v = |\mathbf{v}|$$

$$\theta = \tan^{-1} \frac{\sqrt{v_x^2 + v_y^2}}{v_z}$$

$$\phi = \tan^{-1} \frac{v_y}{v_x}$$

For example, `magAngles (vec (-1) (-2) (-3))` should evaluate to:

```
(3.7416573867739413,2.5010703409103687,-2.0344439357957027)
```

Exercise 10.9. The velocity and acceleration of a ball launched from the ground are

$$\mathbf{v}_{\text{Ball}}(t) = \mathbf{v}_0 t + \frac{1}{2}\mathbf{g}t^2$$

$$\mathbf{a}_{\text{Ball}}(t) = \mathbf{g}$$

where $\mathbf{v}_0$ is the initial velocity of the ball and $\mathbf{g}$ is the acceleration of gravity. Suppose a ball is launched from the ground with an initial speed of 25 m/s at an angle 52° above horizontal. Choose a coordinate system and define a constant

```
gEarth :: Vec
gEarth = undefined
```

for the acceleration of gravity near Earth's surface. It should be 9.8 m/s^2 toward the center of the earth. Next, define a function

```
vBall :: R -> Vec
vBall t = undefined t
```

that gives the velocity of the ball as a function of time. Now define a function

```
speedRateChangeBall :: R -> R
speedRateChangeBall t = undefined t
```

that gives the rate of change of the speed of the ball as a function of time. You may want to use `speedRateChange` for this. At what point in the ball's motion is the rate of change of its speed equal to zero? Is its velocity zero at that point? Is its acceleration zero at that point? Use `plotFunc` from Chapter 7 to make a graph of the rate of change of the speed of the ball as a function of time over the four seconds it is in the air.

Exercise 10.10. Consider a particle in uniform circular motion. If we choose our coordinate system so that the motion takes place in the xy-plane with the origin at the center of the circle, we can write the position of the particle as

$$\mathbf{r}_{\text{UCM}}(t) = R(\cos \omega t\hat{\mathbf{i}} + \sin \omega t\hat{\mathbf{j}})$$

where R is the radius of the circle and ω is the angular velocity of the motion. The velocity of the particle can be found by taking the derivative of position with respect to time.

$$\mathbf{v}_{\text{UCM}}(t) = \omega R(-\sin \omega t\hat{\mathbf{i}} + \cos \omega t\hat{\mathbf{j}})$$

The acceleration of the particle can be found by taking the derivative of velocity with respect to time.

$$\mathbf{a}_{\text{UCM}}(t) = -\omega^2 R(\cos \omega t\hat{\mathbf{i}} + \sin \omega t\hat{\mathbf{j}})$$

This particle in uniform circular motion has a speed $v_{\text{UCM}}(t) = \omega R$, which does not depend on time. The constant speed is what we mean by the word "uniform."

For a uniform circular motion with $R = 2$ m and $\omega = 6$ rad/s, encode the position, velocity, and acceleration of the particle in Haskell. Use `aParallel` to confirm that the tangential component of acceleration is 0 at several different times. Use `aPerp` to confirm that the magnitude of the radial component of acceleration is $[v_{\text{UCM}}(t)]^2/R = \omega^2 R$, again at several different times.

Exercise 10.11. Consider a particle in nonuniform circular motion on a circle with radius R. If we choose our coordinate system so that the motion takes place in the xy-plane with the origin at the center of the circle, we can write the position of the particle as

$$\mathbf{r}_{\text{NCM}}(t) = R[\cos \theta(t)\hat{\mathbf{i}} + \sin \theta(t)\hat{\mathbf{j}}]$$

where $\theta(t)$ describes the angle the particle makes with the x-axis as a function of time. The velocity of the particle can be found by taking the derivative of position with respect to time.

$$\mathbf{v}_{\text{NCM}}(t) = RD\theta(t)[-\sin \theta(t)\hat{\mathbf{i}} + \cos \theta(t)\hat{\mathbf{j}}]$$

The acceleration of the particle can be found by taking the derivative of velocity with respect to time.

$$\mathbf{a}_{\text{NCM}}(t) = -R[D\theta(t)]^2[\cos \theta(t)\hat{\mathbf{i}} + \sin \theta(t)\hat{\mathbf{j}}] + RD^2\theta(t)[-\sin \theta(t)\hat{\mathbf{i}} + \cos \theta(t)\hat{\mathbf{j}}]$$

This particle in circular motion has speed $v_{\text{NCM}}(t) = R\left|D\theta(t)\right|$, which will depend on time unless $D\theta(t)$ is constant. The magnitude of the particle's tangential acceleration is $R\left|D^2\theta(t)\right|$ and the magnitude of its radial acceleration is $[v_{\text{NCM}}(t)]^2/R = R[D\theta(t)]^2$.

Write a function

```
rNCM :: (R, R -> R) -> R -> Vec
rNCM (radius, theta) t = undefined radius theta t
```

that takes a radius R, a function θ, and a time t as input and returns a position vector as output.

The purpose of this exercise is to confirm that, even in nonuniform circular motion, the magnitude of a particle's radial component of acceleration is equal to the square of its speed divided by the radius of the circle. The following function finds the radial component of acceleration for any particle whose position can be given as function of time. Its first input is a small time interval to use for numerical derivatives. Its second input is the position function for the particle, and its third input is time.

```
aPerpFromPosition :: R -> (R -> Vec) -> R -> Vec
aPerpFromPosition epsilon r t
    = let v = vecDerivative epsilon r
          a = vecDerivative epsilon v
      in aPerp (v t) (a t)
```

For a radius $R = 2$ m and

$$\theta(t) = \frac{1}{2}(3 \text{ rad/s}^2)t^2$$

use `aPerpFromPosition` to find the radial component of acceleration at $t = 2$ seconds. Then find the speed of the particle at that time. Finally, show that the magnitude of the radial component is equal to the square of its speed divided by the radius of the circle.

11

CREATING GRAPHS

When you make a graph for a formal report, you want to have titles, axis labels, and perhaps other features that will help the reader understand what you are trying to say. In this chapter, we'll show you how to create such a graph using Haskell. We'll look at titles, axis labels, and other labels. We'll see how to plot data given in the form of a list of pairs. Then we'll show how to plot multiple functions or multiple datasets on the same set of axes, how to control the axis ranges, and how to produce your graph as a file that can be imported into some other document.

Title and Axis Labels

The following code produces a graph with a title and axis labels:

```
{-# OPTIONS_GHC -Wall #-}

import Graphics.Gnuplot.Simple

type R = Double

tRange :: [R]
tRange = [0,0.01..5]

yPos :: R  -- y0
     -> R  -- vy0
     -> R  -- ay
     -> R  -- t
     -> R  -- y
yPos y0 vy0 ay t = y0 + vy0 * t + ay * t**2 / 2

plot1 :: IO ()
plot1 = plotFunc [Title "Projectile Motion"
                 ,XLabel "Time (s)"
                 ,YLabel "Height of projectile (m)"
                 ,PNG "projectile.png"
                 ,Key Nothing
                 ] tRange (yPos 0 20 (-9.8))
```

As in the last chapter, we turn on warnings to catch any poor programming we might not have intended. Then we import the `Graphics.Gnuplot.Simple` module, which we use to make plots. Next, we set up `R` as a *type synonym* for `Double`. This lets us think of `Double`s as real numbers and call them by the short name `R`. We then define a list `tRange` of time values that we will use in our plot, and we define a function `yPos` for the height of a projectile.

Finally, we define `plot1` to make a plot. Recall that `plotFunc` has type

```
[Attribute] -> [a] -> (a -> a) -> IO ()
```

where a is a type in some specialized type classes. The `Attribute` type is defined in the `Graphics.Gnuplot.Simple` module. If you type `:i Attribute` at the GHCi prompt (`:i` is short for `:info`), you'll see some options for what you can do with these `Attribute`s. In `plot1`, we pass a list of five `Attribute`s to `plotFunc`. The first creates the title, the second and third produce axis labels, the fourth specifies the filename to use for the output, and the last requests that a legend not appear.

Notice the type `IO ()` (pronounced "eye oh unit") of `plot1`. `IO` is a type constructor, like `Maybe`, but it's a special type constructor that's designed to signal an effect, which is a computation that is not purely functional. An effect changes the world in some way (for example, changing a file on the hard drive or showing a picture on the screen).

The type `()`, called *unit*, is a type that contains only one value, which is also written `()` and also called unit. A type with only one value can't convey any information because there is no choice about what the value might be. Since it cannot convey any information, the unit type by itself is not too useful. However, coupled with the `IO` type constructor, the type `IO ()` comes to represent an effect without a value, which is a very useful type.

The `Attribute` of `Key Nothing` omits the key that is included with the graph by default. Since the key makes reference to a temporary file that we don't care about, it is generally uninformative to include the default key. The reader should be warned that the `Graphics.Gnuplot.Simple` module is not merely simple, but a bit simple-minded. In particular, if an invalid `gnuplot` keyword is passed through a Haskell `String`, the result is no output at all, not even an error. (For example, if you want to move the legend key to the bottom of the plot instead of the top, the attribute `Key (Just ["bottom"])` works, but `Key (Just ["Bottom"])` fails with no output because `gnuplot` keywords are case sensitive.) The reader is encouraged to consult the online documentation for the `Graphics.Gnuplot.Simple` module as well as the documentation for the `gnuplot` program itself.

If you load the code just shown into GHCi and enter `plot1` at the prompt, it will produce a file called *projectile.png* on your hard drive that you can include in a document. Figure 11-1 shows what it looks like.

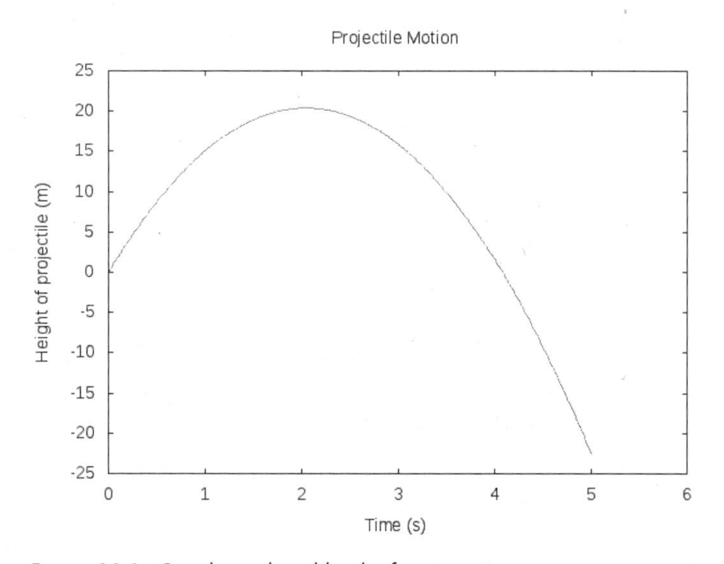

Figure 11-1: Graph produced by the function `plot1`

Other Labels

You may want to put other labels on a plot. Here is how you can do so:

```
plot1Custom :: IO ()
plot1Custom
    = plotFunc [Title "Projectile Motion"
               ,XLabel "Time (s)"
               ,YLabel "Height of projectile (m)"
               ,PNG "CustomLabel.png"
               ,Key Nothing
               ,Custom "label" ["\"Peak Height\" at 1.5,22"]
               ] tRange (yPos 0 20 (-9.8))
```

Note the `Custom` attribute we added. The backslash in front of the quotes is because we need to pass quotes inside of quotes. The coordinates `1.5,22` are the horizontal and vertical coordinates on the graph where we want the label to appear. Figure 11-2 shows what this looks like.

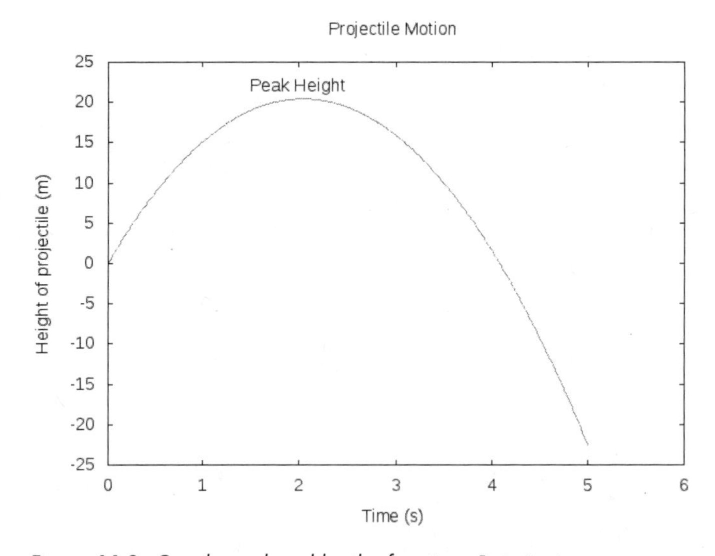

Figure 11-2: Graph produced by the function `plot1Custom`

The syntax for including a custom label is sufficiently awkward and difficult to remember that it makes sense to write a new function that takes its arguments in a simpler way.

```
customLabel :: (R,R) -> String -> Attribute
customLabel (x,y) label
    = Custom "label" ["\"" ++ label ++ "\"" ++ " at "
                                  ++ show x ++ "," ++ show y]
```

We are passing two pieces of information to the custom label function: the coordinates of the location of the label and the name of the label. The first piece of information has type `(R,R)`, and the second has type `String`. Our function customLabel will produce an Attribute that can be included in the attribute list of the function plotFunc. We use the show function to convert an R to a String, and we use the ++ operator to concatenate strings.

We refer to the double-quote character in Haskell by prefixing it with a backslash. The backslash tells the compiler that we mean to write the double-quote character itself rather than to signal the beginning of a string. Having done this, we can treat the double quote character as any other character.

```
Prelude> :t 'c'
'c' :: Char
Prelude> :t '\"'
'\"' :: Char
Prelude> :t "c"
"c" :: [Char]
Prelude> :t "\""
"\"" :: [Char]
```

Having defined the function customLabel, we can use the following nicer syntax to make our graph:

```
plot2Custom :: IO ()
plot2Custom
    = plotFunc [Title "Projectile Motion"
               ,XLabel "Time (s)"
               ,YLabel "Height of projectile (m)"
               ,Key Nothing
               ,customLabel (1.5,22) "Peak Height"
               ] tRange (yPos 0 20 (-9.8))
```

Plotting Data

There will be times when we want to plot points of (x, y) pairs rather than functions. We can use the plotPath function for this (also defined in the package Graphics.Gnuplot.Simple). Let's take a look at the type of the plotPath function to better understand its use.

```
Prelude> :m Graphics.Gnuplot.Simple
Prelude Graphics.Gnuplot.Simple> :t plotPath
plotPath
  :: Graphics.Gnuplot.Value.Tuple.C a =>
     [Attribute] -> [(a, a)] -> IO ()
```

After a list of attributes, plotPath takes a list of pairs containing the data we want to plot. Here is code to produce the same graph as in Figure 11-2 but using plotPath instead of plotFunc:

```
plot3Custom :: IO ()
plot3Custom
```

```
  = plotPath [Title "Projectile Motion"
             ,XLabel "Time (s)"
             ,YLabel "Height of projectile (m)"
             ,Key Nothing
             ,customLabel (1.5,22) "Peak Height"
             ] [(t, yPos 0 20 (-9.8) t) | t <- tRange]
```

We used a list comprehension to produce the list of pairs that `plotPath` requires.

Multiple Curves on One Set of Axes

You can plot multiple curves on a single set of axes. This is particularly useful if you want to compare two functions that have the same independent and dependent variables. The function `plotFuncs` from `Graphics.Gnuplot.Simple` enables us to plot a list of functions.

```
Prelude Graphics.Gnuplot.Simple> :t plotFuncs
plotFuncs
  :: (Graphics.Gnuplot.Value.Atom.C a,
      Graphics.Gnuplot.Value.Tuple.C a) =>
     [Attribute] -> [a] -> [a -> a] -> IO ()
```

Notice that the `plotFuncs` function takes a list of functions as one of its arguments. We promised back in Chapter 5 that we would find a use for a list of functions, and now we have! Here is an example of how to use `plotFuncs`:

```
xRange :: [R]
xRange = [0,0.02..10]

f3 :: R -> R
f3 x = exp (-x)

usePlotFuncs :: IO ()
usePlotFuncs = plotFuncs [] xRange [cos,sin,f3]
```

The range of x-values does not have to be the same for the two plots. Consider the following example, which introduces the new function `plotPaths`.

```
xs1, xs2 :: [R]
xs1 = [0,0.1..10]
xs2 = [-5,-4.9..5]

xys1, xys2 :: [(R,R)]
xys1 = [(x,cos x) | x <- xs1]
xys2 = [(x,sin x) | x <- xs2]

usePlotPaths :: IO ()
usePlotPaths = plotPaths [] [xys1,xys2]
```

The plotPaths function takes a list of lists of pairs where the plotPath function takes a list of pairs.

Controlling the Plot Ranges

By default, gnuplot (the program that is making the graphs behind the scenes) will make plots based on the x-ranges you provide and the corresponding calculated y-ranges. Sometimes, you may want more control over the x-range or the y-range.

Revisiting the previous example of three plots, try the following:

```
usePlotFuncs' :: IO ()
usePlotFuncs' = plotFuncs [ XRange (-2,8)
                          , YRange (-0.2,1)
                          ] xRange [cos,sin,f3]
```

By specifying XRange (-2,8), we produce a graph that runs from $x = -2$ to $x = 8$. Since xRange runs from 0 to 10, no data is calculated in the region from $x = -2$ to $x = 0$, so this region is blank on the graph. Although we ask for data to be calculated up to $x = 10$, it is only shown up to $x = 8$. Because we specify YRange (-0.2,1), values of the cosine and sine functions that fall in the region from $y = -1$ to $y = -0.2$ are not shown.

Notice the funny stylistic way in which I made the list [XRange (-2,8), YRange (-0.2,1)]. People who code in Haskell sometimes put the comma first on the second line of the list, but you don't have to. You could put this all on one line, or put the comma at the end of the first line. It's a matter of style.

Making a Key

The default key that gnuplot provides with a graph is not very useful. It gives the name of a temporary file we are not interested in. It is not a trivial thing to produce a handsome key, but it can be done. The following code gives an example:

```
xRange' :: [R]
xRange' = [-10.0, -9.99 .. 10.0]

sinPath :: [(R,R)]
sinPath = [(x, sin x) | x <- xRange' ]

cosPath :: [(R,R)]
cosPath = [(x, cos x) | x <- xRange' ]

plot4 :: IO ()
plot4 = plotPathsStyle [ Title "Sine and Cosine"
                       , XLabel "x"
                       , YLabel "Function Value"
                       , YRange (-1.2,1.5)
```

```
                       ] [ (defaultStyle {lineSpec = CustomStyle
                                          [LineTitle "sin x"]}, sinPath)
                         , (defaultStyle {lineSpec = CustomStyle
                                          [LineTitle "cos x"]}, cosPath) ]
```

Here we use the function `plotPathsStyle`, which is an extended version of `plotPaths` that allows stylistic alterations. Instead of the list of lists of pairs that `plotPaths` requires, `plotPathsStyle` requires a list of pairs, with each pair consisting of a `PlotStyle` and a list of pairs with the data to be plotted. In this way, we can give a title to each curve that shows up in the key.

Summary

In this chapter, we added plotting tools to our toolkit. We learned how to provide a title, axis labels, and other labels to a graph. We learned how to plot data given in the form of a list of pairs. We saw how to plot multiple functions or multiple lists of pairs on a single set of axes. We learned how to manually control the axis ranges and how to produce the graph as a file that can be imported into another document. In the next chapter, we'll learn how to make stand-alone programs in Haskell.

Exercises

Exercise 11.1. Make a plot of $y = x^2$ from $x = -3$ to $x = 3$ with a title and axis labels.

Exercise 11.2. Make a plot of the cosine and sine functions, together on a single set of axes, from $x = 0$ to $x = 10$.

Exercise 11.3. Take a look at the type signature for `plotPath`, and figure out how to plot the list of points `txPairs` below:

```
ts :: [R]
ts = [0,0.1..6]

txPairs :: [(R,R)]
txPairs = [(t,30 * t - 4.9 * t**2) | t <- ts]
```

Make a plot with a title and axis labels (with units).

Exercise 11.4. Write a function

```
approxsin :: R -> R
approxsin = undefined
```

that approximates the sine function by the first four terms in its Taylor expansion.

$$x - \frac{x^3}{3!} + \frac{x^5}{5!} - \frac{x^7}{7!}$$

(Depending on how you do this, you may or may not run into the issue that you cannot divide an `R` by an `Int` or an `Integer` in Haskell. You can only divide a numeric type by the same numeric type. If you run into this problem, you can use the function `fromIntegral` to convert an `Int` or an `Integer` to some other type, like `R`.)

Test your function by trying the following command in GHCi:

```
plotFuncs [] [-4,-3.99..4] [sin,approxsin]
```

Make a nice version of this plot (with a title, axis labels, labels to indicate which curve is which, and so on).

12

CREATING STAND-ALONE PROGRAMS

Up to this point, we've used GHCi to do all of our calculations and to show us the results. We've written fairly sophisticated source code files, but we've always loaded them into GHCi to use their functions. Haskell, however, is a full-featured, production-ready computer language, and it's completely capable of compiling stand-alone programs that don't require any GHCi involvement. The animations in Chapter 13 and later chapters are best carried out using stand-alone programs rather than GHCi.

This chapter explains three different ways to make a stand-alone (executable) program. The most basic way uses `ghc` to produce the executable program. With this method, you are responsible for installing any library packages that your program needs. The second way uses `cabal`, which will automatically install the library packages your program needs, but these packages must be listed in the appropriate place in a configuration file. The third way uses `stack`, which does even more things automatically, such as installing a version of the GHC compiler compatible with versions of packages you request. To make a stand-alone program, you need to use only one of these three methods. If you are new to Haskell, you may find the `stack` method to be the easiest to use.

For each of the three methods, we'll go through the steps required to produce an executable program (a) for a very simple program and (b) for a program that uses both modules we have written and modules other people have written.

Using GHC to Make a Stand-Alone Program

In this section, we use GHC directly to make a stand-alone program. We do this first for a very simple program called "Hello, world!" and then for a more complex program that imports modules.

Hello, World!

The simplest stand-alone program people often write when learning a new language is called "Hello, world!" All this program does is print the words "Hello, world!" and exit. For many computer languages, it makes sense to learn how to write the "Hello, world!" program very early in the process of learning the language. However, in Haskell, it doesn't make sense to learn "Hello, world!" early because the "Hello, world!" program is all about producing an effect, namely printing something on the screen, while the core of Haskell programming, and functional programming in general, is about pure functions that have no effect.

The "Hello, world!" program in Haskell consists of two lines of code:

```
main :: IO ()
main = putStrLn "Hello, world!"
```

Every stand-alone program needs a function called `main`, which usually has type `IO ()`. We first introduced `IO ()` in Chapter 7 as the type of an impure, or effectful, function that returns no meaningful value but produces an effect. In general, the type `IO a` represents a value of type `a` along with an effect.

The `main` function needs to produce some effect; otherwise, we wouldn't be able to tell that the program actually ran. The purpose of the effectful function `main` is to describe to the compiler what we want the computer to *do*, and the type `IO ()` is the perfect type for this because it represents an effect without a meaningful value.

The function `putStrLn` is a Prelude function that takes a string as input, prints it on the screen, and advances to the next line so that any further printing occurs there. There is also a function called `putStr`, with the same type as `putStrLn`, that prints a string without advancing to the next line so that further printing occurs directly after the printed string. The `Ln` at the end of the name reminds us that the function advances to the next line after printing. The type of putStrLn shows us that it takes a string as input and produces an effect.

```
Prelude> :t putStrLn
putStrLn :: String -> IO ()
```

Suppose we put these two lines in a source code file named *hello.hs*. If your operating system offers a command line, the command

```
$ ghc hello.hs
```

will compile the source code file *hello.hs* to produce an executable file, called *hello*, that you can run. On a Linux system, you can run the program *hello* from the command line with the command

```
$ ./hello
```

The dot-slash in front of the program name tells the operating system to execute the program called *hello* that is in the current working directory. Omitting the dot-slash will cause the operating system to search its standard search path for a program called *hello*, which it may not find if the current working directory is not included in the search path.

A Program That Imports Modules

Now we look at compiling a stand-alone program that uses functions from the SimpleVec module we wrote in Chapter 10 and functions from the Graphics.Gnuplot.Simple module from the gnuplot package. The file *SimpleVec.hs* containing the source code for the SimpleVec module is available at *https://lpfp.io*. Listing 12-1 shows the stand-alone program we want to compile.

```
{-# OPTIONS -Wall #-}

import SimpleVec ( iHat, kHat, xComp, zComp, projectilePos, (^+^), (*^) )
import Graphics.Gnuplot.Simple ( Attribute(..), plotPath )

main :: IO ()
main = let posInitial = 10 *^ kHat
           velInitial = 20 *^ cos (pi/6) *^ iHat ^+^ 20 *^ sin (pi/6) *^ kHat
           posFunc = projectilePos posInitial velInitial
           pairs = [(xComp r, zComp r) | t <- [0, 0.01 ..], let r = posFunc t]
           plottingPairs = takeWhile (\(_,z) -> z >= 0) pairs
       in plotPath [Title "Projectile Motion"
                   ,XLabel "Horizontal position (m)"
                   ,YLabel "Height of projectile (m)"
                   ,PNG "projectile.png"
                   ,Key Nothing
                   ] plottingPairs
```

Listing 12-1: The stand-alone program MakeTrajectoryGraph.hs*, which uses functions from the* SimpleVec *module and the* Graphics.Gnuplot.Simple *module*

The program in Listing 12-1 produces a graph of the trajectory of a ball thrown from the top of a building 10 m above the ground with an initial speed of 20 m/s at an angle 30° above horizontal. The program produces a file named *projectile.png* containing the graph. To do its work, this program imports functions such as `projectilePos`, `xComp`, `zComp`, `iHat`, and `kHat` from the `SimpleVec` module of Chapter 10. The program also uses the `plotPath` function from the `Graphics.Gnuplot.Simple` module. Because the data constructors `Title`, `XLabel`, and so on of the `Attribute` data type are used, we import the `Attribute` data type with its constructors by appending the name of the type `Attribute` with two dots enclosed in parentheses.

We'll assume that the code in Listing 12-1 is contained in a source code file called *MakeTrajectoryGraph.hs*. To use `ghc` to compile the program, two things must be true:

- The file *SimpleVec.hs* containing the `SimpleVec` module must be present in the same directory as the file *MakeTrajectoryGraph.hs* containing the main program. We'll call this directory the *working directory*.
- The working directory must have access to the `Graphics.Gnuplot.Simple` module. This requires that the `gnuplot` package be installed either (a) globally, so it can be accessed from any directory, or (b) locally, so it can be accessed from the working directory. To install the `gnuplot` package globally, issue the following command:

  ```
  $ cabal install --lib gnuplot
  ```

  On my computer, this command creates or changes the file */home/walck/.ghc/x86_64-linux-8.10.5/environments/default* that contains the list of globally installed Haskell packages. To install the `gnuplot` package locally (in the working directory), issue the following command:

  ```
  $ cabal install --lib gnuplot --package-env .
  ```

  This command creates or changes a file with a name such as *.ghc.environment.x86_64-linux-8.10.5* in the current working directory. This file contains a list of packages installed locally (in the current working directory). See the appendix for more information about installing Haskell packages.

Once these two criteria are met, we compile the source code file *MakeTrajectoryGraph.hs* into an executable program by issuing the following command:

```
$ ghc MakeTrajectoryGraph.hs
```

This command must be issued from the same working directory that contains the file *MakeTrajectoryGraph.hs*, the file *SimpleVec.hs*, and access to the `Graphics.Gnuplot.Simple` module.

If the compiler cannot find the `Graphics.Gnuplot.Simple` module, you will see an error like the following:

```
MakeTrajectoryGraph.hs:4:1: error:
    Could not load module 'Graphics.Gnuplot.Simple'
```

In this case, you must install the gnuplot package, either globally or locally, so it is accessible from the working directory.

If all goes well, the compiler will produce an executable file called *MakeTrajectoryGraph* in the current working directory. The executable file is not installed in any global location, so to run the program, you'll need to give the full pathname of the executable file or run it from the directory in which it lives by prefixing the executable filename with ./, as shown here:

```
$ ./MakeTrajectoryGraph
```

The advantage of using ghc to make your executable program is there are no configuration files to worry about. The disadvantage is that any modules your program needs, whether written by you or someone else, must be accessible from the directory in which your program resides. As the number of library packages your program depends on increases, the burden of this installation increases, especially because versions of packages that are acceptable to your program may conflict with versions that are acceptable to other programs you write or to other library packages you want to use. The cabal and stack tools we'll describe next were designed to manage this complexity, so you don't need to deal with it yourself.

Using Cabal to Make a Stand-Alone Program

We used cabal to install a package in the previous section. But the cabal tool can play a larger role in your Haskell ecosystem, managing the modules and packages your stand-alone program needs and using versions that work together, even if they conflict with packages used by other projects you might have. To get basic information about what the cabal tool can do, issue the command

```
$ cabal help
```

at your command prompt.

The first step in using cabal to manage the dependencies of your project is to make a fresh subdirectory that will contain the source code of your stand-alone program as well as some files that cabal needs to do its work. We create a new directory called *Trajectory* under the current directory using the following command. Use a unique name for this directory because the name will be the default name for the executable program as well as the project generally.

```
$ mkdir Trajectory
```

We enter this new directory, and make it the working directory, by issuing the command

```
$ cd Trajectory
```

where `cd` stands for "change directory." Inside this fresh directory, we issue the following command:

```
$ cabal init
```

This creates a file called *Trajectory.cabal* and a subdirectory called *app*, which contains a file called *Main.hs*. Older versions of `cabal` create *Main.hs* in the current directory rather than in the *app* subdirectory.

Imagining that you might want to share your code with others at some point, `cabal` wants you to have a file called *LICENSE* that contains the terms for which others may use your code. The `cabal` tool may demand that you have such a file before it will compile your code, so be prepared to produce one. The `cabal` program does not care about the contents of the *LICENSE* file, only that it exists.

The file *Main.hs* is a default source code file that contains a very simple program. To compile it, type

```
$ cabal install
```

at your command prompt while the current working directory is *Trajectory*. If everything goes smoothly, `cabal` will compile the code in *Main.hs*, produce an executable file called *Trajectory*, and make that executable file available globally, meaning it can be run by giving its name, *Trajectory*, rather than its full pathname containing the directory structure leading to the file.

We can test the executable with

```
$ Trajectory
```

and we should get a short welcome message.

Moving on to using `cabal` to produce a stand-alone program for the code in Listing 12-1, we take a look at the file *Trajectory.cabal*. This is `cabal`'s configuration file that tells it how to go about compiling source code into executable code for the project in the current directory. The command `cabal init` shown earlier selected default values for several options when it created *Trajectory.cabal*. The lines we are interested in right now look something like the following:

```
executable Trajectory
  main-is:          Main.hs

  -- Modules included in this executable, other than Main.
  -- other-modules:

  -- LANGUAGE extensions used by modules in this package.
  -- other-extensions:
```

```
    build-depends:    base ^>=4.14.2.0
    hs-source-dirs:   app
```

The first line indicates that the name of the executable program will be *Trajectory*. This default name matches the name of the project directory; however, we could change it to something else if we wanted to. The second line gives the name of the source code file that has the function `main` in it. By default, this file is called *Main.hs* and is located in the *app* subdirectory of the *Trajectory* directory. The lines preceded by double hyphens are comments. The line beginning with `build-depends:` is a list of packages that the main program depends on. By default, the *.cabal* file includes only a dependence on the package `base`. The `base` package makes all of the Prelude functions and types available. The line beginning with `hs-source-dirs:` is a list of subdirectories that contain source files.

To compile the code in Listing 12-1, which is contained in the file *MakeTrajectoryGraph.hs*, we need to do three things:

1. Copy or move the file *MakeTrajectoryGraph.hs* into the *app* subdirectory of the *Trajectory* directory. Then edit *Trajectory.cabal* to change the name of the main source code file from *Main.hs* to *MakeTrajectoryGraph.hs*. The modified line in *Trajectory.cabal* looks like this:

```
    main-is:          MakeTrajectoryGraph.hs
```

2. Copy or move the file *SimpleVec.hs* containing the `SimpleVec` module into the *app* subdirectory of the *Trajectory* directory. (This file, along with all the other modules in this book, is available at *https://lpfp.io*.) Then edit *Trajectory.cabal* to uncomment (remove the double hyphen) the `other-modules:` line and add the `SimpleVec` module (without the *.hs* extension).

```
    other-modules:    SimpleVec
```

3. Edit *Trajectory.cabal* to include `gnuplot` in the `build-depends:` line. This allows us to import the module `Graphics.Gnuplot.Simple` in our main program. With all three changes, the modified lines in *Trajectory.cabal* look like this:

```
executable Trajectory
    main-is:          MakeTrajectoryGraph.hs

    -- Modules included in this executable, other than Main.
    other-modules:    SimpleVec

    -- LANGUAGE extensions used by modules in this package.
    -- other-extensions:
    build-depends:    base ^>=4.14.2.0, gnuplot
    hs-source-dirs:   app
```

While the base package has bounds on the allowed versions of base, we have not given version bounds on the gnuplot package. The purpose of version bounds is to allow code that is still in development to evolve in ways that are not compatible with previous versions. Library package writers follow conventions stating that minor changes and bug fixes are indicated by small changes in the version number, while major changes are indicated by bigger changes in the version number. Using version bounds, like those with base just shown, is a technique designed to ensure you are getting the functionality you expect.

Adding gnuplot to the list of build dependencies causes cabal to install the gnuplot package, but in a way that makes it private to this project, the project in the *Trajectory* directory. The gnuplot package will not be available in GHCi, for example, as a result of adding it to the build dependencies for this project. To make gnuplot available in GHCi, follow the instructions in the appendix.

Now reissue

```
$ cabal install
```

to recompile the program called *Trajectory*. We can test the executable with

```
$ Trajectory
```

and the executable should create a file called *projectile.png*.

The packages, such as gnuplot, that cabal installs reside at *https://hackage.haskell.org*. You can go there to search for, browse, and read documentation about any of the packages cabal can install.

Using Stack to Make a Stand-Alone Program

The stack tool can manage the modules and packages your stand-alone program needs, using versions that work together, even if they conflict with packages used by other projects you might have. To get basic information about what the stack tool can do, issue the command

```
$ stack --help
```

at your command prompt.

The first step in using stack to manage the dependencies of a new project called *Trajectory* is to issue the command

```
$ stack new Trajectory
```

This will make a subdirectory with the name *Trajectory*. We enter this new directory, and make it the current directory, by issuing the following command:

```
$ cd Trajectory
```

Inside this directory, we find several files and subdirectories that `stack` has created for us. The most important file is *Trajectory.cabal*, which contains important information about how your program gets compiled. The `stack` tool is built on top of the `cabal` tool and uses its configuration file. The most important lines from *Trajectory.cabal* look like this:

```
library
  exposed-modules:
      Lib
  other-modules:
      Paths_Trajectory
  hs-source-dirs:
      src
  build-depends:
      base >=4.7 && <5
  default-language: Haskell2010

executable Trajectory-exe
  main-is: Main.hs
  other-modules:
      Paths_Trajectory
  hs-source-dirs:
      app
  ghc-options: -threaded -rtsopts -with-rtsopts=-N
  build-depends:
      Trajectory
    , base >=4.7 && <5
  default-language: Haskell2010
```

Here we see two stanzas: one beginning with `library` and the other beginning with `executable`. The `library` stanza is in charge of the names, locations, and dependencies of modules we have written, such as `SimpleVec`. The names of modules we have written that we want `stack` to manage go under the heading `exposed-modules:`, and they are separated by commas if there is more than one module. The default program that comes with a new stack project uses only one module, named `Lib`. Here, we do not need to use the `other-modules:` heading; we can leave it alone. The directories in which our modules are located go under the heading `hs-source-dirs:`. By default, the subdirectory *src* under the directory *Trajectory* is the location for modules, and we do not need to change this. We will simply copy or move our modules into the *src* directory that `stack` created for us. Packages that we have not written but upon which our modules depend, such as `gnuplot`, are listed under the `build-depends:` heading.

The first line of the `executable` stanza indicates that the name of the executable program will be *Trajectory-exe*. This default name matches the name we gave the project; however, we could change it to something else if we wanted to. The heading `main-is:` is followed by the name of the source code file that has the function main in it. The default value is *Main.hs*. In the executable stanza, as in the library stanza, we do not need to use the other-modules: heading; we can leave it alone. The directories in which the source code for our executable (stand-alone) programs are located go under the heading hs-source-dirs:. By default, the subdirectory *app* under the directory *Trajectory* is the location for main program source code, and we do not need to change this. We will simply copy or move our code into the *app* directory that stack created for us. At present, the *app* subdirectory contains the *Main.hs* source code file.

The file *Main.hs* is a default source code file that contains a very simple program. To compile it, type

```
$ stack install
```

at your command prompt while the current working directory is *Trajectory* (the directory containing the *.cabal* file). If everything goes smoothly, stack will compile the code in *Main.hs*, produce an executable file called *Trajectory-exe*, and make that executable file available globally, so it can be run, even from other directories, by giving its name, *Trajectory-exe*.

We can test the executable with

```
$ Trajectory-exe
```

and we should see a short text string appear on the screen.

Moving on to using stack to produce a stand-alone program for the code in Listing 12-1, contained in the file *MakeTrajectoryGraph.hs*, we need to do three things:

1. Copy or move the file *MakeTrajectoryGraph.hs* into the *app* subdirectory of the *Trajectory* directory. Then edit *Trajectory.cabal* to change the name of the main source code file from *Main.hs* to *MakeTrajectoryGraph.hs*. The modified line in *Trajectory.cabal* looks like this:

```
  main-is: MakeTrajectoryGraph.hs
```

2. Copy or move the file *SimpleVec.hs* containing the SimpleVec module into the *src* subdirectory of the *Trajectory* directory. This file, along with all the other modules in this book, is available at *https://lpfp.io*. Then edit *Trajectory.cabal* to include the SimpleVec module in the exposed-modules: field of the library stanza.

```
library
  exposed-modules:
      SimpleVec
```

3. Edit *Trajectory.cabal* to include the gnuplot package under the build-depends: heading of the executable stanza. This allows us to import the module Graphics.Gnuplot.Simple in our main program. With all three changes, the modified lines in *Trajectory.cabal* look like this:

```
library
  exposed-modules:
      SimpleVec
  other-modules:
      Paths_Trajectory
  hs-source-dirs:
      src
  build-depends:
      base >=4.7 && <5
  default-language: Haskell2010

executable Trajectory-exe
  main-is: MakeTrajectoryGraph.hs
  other-modules:
      Paths_Trajectory
  hs-source-dirs:
      app
  ghc-options: -threaded -rtsopts -with-rtsopts=-N
  build-depends:
      Trajectory
    , base >=4.7 && <5
    , gnuplot
  default-language: Haskell2010
```

Keep in mind package names, not module names, need to be included in the list of build dependencies. When using `stack`, accidentally substituting the module name `Graphics.Gnuplot.Simple` for the package name `gnuplot` gives a parse error with no hint of what the real trouble is.

Now reissue

```
$ stack install
```

to recompile the program called *Trajectory-exe*. We can test the executable with

```
$ Trajectory-exe
```

and the executable should create a file called *projectile.png*.

The packages, such as `gnuplot`, that `stack` installs reside at *https://hackage.haskell.org*. You can go there to search for, browse, and read documentation about any of the packages `stack` can install.

Summary

This chapter showed three ways to produce a stand-alone Haskell program. The first uses `ghc`, and you must install any needed library packages yourself. The second uses `cabal`, which can help manage library package dependencies. The third uses `stack`, which can also help manage library package dependencies. In the next chapter, we will put these techniques to use in making animations.

Exercises

Exercise 12.1. The `print` function is useful inside a stand-alone program. Ask GHCi for the type of `print`, and GHCi will tell you that `print` is a function whose input can have any type that is an instance of `Show` and whose output is `IO ()`, meaning that it *does* something. What `print` does is send the value of its input to your screen. You can `print` numbers, lists, strings, and anything that can be shown. You can use `print` inside of GHCi, but it is not needed there because GHCi automatically prints the value of whatever you give it.

Write a stand-alone program that prints the first 21 powers of 2, starting with 2^0 and ending with 2^{20}. When you run your program, the output should look like this:

```
[1,2,4,8,16,32,64,128,256,512,1024,2048,4096,8192,16384,32768,65536,131072,
262144,524288,1048576]
```

13

CREATING 2D AND 3D ANIMATIONS

A picture that changes over time can be a good way to visualize many situations. The Haskell Prelude itself does not have any support for animation, but some good library packages are available at *https://hackage.haskell.org*. For two-dimensional pictures and animations, we'll use the `gloss` package. For three-dimensional pictures and animations, we'll use a package named `not-gloss`.

2D Animation

The `gloss` package supplies the `Graphics.Gloss` module, which provides four main functions: `display`, `animate`, `simulate`, and `play`. The first is for still pictures, the second and third are for pictures that change with time, and the fourth is for pictures that change with time and user input. We are interested primarily in the first three functions. We'll describe these functions in the next few sections.

Displaying a 2D Picture

The function display produces a static picture. Let's ask GHCi for the type of display. Since display is not part of the Prelude, we must first load the module.

```
Prelude> :m Graphics.Gloss
Prelude Graphics.Gloss> :t display
display :: Display -> Color -> Picture -> IO ()
```

The types Display, Color, and Picture are defined by the Graphics.Gloss module, or perhaps by another module in the gloss package that is imported by Graphics.Gloss. The types Display and Color are for display mode and background color, respectively. The most interesting type is Picture, which represents the type of things that can be displayed. The gloss documentation on Picture describes the pictures we can make (lines, circles, polygons, and so on). You can find the documentation at *https://hackage.haskell.org/package/gloss* by clicking on Graphics.Gloss.

GHCi is not so good at showing the pictures that gloss creates, so it's better to make a stand-alone program. Let's write a program to get us familiar with the default coordinate system the Graphics.Gloss module uses. We'll draw a red line segment from the origin to the point (100, 0) and a green line segment from the origin to the point (0, 100). Because gloss measures distance in pixels, we'll use 100 so that the lines we produce will be long enough to see on the screen.

```
{-# OPTIONS -Wall #-}

import Graphics.Gloss

displayMode :: Display
displayMode = InWindow "Axes" (1000, 700) (10, 10)

axes :: Picture
axes = Pictures [Color red   $ Line [(0,0),(100,  0)]
                ,Color green $ Line [(0,0),(  0,100)]
                ]

main :: IO ()
main = display displayMode black axes
```

As usual, we turn on warnings. Then we import the Graphics.Gloss module. We need to do this because the code that follows uses the types Display and Picture; the data constructors InWindow, Pictures, Color, and Line; the constants red, green, and black; and the function display. These names are all defined in the Graphics.Gloss module. Without the import statement, we would get "Variable not in scope" errors every time we used each of those names.

We define a constant displayMode to hold the value of type Display that the function display requires. We give the name "Axes" to the window that will be opened by the display function. We ask for the window to be 1,000 pixels wide and 700 pixels high, and we ask for it to be placed 10 pixels up and 10 pixels over from whatever the window system thinks is the origin.

We define a constant called axes to hold the Picture we want to make. We produce the picture using the data constructor Pictures, which gives a way to combine a list of pictures into a single picture. We can ask GHCi for the types of some of the things that didn't get explicit types in our code.

```
Prelude Graphics.Gloss> :t Line [(0,0),(100,0)]
Line [(0,0),(100,0)] :: Picture
Prelude Graphics.Gloss> :t Color green $ Line [(0,0),(0,100)]
Color green $ Line [(0,0),(0,100)] :: Picture
```

The main function here uses the function display to make the picture. We pass display our displayMode, the background color black, and our picture axes.

When we compile and run the program above using one of the three methods described in Chapter 12, we should see a red horizontal line and a green vertical line. The default gloss orientation has the x-axis going toward the right and the y-axis going upward. Depending on your operating system, you might need to hit CTRL-C twice to close the graphics window.

The gloss package does not have a primitive disk, or filled circle, picture. As a second example of the display function, let's make a picture of a blue circle and a red disk, side by side.

```
{-# OPTIONS -Wall #-}

import Graphics.Gloss

displayMode :: Display
displayMode = InWindow "My Window" (1000, 700) (10, 10)

blueCircle :: Picture
blueCircle = Color blue (Circle 100)

disk :: Float -> Picture
disk radius = ThickCircle (radius / 2) radius

redDisk :: Picture
redDisk = Color red (disk 100)

wholePicture :: Picture
wholePicture = Pictures [Translate (-120) 0 blueCircle
                        ,Translate   120  0 redDisk
                        ]

main :: IO ()
main = display displayMode black wholePicture
```

Here we used the same warnings, import, and displayMode lines as before. The constant blueCircle is a blue circle with a radius of 100 pixels.

Since gloss does not provide a function to make a disk, we'll write our own. Our disk function uses gloss's built-in ThickCircle function to make a disk. ThickCircle takes a radius and a thickness as inputs. Here we choose the radius of the thick circle to be half of the desired radius of the disk, and we choose the thickness to be the full desired radius of the disk. This circle is so thick that there is no hole left in the middle, making a disk.

The constant redDisk is a red disk with a radius of 100 pixels. The constant wholePicture uses the Translate data constructor of the Picture type to shift the circle to the left and the disk to the right. The main function is very similar to that of the last program, except now we are displaying wholePicture.

When we run the program, we should see a blue circle to the left of a red disk of the same size.

Making a 2D Animation

Given a picture as a function of time, the function animate produces an animation. Let's ask GHCi the type of animate.

```
Prelude> :m Graphics.Gloss
Prelude Graphics.Gloss> :t animate
animate :: Display -> Color -> (Float -> Picture) -> IO ()
```

The difference in type compared to display is that Picture in display has been replaced by Float -> Picture in animate. The animate function uses a Float to describe time, so an expression of type Float -> Picture is a function from time to a picture, or a picture as a function of time.

Here is an example of how to use animate:

```
{-# OPTIONS -Wall #-}

import Graphics.Gloss

displayMode :: Display
displayMode = InWindow "My Window" (1000, 700) (10, 10)

disk :: Float -> Picture
disk radius = ThickCircle (radius / 2) radius

redDisk :: Picture
redDisk = Color red (disk 25)

projectileMotion :: Float -> Picture
projectileMotion t = Translate (xDisk t) (yDisk t) redDisk

xDisk :: Float -> Float
xDisk t = 40 * t
```

```
yDisk :: Float -> Float
yDisk t = 80 * t - 4.9 * t**2

main :: IO ()
main = animate displayMode black projectileMotion
```

The function `projectileMotion` takes a `Float` (the time) as input and produces a `Picture` by translating the red disk to the right by `xDisk t` and upward by `yDisk t`. The functions `xDisk` and `yDisk` are given explicitly as functions of time.

When we compile and run this code, we'll see a red disk experiencing projectile motion. One meter is represented by one pixel, and one second of time in the real world is one second of time in the animation. The projectile starts with an initial x-component of velocity of 40 m/s and an initial y-component of velocity of 80 m/s.

Making a 2D Simulation

The `gloss` package's `simulate` function allows the user to make an animation when an explicit function describing a picture as a function of time is not available. Let's ask GHCi the type of `simulate`.

```
Prelude> :m Graphics.Gloss
Prelude Graphics.Gloss> :t simulate
simulate
  :: Display
     -> Color
     -> Int
     -> model
     -> (model -> Picture)
     -> (Graphics.Gloss.Data.ViewPort.ViewPort
         -> Float -> model -> model)
     -> IO ()
```

The `simulate` function asks for six pieces of information. The first two, display mode (type `Display`) and background color (type `Color`), are the same as in `display` and `animate`. The third piece of information (an `Int`) is the rate, in updates per second, at which the simulation should run. The fourth piece of information has the type variable `model` instead of a concrete type. We can tell that `model` is a type variable because it starts with a lowercase letter; it cannot be a constant or a function because it sits in a type signature in the place where a type needs to sit. We, the users of the `simulate` function, get to decide what type to use for `model`. We need a type that can hold the *state* of the system we are simulating, which is the collection of information necessary to (1) produce a picture at any given moment of time and (2) determine what will happen next as time evolves. This notion of state will play a large role in the physics we describe in Parts II and III of the book. The value of type `model` that `simulate` needs is the initial state of the situation to be displayed.

The fifth piece of information (type `model -> Picture`) is a function that describes what picture to produce given a value of type `model`. The sixth piece of information `simulate` needs is a function (type `Viewport -> Float -> model -> model`) that describes how the state of the system should advance in time. The `Float` represents a time step here, and we will not use the `Viewport`.

Listing 13-1 gives a complete program that shows how to use the `simulate` function.

```
{-# OPTIONS -Wall #-}

import Graphics.Gloss

displayMode :: Display
displayMode = InWindow "My Window" (1000, 700) (10, 10)

-- updates per second of real time
rate :: Int
rate = 2

disk :: Float -> Picture
disk radius = ThickCircle (radius / 2) radius

redDisk :: Picture
redDisk = Color red (disk 25)

type State = (Float,Float)

initialState :: State
initialState = (0,0)

displayFunc :: State -> Picture
displayFunc (x,y) = Translate x y redDisk

updateFunc :: Float -> State -> State
updateFunc dt (x,y) = (x + 10 * dt, y - 5 * dt)

main :: IO ()
main = simulate displayMode black rate initialState displayFunc
       (\_ -> updateFunc)
```

Listing 13-1: Sample use of the `simulate` function from the `gloss` package

The display mode and background color are the same as before. We define a constant `rate` to hold the simulation rate. For the type variable `model`, we've chosen `(Float,Float)` and given it the type synonym `State`. This state is meant to represent the (x, y) coordinates of the current location of the red disk. The initial value of the state is defined in `initialState`.

The heart of the simulation is contained in the two functions `displayFunc` and `updateFunc`. The first tells how to make a picture from a state. In this case, we use the (x, y) coordinates in the state to translate the red disk over *x* and up *y*. The display function only cares about the current state of affairs (the current values of *x* and *y*). It has nothing to do with how the picture changes with time.

The update function `updateFunc` explains how the state changes with time. We need to give a rule for what the new state will be in terms of the old state and a time step `dt`. In this case, we add 10 pixels/second to the x-value and subtract 5 pixels/second from the y-value.

When we run the program, we should see the red disk move to the right and downward as the simulation evolves. The simulation will be jerky because we chose a rate of 2 updates/second, so you see each update as a discrete motion. Try increasing the rate to get a smoother animation. High definition TV uses 24–60 frames per second, so you shouldn't need to go higher than that. If the lights in your building dim, you've picked a frame rate that's too high.

For one more example of the `simulate` function, let's see how the projectile motion we did before with `animate` would look when done with `simulate`. The difference between constant velocity motion and projectile motion is that the velocity changes in projectile motion. To allow the velocity to change, we need to expand the information in the state to include velocity as well as the position of the red disk. For this purpose, we define the type synonyms `Position`, `Velocity`, and `State` in the code in Listing 13-2.

Our `initialState` now needs to contain both the initial position `(0,0)` and an initial velocity `(40,80)`. The initial x-component of velocity is 40 m/s, and the initial y-component is 80 m/s.

Our display function doesn't need to change in meaning from the previous simulation. The display of the red disk still depends only on the position of the disk and not on the current velocity. However, the `displayFunc` function needs a little bit of syntactic revision since the type of the state has changed. The syntactic revision entails replacing the argument `(x,y)` with `((x,y),_)` to reflect the new type of state. If we left the function entirely unchanged, the compiler would think that the argument `(x,y)` meant `x` for position and `y` for velocity. We would get a type error complaining that the expected type of `x` is `Float` but its actual type is `Position`. The "actual type" of `Position` comes from the type signature of `displayFunc`, while the "expected type" of `Float` comes from the way `x` is used, as an argument to `Translate`, in the definition of `displayFunc`.

Let's implement these changes; the result is in Listing 13-2.

```
{-# OPTIONS -Wall #-}

import Graphics.Gloss

displayMode :: Display
displayMode = InWindow "My Window" (1000, 700) (10, 10)

-- updates per second of real time
rate :: Int
```

```
rate = 24

disk :: Float -> Picture
disk radius = ThickCircle (radius / 2) radius

redDisk :: Picture
redDisk = Color red (disk 25)

type Position = (Float,Float)
type Velocity = (Float,Float)
type State = (Position,Velocity)

initialState :: State
initialState = ((0,0),(40,80))

displayFunc :: State -> Picture
displayFunc ((x,y),_) = Translate x y redDisk

updateFunc :: Float -> State -> State
updateFunc dt ((x,y),(vx,vy))
    = (( x + vx * dt, y +  vy * dt)
      ,(vx          ,vy - 9.8 * dt))

main :: IO ()
main = simulate displayMode black rate initialState displayFunc
       (\_ -> updateFunc)
```

Listing 13-2: Sample use of the `simulate` function to produce projectile motion

The update function is where all of the action happens. The components x and y of position get updated based on the current velocity. The velocity components vx and vy get updated based on the components of acceleration. The x-component of acceleration is 0, so the x-component of velocity stays the same. The y-component of acceleration is -9.8 m/s^2, so we update the y- component of velocity using that, assuming that 1 meter represents 1 pixel in our simulation.

When we run this program, the results should be the same as the projectile program we wrote with animate.

Notice the difference in the information required to produce the projectile motion animation using animate compared to simulate. To use animate, we need to have explicit expressions for the position as a function of time. To use simulate, we provide equivalent information, but it seems like we're providing less. The state update procedure is a powerful tool in numerically solving equations of motion. We'll exploit this tool much more in Parts II and III of the book.

3D Animation

The not-gloss package provides four main functions whose names are identical to those in gloss: display, animate, simulate, and play. As in gloss, the first is for still pictures, the second and third are for pictures that change with time, and the fourth is for pictures that change with time and user input. We're interested primarily in the first three functions. The types of these functions are different from those of the corresponding gloss functions, in part because the not-gloss package has a different author from the gloss package. There are similarities between the two packages, but there are also differences that we will point out.

Displaying a 3D Picture

Let's check the type of display. Just as the gloss package has a module named Graphics.Gloss that must be imported before its functions can be used, the not-gloss package has a module named Vis that we must import.

```
Prelude> :m Vis
Prelude Vis> :t display
display :: Real b => Options -> VisObject b -> IO ()
```

If we ask about the type class Real, we learn that Real is for numeric types that can be converted to rational numbers:

```
Prelude Vis> :i Real
class (Num a, Ord a) => Real a where
  toRational :: a -> Rational
  {-# MINIMAL toRational #-}
   -- Defined in 'GHC.Real'
instance Real Word -- Defined in 'GHC.Real'
instance Real Integer -- Defined in 'GHC.Real'
instance Real Int -- Defined in 'GHC.Real'
instance Real Float -- Defined in 'GHC.Float'
instance Real Double -- Defined in 'GHC.Float'
```

Our favorite instance of the Real type class is R (or Double). This will be our default choice unless there is a reason to choose something else. If the type variable b in the type of display is R, the type of display is the following:

```
display :: Options -> VisObject R -> IO ()
```

The display function is asking us to provide two things: a thing with type Options, and the object to be displayed (type VisObject R). The return type IO () means that the computer will *do* something (in this case display the object).

What kinds of things are there that have type VisObject R? The not-gloss package provides a long list of possibilities, including spheres, cubes, lines, text, and more. You can find documentation at the *https://hackage.haskell.org* site by searching for not-gloss.

Here is an example that produces a blue cube:

```
{-# OPTIONS -Wall #-}

import Vis

type R = Double

blueCube :: VisObject R
blueCube = Cube 1 Solid blue

main :: IO ()
main = display defaultOpts blueCube
```

The constant `defaultOpts` is provided by the `Vis` module as a set of default options. You can compile this code into a stand-alone program as before. When you run the program, a display window containing a blue cube will open. After the display window opens, press **e** to zoom in and **q** to zoom out. You can also use the mouse to rotate the cube. These are standard features of `not-gloss` that we don't need to program.

The next program will get us familiar with the default coordinate system the `Vis` module uses. We'll draw a red line segment from the origin to the point $(1, 0, 0)$, a green line segment from the origin to the point $(0, 1, 0)$, and a blue line segment from the origin to the point $(0, 0, 1)$.

```
{-# OPTIONS -Wall #-}

import Vis
import Linear

type R = Double

axes :: VisObject R
axes = VisObjects [Line Nothing [V3 0 0 0, V3 1 0 0] red
                  ,Line Nothing [V3 0 0 0, V3 0 1 0] green
                  ,Line Nothing [V3 0 0 0, V3 0 0 1] blue
                  ]

main :: IO ()
main = display defaultOpts axes
```

Here we import the `Linear` module to have access to the `V3` constructor. The `Linear` module defines several kinds of vector; `V3` is the one used by the `Vis` module. The `Nothing` means to use the default line width (try `(Just 5)` in place of `Nothing` to get a thicker line width).

When we compile and run the program just shown, we see axes for a three-dimensional coordinate system. We see that `not-gloss`'s default orientation has the x-axis going toward the right and toward the viewer, the y-axis going toward the left and toward the viewer, and the z-axis going downward.

Personally, I think having the positive z-axis point downward is disturbing and abhorrent. I like to think of myself as a flexible person, but this crosses a line. (Greg Horn, the author of `not-gloss`, tells me that the z-down convention is standard in much of the aerospace industry.) Fortunately, `not-gloss` has tools that allow us to rotate things the way we want. I like to have the x-axis coming mostly out of the page, the y-axis pointing toward the right, and the z-axis pointing upward. Here is a program that will do just that:

```
{-# OPTIONS -Wall #-}

import Vis
import Linear
import SpatialMath

type R = Double

axes :: VisObject R
axes = VisObjects [Line Nothing [V3 0 0 0, V3 1 0 0] red
                  ,Line Nothing [V3 0 0 0, V3 0 1 0] green
                  ,Line Nothing [V3 0 0 0, V3 0 0 1] blue
                  ]

orient :: VisObject R -> VisObject R
orient pict = RotEulerDeg (Euler 270 180 0) $ pict

main :: IO ()
main = display defaultOpts (orient axes)
```

We import `Vis` and `Linear` as before, but here we also import `SpatialMath` so we can use `Euler` to perform three-dimensional rotations using Euler angles. The `axes` picture has not changed. We define a function `orient` that takes a picture as input and gives back a reoriented picture as output. To do this, we perform a rotation specified by Euler angles using the `RotEulerDeg` data constructor of the `VisObject` type. In this case, the Euler angles mean we rotate first by 0° about the x-axis, then by 180° about the y-axis, and then by 270° about the z-axis. Equivalently, we can view this as a rotation first about the z-axis by 270°, then about the *rotated* y-axis by 180°, and then about the rotated x-axis by 0°.

Finally, we pass `orient axes` to `display` as the picture to be displayed. If you like this way of orienting the coordinate system, you could pass any picture to the function `orient` before displaying it as a way of using this coordinate system. You could even define your own display function that does the reorientation for you.

```
myDisplay :: VisObject R -> IO ()
myDisplay pict = display defaultOpts (orient pict)
```

Making a 3D Animation

Let's look at the type of animate.

```
Prelude> :m Vis
Prelude Vis> :t animate
animate :: Real b => Options -> (Float -> VisObject b) -> IO ()
```

The type of animate is the same as the type of display, except that the VisObject b of display is replaced by Float -> VisObject b in animate. Instead of asking us to provide a picture, animate is asking us to provide a function from time to a picture. The animate function demands that we use Float for the real number representing time.

The following animation of a rotating blue cube has the cube rotating counterclockwise about the x-axis in my favorite coordinate system (x-axis out of the screen, y-axis to the right, and z-axis up the screen):

```
{-# OPTIONS -Wall #-}

import Vis
import SpatialMath

rotatingCube :: Float -> VisObject Float
rotatingCube t = RotEulerRad (Euler 0 0 t) (Cube 1 Solid blue)

orient :: VisObject Float -> VisObject Float
orient pict = RotEulerDeg (Euler 270 180 0) $ pict

main :: IO ()
main = animate defaultOpts (orient . rotatingCube)
```

Note the use of function composition between rotatingCube and orient. The function rotatingCube takes a number as input and produces a picture as output. The function orient takes a picture as input and produces a (re-oriented) picture as output. The composition orient . rotatingCube takes a number as input and produces a picture as output, which is just the type of function that animate wants.

Making a 3D Simulation

The not-gloss function simulate allows the user to make an animation when an explicit function describing a picture as a function of time is not available. Let's ask GHCi the type of simulate.

```
Prelude> :m Vis
Prelude Vis> :t simulate
simulate
  :: Real b =>
     Options
     -> Double
```

```
-> world
-> (world -> VisObject b)
-> (Float -> world -> world)
-> IO ()
```

The simulate function asks for five pieces of information. The first one (type Options) is the same as in display and animate. The second piece of information (a Double) is the time step, in seconds per update, between successive frames of the animated display. Note the difference with the gloss library: gloss asks for a rate in updates per second, but not-gloss asks for a time step in seconds per update.

The third piece of information is the initial state of the situation to be displayed. The type variable world stands for a type that the user chooses to describe the state of the situation, very much like the type variable model used in the gloss function simulate.

The fourth piece of information (type world -> VisObject b) is a display function that describes what picture to produce given a value of type world. This display function is very much like the display function of gloss.

Finally, the fifth piece of information needed by simulate is a function (type Float -> world -> world) that describes how the state of the system should advance in time. The Float in this type represents the total time elapsed since the beginning of the simulation. This is different from gloss, where the Float in the analogous term describes the time step since the previous frame.

Listing 13-3 demonstrates how the simulate function uses the update function that we provide as the fifth piece of information. The purpose of the code is to determine, experimentally, what the simulate function does with the function of type Float -> world -> world that we provide. If you're not used to working with higher-order functions, this can seem like a weird question. Normally, we write functions for our own use, or we use functions that others have written. But when someone else writes a higher-order function for us to use, and that higher-order function takes a user-defined function as input, we might well wonder how the higher-order function intends to use the user-defined function that we provide. (We might read the code or the documentation of the higher-order function, but here we're going to figure it out experimentally.)

What's weird is that we're writing a function, updateFunc in Listing 13-3, but we are not going to use that function directly. We don't decide what Float to send to updateFunc; another function, simulate, decides that.

```
{-# OPTIONS -Wall #-}

import Vis

type State = (Int,[Float])

-- seconds / update
dt :: Double
```

```
dt = 0.5

displayFunc :: State -> VisObject Double
displayFunc (n,ts) = Text2d (show n ++ " " ++ show (take 4 ts))
                     (100,100) Fixed9By15 orange

updateFunc :: Float -> State -> State
updateFunc t (n,ts) = (n+1,t:ts)

main :: IO ()
main = simulate defaultOpts dt (0,[]) displayFunc updateFunc
```

Listing 13-3: Use of the `simulate` function from the `not-gloss` library. The purpose of this code is to experimentally determine what `simulate` is doing with `updateFunc`.

We begin the code by importing the `Vis` module. For the type variable `world`, we choose a pair `(Int,[Float])` in which the `Int` is designed to hold the number of updates performed since the beginning of the simulation and the list of floats is intended to be a list of the time values passed to the update function `updateFunc`. We don't choose those time values; `simulate` does.

We set a time step called `dt` to be half of a second. The `displayFunc` tells how to produce a picture from a `State`. It uses the `Text2d` data constructor of the `VisObject` type, which you will find if you check out the documentation on `not-gloss`'s `simulate` function.

The update function `updateFunc` keeps track of two things: the number of times it has been called and the `Float` values it has been called with. Each time `updateFunc` is called, it increases the call count by one and tacks the most recent `Float` onto the front of the list.

When we run this program, we can watch the number of updates increase at the rate of two per second and see that the times being passed in are increasing, confirming the assertion that the update function takes the time since simulation start as input.

Summary

In this chapter, we explored several ways to produce two-dimensional and three-dimensional graphics and animation. We gave code for programs exhibiting each of the graphics functions that we will use later in the book to provide visual support to our thinking and writing about physics.

With this chapter, we have completed Part I of the book—an introduction to functional programming ideas in general and to the Haskell programming language in particular. In Part II, we explore Newtonian mechanics, in which our goal is to predict the motion of one or more objects experiencing forces. The central principle of Newtonian mechanics is Newton's second law, which is the subject of the next chapter.

Exercises

Exercise 13.1. Consult the `gloss` documentation on the `Picture` type and make an interesting picture using the `display` function. Combine lines, circles, text, colors, and whatever you like. Be creative.

Exercise 13.2. Use `animate` to make a simple animation. Be creative.

Exercise 13.3. Use `animate` to make the red disk oscillate left and right. Then, change your code a little bit to make the red disk orbit in a circle. Can you make the red disk move in an ellipse?

Exercise 13.4. Use `animate` to produce the same motion of the red disk that we achieved with `simulate` in Listing 13-1.

Exercise 13.5. Use `simulate` to do something you think is interesting. Be creative.

Exercise 13.6. In the 2D projectile motion example of Listing 13-2, one meter in the real world is represented by one pixel in the animation. Modify the code so that one meter is represented by 10 pixels. Feel free to change the initial velocity components so the projectile doesn't speed off the screen right away.

Exercise 13.7. Challenging exercise: Try to use `simulate` to make the red disk oscillate left and right without explicitly giving it an oscillating function like sin or cos. We will show how to do this in Part II of the book.

Exercise 13.8. Rewrite the 3D axes code so that the x-axis points to the right, the y-axis points upward, and the z-axis points out of the page. This is my second-favorite coordinate system.

Exercise 13.9. Modify the rotating cube animation to make the rotation occur clockwise about the x-axis instead of counterclockwise.

Exercise 13.10. Write an experimental program, similar to Listing 13-3, using the `gloss` function `simulate` to understand how `gloss`'s `simulate` uses the update function. Use the same expressions for `updateFunc` and `State` that we used in Listing 13-3. You will need to change the values of `displayFunc` and `main`. Use a `rate` of `2` instead of a `dt` of `0.5`. When you run this, you should see that the times passed in by `gloss`'s `simulate` are time steps that are all close to 0.5.

PART II

EXPRESSING NEWTONIAN MECHANICS AND SOLVING PROBLEMS

14

NEWTON'S SECOND LAW AND DIFFERENTIAL EQUATIONS

Isaac Newton accomplished a lot. Among the numerous physical and mathematical insights he left us are three numbered laws that bear his name. Newton's second law is the most important of these; it provides a method for understanding the motion of an object if we know the forces that act on it. Newton's third law is almost as significant; it's a rule about how two objects interact. Newton's first law, from a mathematical standpoint, is a corollary to Newton's second law, so it seems the smallest of the three. But since Newton's second law is sufficiently intuition shattering, it's helpful to get our minds around something simpler before trying to grasp it. Newton's first law serves well in this capacity; it makes a bold claim that seems obviously false.

In this chapter, we'll discuss Newton's first law and then turn our attention to Newton's second law in one linear dimension, such as a horizontal line or a vertical line. We'll show how to think about Newton's second law in a sequence of settings of increasing complexity, organized by what the forces depend on. We'll start with constant forces, the simplest situation, before moving on to forces that depend only on time. Then we'll turn to forces that depend on the velocity of the particle they act on, followed by forces that depend on both time and velocity. The techniques for solving Newton's second law change as the forces involved depend on different physical quantities. We'll introduce the Euler method for solving a differential equation and explore a number of situations in which Newton's second law is the central principle that allows us some traction in understanding the motion of an object.

Newton's First Law

Let's return to the air track of Chapter 4. If you give the car a little push on the air track and then let it go, it will travel at a constant speed until it hits the end of the track. After we stop pushing the car, it continues to move at some speed even with no force applied in the direction of motion. This tendency for moving objects to keep moving is called *inertia*. The idea of inertia is relevant in the one-dimensional spatial setting of the air track, and it's also relevant in the unconstrained three-dimensional spatial setting of the world in which we live. The idea is important enough to be enshrined in a principle of physics called *Newton's first law*. Here are three versions:

> **Newton's first law, Newton's words [15]**
> *Every body perseveres in its state of being at rest or of moving uniformly straight forward, except insofar as it is compelled to change its state by forces impressed.*
>
> **Newton's first law, poetic version**
> *A body in motion stays in motion. A body at rest stays at rest.*
>
> **Newton's first law, modern version**
> *In the absence of applied forces, an object maintains the same velocity.*

Recall that velocity is a vector, so maintaining the same velocity means keeping the same speed as well as the same direction. Since acceleration is change in velocity per unit of time, an equivalent way of expressing Newton's first law is that in the absence of forces, an object experiences no acceleration.

Notice that Newton's first law makes no mention of forces that were applied *in the past*. The point is that if there are no forces acting *now*, the velocity will stay constant now. Any time there are no forces present, the velocity will stay constant.

Why does Newton's first law seem obviously false? Because we're stuck on the surface of the earth, a place that is rife with forces we might fail to consider, friction and air resistance not least among them. Things are a bit simpler out in space. We can imagine one astronaut tossing a small wrench to another at slow speed. The wrench just glides straight across the ship, perhaps rotating slowly about its center. That wrench is a great example of Newton's first law.

Perhaps you've been in a car when the driver slams on the brakes so that books, papers, and toys go flying forward (with respect to the car's seats). In my family, we celebrate these moments by shouting "Newton's first law!" From a perspective outside the (decelerating) car, the books, papers, and toys are doing their best to travel in a straight line, at least for the short period of time before gravity and other objects put an end to their line-like motion.

Newton's first law tells us that objects naturally go steady and straight. In practice, though, they don't. Newton's second law explains how and why.

Newton's Second Law in One Dimension

Newton's first law tells us that when no forces are present, an object does not accelerate. Newton's second law claims that acceleration is caused by forces.

> **Newton's second law, Newton's words [15]**
> *A change in motion is proportional to the motive force impressed and takes place along the straight line in which that force is impressed.*

> **Newton's second law, poetic version**
> *An object's acceleration is directly proportional to the net force acting on the object and inversely proportional to its mass.*

Modern versions of Newton's second law are expressed by Equation 14.1 for Newton's second law in one dimension, and Equation 16.1 for Newton's second law in three dimensions. In the remainder of this chapter, we'll treat Newton's second law in one dimension, which allows us to keep things simple by using numbers rather than vectors for velocity, acceleration, and force. In Chapter 16, we'll treat Newton's second law in full generality with vectors.

To discuss force and mass in a quantitative way, we need units of measure. In the SI system, force is measured in Newtons (N). A 100-N force has a different effect on a golf ball than it has on a bowling ball. According to Newton, each object has a *mass*, which determines the readiness of an object to accelerate in response to a force. A large-mass object experiences small acceleration compared to a small-mass object exposed to the same force. The SI unit of mass is the kilogram (kg).

Newton's second law expresses a relationship between the following three quantities:

- The forces that act on an object
- The mass of the object
- The acceleration of the object

Newton's second law says that the acceleration of an object can be found by dividing the net force acting on the object by the mass of the object. The *net force* acting on an object is the sum of all the forces acting on the object. In one dimension, some forces may be negative and some may be positive.

Newton's second law is usually written as $F_{net} = ma$. Unlike the one-dimensional equations for velocity and acceleration (Equations 4.5 and 4.12), this equation is not an equality of functions. The acceleration of the object is only a function of time, but the net force generally depends on the time, the position of the object, and the velocity of the object. The net force at time t is $F_{net}(t, x(t), v(t))$. A better way to write Newton's second law is:

Newton's second law in one dimension

$$F_{net}(t, x(t), v(t)) = ma(t) \tag{14.1}$$

There is a chicken-and-egg issue going on with Newton's second law. We know from Equations 4.5 and 4.12 that $v = Dx$ and $a = Dv$. If we know the function a (meaning we know its value at all times), we can find the function v given an initial velocity. (See Equation 6.1 and the corresponding function `velFromAcc`.) We can then go on and find the function x given an initial position. (See Equation 6.5 and the corresponding function `posFromVel`.) But Newton's second law is telling us that acceleration depends on the forces, which depend on the position and the velocity. To find the position of my object, it seems that I need to find the velocity, and for that I need the acceleration. However, the acceleration depends on both the position and the velocity.

There is a name for this particular kind of chicken-and-egg problem. Newton's second law is an example of a *differential equation*. A differential equation is a relationship between derivatives of an unknown function, with the unknown function itself often regarded as the zeroth derivative. The unknown function in the case of Newton's second law is usually either the position x or the velocity v. Velocity can be written as the first derivative of position ($v = Dx$), and acceleration can be written as the second derivative of position ($a = Dv = D^2x$).

Newton's second law looks more like a differential equation if we write it in terms of an unknown position function.

$$F_{net}(t, x(t), Dx(t)) = mD^2x(t) \tag{14.2}$$

This is a second-order differential equation because it is a relationship between the position function x, its first derivative Dx, and its second derivative D^2x. The relationship for a particular physical object depends on the function F_{net}, which depends on the nature of the forces acting on the object.

In simple situations, the net force on an object may not depend on time, position, and velocity, but rather only on zero, one, or two of these physical quantities. In these simple situations, Newton's second law may appear as something simpler than a second-order differential equation. Table 14-1 lists situations by the physical quantities that the forces depend on and gives the mathematical technique needed to solve Newton's second law.

Table 14-1: The Technique for Solving Newton's Second Law, Based on Which Physical Quantities the Forces Depend On

Forces depend only on	Solution technique
Nothing	Algebra
Time	Integration
Velocity	First-order differential equation
Time and velocity	First-order differential equation
Time, position, and velocity	Second-order differential equation

A net force that depends on nothing is a constant net force. Its value remains constant over time, independent of time, position, or velocity. In the next several sections, we'll look at constant forces, forces that depend only on time, forces that depend only on velocity, and forces that depend on both time and velocity. This restriction allows us to limit our attention in this chapter to first-order differential equations. In Chapter 15, we'll look at the more general case of one-dimensional motion in which the net force can depend on time, position, *and* velocity.

Second Law with Constant Forces

The simplest situation for Newton's second law is when the net force is constant, independent of time, position, and velocity. Most problems in an introductory physics course are like this because they can be solved without differential equations and without a computer.

Let's consider an example problem with constant forces.

Example 14.1. Suppose we have a car with mass 0.1 kg on an air track. The car is initially moving east at a speed of 0.6 m/s. Starting at time $t = 0$, we apply to this car a constant force of 0.04 N to the east. At the same time, our friend applies to the same car a constant force of 0.08 N to the west. What will the subsequent motion of the car look like? In particular, how will the velocity and the position of the car change in time?

Figure 14-1 shows the schematic diagram.

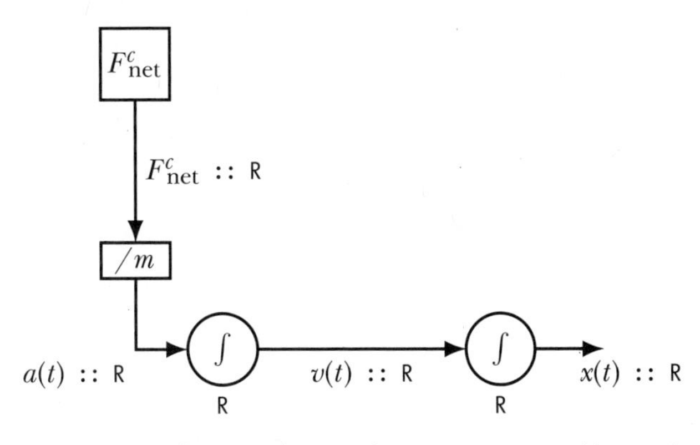

Figure 14-1: Schematic diagram for Newton's second law with constant forces

The constant net force F^c_{net} (superscript c for constant) acting on the object needs to be divided by the mass of the object to obtain the acceleration of the object. Because the net force is constant, the acceleration is also constant.

$$a(t) = \frac{F^c_{\text{net}}}{m} \tag{14.3}$$

We write $a(t)$ rather than a for acceleration, not because acceleration changes with time, but because a is the acceleration function (type `R -> R`) and $a(t)$ is the acceleration (type `R`). We then integrate acceleration to obtain the velocity.

$$v(t) = v(0) + \frac{F^c_{\text{net}}t}{m} \tag{14.4}$$

The integrator that produces velocity contains a real number (type `R`) as state. This type is shown below the integrator in Figure 14-1. This integrator remembers the current velocity so that it can be updated using the acceleration.

We then integrate the velocity to obtain the position.

$$x(t) = x(0) + v(0)t + \frac{F^c_{\text{net}}t^2}{2m} \tag{14.5}$$

The wires of the diagram represent quantities that are continuously changing in time. Each wire in the diagram is labeled with a name and a type. For this diagram, all of the wire types are real numbers.

Rectangular boxes represent purely functional constants and functions. In other words, they are constants and functions that do not contain any state, so that the output is a function only of the input. The circular integrators contain states that must be combined with the input to produce the output. The integrators are labeled with the type of state they contain, which is the same as the type of the output from the integrator.

Before we write Haskell code to solve Newton's second law for constant forces, we are going to write a few lines of code that need to be at the top of the source code file we build throughout this chapter. The first line turns on warnings, which I recommend doing because the compiler will warn you of things that are legal but unusual enough that they may not be what you intended. The second line gives the code in this chapter the module name `Newton2`. If we want to use functions we write here in later chapters, we'll refer to the current code using its module name. A module name is optional, but if you use one, it must match the filename; in this case, the filename should be *Newton2.hs*. The third line loads the `gnuplot` graphics library so that we can make a graph. Imports like this must occur before any function definitions or type signatures.

```
{-# OPTIONS -Wall #-}

module Newton2 where

import Graphics.Gnuplot.Simple
```

Example 14.1 is typical of situations in which Newton's second law applies. Given a mass, an initial velocity, and some forces, we are asked to produce velocity as a function of time. In the Haskell language, a solution to this example situation would be a (higher-order) function `velocityCF` (`CF` for constant forces) with the following type:

```
velocityCF :: Mass
           -> Velocity          -- initial velocity
           -> [Force]           -- list of forces
           -> Time -> Velocity  -- velocity function
```

Recall that there are (at least) two ways to read this type signature. On one reading, `velocityCF` takes four inputs—mass, initial velocity, a list of forces, and a time—and produces as output a real number representing velocity. An alternative reading is that `velocityCF` takes three inputs—mass, initial velocity, and a list of forces—and produces as output a *function* for how velocity changes with time. If we wanted to emphasize the latter viewpoint, we could write

```
velocityCF :: Mass -> Velocity -> [Force] -> (Time -> Velocity)
```

but it means the same thing as the original type signature.

We used the types `Time`, `Mass`, `Velocity`, and `Force`. These are not built-in types in Haskell, so we'd better define what they mean. In one-dimensional mechanics, all of these quantities can be represented with real numbers, so we can write some type synonyms to define these types. Using a type synonym in which `R` stands for `Double`,

```
type R = Double
```

we can write type synonyms for all of the other types:

```
type Mass     = R
type Time     = R
type Position = R
type Velocity = R
type Force    = R
```

The definitions for types `Mass`, `Time`, and so on, need not appear before their use in a type signature. Haskell allows definitions of constants, functions, and types before or after their use.

If we can write a function `velocityCF` with the type signature above, we will have solved not just Example 14.1, but all others like it. Our strategy in writing such a function is:

- Find the net force by adding all of the forces
- Find the acceleration using Newton's second law (Equation 14.3)
- Find the velocity from the acceleration (Equation 4.14 or 14.4)

Here's a definition for `velocityCF` that expresses these three steps and has the type we claimed earlier.

```
velocityCF m v0 fs
    = let fNet = sum fs         -- net force
          a0   = fNet / m       -- Newton's second law
          v t  = v0 + a0 * t    -- constant acceleration eqn
      in v
```

To write the function velocityCF, we begin by naming the three inputs: mass m, initial velocity v0, and list of forces fs. We then use a let construction to define three local names for net force, acceleration, and velocity. To find the net force, we sum up the forces in the list using the built-in sum function. To find the acceleration, we divide the net force on the object by the mass of the object, as Newton's second law prescribes.

The third equation in the let construction defines a local function v to represent the velocity function. We use Equation 4.14, one of the constant acceleration equations introduced in standard introductory physics textbooks, but we could just as easily have used Equation 14.4 in place of the second and third lines of the let construction. Notice that we have written the definition of velocityCF using the "three-input thinking" mentioned earlier. Exercise 14.1 asks you to rewrite the function using four-input thinking.

We can write a function positionCF that produces a position function given mass, initial position, initial velocity, and a list of constant forces.

```
positionCF :: Mass
           -> Position          -- initial position
           -> Velocity          -- initial velocity
           -> [Force]           -- list of forces
           -> Time -> Position  -- position function
positionCF m x0 v0 fs
    = let fNet = sum fs
          a0   = fNet / m
          x t  = x0 + v0 * t + a0*t**2 / 2
      in x
```

Here, we have used Equation 4.15 or 14.5. Returning to Example 14.1, the velocity of the car as a function of time is

```
velocityCF 0.1 0.6 [0.04, -0.08]
```

because 0.1 kg is the mass of the car, 0.6 m/s is its initial velocity, and the square-bracketed list contains the forces in Newtons. We can ask for the type of this function in GHCi, and we can ask for values of the velocity at specific times.

```
Prelude> :l Newton2
[1 of 1] Compiling Newton2          ( Newton2.hs, interpreted )
Ok, one module loaded.
*Newton2> :t velocityCF 0.1 0.6 [0.04, -0.08]
velocityCF 0.1 0.6 [0.04, -0.08] :: Time -> Velocity
*Newton2> velocityCF 0.1 0.6 [0.04, -0.08] 0
```

```
0.6
*Newton2> velocityCF 0.1 0.6 [0.04, -0.08] 1
0.2
```

Since we have the velocity function in hand, we can graph it. Let's write the code to do so first. Most of the code below is for setting up a title, axis labels, and the name of the file we want produced. The interesting stuff is at the end, where we give a list of times at which to evaluate the function and the function itself.

```
carGraph :: IO ()
carGraph
    = plotFunc [Title "Car on an air track"
               ,XLabel "Time (s)"
               ,YLabel "Velocity of Car (m/s)"
               ,PNG "CarVelocity.png"
               ,Key Nothing
               ] [0..4 :: Time] (velocityCF 0.1 0.6 [0.04, -0.08])
```

This code produces the graph in Figure 14-2.

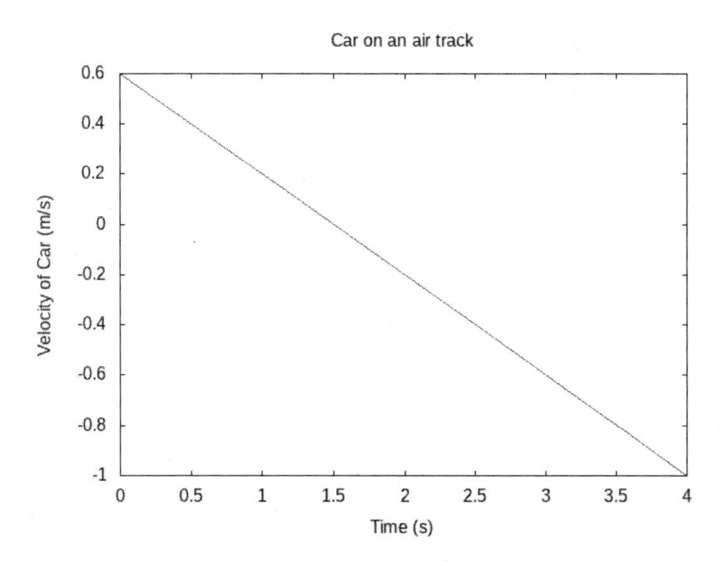

Figure 14-2: Car velocity as a function of time in Example 14.1

If you load this chapter's module, `Newton2`, into GHCi and enter `carGraph`,

```
*Newton2> carGraph
```

you will not get any return value, but the function will produce a Portable Network Graphics (PNG) file named *CarVelocity.png* on your hard drive. Without the `PNG "CarVelocity.png"` option, the `carGraph` function would produce a graph on the screen.

Note that the negative acceleration in the graph in Figure 14-2 (which exists over the entire time interval from $t = 0$ to $t = 4$ s) does not mean that the car is always slowing down. Rather, a negative acceleration means an acceleration to the west. The car slows down during the first 1.5 s as it is moving east but then begins to speed up as it moves west. When the acceleration and velocity of an object point in the same direction, the object speeds up. When the acceleration and velocity of an object point in opposite directions, the object slows down.

With the functions `velocityCF` and `positionCF`, we have general-purpose ways of solving any Newton's second law type problem in one spatial dimension with constant forces. Next we'll consider forces that change in time.

Second Law with Forces That Depend Only on Time

The next situation for Newton's second law is when the net force depends on time but not on position or velocity. Figure 14-3 shows a schematic diagram for Newton's second law with forces that depend only on time.

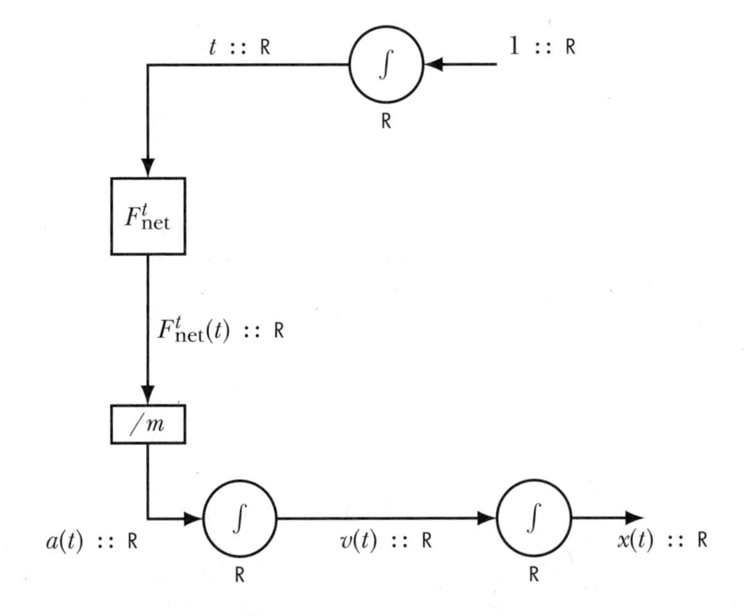

Figure 14-3: Schematic diagram for Newton's second law with forces that depend only on time

The constant number 1 is fed into an integrator to produce a value for time. (The time changes at a rate of 1 second per second.) As usual, wires are labeled with names and types. Integrators are labeled with the type of state they hold. Time is fed into the net force function F^t_{net} (superscript t for time-dependent), which produces net force as output. To obtain the acceleration of the object, we need to divide the net force acting on the object by the object's mass.

$$a(t) = \frac{F^t_{\text{net}}(t)}{m} \tag{14.6}$$

We then integrate the acceleration to obtain the velocity,

$$v(t) = v(0) + \int_0^t a(t')\, dt' = v(0) + \frac{1}{m} \int_0^t F^t_{\text{net}}(t')\, dt' \tag{14.7}$$

and we integrate the velocity to obtain the position:

$$\begin{aligned} x(t) &= x(0) + \int_0^t v(t'')\, dt'' \\ &= x(0) + \int_0^t \left[v(0) + \frac{1}{m} \int_0^{t''} F^t_{\text{net}}(t')\, dt' \right] dt'' \\ &= x(0) + v(0)t + \frac{1}{m} \int_0^t \left[\int_0^{t''} F^t_{\text{net}}(t')\, dt' \right] dt'' \end{aligned}$$

The wires of the diagram represent quantities that are continuously changing in time. Rectangular boxes represent pure functions, whereas circular elements contain state.

To solve Newton's second law problems with forces that depend on time, we'd like a higher-order function that produces a velocity function, similar to `velocityCF` in the previous section. One difference is that now we need to provide a list of force *functions* rather than a list of numerical forces. We want a function `velocityFt` (the `Ft` suffix denotes that forces depend only on time) with the following type signature:

```
velocityFt :: Mass -> Velocity -> [Time -> Force] -> Time -> Velocity
```

Given the mass of our object, its initial velocity, and a list of force functions, we want to produce a velocity function.

Because we're going to do numerical integration to get the velocity function, we'll add one additional parameter to this type signature, namely the time step for numerical integration. Thus, we arrive at the following definition for `velocityFt`:

```
velocityFt :: R                    -- dt for integral
           -> Mass
           -> Velocity             -- initial velocity
           -> [Time -> Force]      -- list of force functions
           -> Time -> Velocity     -- velocity function
velocityFt dt m v0 fs
    = let fNet t = sum [f t | f <- fs]
          a t = fNet t / m
      in antiDerivative dt v0 a
```

In this definition, we begin by naming the inputs: `dt` for an integration time step, `m` for the mass of the object we are attending to, `v0` for the initial velocity of this object, and `fs` for a list of force functions. Note that the local variable for forces, `fs`, had type `[Force]` (or `[R]`) when used in `velocityCF` and `positionCF` for situations with constant forces, but it now has type `[Time -> Force]` (or `[R -> R]`) for situations with forces that depend on time.

We again use a let construction to define local functions, a net force function, and an acceleration function. The net force function adds together the forces provided in the list fs. We might have hoped we could use the same line of code we used in velocityCF, namely fNet = sum fs, to sum the forces. After all, fs is still a list. The trouble is that sum works only with types that are instances of Num, as you can see if you look at the type of sum. So while it is happy to add numbers (type R), it is not happy to add functions (type R -> R). Fortunately, we can evaluate the force functions at a time t introduced as an argument to fNet and then add the resulting numbers.

The acceleration function comes from Newton's second law. Here, we might have hoped that we could divide the net force function by the mass to obtain the acceleration function, perhaps writing a = fNet / m. But recall that the division operator insists that it work with two values that have the same type and that this type be an instance of Fractional. The division operator does not want to work with functions. Again, we address this by evaluating the fNet function at the time t introduced as the argument to the acceleration function a.

Finally, the velocity comes from taking an antiderivative of the acceleration function. We defined the functions antiDerivative and integral in Chapter 6, but we'll repeat their definitions here:

```
antiDerivative :: R -> R -> (R -> R) -> (R -> R)
antiDerivative dt v0 a t = v0 + integral dt a 0 t

integral :: R -> (R -> R) -> R -> R -> R
integral dt f a b
    = sum [f t * dt | t <- [a+dt/2, a+3*dt/2 .. b - dt/2]]
```

Note that velocityFt dt m v0 fs has type R -> R and is the velocity function for an object with mass m, initial velocity v0, and list of force functions fs. This velocity function is part of the solution to the mechanics problem. Another part of the solution is a position function. We can write a function positionFt that produces a position function given mass, initial position, initial velocity, and a list of force functions.

```
positionFt :: R                  -- dt for integral
           -> Mass
           -> Position           -- initial position
           -> Velocity           -- initial velocity
           -> [Time -> Force]    -- list of force functions
           -> Time -> Position   -- position function
positionFt dt m x0 v0 fs
    = antiDerivative dt x0 (velocityFt dt m v0 fs)
```

This function works by taking an antiderivative of the velocity function, which we find using velocityFt.

As an example of solving Newton's second law with a time-dependent force, consider a child riding a bike. By working the pedals, the child arranges for the ground to apply a constant forward force of 10 N on the bike for 10 seconds, after which the child coasts for the next 10 seconds. Following the coasting, the child returns to the 10-N force for another 10 seconds, and so on, as illustrated in Figure 14-4.

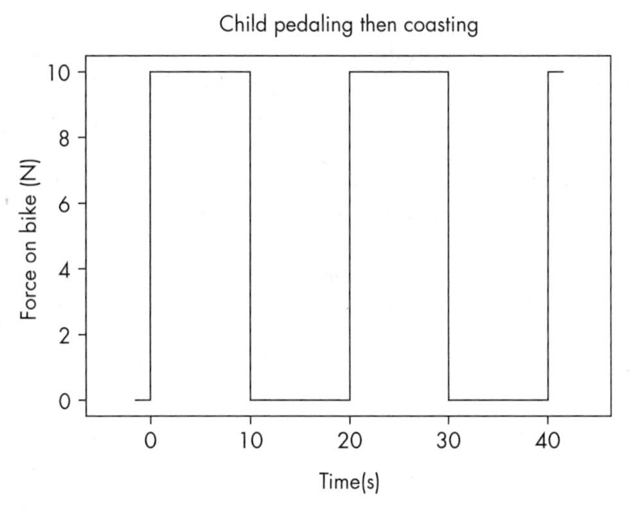

Figure 14-4: Force as a function of time for a child on a bike

In this example, we'll assume that air resistance is not important and that there is only one force on the bike.

Here is the equation for the time-dependent force of pedaling and coasting:

$$F_{\text{pc}}(t) = \begin{cases} 10 \text{ N}, & (20 \text{ s})n \leq t < (20 \text{ s})n + 10 \text{ s for some integer } n \\ 0 \text{ N}, & (20 \text{ s})n + 10 \text{ s} \leq t < (20 \text{ s})n + 20 \text{ s for some integer } n \end{cases} \tag{14.8}$$

The force is either 0 N or 10 N, depending on where the time falls in a 20-second cycle. If the time falls in the first 10 seconds of the cycle, the force is 10 N. If, on the other hand, the time falls in the last 10 seconds of the cycle, the force is 0 N.

Here is the time-dependent force of Equation 14.8 in Haskell:

```
pedalCoast :: Time -> Force
pedalCoast t
    = let tCycle = 20
          nComplete :: Int
          nComplete = truncate (t / tCycle)
          remainder = t - fromIntegral nComplete * tCycle
      in if remainder < 10
         then 10
         else 0
```

The local variable `tCycle` is the number of seconds for a full cycle. The variable `nComplete` uses the Prelude function `truncate` to calculate the number of complete cycles from the time `t`. The `truncate` function produces a type with type class Integral (recall Integer and Int are instances of Integral). We provide a local type signature to say that we want nComplete to have type Int. The local type signature is optional, but the compiler will give us a warning that it chose a default type if we don't specify something. Remove the local type signature to see what the warning looks like. This is a mild warning. We don't mind that the compiler chooses Integer instead of Int. You can feel free to ignore this warning and use the code without the type signature if you wish.

The remainder is the number of seconds, between 0 and 20, that have elapsed since the beginning of the most recent cycle. We want remainder to be a real number, so we must use fromIntegral to convert nComplete :: Int into a real number.

Figure 14-5 shows the position of the child as a function of time.

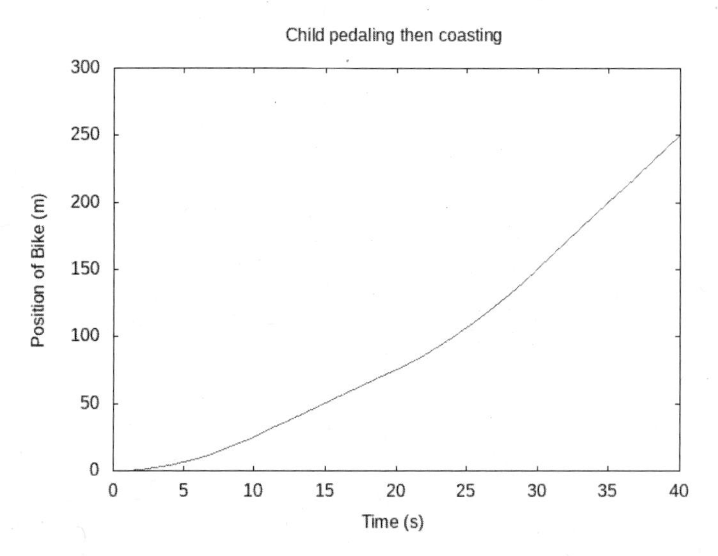

Figure 14-5: Position as a function of time for the child on a bike

Here is the Haskell code that produced Figure 14-5:

```
childGraph :: IO ()
childGraph
    = plotFunc [Title "Child pedaling then coasting"
               ,XLabel "Time (s)"
               ,YLabel "Position of Bike (m)"
               ,PNG "ChildPosition.png"
               ,Key Nothing
               ] [0..40 :: R] (positionFt 0.1 20 0 0 [pedalCoast])
```

The most interesting part of the code is the last line, where we specify the function we want plotted. This function, positionFt 0.1 20 0 0 [pedalCoast], uses the positionFt function we developed earlier in the chapter with a time step of 0.1 s, a mass of 20 kg, 0s for initial position and initial velocity, and a list of forces that includes only the force of pedaling and coasting. All of the relevant physical information is included in the "name" of the function we are plotting.

You can see from the graph in Figure 14-5 that during the first 10 seconds, the child's position curve is parabolic, as we'd expect from constant acceleration. From 10 to 20 seconds, the position shows constant velocity while the child is coasting. From 20 to 30 seconds, there is another period of acceleration in which the position curve is parabolic, followed by a second period of coasting.

With the functions `velocityFt` and `positionFt`, we have general-purpose ways of solving any Newton's second law type problem in one spatial dimension with forces that depend only on time. We're now ready to look at forces that depend on velocity, the most common of which is air resistance.

Air Resistance

In this section, we'll make a short diversion from our path of considering Newton's second law in the presence of forces that depend on time, velocity, neither, or both to develop an expression for the force of air resistance on an object in one-dimensional motion. Air resistance is a force that depends only on velocity, and we'll use it in the next several sections as we develop ways to solve Newton's second law with forces that depend on velocity.

Introductory physics classes typically ignore air resistance or treat it very lightly, because the presence of air resistance turns Newton's second law into a differential equation, which is considered beyond the scope of an introductory physics course. In this chapter and the next, we'll develop numerical methods for solving differential equations, meaning that air resistance is not something we want to avoid; in fact, it showcases the power of our tools.

To develop a model of air resistance, let's think of the interaction between an object and the air around it as a collision. Suppose the object is moving with velocity v. In this section, v represents the real-valued, one-dimensional velocity of the object (a quantity with type `R`) and not the velocity function or the speed.

Let the cross-sectional area of the object be A and the density of air be ρ. We analyze the motion of the object over a small time interval dt. We assume that the initial velocity of the air is 0, and that the final velocity of the air is v (in other words, after the collision, the air is traveling at the same speed as the object).

The distance the object travels in time dt is $v\,dt$. The volume of air swept out by the object in time dt is $Av\,dt$. The mass of air disturbed by the object in time dt is $\rho Av\,dt$. The momentum imparted to the air by the object in time dt is the product of the mass of the air, $\rho Av\,dt$, and the change in velocity of the air, which is v, as we assume that the air starts from rest and ends the short time interval with velocity v. The momentum imparted to the air is $\rho Av^2\,dt$. The force felt by the air is this change in momentum per unit time, or ρAv^2. The force felt by the object from the air is equal and opposite to this following Newton's third law, which we will discuss in Chapter 19.

Our derivation was really quite approximate because we don't know that the air molecules really end up with velocity v, and we haven't even tried to account for the forces of air molecules on each other as the air compresses. Nevertheless, the form of our result is quite useful and approximately correct. Objects with different shapes respond a bit differently though, so it is useful to introduce a *drag coefficient* C to account for these differences. The drag coefficient is a dimensionless constant that is a property of the object that is flying through the air. It is also conventional to include a factor of $1/2$ so that the magnitude of the force of air resistance on the object is $C\rho Av^2/2$. This expression is never negative. We would prefer an expression in which the force is negative when the velocity is positive and positive when the velocity is negative. Our final expression for the one-dimensional force of air resistance is

$$F_{\text{air}}(v) = -\frac{1}{2}C\rho A\,|v|\,v \tag{14.9}$$

where the minus sign and the absolute value ensure that the force acts in a direction opposite the velocity. Air resistance is acting to slow the object. In Haskell, we'll write Equation 14.9 for air resistance as follows:

```
fAir :: R  -- drag coefficient
     -> R  -- air density
     -> R  -- cross-sectional area of object
     -> Velocity
     -> Force
fAir drag rho area v = -drag * rho * area * abs v * v / 2
```

In the mathematical notation of Equation 14.9, we're treating F_{air} as a function of one variable. The parameters C, ρ, and A are not listed explicitly as variables that F_{air} depends on. Eliding parameters like this is standard practice in physics, but in some sense it's an abuse of notation. In the Haskell notation, we must include all of the variables that the force of air resistance depends on. We list the three parameters first, before the `Velocity`, so that an expression like `fAir 1 1.225 0.6` is a fully legitimate function that takes only velocity as input. The function `fAir 1 1.225 0.6` has already chosen `drag = 1`, `rho = 1.225`, and `area = 0.6`.

With this brief foray into air resistance, and particularly the development of Equation 14.9, we're now ready to look at Newton's second law in the case where forces on our object depend only on its velocity.

Second Law with Forces That Depend Only on Velocity

The next situation for Newton's second law is when the net force depends on velocity but not on time or position. What we really mean here is that the forces do not depend *explicitly* on time. Velocity is a function that depends on time, and forces are allowed to depend on the velocity in this section, so there is a sense in which the forces depend on time. The constraint in this section is that the forces can depend on time *only through the velocity*.

The force functions may depend only on one variable, the velocity. We use F^v_j to denote the *j*th force function of one variable that gives force when supplied with velocity and we use F^v_{net} to denote the function of one variable that gives net force when supplied with velocity.

$$F^v_{\text{net}}(v_0) = \sum_j F^v_j(v_0)$$

We use v_0 as a local variable for velocity (type `R`) rather than v in this section because we want v to stand for the velocity function of our object (type `R -> R`).

Figure 14-6 shows a schematic diagram for Newton's second law with forces that depend only on velocity.

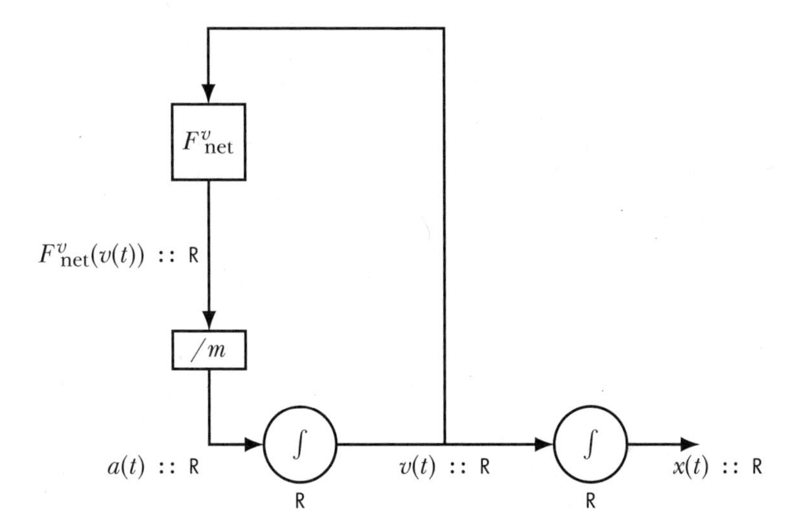

Figure 14-6: Newton's second law with forces that depend only on velocity

This diagram, unlike previous diagrams, contains a loop. The velocity produced by the integrator of acceleration serves as the input to the net force function. The loop in the diagram indicates that Newton's second law produces a differential equation. Because the loop contains one integrator, we get a first-order differential equation. A differential equation is a more difficult mathematical problem than a mere integral or antiderivative, as we had when forces depended only on time.

Newton's second law is given by the following equation:

$$\frac{d}{dt}\left[v(t)\right] = \frac{1}{m}\sum_j F^v_j(v(t)) \tag{14.10}$$

The information this equation represents is the same as the information in the schematic diagram of Figure 14-6. The equation describes how the rate of change of velocity depends on velocity itself through the forces that act on the object. The function `newtonSecondV`, presented next, is yet a third way to express Newton's second law; this function returns the rate of change of velocity when given the current value of velocity along with the forces that act on the object.

```
newtonSecondV :: Mass
              -> [Velocity -> Force]  -- list of force functions
              -> Velocity             -- current velocity
              -> R                    -- derivative of velocity
newtonSecondV m fs v0 = sum [f v0 | f <- fs] / m
```

We can integrate the acceleration to obtain the velocity.

$$v(t) = v(0) + \int_0^t a(t')\, dt' = v(0) + \frac{1}{m} \int_0^t F^v_{\text{net}}(v(t'))\, dt'$$

Unlike the case with time-dependent forces, we cannot simply perform the integral here because the velocity function we are trying to find appears under the integral. How to proceed?

To solve the differential equation, Equation 14.10, we will discretize time, which is something we have been doing with our numerical derivatives and integrals when we chose a time step. As long as our time step Δt is smaller than any important time scales in the situation we are addressing, the slope of the line connecting points $(t, v(t))$ and $(t + \Delta t, v(t + \Delta t))$ will be approximately equal to the derivative of velocity at time t.

$$\frac{v(t + \Delta t) - v(t)}{\Delta t} \approx \frac{dv(t)}{dt}$$

Rearranging this equation leads to the *Euler method* for solving a first-order differential equation.

$$v(t + \Delta t) \approx v(t) + \frac{dv(t)}{dt} \Delta t \tag{14.11}$$

The Euler method approximates the velocity at $t + \Delta t$ by the sum of the velocity at t and the product of the derivative at t with the time step Δt. The Euler method gives a way to find velocity at a later time from velocity at an earlier time if we know the derivative of velocity at the earlier time.

Figure 14-7 pictorially describes the Euler method for solving Newton's second law.

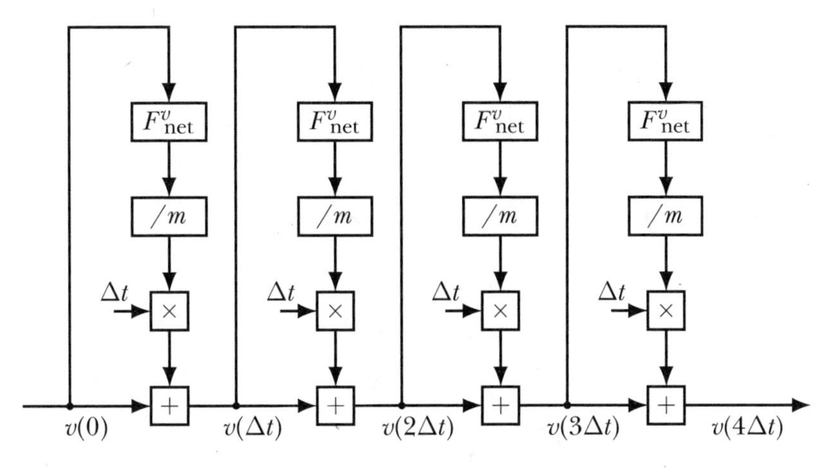

Figure 14-7: Euler method for Newton's second law in one dimension, for the special case in which net force depends only on velocity

The diagram shows how data is acted on by pure functions to compute the velocity of the object at different times. Because the diagram employs only pure functions (functions that do nothing but return an output from inputs and unchanging global values), we refer to this as a *functional diagram*. Whereas the schematic diagram in Figure 14-6 presents time as continuous, this diagram shows time as discrete. And whereas the schematic diagram has wires with values that are continuously changing in time, the functional diagram has wires with values that do not change. Different points in time have different wires in the functional diagram. While a schematic diagram may contain the stateful integrator from Figure 6-5, a functional diagram uncoils and replaces the integrator with a discrete, functional model like the one in Figure 6-7. We can see from Figure 14-7 that the same set of computations occurs at each time step to produce a new velocity from an old velocity. We call the set of computations that occurs at each time step the *velocity-update function*.

Figure 14-8 shows the velocity-update function, which is based on the application of the Euler method to one small time step.

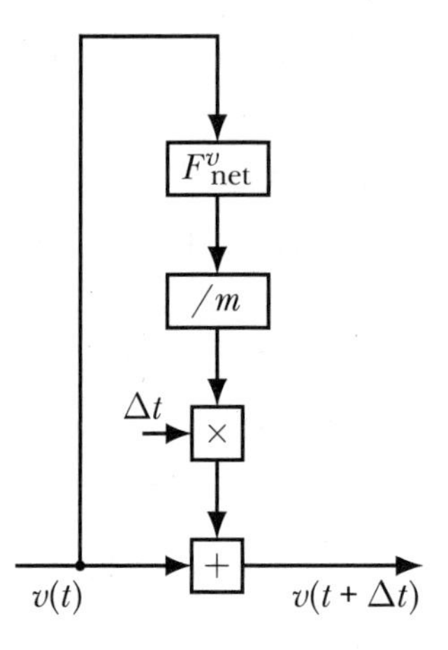

Figure 14-8: Velocity-update function used in the Euler method for solving Newton's second law with forces that depend only on velocity

Figure 14-8 shows a functional diagram for velocity update, visually describing how velocity at $t + \Delta t$ is computed from velocity at t and the forces.

Here is the velocity-update equation showing how a new velocity is obtained from an old velocity:

$$v(t + \Delta t) = v(t) + \frac{F^v_{\text{net}}(v(t))}{m} \Delta t \tag{14.12}$$

Lastly, we have the Haskell function `updateVelocity`, which advances the value of the velocity by one time step.

```
updateVelocity :: R                         -- time interval dt
               -> Mass
               -> [Velocity -> Force]  -- list of force functions
               -> Velocity              -- current velocity
               -> Velocity              -- new velocity
updateVelocity dt m fs v0
    = v0 + (newtonSecondV m fs v0) * dt
```

The functional diagram in Figure 14-8, the velocity-update equation (Equation 14.12), and the function `updateVelocity` express the same information in different forms, namely how to take one step in time with the Euler method.

Now we want to write a function `velocityFv`, similar to `velocityCF` and `velocityFt`, but for the case of forces that depend on velocity. To think of `updateVelocity` as a function that takes `Velocity` as input and gives `Velocity` as output, we want to think of the time step, mass, and list of force functions as parameters. The function `updateVelocity dt m fs` has type `Velocity -> Velocity` and plays the role of the iterable function `f` in Figure 6-4 on page 76.

```
velocityFv :: R                        -- time step
           -> Mass
```

```
          -> Velocity              -- initial velocity v(0)
          -> [Velocity -> Force]   -- list of force functions
          -> Time -> Velocity      -- velocity function
velocityFv dt m v0 fs t
    = let numSteps = abs $ round (t / dt)
      in iterate (updateVelocity dt m fs) v0 !! numSteps
```

We define a local variable `numSteps` to be the number of time steps we need to take to get as close as possible to the desired time `t`. We iterate the function `updateVelocity dt m fs`, starting at the initial velocity `v0`, and then select the single value of velocity from this infinite list that is closest to the desired time.

As an example of a situation with forces that depend only on velocity, let's consider a bicycle rider heading north on a flat, level road. We'll consider two forces in this situation. First, there is the northward force that the road exerts on the tires of the bicycle because the rider is working the pedals. Let us call this force F_{rider} (it is directly produced by the road on the bike, but it is indirectly produced by the rider), and assume that this force is a constant 100 N. Second, there is the southward force of air resistance that impedes the northward progress of the rider, especially when she is traveling quickly. We'll use the expression for air resistance that we developed in the previous section with Equation 14.9. The net force is

$$\begin{aligned} F^v_{\text{net}}(v_0) &= F_{\text{rider}} + F_{\text{air}}(v_0) \\ &= F_{\text{rider}} - \frac{1}{2} C \rho A \left| v_0 \right| v_0 \end{aligned}$$

Let's take the mass of the bike plus rider to be $m = 70$ kg. We'll choose a drag coefficient of $C = 2$, take the density of air to be $\rho = 1.225\ \text{kg/m}^3$, and approximate the cross-sectional area of bike and rider to be $0.6\ \text{m}^2$. Starting from rest, our mission is to find the velocity of the bike as a function of time.

Before we use our Haskell functions to investigate the motion of the bike, we're going to show how to use the Euler method by hand.

Euler Method by Hand

Let's use the Euler method by hand to compute the first several values of velocity for the bike. The purpose in doing this is to get a clear understanding of what is happening in the Euler method, so the code we write will be meaningful and not just a formal representation of some abstract vague process. We choose a time step of 0.5 s. Our mission is to complete the following table. We can fill in all of the time values because they are simply spaced at 0.5 s intervals. The initial velocity is 0, so we fill that in as well.

t (s)	v(t) (m/s)
0.0	0.0000
0.5	
1.0	
1.5	

We will complete the table by using Equation 14.12 to update the velocity over and over again. To compute the velocity at 0.5 s, we choose $t = 0$ in Equation 14.12.

$$
\begin{aligned}
v(0.5\text{ s}) &= v(0.0\text{ s}) + \frac{F^v_{\text{net}}(v(0.0\text{ s}))(0.5\text{ s})}{70\text{ kg}} \\
&= 0.0000\text{ m/s} + \frac{F^v_{\text{net}}(0.0000\text{ m/s})(0.5\text{ s})}{70\text{ kg}} \\
&= 0.0000\text{ m/s} + \frac{(100\text{ N})(0.5\text{ s})}{70\text{ kg}} \\
&= 0.7143\text{ m/s}
\end{aligned}
$$

We update the table with

t (s)	v(t) (m/s)
0.0	0.0000
0.5	0.7143
1.0	
1.5	

and then we calculate $v(1.0\text{ s})$ using Equation 14.12 with $t = 0.5$ s:

$$
\begin{aligned}
v(1.0\text{ s}) &= v(0.5\text{ s}) + \frac{F^v_{\text{net}}(v(0.5\text{ s}))(0.5\text{ s})}{70\text{ kg}} \\
&= 0.7413\text{ m/s} + \frac{F^v_{\text{net}}(0.7143\text{ m/s})(0.5\text{ s})}{70\text{ kg}} \\
&= 0.7413\text{ m/s} \\
&\quad + \frac{[100\text{ N} - (1)(1.225\text{ kg/m}^3)(0.6\text{ m}^2)(0.7143\text{ m/s})^2](0.5\text{ s})}{70\text{ kg}} \\
&= 1.4259\text{ m/s}
\end{aligned}
$$

We add this to the appropriate row of the table and continue.

$$
\begin{aligned}
v(1.5\text{ s}) &= v(1.0\text{ s}) + \frac{F^v_{\text{net}}(v(1.0\text{ s}))(0.5\text{ s})}{70\text{ kg}} \\
&= 1.4259\text{ m/s} + \frac{F^v_{\text{net}}(1.4259\text{ m/s})(0.5\text{ s})}{70\text{ kg}} \\
&= 1.4259\text{ m/s} \\
&\quad + \frac{[100\text{ N} - (1)(1.225\text{ kg/m}^3)(0.6\text{ m}^2)(1.4259\text{ m/s})^2](0.5\text{ s})}{70\text{ kg}} \\
&= 2.1295\text{ m/s}
\end{aligned}
$$

The completed table looks like this:

t (s)	v(t) (m/s)
0.0	0.0000
0.5	0.7143
1.0	1.4259
1.5	2.1295

Euler Method in Haskell

Now we'll use the `velocityFv` function to calculate velocity for the bike. Here is a velocity function for the bike with a time step of 1 s:

```
bikeVelocity :: Time -> Velocity
bikeVelocity = velocityFv 1 70 0 [const 100,fAir 2 1.225 0.6]
```

The higher-order function `const` can be used to make a constant function. The function `const 100` takes one input, ignores it, and returns 100 as output. It is equivalent to the anonymous function `\_ -> 100`. We're using it here to represent the constant force of 100 N.

Notice the data that must be supplied to solve the bike problem. We provide the 70-kg mass, the 0 m/s initial velocity of the bike, and the two forces: `const 100`, a constant force of 100 N, and `fAir 2 1.225 0.6`, which is the force of air resistance with a drag coefficient of 2, an air density of 1.225 kg/m^3, and a cross-sectional area of 0.6 m^2.

Here is the code to produce a graph of velocity versus time:

```
bikeGraph :: IO ()
bikeGraph = plotFunc [Title "Bike velocity"
                     ,XLabel "Time (s)"
                     ,YLabel "Velocity of Bike (m/s)"
                     ,PNG "BikeVelocity1.png"
                     ,Key Nothing
                     ] [0,0.5..60] bikeVelocity
```

The code plots the `bikeVelocity` function, including a title and axis labels, and makes a PNG file that can be included in another document. Figure 14-9 contains the graph itself.

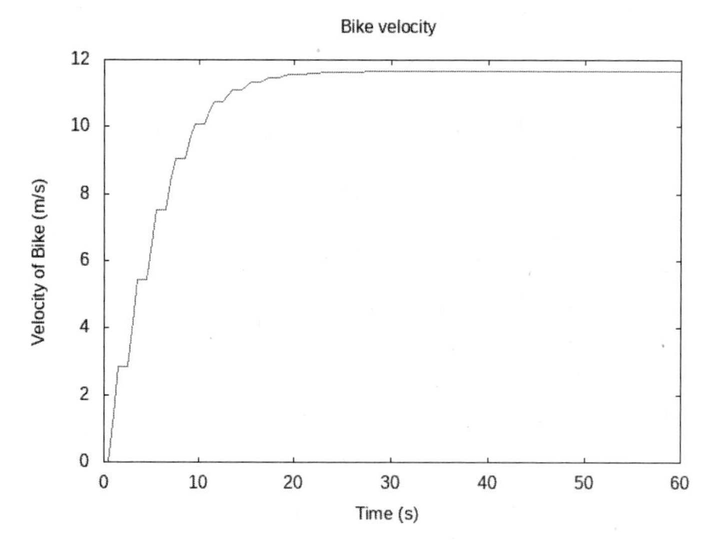

Figure 14-9: Bike velocity as a function of time. The stair-stepping look can be fixed and is discussed in the text.

A phenomenon occurs in Figure 14-9 that does not occur in constant acceleration situations: the establishment of a terminal velocity. After 20 seconds or so, the forward force of the road (from the pedaling) matches the backward force of the air. At this point we have no net force (or a very small net force), and the velocity stays at the terminal velocity.

Why the stair-stepping look to Figure 14-9? We used a time step of one second to do the calculation of the velocity function `bikeVelocity`, but then we asked the `plotFunc` function to give us a plot of that function every half a second. If we want a smooth plot, we have a couple of options. The simplest would be to ask for a plot with time values spaced at least one second apart. Alternatively, we could calculate the `bikeVelocity` function using a smaller time step. In any case, we shouldn't ask for more resolution in the graph than we asked for in the function we are graphing.

With the functions `velocityFv` and `positionFv`, the latter of which you are asked to write in Exercise 14.4, we have general-purpose tools for solving any Newton's second law type problem in one spatial dimension with forces that depend only on velocity. Before we turn to the case in which forces depend on both time and velocity, let's take a moment to view what we've just done from a broader perspective.

The State of a Physical System

A fruitful way to structure our thinking about Newton's second law, and also later about the Maxwell equations, revolves around the concept of the *state* of a physical system, which is the collection of information needed to say precisely what is going on with the system *at a particular instant of time*.

The state represents the current "state of affairs" of the system, containing enough information that future prediction can be based on the current state instead of past information about the system. The state evolves in time, changing according to some rule.

Given a physical system that we wish to understand, the state-based paradigm suggests the following conceptual division:

1. What information is required to specify the state of the system?
2. What is the state at some initial time?
3. By what rule does the state change with time?

When we treated Newton's second law with constant forces and forces that depend only on time, we did not use a state-based method because we did not need one. In those cases, we could use algebra or integration to find how the position and velocity of our object changed in time. When we looked at forces that depend on velocity, we had a schematic diagram with a loop that corresponded to a differential equation, shown in Figure 14-6. The state-based method is particularly useful for differential equations.

There are three things to notice about Figure 14-6 that relate to the state-based method. First, notice that there is one integrator in the loop and that this integrator holds the value of velocity as state. Second, notice that the differential equation, Equation 14.10, gives an expression for the rate of change of velocity. Lastly, notice that the forces depend on velocity. For these three reasons, in the case where forces depend only on velocity, the state of the object consists of the velocity of the object.

In general, the answer to question 1 is a data type. The state of an object experiencing forces that depend only on the object's velocity is a value of the data type `Velocity`. In the next section, where forces depend on time and velocity, the data type we will use for state is the pair `(Time,Velocity)`. As we consider more complex physical situations, the data type we use to hold the state of our physical system will contain more information.

Question 2 above is, in some sense, the smallest question. It may even be possible to do some analysis without an answer to question 2. But if we wish to know properties of a system at a later time, then we wish to know the state at a later time, and this typically requires knowing the state at some earlier time. The answer to question 2 is a value of the data type from question 1.

Question 3 requires a physical theory to answer. In the case of mechanics, Newton's second law gives the rule by which the state changes in time.

Let's see how the state-based method applies in the case where the forces on an object depend only on time and the velocity of the object.

Second Law with Forces That Depend on Time and Velocity

The next situation for Newton's second law is when the forces depend on both time and velocity but not on position. The force functions depend on two variables, time and velocity. We use F^{tv}_j to denote the jth function of two variables that gives a force when supplied with time and velocity, and we use F^{tv}_{net} to denote the function of two variables that gives net force when supplied with time and velocity.

$$F^{tv}_{net}(t, v_0) = \sum_j F^{tv}_j(t, v_0)$$

Figure 14-10 shows a schematic diagram for Newton's second law with forces that depend on time and velocity.

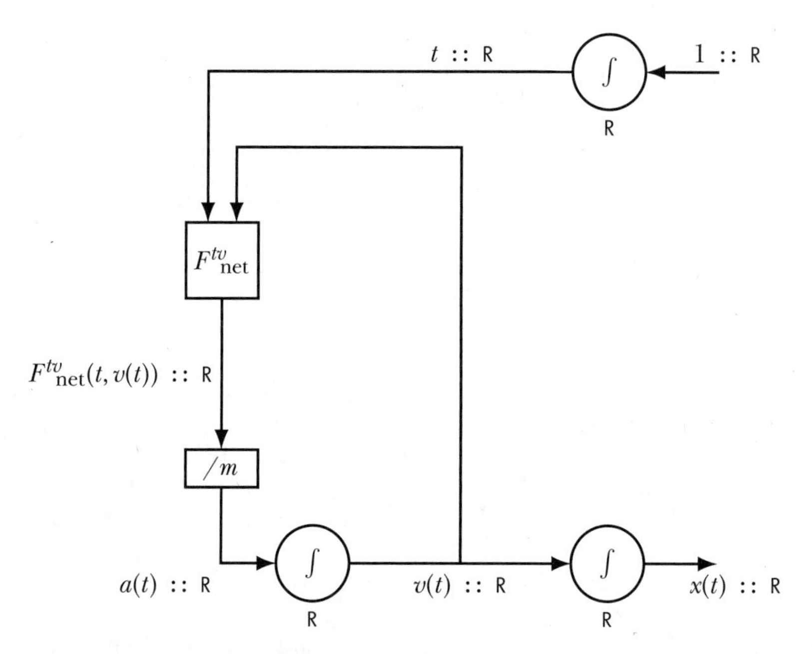

Figure 14-10: Newton's second law with forces that depend on time and velocity

The schematic diagram contains a loop, so Newton's second law is a differential equation, given in Equation 14.14.

$$\frac{d}{dt}[t] = 1 \tag{14.13}$$

$$\frac{d}{dt}[v(t)] = \frac{1}{m}\sum_j F^{tv}_j(t, v(t)) \tag{14.14}$$

Notice that there is one integrator in the loop in Figure 14-10, which holds the value of velocity as state. There is a way to solve this differential equation using only velocity as the state of the object. However, since the rate of change of velocity in Equation 14.14 depends on both time and velocity (because the forces depend on time and velocity), the state-based method is simpler to apply if we allow both time and velocity to be *state variables*. This is to say that the data type we will use for state is `(Time,Velocity)`. The difference between Equation 14.10, which expresses Newton's second law with forces that depend only on velocity, and Equation 14.14, which expresses Newton's second law with forces that depend on time and/or velocity, is simply that we need to know the current value of time in the latter case but not in the former. Including time in the state `(Time,Velocity)` is a simple way to gain convenient access to the current time.

Which quantities deserve to be called state variables? Say I have a particle in space acted on by a known (time-independent) force law. The state variables are the position and velocity because we can calculate the position and velocity at the next time instant from them. Why is acceleration not a state variable? To use the terminology of earlier sections in this chapter, state variables are numbers that identify a particular solution to the differential equation—they are the initial values that convert integrals into antiderivatives. Time is usually not considered a state variable, but taking it as one makes it easier to think about time-dependent forces. Readers interested in a more in-depth discussion of state variables and their uses are encouraged to see [16] and [17].

The Haskell function `newtonSecondTV`, shown below, expresses Newton's second law in the case where forces depend on time and velocity.

```
newtonSecondTV :: Mass
               -> [(Time,Velocity) -> Force]  -- force funcs
               -> (Time,Velocity)             -- current state
               -> (R,R)                       -- deriv of state
newtonSecondTV m fs (t,v0)
    = let fNet = sum [f (t,v0) | f <- fs]
          acc = fNet / m
      in (1,acc)
```

Given the mass of an object and a list of forces that act on the object, now expressed as functions of the state `(Time,Velocity)`, `newtonSecondTV` gives instructions for computing the time derivatives of the state variables from the state variables themselves. The return type `(R,R)` is meant to stand for time derivative of time, which is always the dimensionless number 1, and time derivative of velocity, which is acceleration. The acceleration is computed from Newton's second law by finding the net force and dividing by the mass.

To solve Equation 14.14, we will discretize time and use the Euler method. We'll continue to use Equation 14.11 for the Euler method. Figure 14-11 pictorially describes the Euler method for solving Newton's second law when forces depend on time and/or velocity.

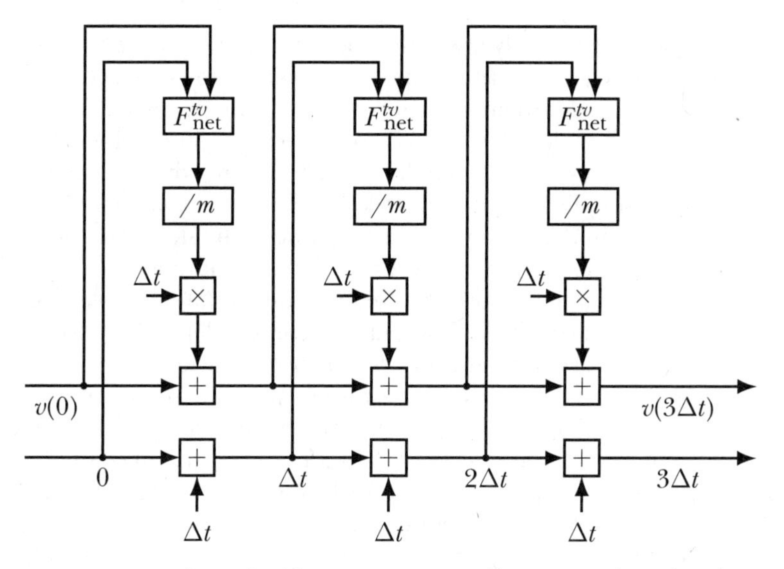

Figure 14-11: Euler method for Newton's second law in one dimension, for the special case in which net force depends only on time and/or velocity

The diagram shows how functions act on the state variables at one point in time to compute the state variables at the next point in time. The same set of computations reoccurs at each time step to produce a new state from an old state. We call the set of computations that occurs at each time step the *state-update function*.

The state-update function is shown pictorially in Figure 14-12. The figure shows a functional diagram for state update, visually describing how time and velocity at $t + \Delta t$ are computed from time and velocity at t, given the force functions.

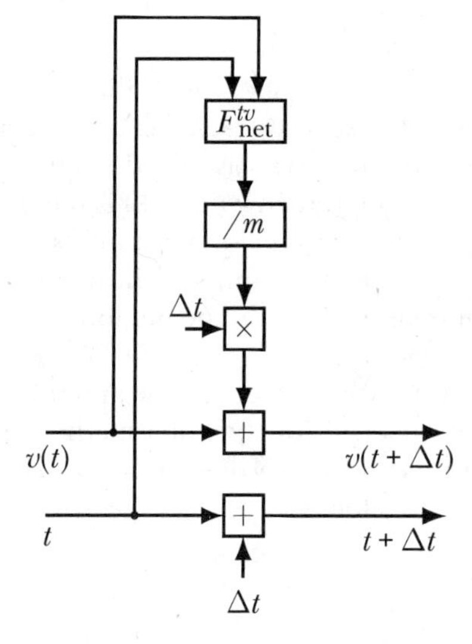

Figure 14-12: Euler method update for Newton's second law with forces that depend only on time and velocity

Here are the state-update equations showing how the new state variables are obtained from the old state variables:

$$t + \Delta t = t + 1\Delta t \tag{14.15}$$

$$v(t + \Delta t) = v(t) + \frac{F^{tv}_{\text{net}}(t, v(t))}{m}\Delta t \tag{14.16}$$

Equations 14.15 and 14.16 are state-update equations for an object exposed to forces that depend on time and velocity. The state-update equations tell us how the state variables time and velocity must be updated to advance to the next time step. The time update in Equation 14.15 is easy: we just add Δt to the old time to get the new time. To update the velocity in Equation 14.16, we compute an acceleration, multiply by a time step to get a change in velocity, and add that change to the old velocity. Applying these state-update equations is how we carry out the Euler method for solving a differential equation. This state-update procedure is the main tool we will use to solve problems in Newtonian mechanics.

The following Haskell function `updateTV`, named because it updates both time and velocity, advances the value of the state by one time step.

```
updateTV :: R                                 -- time interval dt
         -> Mass
         -> [(Time,Velocity) -> Force]        -- list of force funcs
         -> (Time,Velocity)                   -- current state
         -> (Time,Velocity)                   -- new state
updateTV dt m fs (t,v0)
    = let (dtdt, dvdt) = newtonSecondTV m fs (t,v0)
      in (t  + dtdt * dt
         ,v0 + dvdt * dt)
```

The function `updateTV` takes a few parameters and produces a function with type `(Time,Velocity) -> (Time,Velocity)`. The third input of `updateTV`, named `fs` with type `[(Time,Velocity) -> Force]`, could have been an input with type `[Time -> Velocity -> Force]`; it's a matter of style, and either choice will work just fine. Here I chose the former, as time and velocity are already paired in the function output.

The time-velocity pair we are passing around in this function represents the state of the object to which we are applying Newton's second law. The function `updateTV` is then an example of a state-update function. In an earlier section, when forces depended only on velocity, the velocity alone acted as state, and the function `updateVelocity` was the appropriate state-update function.

The functional diagram in Figure 14-12, Equations 14.15 and 14.16, and the function `updateTV` express the same information in different forms, namely how to take one step in time with the Euler method.

Depending on what we want to calculate, there are two things we might do with the `updateTV` function, corresponding to two types of representation of the time-velocity data. First, we may wish to produce a list of time-velocity pairs. Second, we may wish to produce velocity as a function of time. We'll develop functions for these two purposes in the next two subsections.

Method 1: Produce a List of States

A list of time-velocity pairs can be regarded as a solution to a Newton's second law problem with forces that depend on time and velocity because a time-velocity pair gives the state. The list of states contains a time-velocity pair for each time that has been probed by the Euler method in Figure 14-11. The function statesTV produces a list of time-velocity pairs when given a time step, a mass, an initial state, and a list of force functions.

```
statesTV :: R                                -- time step
         -> Mass
         -> (Time,Velocity)                  -- initial state
         -> [(Time,Velocity) -> Force]       -- list of force funcs
         -> [(Time,Velocity)]                -- infinite list of states
statesTV dt m tv0 fs
    = iterate (updateTV dt m fs) tv0
```

We use iterate to achieve the repeated composition in Figure 14-11. But which function do we want to iterate? It's not simply updateTV because updateTV takes three parameters as input before the time-velocity pair. The function we iterate must have type `a -> a`, or in this case `(Time,Velocity) -> (Time,Velocity)`. The solution is to give updateTV its first three parameters to form the function we send to iterate. The function we want to iterate is `updateTV dt m fs`, starting with the initial time-velocity pair tv0.

The function statesTV gives a general-purpose way of solving any Newton's second law type problem in one spatial dimension with forces that depend only on time and velocity. By a solution, we mean an infinite list of states (time-velocity pairs) of the object, spaced one time step apart from each other.

Method 2: Produce a Velocity Function

Now we want to write a function, velocityFtv, that is similar to velocityCF, velocityFt, and velocityFv, but for the case of forces that depend on time and velocity. We'll use the infinite list produced by statesTV, picking out the particular time-velocity pair that comes closest to our desired time and using the Prelude function snd to return the velocity, unpaired from the time.

```
velocityFtv :: R                             -- time step
            -> Mass
            -> (Time,Velocity)               -- initial state
            -> [(Time,Velocity) -> Force]    -- list of force funcs
            -> Time -> Velocity              -- velocity function
velocityFtv dt m tv0 fs t
    = let numSteps = abs $ round (t / dt)
      in snd $ statesTV dt m tv0 fs !! numSteps
```

With the functions `velocityFtv` and `positionFtv`, the latter of which you will be asked to write in Exercise 14.9, we have general-purpose ways of solving any Newton's second law type problem in one spatial dimension with forces that depend only on time and velocity. Let's now take a look at a situation that involves just such forces.

Example: Pedaling and Coasting with Air Resistance

As an example of a situation with forces that depend on time and velocity, let's reconsider our child bicycle rider who is pedaling and coasting, but now in the presence of air resistance. We'll consider two forces in this situation. First, there's the time-dependent force $F_{\text{pc}}(t)$ of pedaling from Equation 14.8. Second, there's the force of air resistance $F_{\text{air}}(v_0)$ that impedes the motion of the child, for which we'll use Equation 14.9. The net force is

$$F^{tv}_{\text{net}}(t, v_0) = F_{\text{pc}}(t) + F_{\text{air}}(v_0)$$

The mass of the bike plus child is $m = 20$ kg. We'll choose a drag coefficient of $C = 2$, take the density of air to be $\rho = 1.225\ \text{kg/m}^3$, and approximate the cross-sectional area of bike and rider to be $0.5\ \text{m}^2$. Starting from rest, our mission is to find the velocity of the bike as a function of time.

We update the velocity with Equation 14.16. Before we use our Haskell functions to investigate the motion of the bike, we'll show how to use the Euler method by hand.

Euler Method by Hand

Let's use the Euler method by hand to compute several values of velocity for the bike. Again, the purpose in doing the Euler method by hand is simply to get a clear picture of how the state variables get updated in the Euler method. We'll pick a time step of 6 s, even though this is too big to get accurate results, as it is not small compared to relevant time scales, such as the 20-second cycle time. We choose a time step of 6 s so we can sample both pedaling and coasting over the first few time steps. Our mission is to complete the following table. We can fill in all of the time values because they are simply spaced at six-second intervals. The initial velocity is 0, so we'll fill that in as well.

t (s)	v(t) (m/s)
0	0.0000
6	
12	
18	

The force of pedaling is either 10 N or 0 N, depending on the value of the time.

$$F_{\text{pc}}(0\ \text{s}) = F_{\text{pc}}(6\ \text{s}) = 10\ \text{N}$$
$$F_{\text{pc}}(12\ \text{s}) = F_{\text{pc}}(18\ \text{s}) = 0\ \text{N}$$

Repeatedly applying Equation 14.16, we obtain the following:

$$\begin{aligned}
v(6\text{ s}) &= v(0\text{ s}) + \frac{F_{\text{net}}^{tv}(0\text{ s}, v(0\text{ s}))(6\text{ s})}{20\text{ kg}} \\
&= 0.0000\text{ m/s} + \frac{(10\text{ N})(6\text{ s})}{20\text{ kg}} \\
&= 3.0000\text{ m/s} \\
v(12\text{ s}) &= v(6\text{ s}) + \frac{F_{\text{net}}^{tv}(6\text{ s}, v(6\text{ s}))(6\text{ s})}{20\text{ kg}} \\
&= 3.0000\text{ m/s} + \frac{[(10\text{ N}) - (1)(1.225\text{ kg/m}^3)(0.5\text{ m}^2)(3.0000\text{ m/s})^2](6\text{ s})}{20\text{ kg}} \\
&= 4.3463\text{ m/s} \\
v(18\text{ s}) &= v(12\text{ s}) + \frac{F_{\text{net}}^{tv}(12\text{ s}, v(12\text{ s}))(6\text{ s})}{20\text{ kg}} \\
&= 4.3463\text{ m/s} + \frac{[(0\text{ N}) - (1)(1.225\text{ kg/m}^3)(0.5\text{ m}^2)(4.3463\text{ m/s})^2](6\text{ s})}{20\text{ kg}} \\
&= 0.8752\text{ m/s}
\end{aligned}$$

The completed table looks like this:

t (s)	v(t) (m/s)
0	0.0000
6	3.0000
12	4.3463
18	0.8752

Let's turn now to Haskell, using each of the two methods we discussed earlier.

Method 1: Produce a List of States

Here we'll use the function `statesTV` to produce an infinite list of velocity-time pairs called `pedalCoastAir` for the child on the bike.

```
pedalCoastAir :: [(Time,Velocity)]
pedalCoastAir = statesTV 0.1 20 (0,0)
                [\(t,_) -> pedalCoast t
                ,\(_,v) -> fAir 2 1.225 0.5 v]
```

Notice the data that must be supplied to solve this problem. We provide a 0.1-s time step, the 20-kg mass, an initial state consisting of 0 for the time and 0 for the velocity, and the two forces, expressed here as anonymous functions. The function `pedalCoast` is a function only of time, so it cannot be listed directly as a force function because a force function for `statesTV` takes a time-velocity pair as input. The underscores are present because the pedaling function does not depend on the second item in the state, which happens to be velocity, and because air resistance does not depend on the first item in the state, which happens to be time.

A list of pairs is something we can plot with the plotPath function from the gnuplot library, but we need to truncate the list to a finite list before plotting, or plotPath will hang while trying to finish calculating an infinite list. In pedalCoastAirGraph below, we use the takeWhile function to extract the states with times less than or equal to 100 seconds.

```
pedalCoastAirGraph :: IO ()
pedalCoastAirGraph
    = plotPath [Title "Pedaling and coasting with air"
               ,XLabel "Time (s)"
               ,YLabel "Velocity of Bike (m/s)"
               ,PNG "pedalCoastAirGraph.png"
               ,Key Nothing
               ] (takeWhile (\(t,_) -> t <= 100)
                  pedalCoastAir)
```

This code produces Figure 14-13, which shows the velocity as a function of time for the child pedaling and coasting in the presence of air resistance.

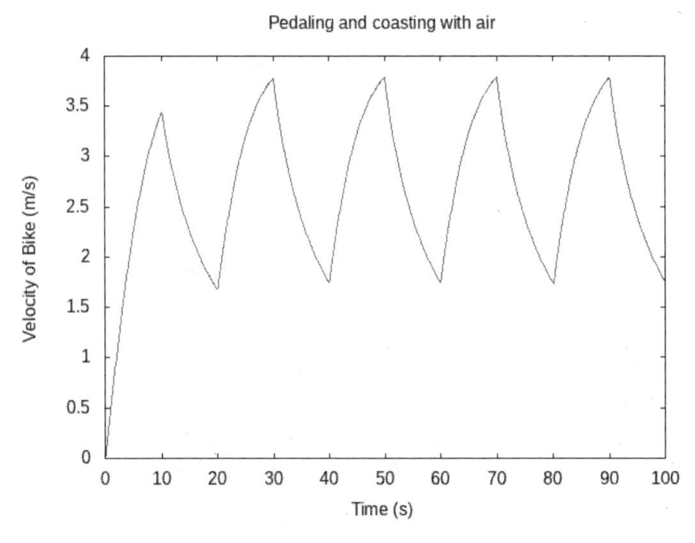

Figure 14-13: Pedaling and coasting with air resistance

As expected, the child's speed increases during the pedaling intervals and decreases during the coasting intervals.

Method 2: Produce a Velocity Function

Now let's use the function velocityFtv to produce a velocity function for the child on the bike.

```
pedalCoastAir2 :: Time -> Velocity
pedalCoastAir2 = velocityFtv 0.1 20 (0,0)
                 [\( t,_v) -> pedalCoast t
                 ,\(_t, v) -> fAir 1 1.225 0.5 v]
```

The data we give to `pedalCoastAir2` is the same data we gave to `pedalCoastAir`. Because `pedalCoastAir2` is a function `R -> R`, it can be plotted with the `plotFunc` function from the `gnuplot` package. It would produce the same graph as that in Figure 14-13.

Summary

This chapter discussed Newton's first law and introduced Newton's second law in the context of one-dimensional motion. The chapter presented a sequence of increasingly sophisticated settings for Newton's second law. Easiest among them is when the forces on an object are constant, that is, unchanging in time. Next is when the forces on an object depend only on time, in which case we can apply integration to find the velocity and the position of the object. Forces that depend on velocity, such as the air resistance introduced in this chapter, require that we solve a differential equation, which is a more complex task than integration. The chapter also introduced the Euler method for solving a first-order differential equation. The Euler method, along with Newton's second law, provides a rule for updating the state of the object we are tracking, allowing us to predict its future motion. The choice of state variables, or physical quantities contained in the state, is determined by what the forces depend on. If forces depend only on velocity, then velocity alone can serve as the particle state. If forces depend on time and velocity, we use time and velocity as state variables.

In the next chapter, we allow the forces to depend on position as well as time and velocity. This produces a second-order differential equation and requires that time, position, and velocity all be state variables.

Exercises

Exercise 14.1. Write a function `velocityCF'` that does the same thing and has the same type signature as `velocityCF`, but in which the time `t :: Time` is listed explicitly on the left of the equal sign in the definition.

```
velocityCF' :: Mass
            -> Velocity             -- initial velocity
            -> [Force]              -- list of forces
            -> Time -> Velocity     -- velocity function
velocityCF' m v0 fs t = undefined m v0 fs t
```

Exercise 14.2. Using the `positionCF` function, make a graph for the position of the car on the air track in Example 14.1 as a function of time. Assume the initial position of the car is −1 m.

Exercise 14.3. Write a function

```
sumF :: [R -> R] -> R -> R
sumF = undefined
```

that adds a list of functions to produce a function that represents the sum. Replace `undefined` with your code, and feel free to include one or two variables to the left of the equal sign in the definition. Using `sumF`, we could write the first line in the `let` construction of `velocityFt` as `fNet = sumF fs`.

Exercise 14.4. Write a Haskell function

```
positionFv :: R                        -- time step
           -> Mass
           -> Position                 -- initial position x(0)
           -> Velocity                 -- initial velocity v(0)
           -> [Velocity -> Force]      -- list of force functions
           -> Time -> Position         -- position function
positionFv = undefined
```

that returns a position function for a Newton's second law situation in which the forces depend only on the velocity. Replace the `undefined` with your code, and feel free to include variables to the left of the equal sign in the definition.

Exercise 14.5. Any Newton's second law problem that can be solved with `velocityFv` can also be solved with `velocityFtv`. Rewrite the `bikeVelocity` function so that it uses `velocityFtv` instead of `velocityFv`.

Exercise 14.6. Doing the Euler method by hand on page 225, we found the velocity after 1.5 s to be $v(1.5\text{ s}) = 2.1295\text{ m/s}$. Use the `velocityFv` function to calculate this same number.

Exercise 14.7. Doing the Euler method by hand on page 235, we found the velocity after 18 s to be $v(18\text{ s}) = 0.8752\text{ m/s}$. Use `statesTV` or `velocityFtv` to calculate this same number.

Exercise 14.8. Fix the stair-stepping issue in Figure 14-9 so that a smooth plot appears.

Exercise 14.9. Write a Haskell function

```
positionFtv :: R                        -- time step
            -> Mass
            -> Position                 -- initial position x(0)
            -> Velocity                 -- initial velocity v(0)
            -> [(Time,Velocity) -> Force]  -- force functions
            -> Time -> Position         -- position function
positionFtv = undefined
```

that returns a position function for a Newton's second law situation in which the forces depend only on time and velocity. Replace the `undefined` with your code, and feel free to include variables to the left of the equal sign in the definition.

Exercise 14.10. Produce a graph of position versus time for the situation in Figure 14-13.

Exercise 14.11. To deepen our understanding of the Euler method, we'll do a calculation by hand (using only a calculator and not the computer).

Consider a 1-kg mass exposed to two forces. The first force is an oscillatory force, pushing first one way and then the other. With t in seconds, the force in Newtons is given by

$$F_1(t) = 4\cos 2t$$

The second force is an air resistance force in Newtons, given by

$$F_2(v_0) = -3v_0$$

where v_0 is the current velocity of the mass in meters per second.

The net force is

$$F^{tv}_{\text{net}}(t, v_0) = F_1(t) + F_2(v_0) = 4\cos 2t - 3v_0$$

Suppose the mass is initially moving 2 m/s so that

$$v(0\text{ s}) = 2\text{ m/s}$$

Use the Euler method with a time step of $\Delta t = 0.1$ s to approximate the value of $v(0.3\text{ s})$. Keep at least four figures after the decimal point in your calculations. Show your calculations in a small table.

Exercise 14.12. Write a Haskell function

```
updateExample :: (Time,Velocity)  -- starting state
              -> (Time,Velocity)  -- ending state
updateExample = undefined
```

that takes a time-velocity pair (t_0, v_0) and returns an updated time-velocity pair (t_1, v_1) for a single step of the Euler method for a 1-kg object experiencing a net force of

$$F^{tv}_{\text{net}}(t, v_0) = F_1(t) + F_2(v_0) = 4\cos 2t - 3v_0$$

Use a time step of $\Delta t = 0.1$ s. Show how to use the function `updateExample` to calculate the value $v(0.3\text{ s})$ that you calculated by hand in Exercise 14.11.

Exercise 14.13. Consider a 1-kg object experiencing a net force

$$F^{tv}_{\text{net}}(t, v_0) = -\alpha v_0$$

where $\alpha = 1$ N s/m, subject to the initial condition $v(0\text{ s}) = 8$ m/s. Use the Euler method to find the velocity of the object over the time interval $0\text{ s} \leq t \leq 10$ s. Plot velocity as a function of time to see what it looks like. Compare your results to the exact solution:

$$v(t) = (8\text{ m/s})e^{-\alpha t}$$

Try out different time steps to see what happens when the time step gets too big.

Find a time step that is small enough that the Euler solution and the exact solution nicely overlap on a plot. Find another time step that is big enough that you can see the difference between the Euler solution and the exact solution on a plot.

Make a nice plot (with title, axis labels, and so on) with these three solutions on a single graph (bad Euler, good Euler, and exact). Label the Euler results with the time step you used and label the exact result "Exact."

Exercise 14.14. Consider the differential equation

$$\frac{dv(t)}{dt} = \cos(t + v(t))$$

subject to the initial condition $v(0) = 0$. This differential equation has no exact solution. Use the Euler method with a step size of $\Delta t = 0.01$ to find $v(t)$ over the interval $0 \leq t \leq 3$. Make a nice plot of the resulting function and include the value $v(3)$ to five significant figures.

Exercise 14.15. Each wire in a functional diagram can be labeled with a type. Label each wire in Figure 14-11 with a type.

15

MECHANICS IN ONE DIMENSION

In this chapter, we'll complete the story of mechanics in one dimension by developing tools for situations in which the forces on an object can depend on position as well as on time and velocity. This will require thinking about second-order differential equations, which we'll convert into systems of first-order differential equations.

As before, we'll transform information about the physical situation through several different forms, beginning with the object's mass and the forces that act on it, and ending with functions that give the position and velocity of the object with time. A functional language like Haskell helps organize our thoughts about what it means to solve a mechanics problem by giving names and types to each form of information and by allowing that information to exist naturally in the form of a function when appropriate.

To apply the tools we're developing, we'll look at an example situation: the motion of a Ping-Pong ball attached to the end of a Slinky in the presence of air resistance—in other words, a damped harmonic oscillator. We'll show how to apply the Euler method for a second-order differential equation, and we'll see how either a list of states or position and velocity functions for an object can be regarded as a solution to a mechanics problem. We'll then introduce the Euler-Cromer method, an improvement on the

Euler method for second-order differential equations. Since we'll continue to modify the state variables and the data type used for state as we introduce vectors and multiple particles in upcoming chapters, at the end of this chapter, we'll generalize our differential equation solving methods to allow arbitrary data types for state and allow a choice of numerical methods.

Introductory Code

Let's begin with the code that needs to appear at the beginning of our source code file. I always like to start by turning on warnings. I also include two language settings that we will use in "Solving Differential Equations" later in this chapter.

```
{-# OPTIONS_GHC -Wall #-}
{-# LANGUAGE FlexibleInstances, MultiParamTypeClasses #-}
```

Let's make the code in this chapter into a module named `Mechanics1D`.

```
module Mechanics1D where
```

We'll also want to make a plot later, so let's import the necessary module.

```
import Graphics.Gnuplot.Simple
```

For our example of a Ping-Pong ball on the end of a Slinky, we'll use the air resistance function `fAir` from the previous chapter. To access this function, we import the module called `Newton2` we wrote in Chapter 14.

```
import Newton2 ( fAir )
```

The Haskell compiler will look for a file called *Newton2.hs* in the current working directory (the same directory that contains the *Mechanics1D.hs* file we are writing in this chapter). The file *Newton2.hs* contains the code we wrote in the previous chapter and is available at *https://lpfp.io*.

If we include a comma-separated list of types and functions in parentheses after the module name, we'll import only those types and functions. If we omit such a list, as we did when importing the `Graphics.Gnuplot.Simple` module, we'll import all types and functions provided by the module.

As usual, we'll use the type synonym `R` instead of `Double`. We made this type synonym in the `SimpleVec` module of Chapter 10, and we import it from there.

```
import SimpleVec ( R )
```

The warnings option, the language setting, the module name, and the import statements need to be at the beginning of the source code file. Type synonyms can appear anywhere.

When we do mechanics in one dimension, time, a time step, mass, position, velocity, and force are each represented by a real number.

```
type Time     = R
type TimeStep = R
type Mass     = R
type Position = R
type Velocity = R
type Force    = R
```

Forces That Depend on Time, Position, and Velocity

When the forces on an object depend on time, position, and velocity, the force functions depend on three variables. We'll use F_j^{txv} to denote the jth function of three variables that gives a force when supplied with time, position, and velocity; we'll use F_{net}^{txv} to denote the function of three variables that gives net force. Figure 15-1 shows a schematic diagram for Newton's second law with forces that depend on time, position, and velocity.

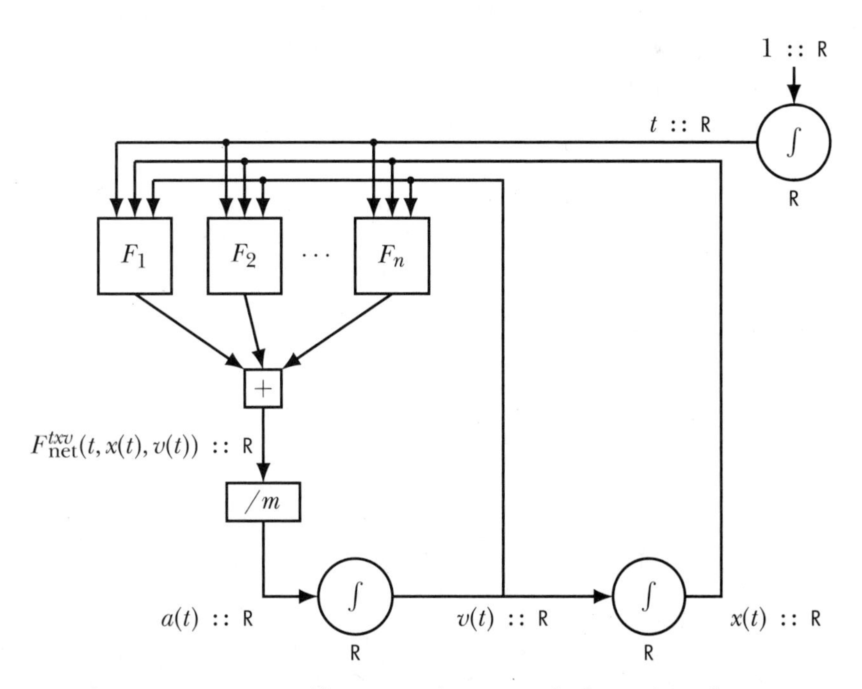

Figure 15-1: Newton's second law in one dimension. The forces depend on time, position, and velocity.

The rectangular boxes in the diagram are pure functions whose output depends only on the input. The integrators are contained in circles to remind us that each integrator contains some state. The integrator whose output is $x(t)$, for example, must contain the current position value. The schematic diagram in Figure 15-1 is continuous and stateful.

Newton's second law appears as the following differential equation:

$$F_{\text{net}}^{txv}(t, x(t), Dx(t)) = mD^2x(t) \tag{15.1}$$

We've come to the full generality of Newton's second law in one dimension, as given in the previous chapter by Equation 14.2. Position is the unknown function we want to find in this differential equation. This is a second-order differential equation because a second derivative of position appears in the equation. To solve a second-order differential equation using a state-based approach, we'll choose some state variables and write first-order differential equations for each of the state variables.

What should our state variables be? Figure 15-1 has two integrators contained in loops. These integrators hold position and velocity as state, so position and velocity are required to be state variables. While it is possible to solve Equation 15.1 using only position and velocity as state variables, it is easier if we also allow time to serve as a state variable because the forces could also depend on time. Our data type describing state for the mechanics of a single object moving in one dimension is a time-position-velocity triple.

```
type State1D = (Time,Position,Velocity)
```

We use the name *state space* for data types like `State1D` that describe state. The types `Velocity` and `(Time,Velocity)` from the previous chapter are other examples of state spaces.

Equations 15.2, 15.3, and 15.4 show the three first-order differential equations obtained by writing expressions for the time derivatives of each state variable. What appears on the right side of these equations can involve the state variables but may not have any derivatives. Because the derivative of the position function depends on the velocity function, and vice versa, we refer to this as a set of *coupled* differential equations.

$$\frac{d}{dt}\left[t\right] = 1 \tag{15.2}$$

$$\frac{d}{dt}\left[x(t)\right] = v(t) \tag{15.3}$$

$$\frac{d}{dt}\left[v(t)\right] = \frac{1}{m}\sum_j F_j^{txv}(t, x(t), v(t)) \tag{15.4}$$

The function `newtonSecond1D` gives expressions for the time derivatives of the state variables in terms of the state variables themselves. Note that the state variables in this function are real numbers rather than functions of time. By expressing the differential Equations 15.2, 15.3, and 15.4, the function `newtonSecond1D` expresses Newton's second law in one dimension. The schematic diagram in Figure 15-1, the differential equations, and the Haskell function `newtonSecond1D` each contain the essential information for understanding and solving Newton's second law in one dimension.

```
newtonSecond1D :: Mass
               -> [State1D -> Force]  -- force funcs
               -> State1D             -- current state
               -> (R,R,R)             -- deriv of state
newtonSecond1D m fs (t,x0,v0)
    = let fNet = sum [f (t,x0,v0) | f <- fs]
          acc = fNet / m
      in (1,v0,acc)
```

A General Strategy for Solving Mechanics Problems

Our strategy for constructing and solving Newton's second law consists of transforming information about the physical situation through a sequence of five different forms:

1. Mass and force functions
2. Differential equation
3. State-update function
4. List of states
5. Position and velocity functions

The information begins with the mass of the object under consideration and the forces that act on it, expressed as functions of the state variables. The function `newtonSecond1D` transforms this mass and force information into a differential equation. A differential equation is a function `State1D -> (R,R,R)` that gives the derivatives of the state variables time, position, and velocity in terms of the state variables themselves. The Euler method transforms a differential equation into a state-update function, a function `State1D -> State1D` that computes the state variables at a later time from the state variables at an earlier time. From a state-update function and an initial state, we can then compute an infinite list of states, the fourth representation of the data describing our physical situation. Finally, we can extract position as a function of time and velocity as a function of time from the list of states. Predicting the position and velocity of the object as a function of time is the final data representation, which we regard as a solution to the problem of understanding the motion of an object.

Figure 15-2 is a functional diagram showing the five data representations above and the functions that transform the data from one representation to another.

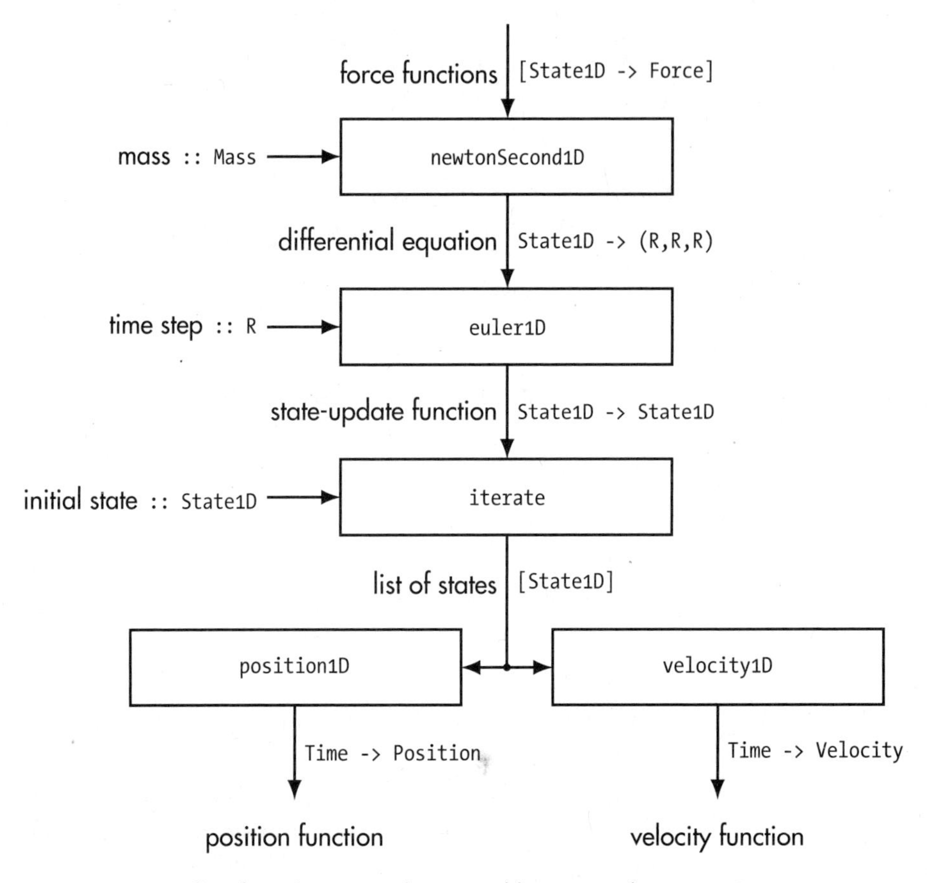

Figure 15-2: Data flow for solving a mechanics problem in one dimension. Functions transform the data through a sequence of five representations.

We have already discussed how `newtonSecond1D` transforms mass and force data into a differential equation. We'll discuss the other transformations in detail later. Figure 15-2 is the first of several figures in this chapter and the next few chapters that serve to give an overview of the process of solving a mechanics problem. As we expand and generalize the ideas of state and numerical method, these figures will become more general, more streamlined, and a bit more abstract. Their purpose, central to a deep understanding of Newtonian mechanics, is to suggest a high-level way of thinking about the meanings and solution techniques of Newton's laws. The figures organize the many functions we've written into a coherent, operational description of how to make predictions with Newtonian mechanics. Understanding what it takes to solve Newton's second law is a prerequisite to a more profound insight into what the law means.

Note that there are two places in Figure 15-2 where additional information comes in. A time step is required for the Euler method that transforms a differential equation into a state-update function, and an initial state is required to transform a state-update function into a list of states. In this figure,

I regard the object's mass, along with the forces, to be part of the initial information that specifies the problem.

Solving with Euler's Method

To solve differential Equations 15.2, 15.3, and 15.4, we will discretize time, choosing a time step Δt that is smaller than any important time scales in the problem we are solving. We'll use the Euler method, which approximates the time-slope of each state variable over the course of one time step by the time derivative of that variable at the beginning of the time step. Figure 15-3 shows the Euler method for solving Newton's second law in one dimension.

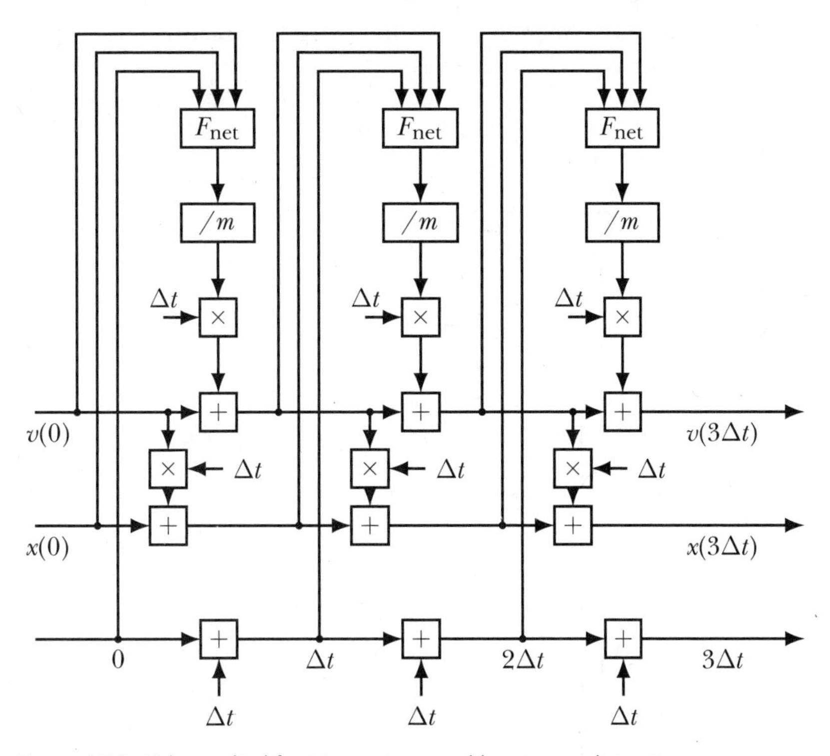

Figure 15-3: Euler method for Newton's second law in one dimension

Unlike the schematic diagram of Figure 15-1, with values that continuously change with time, the functional diagram in Figure 15-3 has values that do not change. Change in a quantity, such as position, over time is represented by a sequence of values on different wires.

The function `euler1D`, shown below, transforms a differential equation into a state-update function, as shown in Figure 15-2. To do this, it takes a time step in addition to the differential equation as input. Each state variable is updated by changing its value by its derivative (calculated from the differential equation) multiplied by the time step.

```
euler1D :: R                            -- time step dt
        -> (State1D -> (R,R,R))  -- differential equation
        -> State1D -> State1D    -- state-update function
euler1D dt deriv (t0,x0,v0)
    = let (_, _, dvdt) = deriv (t0,x0,v0)
          t1 = t0 + dt
          x1 = x0 + v0 * dt
          v1 = v0 + dvdt * dt
      in (t1,x1,v1)
```

Figure 15-3 composes a state-update function over and over again. We call the state-update function `updateTXV` because it updates time, position, and velocity. As before, let's go through three expressions of this state-update function. Figure 15-4 shows the state-update function as a functional diagram, indicating how a new time-position-velocity triple is formed from an old triple.

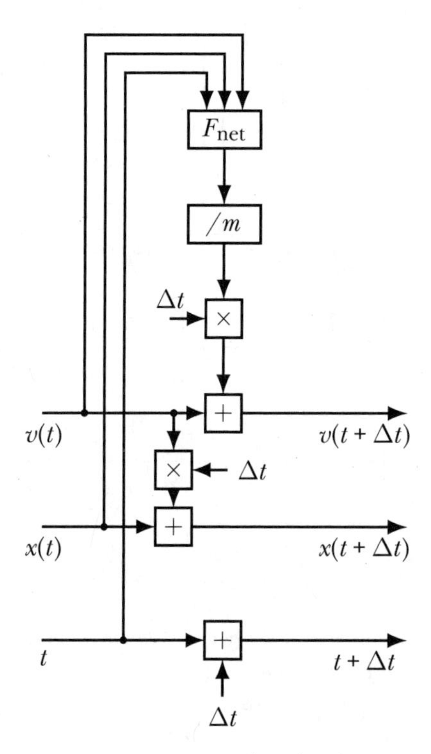

Figure 15-4: How to update the state variables time, position, and velocity when a small time interval Δt has elapsed

Now we'll give state-update equations in mathematical notation. The state-update equations tell us how the state variables time, position, and velocity must be updated to advance to the next time step.

$$t + \Delta t = t + 1\Delta t \tag{15.5}$$

$$x(t + \Delta t) = x(t) + v(t)\Delta t \tag{15.6}$$

$$v(t + \Delta t) = v(t) + \frac{F_{\text{net}}^{txv}(t, x(t), v(t))}{m}\Delta t \tag{15.7}$$

Finally, we define the Haskell function `updateTXV`, which produces a new state triple from an old state triple.

```
updateTXV :: R                       -- time interval dt
          -> Mass
          -> [State1D -> Force]  -- list of force funcs
          -> State1D -> State1D  -- state-update function
updateTXV dt m fs = euler1D dt (newtonSecond1D m fs)
```

Note that `updateTXV` is essentially the composition of `newtonSecond1D` and `euler1D`, so it transforms the mass and force data (the first of our five data representations) into a state-update function (the third of the five).

Producing a List of States

The next step in the data flow of Figure 15-2 is to produce an infinite list of states from the state-update function and an initial state. The Haskell Prelude function `iterate` can do this, essentially transforming Figure 15-4 into Figure 15-3. A list of time-position-velocity triples can be regarded as a solution to a Newton's second law problem. The list of states contains a time-position-velocity triple for every time that has been probed by the Euler method of Figure 15-3. The function `statesTXV` produces a list of states when given a time step, a mass, an initial state, and a list of force functions.

```
statesTXV :: R                       -- time step
          -> Mass
          -> State1D             -- initial state
          -> [State1D -> Force]  -- list of force funcs
          -> [State1D]           -- infinite list of states
statesTXV dt m txv0 fs = iterate (updateTXV dt m fs) txv0
```

All we are doing here is iterating the update function to produce an infinite list of states. Note that we need to pass the time step, mass, and list of forces as parameters to `updateTXV` before it can serve as an iterable function passable to `iterate`. Iterate requires a function with type `a -> a` that can be applied again and again.

The function `statesTXV` transforms the mass and force data (the first of our five data representations), along with a time step and an initial state, into an infinite list of states (the fourth of our five representations). With this function, we have a general-purpose way of solving any Newton's second law type problem in one spatial dimension. By a solution, we mean an infinite list of states (time-position-velocity triples) of the object, spaced one time step apart from each other. From this infinite list of states, we can extract whatever data we are most interested in and graph it or animate it.

Position and Velocity Functions

For our fifth and last data representation, we wish to write a function velocityFtxv, similar to velocityCF, velocityFt, velocityFv, and velocityFtv, but for the case of forces that depend on time, position, and velocity. This function will transform the mass and force data (the first of our five data representations), along with a time step and an initial state, into a velocity function (part of the fifth of our five representations).

To help us do this, we want to write a function velocity1D that transforms from an infinite list of states (the fourth of our five data representations) into a velocity function.

```
-- assume that dt is the same between adjacent pairs
velocity1D :: [State1D]             -- infinite list
           -> Time -> Velocity      -- velocity function
velocity1D sts t
    = let (t0,_,_) = sts !! 0
          (t1,_,_) = sts !! 1
          dt = t1 - t0
          numSteps = abs $ round (t / dt)
          (_,_,v0) = sts !! numSteps
      in v0
```

We name the infinite list of states sts and the desired time t. We assume that the time step stays the same throughout the infinite list of states, and we calculate the time step from the time values in the first two states of the list. In the first two lines of the let clause, we use the list index operator (!!) to pick out the first and second states of the list. Since State1D is a type synonym for (R,R,R), we use pattern matching to pick out the time from the state. The local variables t0 and t1 are the times for the first two states in the list.

The third line in the let clause defines a local variable dt for the time step, calculated as the difference between the times of the first and second states. The fourth line in the let clause computes the number of time steps required to reach the state that is closest to the desired time, and it names this number of steps with the local variable numSteps. The fifth line in the let clause uses the list index operator to pick out the state closest to the desired time and then names the velocity of that state v0 using pattern matching.

All of the real work in this function is done in the let clause. The function returns v0, which is the velocity that occurs closest to the desired time t.

For the function velocityFtxv, which does the full transformation, we'll use the infinite list produced by statesTXV, in conjunction with velocity1D.

```
velocityFtxv :: R                    -- time step
             -> Mass
```

```
             -> State1D              -- initial state
             -> [State1D -> Force]  -- list of force funcs
             -> Time -> Velocity    -- velocity function
velocityFtxv dt m txv0 fs = velocity1D (statesTXV dt m txv0 fs)
```

Once we have a velocity function, we could integrate to get a position function, but all of this position information is contained in our list of states, so we might as well extract the position function from that using `position1D`, which is very similar to `velocity1D`.

```
-- assume that dt is the same between adjacent pairs
position1D :: [State1D]             -- infinite list
           -> Time -> Position      -- position function
position1D sts t
    = let (t0,_,_) = sts !! 0
          (t1,_,_) = sts !! 1
          dt = t1 - t0
          numSteps = abs $ round (t / dt)
          (_,x0,_) = sts !! numSteps
      in x0
```

Here is the function `positionFtxv`, which makes the full transition from our initial data representation in terms of mass and forces, along with a time step and an initial state, into our final representation giving the position of our object as a function of time.

```
positionFtxv :: R                   -- time step
             -> Mass
             -> State1D              -- initial state
             -> [State1D -> Force]  -- list of force funcs
             -> Time -> Position    -- position function
positionFtxv dt m txv0 fs = position1D (statesTXV dt m txv0 fs)
```

With the functions `velocityFtxv` and `positionFtxv`, we have a general purpose way of solving any Newton's second law problem in one spatial dimension.

Next let's look at an example of where to use this technology: a Ping-Pong ball oscillating at the end of a Slinky. The restoring force of the Slinky on the ball depends on the position of the Ping-Pong ball, and air resistance depends on the velocity of the ball.

A Damped Harmonic Oscillator

As an example of a situation with forces that depend on position and velocity, let's consider a damped harmonic oscillator. In particular, let's consider a Ping-Pong ball oscillating at the end of a vertically hanging Slinky. A Slinky is a spring, made of metal or plastic, that is sold as a child's toy. We'll choose a coordinate system where up is positive and choose the zero of position to be the place where the lower end of the Slinky hangs without the Ping-Pong ball attached. The Ping-Pong ball has a mass of 2.7 g and a radius of 2 cm.

We'll include three forces on the ball, all acting vertically. The first force is produced by the spring and acts to restore the mass toward the equilibrium position. The spring force is given by *Hooke's law*,

$$F_{\text{spring}}(x_0) = -kx_0 \tag{15.8}$$

which claims that the force produced by the spring is proportional to the displacement x_0 of the mass from its equilibrium position. The constant k is called the *spring constant* of the spring. A spring with a large spring constant is stiff and takes lots of force to extend or compress. The minus sign makes the spring force a *restoring force*, which acts in a direction to restore the object to its equilibrium position.

The equilibrium position is $x_0 = 0$. If x_0 is positive, then $F_{\text{spring}}(x_0)$ is negative, and the force acts toward the equilibrium position. If x_0 is negative, then $F_{\text{spring}}(x_0)$ is positive, and again the force acts toward the equilibrium position.

The spring force depends only on the ball's position and not on its velocity or on the time. But knowing that we want to use a function like `statesTXV` or `positionFtxv`, which requires a list of forces `State1D -> Force` given as functions of the state, we'll write Hooke's law so that it has this type.

```
springForce :: R -> State1D -> Force
springForce k (_,x0,_) = -k * x0
```

The second force is the force of air resistance, a force that damps the natural oscillation of the ball on the spring. We'll use Equation 14.9 for air resistance, repeated here.

$$F_{\text{air}}(v) = -\frac{1}{2}C\rho A\,|v|\,v$$

We use a drag coefficient of 2.

The third force is the force of gravity acting on the ball. Near Earth's surface, the force of gravity on an object with mass m is

$$F_g = -mg$$

where $g = 9.80665 \text{ m/s}^2$ is the acceleration of gravity, and where we employ a coordinate system in which "away from Earth's center" is the positive direction.

We take the Slinky to have a spring constant of 0.8 kg/s^2. We'll release the ball at a position 10 cm above the equilibrium position so that $x(0 \text{ s}) = 0.1 \text{ m}$ and $v(0 \text{ s}) = 0 \text{ m/s}$.

Before we use our Haskell functions to investigate the motion of the ball, let's investigate it using Euler's method by hand.

Euler Method by Hand

We want to have a clear mental picture of what the computer is doing when it applies the Euler method. To this end, we will calculate a few steps of the

Euler method by hand (meaning with a calculator) to see in detail what is happening.

We'll use the state-update Equations 15.6 and 15.7 with a time step of Δt = 0.1 s. A time step of 0.1 seconds is too large to get accurate results for this problem, but it serves to show the essence of the Euler method.

Our mission is to complete the following table. We can fill in all of the time values because they are simply spaced at 0.1 s intervals. We'll also fill in the initial values of position and velocity.

t (s)	x(t) (m)	v(t) (m/s)
0.0	0.1000	0.0000
0.1		
0.2		
0.3		

The net force on the Ping-Pong ball is given by the following expression:

$$F_{\text{net}}^{txv}(t, x_0, v_0) = -kx_0 - \frac{1}{2}C\rho A\left|v_0\right|v_0 - mg$$

Using t = 0.0 s, x(0.0 s) = 0.1000 m and v(0.0 s) = 0.0000 m/s (in other words, the information in the first row of our table) in the state-update Equations 15.6 and 15.7, we can find x(0.1 s) and v(0.1 s) (the information in the second row of our table). The state-update equations are exactly what we need to produce a new row of our table from an existing row.

$$\begin{aligned}
x(0.1\text{ s}) &= x(0\text{ s}) + v(0\text{ s})\Delta t \\
&= 0.1000\text{ m} + (0.0000\text{ m/s})(0.1\text{ s}) \\
&= 0.1000\text{ m}
\end{aligned}$$

$$\begin{aligned}
v(0.1\text{ s}) &= v(0\text{ s}) + \frac{F_{\text{net}}^{txv}(0\text{ s}, x(0\text{ s}), v(0\text{ s}))}{m}\Delta t \\
&= 0.0000\text{ m/s} + \frac{(-0.106478\text{ N})(0.1\text{ s})}{0.0027\text{ kg}} \\
&= -3.9436\text{ m/s}
\end{aligned}$$

Using t = 0.1 s, x(0.1 s) = 0.1000 m, and v(0.1 s) = −3.9436 m/s (the information in the second row of our table) in the state-update equations, we can find x(0.2 s) and v(0.2 s) (the information in the third row).

$$\begin{aligned}
x(0.2\text{ s}) &= x(0.1\text{ s}) + v(0.1\text{ s})\Delta t \\
&= 0.1000\text{ m} + (-3.9436\text{ m/s})(0.1\text{ s}) \\
&= -0.2944\text{ m}
\end{aligned}$$

$$\begin{aligned}
v(0.2\text{ s}) &= v(0.1\text{ s}) + \frac{F_{\text{net}}^{txv}(0.1\text{ s}, x(0.1\text{ s}), v(0.1\text{ s}))}{m}\Delta t \\
&= -3.9436\text{ m/s} + \frac{(-0.082538\text{ N})(0.1\text{ s})}{0.0027\text{ kg}} \\
&= -7.0005\text{ m/s}
\end{aligned}$$

Using $t = 0.2$ s, $x(0.2 \text{ s}) = -0.2944$ m, and $v(0.2 \text{ s}) = -7.0005$ m/s (the information in the third row of our table), we can find $x(0.3 \text{ s})$ and $v(0.3 \text{ s})$. (the information in the fourth row).

$$
\begin{aligned}
x(0.3 \text{ s}) &= x(0.2 \text{ s}) + v(0.2 \text{ s})\Delta t \\
&= -0.2944 \text{ m} + (-7.0005 \text{ m/s})(0.1 \text{ s}) \\
&= -0.9945 \text{ m} \\
v(0.3 \text{ s}) &= v(0.2 \text{ s}) + \frac{F_{\text{net}}^{txv}(0.2 \text{ s}, x(0.2 \text{ s}), v(0.2 \text{ s}))}{m}\Delta t \\
&= -7.0005 \text{ m/s} + \frac{(0.284482 \text{ N})(0.1 \text{ s})}{0.0027 \text{ kg}} \\
&= 3.5359 \text{ m/s}
\end{aligned}
$$

The completed table looks as follows:

t (s)	x(t) (m)	v(t) (m/s)
0.0	0.1000	0.0000
0.1	0.1000	–3.9436
0.2	–0.2944	–7.0005
0.3	–0.9945	3.5359

Since states are time-position-velocity triples, this completed table contains the first four states of the Ping-Pong ball for the Euler method with a time step of 0.1 s. If we ask the computer to produce an infinite list of states using our function statesTXV, the first four states should look like this. We can imagine that the process we used to complete this table is what the computer is doing over and over again to produce a list of states for us.

Let's now turn to the technology we developed earlier in the chapter to find the position of the Ping-Pong ball as a function of time.

Method 1: Producing a List of States

Here we'll use the statesTXV function to produce a list of states and then extract position versus time information to make a graph.

We're not required to give a name to the list of forces that act on the Ping-Pong ball. We could insert the list as the appropriate input to statesTXV, but I think it may organize and help our thinking to give the list of forces a name. Let's call it dampedHOForces.

```
dampedHOForces :: [State1D -> Force]
dampedHOForces = [springForce 0.8
                 ,\(_,_,v0) -> fAir 2 1.225 (pi * 0.02**2) v0
                 ,\_ -> -0.0027 * 9.80665
                 ]
```

We see that this is a list of the three forces we discussed earlier. First we have the spring force, then the force of air resistance, and then the force of gravity. We use anonymous-function notation to express the latter two forces because they need to be expressed as functions of the state. In the case of the force of gravity, we don't care at all about the state, so we don't need to name the state or name the time, position, and/or velocity variables that make up the state. For the force of air resistance, underscores in the time and position entries for the state are reminders that the air resistance function does not need their values.

To produce an infinite list of states, we use the statesTXV function with a time step of 1 ms, or 0.001 s, which is on the threshold of acceptability in the sense that smaller time steps would produce a graph that is only a little bit different in its results.

```
dampedHOStates :: [State1D]
dampedHOStates = statesTXV 0.001 0.0027 (0.0,0.1,0.0) dampedHOForces
```

We need to send dampedHOStates all of the information about our problem. In addition to the time step of 0.001 s, we send the Ping-Pong ball mass of 0.0027 kg, the initial state consisting of time 0 s, position 0.1 m, and velocity 0 m/s, and the list of forces, which we named dampedHOForces.

If you want to look at the raw time-position-velocity data, you can use the list element operator (!!) to select particular states from the list, or you can take the first several elements of this infinite list.

```
Prelude> :l Mechanics1D
[1 of 2] Compiling Newton2          ( Newton2.hs, interpreted )
[2 of 2] Compiling Mechanics1D      ( Mechanics1D.hs, interpreted )
Ok, two modules loaded.
*Mechanics1D> dampedHOStates !! 0
(0.0,0.1,0.0)
*Mechanics1D> dampedHOStates !! 5
(5.0e-3,9.960571335911717e-2,-0.1970379672671094)
*Mechanics1D> take 2 dampedHOStates
[(0.0,0.1,0.0),(1.0e-3,0.1,-3.943627962962963e-2)]
```

A list of pairs is something we can plot with the plotPath function from the gnuplot package, but we need to truncate the list to a finite list before plotting; otherwise, plotPath will hang while trying to finish calculating an infinite list. In the code that follows, we use the take function to extract the first 3,000 states, corresponding to the first three seconds of motion.

```
dampedHOGraph :: IO ()
dampedHOGraph
    = plotPath [Title "Ping Pong Ball on a Slinky"
               ,XLabel "Time (s)"
               ,YLabel "Position (m)"
               ,PNG "dho.png"
               ,Key Nothing
               ] [(t,x) | (t,x,_) <- take 3000 dampedHOStates]
```

Figure 15-5 shows the oscillation of the ball as a function of time. Notice the oscillation is not centered at position zero. This is because zero is the equilibrium position of the spring with no ball attached. When we attach the Ping-Pong ball, its weight will extend the spring downward, forming a new equilibrium position at the place where $-kx_0 - mg = 0$. The upward force of the spring cancels the downward force of gravity; air resistance plays no role in the new equilibrium position because the ball is not moving at equilibrium. The new equilibrium position is $x_0 = -mg/k = -0.033$ m, so this is the position about which the oscillation is centered, as you can see in Figure 15-5.

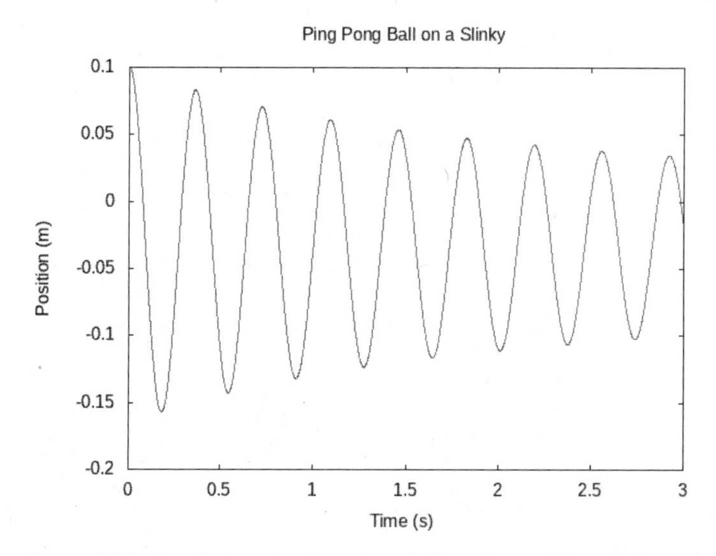

Figure 15-5: Oscillation of a Ping-Pong ball on the end of a Slinky

Having explored one data representation for a solution to Newton's second law, namely a list of states, let's take a look at another representation, that of position and velocity functions.

Method 2: Producing Position and Velocity Functions

We can use the function `positionFtxv` to produce a position function for the Ping-Pong ball on the Slinky.

```
pingpongPosition :: Time -> Velocity
pingpongPosition = positionFtxv 0.001 0.0027 (0,0.1,0) dampedHOForces
```

All of the information necessary to describe the situation is included in the one line that makes up the body of this function: the 0.0027-kg mass; the initial state consisting of time 0 s, position 0.1 m, and velocity 0 m/s; and the list `dampedHOForces` of three forces defined earlier.

The following code will produce a graph very much like that shown in Figure 15-5.

```
dampedHOGraph2 :: IO ()
dampedHOGraph2
    = plotFunc [Title "Ping Pong Ball on a Slinky"
               ,XLabel "Time (s)"
               ,YLabel "Position (m)"
               ,Key Nothing
               ] [0,0.01..3] pingpongPosition
```

We can use the function `velocityFtxv` to produce a velocity function for the Ping-Pong ball on the Slinky.

```
pingpongVelocity :: Time -> Velocity
pingpongVelocity = velocityFtxv 0.001 0.0027 (0,0.1,0) dampedHOForces
```

As before, we can graph our function:

```
dampedHOGraph3 :: IO ()
dampedHOGraph3
    = plotFunc [Title "Ping Pong Ball on a Slinky"
               ,XLabel "Time (s)"
               ,YLabel "Velocity (m/s)"
               ,PNG "dho2.png"
               ,Key Nothing
               ] [0,0.01..3] pingpongVelocity
```

This code produces Figure 15-6, which shows a graph of velocity versus time for the Ping-Pong ball. Velocity starts at zero since we release the ball from rest and then becomes negative as the ball moves downward. The velocity oscillates and experiences the same sort of damping that the position does.

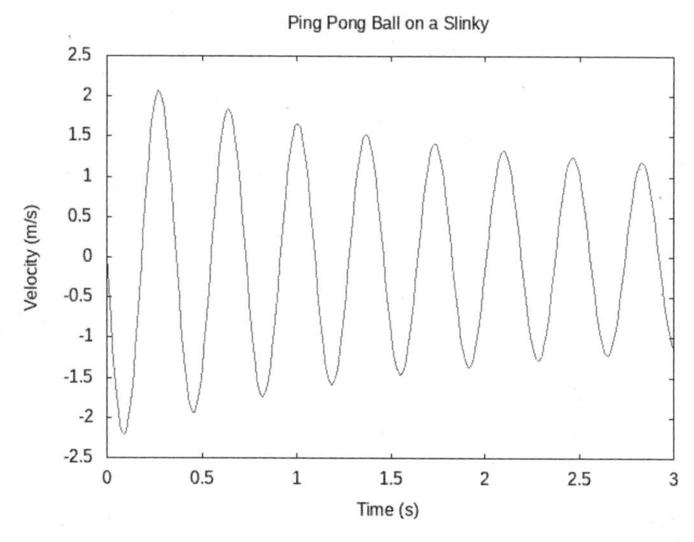

Figure 15-6: Velocity of a Ping-Pong ball on the end of a Slinky

The Euler method is a general-purpose state-based way of solving a system of first-order differential equations (which is more or less equivalent to a single higher-order differential equation). However, the Euler method does not usually get very much bang out of the computational buck, in that the time step often must be very small to get acceptable results. There are many other methods to choose from, and a small modification to the Euler method, which we explore in the next section, typically allows one to get acceptable results with a larger step size, decreasing the computational cost of a calculation.

Euler-Cromer Method

We can make a slight modification to the Euler method that improves the results of Newton's second law calculations in many cases. Instead of the functional diagram that describes the Euler method in Figure 15-3, let's look at a functional diagram for the Euler-Cromer method, shown in Figure 15-7.

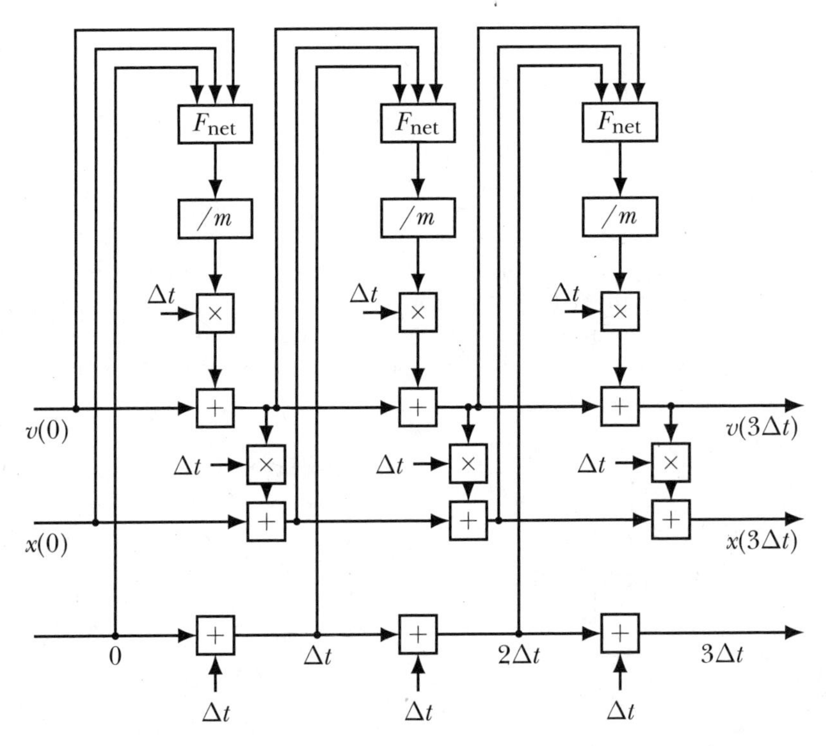

Figure 15-7: The Euler-Cromer method for Newton's second law in one dimension

The difference lies in the value of velocity used to update position. The Euler method used the old velocity to update the position. The Euler-Cromer method calculates a new velocity using the same velocity-update equation as in the Euler method but then uses this new velocity to update the position.

Instead of the state-update Equations 15.6 and 15.7 for the Euler method, the Euler-Cromer method uses the following slightly modified equations:

$$v(t + \Delta t) = v(t) + \frac{F^{txv}_{\text{net}}(t, x(t), v(t))}{m}\Delta t \tag{15.9}$$

$$x(t + \Delta t) = x(t) + v(t + \Delta t)\Delta t \tag{15.10}$$

The velocity-update equation for Euler-Cromer is the same as the velocity-update equation for Euler. The difference is the replacement of $v(t)$ with $v(t + \Delta t)$ in the position-update equation. While the order in which the Euler update equations 15.6 and 15.7 are evaluated is irrelevant, the Euler-Cromer velocity-update equation must be evaluated before the position-update equation because the updated velocity is used in the latter equation.

However, one of the benefits of a functional language with names that refer to objects that don't change is that we don't need to worry about telling the computer to update the velocity first. We can put the equations in whatever order we want, and the compiler will figure out an appropriate order of evaluation. The following function plays the role for Euler-Cromer that `euler1D` played for Euler:

```
eulerCromer1D :: R                          -- time step dt
              -> (State1D -> (R,R,R))  -- differential equation
              -> State1D -> State1D     -- state-update function
eulerCromer1D dt deriv (t0,x0,v0)
    = let (_, _, dvdt) = deriv (t0,x0,v0)
          t1 = t0 + dt
          x1 = x0 + v1 * dt
          v1 = v0 + dvdt * dt
      in (t1,x1,v1)
```

In this code, I use the local variable `v1`, which is the updated value of velocity, before I give the equation for how to find it. The compiler will know to order the evaluations so that `v1` is computed before it is used.

The function `updateTXVEC` is the Euler-Cromer version of the state-update function corresponding to Euler's `updateTXV`.

```
updateTXVEC :: R                        -- time interval dt
            -> Mass
            -> [State1D -> Force]   -- list of force funcs
            -> State1D -> State1D   -- state-update function
updateTXVEC dt m fs = eulerCromer1D dt (newtonSecond1D m fs)
```

The Euler-Cromer method comes closer to conserving energy in cases where it should be conserved, and it's generally better for situations with oscillatory behavior. Both the Euler and Euler-Cromer methods converge to the correct result as the time step is decreased, but Euler-Cromer can often achieve acceptable results with a larger time step than Euler would require.

Figure 15-7 shows an alternative uncoiling of Figure 15-1 corresponding to the Euler-Cromer method. Comparing Figure 15-3 (which describes the Euler method) and Figure 15-7, we see that the only difference is that the updated velocity is used to update the position.

The functions `statesTXV`, `velocityFtxv`, and `positionFtxv` use the function `updateTXV`, which means they use the Euler method. In Exercise 15.13, you are asked to write analogous functions for the Euler-Cromer method.

Note that the Euler-Cromer method is specific to second-order differential equations because there must be a state variable playing the role of velocity that can be updated first and then used to update the main unknown function (position for Newton's second law).

As we expand our study of mechanics in the next several chapters, we'll continue to solve differential equations. The final section of this chapter prepares the way for future chapters by treating the process of differential equation solving in greater generality.

Solving Differential Equations

A typical mechanics problem starts as a physics problem as we use physical information to construct a differential equation, then becomes a mathematics problem as we solve the differential equation, and finally becomes a physics problem again as we interpret the results. This section focuses on the mathematical activity, beginning with a differential equation and ending with knowledge of how the state variables change with the independent variable (which in mechanics represents time).

In Figure 15-2, we showed how information was transformed in the process of solving a mechanics problem. Starting with mass and force information, Newton's second law produces a differential equation. The Euler method then transforms this differential equation into a state-update function. By iterating the state-update function, given some initial state, we arrive at a list of states. This list of states can be regarded as a solution to the problem, or we can proceed to an additional step and produce position and velocity functions for the object.

In the next several chapters, in which we use vectors to describe quantities like velocity and treat multiple interacting particles, we will continue to view the process of solving a mechanics problem as one of information transformation, like that in Figure 15-2. For this purpose we need to generalize Figure 15-2 in two ways. The function `euler1D` will work with any differential equation that uses the state space `State1D`. In the previous chapter, we used `Velocity` and `(Time,Velocity)` as state spaces, and in the coming chapters, we will continue to expand the state spaces we use to include vectors and multiple particles. We want to be able to use the Euler method with the new state spaces we design, and it would be really great if we could write the Euler method once and for all in a way that would work with any state space. Later in this section, we'll identify the commonality among these state spaces that allows us to do this. So, the first generalization is from the state space `State1D` to a broader class of state spaces.

Figure 15-2 uses the Euler method to transform a differential equation into a state-update function. Now that we have introduced the Euler-Cromer method, we have two numerical methods, each of which can perform this transformation. The second generalization is from the Euler method to other numerical methods. We want our information transformation process to allow whatever numerical method we might wish to use.

Let's turn to the question of how to generalize the state space.

Generalizing the State Space

To generalize the state space from `State1D` to other possibilities, we will use a type variable `s` for the data type of the state. If we can write functions with types expressed in terms of `s`, rather than `State1D`, we can use these functions with any state space. When we generalize `Velocity`, `(Time,Velocity)`, and `State1D`, type `s` will contain whatever state variables are necessary for the physical system in question.

Where Figure 15-2 has a differential equation based on state space `State1D`, we want differential equations based on state space `s`. Where Figure 15-2 has a state-update function based on state space `State1D`, we want a state-update function based on state space `s`. Where `euler1D` is a numerical method that transforms a differential equation based on `State1D` into a state-update function based on `State1D`, we want to be able to talk and write about numerical methods that transform a differential equation based on state space `s` into a state-update function based on `s`.

To be precise, we're going to give formal definitions for the differential equation, state-update function, and numerical method for state space `s` by writing type synonyms. We'll start with the state-update function because that is the simplest of the three.

Since a state-update function, such as `updateTXV dt m fs`, produces a new state with the same type as the input state, a state-update function for state space `s` is a function `s -> s`. This definition can be encoded as a type synonym.

```
type UpdateFunction s = s -> s
```

The type `UpdateFunction s` is the type for state-update functions that work with state space `s`.

Table 15-1 shows the state-update functions we have used so far in the last chapter and the current chapter.

Table 15-1: State-Update Functions That Work with Different State Spaces

State-update function	Type
`updateVelocity dt m fs`	`UpdateFunction Velocity`
`updateTV dt m fs`	`UpdateFunction (Time,Velocity)`
`updateTXV dt m fs`	`UpdateFunction State1D`

Note that for each function we must supply a time step, a mass, and a list of force functions before the resulting expression has the type shown on the right side of the table.

A differential equation takes a state as input and produces derivatives of each state variable as output. With the state space `State1D`, the input consists of time, position, and velocity, while the output is numbers, velocity, and acceleration.

Time is not the same thing as a dimensionless number, position is not the same thing as velocity, and velocity is not the same as acceleration. However, these are all real numbers for `State1D`, so the state space can also be written as `(R,R,R)`. In the function `euler1D` from earlier, I used the type `State1D -> (R,R,R)` for the differential equation and the type `State1D -> State1D` for the state-update function. To the compiler, these are the same type. I wrote them differently because numbers, velocity, and acceleration aren't the quantities that belong in a `State1D`.

To treat the difference between state variables and their time derivatives, we will use a type variable `ds` for a quantity that represents the time derivatives of the state variables. Just as the type variable `s` is for state, the type variable `ds` is for the time derivative of state.

A differential equation, as represented in Haskell, is a function that returns a set of derivatives of state variables when given a set of state variables. If `s` is the data type for state, and `ds` is the data type for time derivative of state, then the definition of a differential equation can be given as a type synonym.

```
type DifferentialEquation s ds = s -> ds
```

The type `DifferentialEquation s ds` is the type for differential equations that work with state space `s` and time derivative of state space `ds`.

Table 15-2 shows the differential equations we have used so far in the last chapter and the current chapter.

Table 15-2: Differential Equations That Work with Different State Spaces

Differential equation	Type
`newtonSecondV m fs`	`DifferentialEquation Velocity R`
`newtonSecondTV m fs`	`DifferentialEquation (Time,Velocity) (R,R)`
`newtonSecond1D m fs`	`DifferentialEquation State1D (R,R,R)`

Note that for each function we must supply a mass and a list of force functions before the resulting expression has the type shown on the right side of the table.

A numerical method transforms a differential equation into a state-update function. The definition of numerical method for state space `s` and derivative space `ds` can be given as a type synonym.

```
type NumericalMethod s ds = DifferentialEquation s ds -> UpdateFunction s
```

The type `NumericalMethod s ds` is the type for numerical methods that work with the state space `s` and time derivative space `ds`.

While the differential equation itself is a mathematically exact expression, applying a numerical method to solve it necessarily involves approximation. We have seen two numerical methods so far: the Euler method and the Euler-Cromer method. However, there are many numerical methods to choose from in solving a differential equation.

Since there are many numerical methods, which all lead to different approximations, it makes little sense to bake one particular numerical method into the foundational ideas and code that constitute both the structure of how we think about solving mechanics problems and the computational tools to solve them. We should be able to choose a numerical method freely, independently of the differential equation we are solving. We want to separate the numerical method from the differential equation.

Table 15-3 shows the numerical methods we have used so far in this chapter. In the previous chapter, we never wrote an explicit numerical method, instead placing the Euler method code inside the state-update function.

Table 15-3: Numerical Methods That Work with the `State1D` State Space

Numerical method	Type
`euler1D dt`	`NumericalMethod State1D (R,R,R)`
`eulerCromer1D dt`	`NumericalMethod State1D (R,R,R)`

Note that for each function we must supply a time step before the resulting expression has the type shown on the right side of the table.

Given a differential equation and a numerical method, we can solve the differential equation by applying the numerical method, which results in a state-update function, and then iterate the state-update function to produce a list of states. The following function takes a numerical method, a differential equation, and an initial state as input and produces a list of states as output. We can think of this function as a general-purpose differential equation solver.

```
solver :: NumericalMethod s ds -> DifferentialEquation s ds -> s -> [s]
solver method = iterate . method
```

We said earlier that solving a differential equation is the mathematical part of solving a mechanics problem, in which we transform a differential equation through a state-update function and ultimately to a list of states. The function `solver` carries out the entire mathematical process of solving a differential equation. In other words, it handles the mathematical part of solving a mechanics problem.

We've given definitions for differential equation, state-update function, and numerical method, each of which works with a state space `s` and time derivative space `ds`. We now turn to writing a general-purpose function `euler` that can act as a numerical method for an almost-arbitrary state space `s`.

Type Classes for State Spaces

The state space type s and time derivative space ds cannot be just any data types. To carry out the Euler method and other numerical methods, we need to be able to add elements of the time derivative space ds, and we want to be able to scale elements by a time step. A space that supports addition of states and scalar multiplication of states by real numbers is called a *real vector space*.

To express the constraint that our time derivative space be a real vector space, we define a type class.

```
class RealVectorSpace ds where
      (+++) :: ds -> ds -> ds
      scale :: R -> ds -> ds
```

This code defines a new type class RealVectorSpace that owns two functions: (+++) for addition and scale for scalar multiplication. You can see from the type signatures that the addition function takes two state derivatives as input and produces one as output, while the scalar multiplication function takes a real number and a state derivative as input and produces a state derivative as output.

For each derivative space we wish to use, we'll write an instance declaration saying exactly how addition and scalar multiplication are to be defined for that data type. For the type (R,R,R), which is the derivative space associated with state space State1D, here is the instance declaration:

```
instance RealVectorSpace (R,R,R) where
    (dtdt0, dxdt0, dvdt0) +++ (dtdt1, dxdt1, dvdt1)
        = (dtdt0 + dtdt1, dxdt0 + dxdt1, dvdt0 + dvdt1)
    scale w (dtdt0, dxdt0, dvdt0) = (w * dtdt0, w * dxdt0, w * dvdt0)
```

We use the local variable dxdt0 to remind us that this name stands for a quantity that represents the derivative of a position with respect to time. The instance declaration defines adding two triples to mean adding each corresponding pair; it defines scaling a triple by a real number to mean scaling each item in the triple by the real number.

We will also use a type class to claim a relationship that must hold between a state space s and its derivative space ds. The relationship describes how a state variable in s can be advanced in time using knowledge of the state variable's time derivative in ds. We name the type class Diff to remind us of the differentiation (the taking of a time derivative) that relates a state space s to its derivative space ds.

```
class RealVectorSpace ds => Diff s ds where
    shift :: R -> ds -> s -> s
```

This type class definition contains a type class constraint saying that ds must be a RealVectorSpace. The type class Diff owns the function shift. For each pair of types we wish to use as state space and derivative space, we will provide an instance declaration that defines what the function shift does with values of the two spaces. The function `shift` describes how a state can be advanced in time using knowledge of the derivative state. The function `shift` takes a time step, a state derivative, and a state as input and gives a new state as output.

A type class used to relate two types rather than to claim membership of a single type is called a *multi-parameter type class*. The second line of the introductory code for the module we are writing in this chapter turns on a `LANGUAGE` feature called `MultiParamTypeClasses` to allow the use of multi-parameter type classes. Their use is disabled by default.

The following instance declaration claims a differentiation relationship between the types `State1D` and `(R,R,R)` by defining the function `shift` for them:

```
instance Diff State1D (R,R,R) where
    shift dt (dtdt,dxdt,dvdt) (t,x,v)
        = (t + dtdt * dt, x + dxdt * dt, v + dvdt * dt)
```

The function `shift` says that to update time, we should multiply a time step `dt` by the rate `dtdt` at which time changes (which is 1). It says that to update position, we should multiply a time step `dt` by the rate `dxdt` at which position changes and act similarly to update velocity. Time, position, and velocity come from the state, while the rates of change come from the state derivative.

We see that `shift` looks a lot like the Euler method. In fact, it is more basic than that. The `shift` function will be used by the Euler method but also by other numerical methods.

Here is the generic version of the Euler method, which is suitable for any differential equation based on any state space.

```
euler :: Diff s ds => R -> (s -> ds) -> s -> s
euler dt deriv st0 = shift dt (deriv st0) st0
```

As long as data types `s` and `ds` are appropriately related by the `Diff` type class, which is to say that `s` is a state space and `ds` is a derivative space that goes along with `s`, the function `euler` will carry out the Euler method with state variables from `s`, using the differential equation passed in as `deriv`. The Euler method is carried out by a single use of the `shift` function, where the derivatives are evaluated at the current state.

Table 15-4 compares the types of `euler1D` and `euler`.

Table 15-4: Comparison of the Function `euler1D` with the More General Function `euler`

Function		Type
`euler1D`	`::`	`R -> NumericalMethod State1D (R,R,R)`
`euler`	`::`	`Diff s ds => R -> NumericalMethod s ds`

The function euler can be used anywhere that euler1D can be used, and in other places as well. Because it is written with a type variable, we'll be able to use euler for the state spaces we make in future chapters.

Can we do the same thing with the Euler-Cromer method? That is, can we write, once and for all, a numerical method that works with any state space and any differential equation? Sadly, the answer is no. The Euler method is a general-purpose technique for solving any system of first-order differential equations. The Euler-Cromer method is a specialized method for second-order differential equations, or for systems of first-order differential equations where we can identify a quantity to play the role of velocity. We'll need to write a new Euler-Cromer function for each state space we work with.

Next we'll introduce one more general-purpose numerical method that could be used instead of Euler or Euler-Cromer.

One More Numerical Method

In this section, we'll introduce one more general-purpose numerical method, called the fourth-order Runge-Kutta method. Numerical methods are sometimes classified by an order. The order gives an expectation of how the error (the difference between the numerical solution and the exact solution) scales with the step size. When we shrink the step size by a factor of 10, the error for a first-order solver shrinks by about 10, while the error for a second-order solver shrinks by about 10^2, or 100. The Euler and Euler-Cromer methods are first-order methods and have the advantage of simplicity; it is straightforward to see why they work. The fourth-order Runge-Kutta method is substantially more complex. We won't get into why it works well or why it is a fourth-order method. However, it is a popular method for solving differential equations, and it allows us to see a second general-purpose method for solving a differential equation. Fourth-order Runge-Kutta is a general method that will work with any differential equation and state space.

Here is the code for the fourth-order Runge-Kutta method:

```
rungeKutta4 :: Diff s ds => R -> (s -> ds) -> s -> s
rungeKutta4 dt deriv st0
    = let m0 = deriv                      st0
          m1 = deriv (shift (dt/2) m0 st0)
          m2 = deriv (shift (dt/2) m1 st0)
          m3 = deriv (shift  dt    m2 st0)
      in shift (dt/6) (m0 +++ m1 +++ m1 +++ m2 +++ m2 +++ m3) st0
```

You can see that this method is more complex, but it has the same type as euler, and it plays the same role: transforming a differential equation into a state-update function.

Now that we have three numerical methods to use, let's compare them on a differential equation to which we know the solution.

Comparison of Numerical Methods

Let's do a comparison of the three numerical methods we've introduced. The differential equations

$$\frac{d}{dt}[t] = 1$$

$$\frac{d}{dt}[x(t)] = v(t)$$

$$\frac{d}{dt}[v(t)] = x(t)$$

can be written in Haskell as follows:

```
exponential :: DifferentialEquation (R,R,R) (R,R,R)
exponential (_,x0,v0) = (1,v0,x0)
```

These differential equations are exactly solvable and have a solution of

$$x(t) = Ae^{t}$$

$$v(t) = Ae^{t}$$

or

$$x(t) = Ae^{-t}$$

$$v(t) = -Ae^{-t}$$

for some constant A. If we focus on the initial state in which $x(0) = 1$ and $v(0) = 1$, the solution is

$$x(t) = e^{t}$$

$$v(t) = e^{t}$$

Let's compare the exact solution at $t = 8$ to the approximations given by the three numerical methods we've presented using different step sizes.

The `solver` function takes a numerical method, a differential equation, and an initial state to produce a list of states.

```
*Mechanics1D> solver (euler 0.01) exponential (0,1,1) !! 800
(7.999999999999874,2864.8311229272326,2864.8311229272326)
*Mechanics1D> solver (eulerCromer1D 0.1) exponential (0,1,1) !! 80
(7.999999999999988,3043.379244966009,2895.0121485099035)
*Mechanics1D> solver (rungeKutta4 1) exponential (0,1,1) !! 8
(8.0,2894.789038540849,2894.789038540849)
```

In the first use of `solver`, we're using a step size of 0.01, so item number 800 in the list corresponds to $t = 8$. The other two uses of `solver` use different step sizes and consequently different item numbers for $t = 8$.

Table 15-5 compares the three numerical methods we've discussed. You can see where these example calculations fit into the table.

Table 15-5: Comparison of the Euler, Euler-Cromer, and Fourth-Order Runge-Kutta Methods with the Exact Solution of a Differential Equation

	$\Delta t = 1$	$\Delta t = 0.1$	$\Delta t = 0.01$	$\Delta t = 0.001$
Exact	2981	2981	2981	2981
RK4	2895	2981	2981	2981
Euler-Cromer	2584	3043	2988	2982
Euler	256	2048	2865	2969

The exact result does not depend on any step size; it is just e^8. All three numerical methods get closer to the exact result as the step size decreases. The Euler method is within 4 percent of the exact value at a step size of $\Delta t = 0.01$ and within 1 percent at a step size of $\Delta t = 0.001$. Euler-Cromer at $\Delta t = 0.1$ is better than Euler at $\Delta t = 0.01$, and Euler-Cromer at $\Delta t = 0.01$ is better than Euler at $\Delta t = 0.001$. Thus, we can use roughly 10 times the step size with Euler-Cromer compared to Euler and get comparable results for this differential equation. Similarly, we can use roughly 10 times the step size with fourth-order Runge-Kutta compared to Euler-Cromer and get comparable results.

Summary

This chapter completed the treatment of one-dimensional mechanics begun in the previous chapter. We saw how to handle forces that depend on position. With a force that depends on position, Newton's second law is a second-order differential equation that we transform into coupled first-order differential equations for the state variables position and velocity. We can view the solution of a mechanics problem as the transformation of information about the physical situation through a sequence of five representations: mass and force functions, differential equation, state-update function, list of states, and position-velocity functions.

A Ping-Pong ball oscillating on the end of a Slinky in the presence of air resistance was the central example of the chapter. We introduced the Euler-Cromer method, an improvement on the Euler method for second-order differential equations. We also introduced the fourth-order Runge-Kutta method, which, along with the Euler method, is a general-purpose numerical method for solving any differential equation with any set of state variables. In the next chapter, we'll start three-dimensional mechanics by treating position, velocity, and acceleration as vectors.

Exercises

Exercise 15.1. Let's warm up with a basic projectile motion problem where we know what the answer should look like. Suppose someone throws a ball from the ground straight up into the air with an initial velocity of 10 m/s.

Ignoring air resistance, use the function `positionFtxv` to find the height of the ball as a function of time. Make a plot of height as a function of time.

Exercise 15.2. Doing the Euler method by hand on page 254, we arrived at a table of values for position and velocity. Show how to calculate these values using Haskell functions. The values may not match to four decimal places because I rounded intermediate results in doing the Euler method by hand, but the first two digits after the decimal point should match in every case.

Exercise 15.3. (Euler method by hand.) Consider the differential equations

$$\frac{dx(t)}{dt} = v(t)$$
$$\frac{dv(t)}{dt} = -3x(t) + 4\cos 2t - 2v(t)$$

along with the initial conditions

$$x(0) = 2 \qquad\qquad v(0) = 1$$

Use the Euler method with a step size of $\Delta t = 0.1$ to approximate the value of $x(0.3)$. Keep at least four figures after the decimal point in your calculations. Show your calculations in a small table. (The table will have three columns, one each for time, position, and velocity.)

Exercise 15.4. Write a Haskell function

```
update2 :: (R,R,R)  -- starting state
        -> (R,R,R)  -- ending state
update2 = undefined
```

that takes a tuple (t_0, x_0, v_0) and returns a tuple (t_1, x_1, v_1) for a single step of the Euler method for the differential equations

$$\frac{dx(t)}{dt} = v(t)$$
$$\frac{dv(t)}{dt} = -3x(t) + 4\cos 2t - 2v(t)$$

with a step size of $\Delta t = 0.1$ and the same initial conditions as in the previous exercise. Show how to use the function `update2` to calculate the value $x(0.3)$ that you calculated by hand in the previous exercise.

Exercise 15.5. Consider a 3-kg mass connected to a wall by a linear spring with spring constant 100,000 N/m. Ignoring gravity and friction, if the spring is extended by 0.01 m and released, what does the subsequent motion look like? Investigate this motion over several cycles of oscillation. Compare your results to the exact solution. Find a time step that is small enough so that the Euler solution and the exact solution overlap precisely on a plot. Find another time step that is big enough so that you can see the difference between the Euler solution and the exact solution on a plot.

Make a nice plot (with title, axis labels, and so on) with these three solutions on a single graph (bad Euler, good Euler, and exact). Label the Euler results with the time step you used, and label the exact result "Exact."

Exercise 15.6. Let's investigate dropping things from large heights. In particular, let's look at a Ping-Pong ball and a bowling ball. In each case, take $C = 1/2$. You will need to find out good approximations for things like the size and mass of these balls. Let's drop them from heights of 100 m and 500 m. Make graphs of velocity as a function of time and velocity as a function of vertical position. What fraction of terminal velocity is achieved in each case? Assemble your results in some meaningful and understandable way.

Exercise 15.7. Return to the harmonic oscillator of Exercise 15.5. Compare the Euler and Euler-Cromer solutions to the exact solution for a time step of 0.001 s (you will recall that this is not a very good time step for the Euler method). Plot the displacement of the mass as a function of time for the first 0.1 s of motion. Plot Euler, Euler-Cromer, and exact solutions on one set of axes. Also give the value of the position of the mass (to four significant figures) at $t = 0.1$ s for each of the three solutions.

Exercise 15.8. Consider an object with mass m attached to a spring with spring constant k. The other end of the spring is attached to a vertical wall. The object slides horizontally across the floor. There is a coefficient of kinetic friction $\mu_k = 0.3$ between the object and the floor. The weight of the object is mg, so the force of kinetic friction on the object is $\mu_k mg$, directed opposite the velocity of the object.

Let $m = 3$ kg and $k = 12$ N/m.

(a) Write a function with type `State1D -> Force` that gives the horizontal force of kinetic friction. You may want to use the `signum` function.

(b) Use the function `positionFtxv` to find the position of the object as a function of time.

(c) Make a plot of position as a function of time.

Exercise 15.9. In most situations in mechanics, the mass of the object that we care about does not change. There is no need to include a quantity that doesn't change in the state. However, since some forces, such as gravity, depend on mass, there is some motivation to include mass in the state simply for convenience. Several of the functions we developed in this chapter accept a list of force functions `[State1D -> Force]` as input. If we wanted to include Earth's gravity as such a state-dependent force, we would need to write something like the following:

```
earthGravity :: Mass -> State1D -> Force
earthGravity m _ = let g = 9.80665
                   in -m * g
```

Suppose, on the other hand, we include the object's mass in its state by using the following 4-tuple as the data type for state.

```
type MState = (Time,Mass,Position,Velocity)
```

Then we could write an Earth gravity function as

```
earthGravity2 :: MState -> Force
earthGravity2 (_,m,_,_) = let g = 9.80665
                          in -m * g
```

Notice that since mass is included in the state, we no longer need `Mass` as an extra parameter in the function type.

Write definitions for the following functions, using `MState` in place of `State1D`:

```
positionFtxv2 :: R                         -- time step
              -> MState                    -- initial state
              -> [MState -> Force]         -- list of force funcs
              -> Time -> Position          -- position function
positionFtxv2 = undefined

statesTXV2 :: R                            -- time step
           -> MState                       -- initial state
           -> [MState -> Force]            -- list of force funcs
           -> [MState]                     -- infinite list of states
statesTXV2 = undefined

updateTXV2 :: R                            -- dt for stepping
           -> [MState -> Force]            -- list of force funcs
           -> MState                       -- current state
           -> MState                       -- new state
updateTXV2 = undefined
```

Exercise 15.10. The Lennard-Jones potential

$$V_{LJ}(r) = D_e \left[\left(\frac{r_e}{r} \right)^{12} - 2 \left(\frac{r_e}{r} \right)^{6} \right]$$

is sometimes used to model the interaction between atoms. The expression $V_{LJ}(r)$ gives the potential energy for a system of two atoms when the atoms are a distance r apart from each other. As $r \to 0$, the potential energy becomes infinite, expressing the difficulty of having the two atoms very close together. The lowest value of potential energy occurs at an inter-atomic separation of $r = r_e$, meaning that the parameter r_e is the equilibrium separation of the atoms. The parameter D_e represents the dissociation energy for the two-atom molecule, or the amount of energy that must be provided to the molecule to pull the atoms (arbitrarily far) apart.

The Lennard-Jones force

$$F_{LJ}(r) = -\frac{dV_{LJ}(r)}{dr} = \frac{12 D_e}{r_e} \left[\left(\frac{r}{r_e} \right)^{-13} - \left(\frac{r}{r_e} \right)^{-7} \right]$$

gives the force on one of the atoms produced by the other, with positive meaning repulsive and negative meaning attractive. We can think of the

Lennard-Jones force as coming from a nonlinear spring that connects the two atoms. When the atomic separation is greater than r_e, the spring provides an attractive force that attempts to restore equilibrium. When the atomic separation is less than r_e, the spring provides a repulsive force that attempts to restore equilibrium. The spring is nonlinear in that the restoring force is not proportional to the deviation of the atomic separation from equilibrium.

Figure 15-8 shows the Lennard-Jones force as a function of inter-atomic distance r, along with the linear spring force that most closely approximates it.

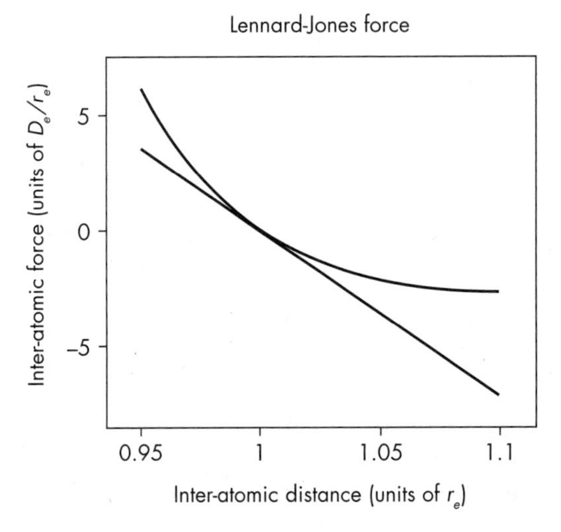

Figure 15-8: The Lennard-Jones force (curved line) and the linear force (straight) that most closely approximates it

The effective spring constant for the Lennard-Jones force is the negative slope of the force as a function of inter-atomic distance.

$$k_{\text{eff}} = -\left.\frac{dF_{LJ}(r)}{dr}\right|_{r = r_e} = \frac{72D_e}{r_e^2}$$

Suppose we have a mass m attached by a Lennard-Jones spring to a wall. We'll ignore gravity and friction in this problem so that the Lennard-Jones force is the only force that acts on the mass. If an oscillation about the equilibrium position is small in amplitude, the angular frequency of the oscillation will be close to

$$\omega = \sqrt{\frac{k_{\text{eff}}}{m}} = \frac{1}{r_e}\sqrt{\frac{72D_e}{m}}$$

and the period will be close to

$$T = \frac{2\pi}{\omega} = 2\pi r_e\sqrt{\frac{m}{72D_e}}$$

Choose any parameters for r_e, D_e, and m.

(a) Make a graph of position versus time when the initial position is $r = 1.01r_e$ and the initial velocity is zero. The graph should be oscillatory. Confirm that the period is close to the value given earlier.

(b) Make a graph of position versus time when the initial position is $r = 5r_e$ and the initial velocity is zero. What is the period now?

This is an example of an anharmonic oscillator, where the period depends on the amplitude of oscillation. It is only the special case of the harmonic oscillator in which period is independent of amplitude.

Exercise 15.11. Applying the Euler method by hand on page 254, we arrived at a table of values for position and velocity. Produce a similar table of values using the Euler-Cromer method by hand.

Exercise 15.12. Write a function `statesTXVEC` that is similar to `statesTXV` but uses the Euler-Cromer method instead of the Euler method. Use this function to check the table of values you calculated by hand in the previous exercise.

Exercise 15.13. Write versions of `statesTXV`, `velocityFtxv`, and `positionFtxv` that use the Euler-Cromer method rather than the Euler method.

Exercise 15.14. Show how to use Haskell functions to calculate the entries of the comparison table for numerical methods.

Exercise 15.15. With the code

```
instance RealVectorSpace (R,R) where
    (dtdt0, dvdt0) +++ (dtdt1, dvdt1) = (dtdt0 + dtdt1, dvdt0 + dvdt1)
    scale w (dtdt0, dvdt0) = (w * dtdt0, w * dvdt0)

instance Diff (Time,Velocity) (R,R) where
    shift dt (dtdt,dvdt) (t,v)
        = (t + dtdt * dt, v + dvdt * dt)
```

we have made the data type `(R,R)` an instance of type class `RealVectorSpace` and written a `Diff` instance for the pair of types `(Time,Velocity)` and `(R,R)`. Now we can use `(Time,Velocity)` as a state space and `(R,R)` as its derivative space for the Euler or fourth-order Runge-Kutta methods. Write a function

```
updateTV' :: R                                    -- dt for stepping
          -> Mass
          -> [(Time,Velocity) -> Force]  -- list of force funcs
          -> (Time,Velocity)                  -- current state
          -> (Time,Velocity)                  -- new state
updateTV' = undefined
```

that does the same thing as `updateTV` from Chapter 14 but uses the function `euler` from this chapter.

Exercise 15.16. Newton's second law generally produces a second-order differential equation (recall Table 14-1). Our `DifferentialEquation s ds` data type is for functions that return the derivatives of state variables when given the state variables themselves. A function with type `DifferentialEquation s ds` expresses a set of coupled first-order differential equations.

In this exercise, we'll rewrite a second-order differential equation as two coupled first-order differential equations. A second-order (ordinary) differential equation has one independent variable and one dependent variable (in other words, one unknown function). A set of two coupled first-order differential equations has one independent variable and two dependent variables (two unknown functions of the independent variable).

Here is the recipe for producing a set of two coupled first-order differential equations from a second-order differential equation:

1. The independent variable of the coupled set is the same as the independent variable of the second-order equation.
2. For the first unknown function of the coupled set, choose the unknown function of the second-order equation.
3. For the second unknown function of the coupled set, choose the derivative of the first unknown function with respect to the independent variable and give this function a new name.
4. The first differential equation in the coupled set expresses that the derivative of the first unknown function is equal to the second unknown function.
5. To form the second differential equation in the set, start with the original second-order differential equation, replace the first derivative of the unknown function with the new second unknown function, replace the second derivative of the unknown function with the derivative of the new second unknown function, and solve for the derivative of the new second unknown function.

Express the differential equation

$$3\frac{d^2x(t)}{dt^2} + 4\left[\frac{dx(t)}{dt}\right]^2 + 5x(t) = 6\sin(2t)$$

as a set of two coupled first-order differential equations.

Exercise 15.17. The Van der Pol oscillator is a generalization of the harmonic oscillator that is often used to explore chaos. It is described by the following differential equation:

$$\frac{d^2x(t)}{dt^2} - \mu\left(1 - [x(t)]^2\right)\frac{dx(t)}{dt} + x(t) = 0$$

We can view this equation as coming from Newton's second law with two forces present: a spring-like linear restoring force and a damping force. We will abandon SI units for this exercise, setting both mass and spring constant to unity. The spring force is then given by

$$F_{\text{spring}}(t, x, v) = -x$$

and the damping force is given by

$$F_{\text{damping}}(t, x, v) = \mu(1 - x^2)v$$

where μ is a parameter that controls how nonlinear the damping force is. If $\mu = 0$, the Van der Pol oscillator reduces to the harmonic oscillator.

When studying chaos, people often like to make phase plane plots, which are graphs of velocity as a function of position. (They can also be graphs of momentum as a function of position, but we will use velocity.) Fill in the undefined parts of the following code to make phase plane plots for $\mu = 0$, $\mu = 2$, $\mu = 4$, and $\mu = 6$, all on the same graph.

```
forces :: R -> [State1D -> R]
forces mu = [\(_t,x,_v) -> undefined x
            ,\(_t,x, v) -> undefined mu x v]

vdp :: R -> [(R,R)]
vdp mu = map (\(_,x,v) -> (x,v)) $ take 10000 $
         solver (rungeKutta4 0.01) (newtonSecond1D 1 $ forces mu) (0,2,0)

vdpPhasePlanePlot :: IO ()
vdpPhasePlanePlot = plotPaths [Title "Van der Pol oscillator"
                              ,XLabel "x"
                              ,YLabel "v"
                              ,PNG "VanderPol.png"
                              ,Key Nothing] (undefined :: [[(R,R)]])
```

The result should look something like Figure 15-9.

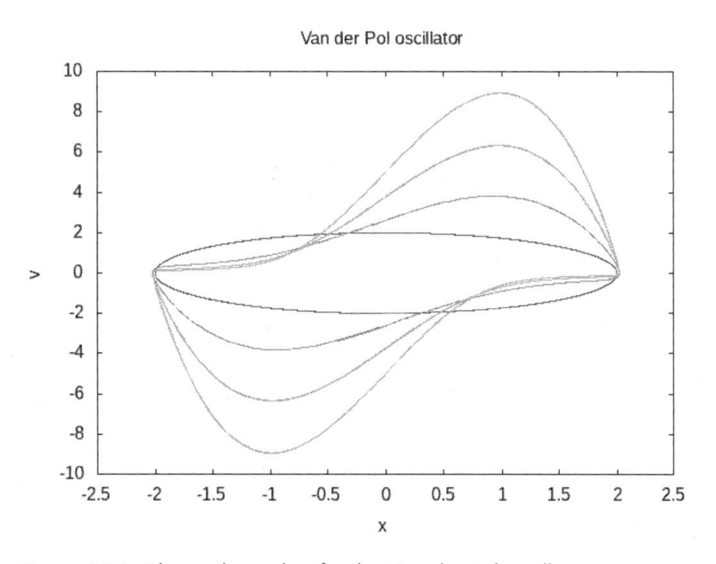

Figure 15-9: Phase plane plots for the Van der Pol oscillator

16

MECHANICS IN THREE DIMENSIONS

To predict the motion of a projectile, a satellite, or any object that can move unconstrained through three-dimensional space, we need to use three-dimensional vectors to describe velocity, acceleration, and force. In this chapter, we'll merge the three-dimensional vectors we described in Chapter 10 with the state-based solution techniques of Chapter 15.

Describing the state of an object or particle remains central to our task of predicting its future motion. We'll develop an appropriate set of state variables for a particle in three dimensions and define a new type called `ParticleState` to hold them.

Acknowledging the dependence that forces have on state variables, we assign the name *one-body force* to a function that returns a force vector when presented with a particle state. We give multiple examples of one-body forces, such as Earth's surface gravity and air resistance.

Solving a mechanics problem is a process of transforming information through a sequence of representations, beginning with a list of one-body forces, then a differential equation, then a state-update function, and finally a list of states. Newton's second law appears as the transformation from forces to differential equation. A numerical method transforms a differential equation into a state-update function. Iteration of the state-update function from an initial state produces a list of states.

This chapter explores the foundational ideas and code that allow us to predict motion by transforming information through a sequence of representations. Let's start with some introductory code.

Introductory Code

In this chapter and the next two, we'll create a module that contains ideas for setting up and solving Newton's second law in three dimensions. Before we start adding type signatures and function definitions, there's some code we need to have at the top of our source code file. This introductory code consists of four parts: a request for warnings, a request to use a language option, a name for the module, and a collection of types and functions we wish to import from other modules.

```
{-# OPTIONS -Wall #-}
{-# LANGUAGE MultiParamTypeClasses #-}

module Mechanics3D where

import SimpleVec
    ( R, Vec, PosVec, (^+^), (^-^), (*^), (^*), (^/), (<.>), (><)
    , vec, sumV, magnitude, zeroV, xComp, yComp, zComp, iHat, jHat, kHat)
import Mechanics1D
    ( RealVectorSpace(..), Diff(..), NumericalMethod
    , Time, TimeStep, rungeKutta4, solver )
import SpatialMath
    ( V3(..), Euler(..) )
import Graphics.Gnuplot.Simple
    ( Attribute(..), Aspect(..), plotFunc, plotPaths )
import qualified Graphics.Gloss as G
import qualified Vis as V
```

As usual, we begin by turning on warnings. We then turn on the language option that allows multi-parameter type classes, just as we did in the last chapter. We give this module the name `Mechanics3D`, which is how we will refer to it when we use any of the types or functions we define in this module in stand-alone programs or in other modules we write in later chapters. The remainder of the code consists of import statements, indicating that we wish to use types, type classes, and functions defined in modules that other people have written, or those defined in modules that we wrote in previous chapters.

In particular, we import vector operations from the `SimpleVec` module we wrote in Chapter 10 and some differential equation solving types, type classes, and functions from the `Mechanics1D` module we wrote in Chapter 15. We've listed the name of each piece we're importing from the `SimpleVec` and `Mechanics1D` modules, as opposed to simply importing the module as a whole. This is my preferred style because it shows where each of the names we use in the module comes from. If you want to import all of the names, you can write a one-line import statement composed of the keyword `import` followed by the module name, as we did in Chapter 15. If you import all of the names from many different modules, one of the names that you use may be defined in multiple modules, causing the compiler to complain. You will then need to clarify which module you want the name imported from.

The parenthetical with two dots, `(..)`, after a data type like `Attribute` means we want to import the data type and all of its constructors. If we omit the two dots, we will only import the name of the data type. Two dots after a type class, such as `RealVectorSpace`, mean we want to import the functions owned by the type class in addition to the name of the type class itself.

Lastly, we have the qualified import of the `Graphics.Gloss` and `Vis` modules. The first qualified import statement assigns the short name `G` to the `Graphics.Gloss` module, allowing us to access any type or function provided by `Graphics.Gloss` as long as it's prefixed by the short name `G` and a dot. The `Picture` type from the `Graphics.Gloss` module, for example, must be referred to as `G.Picture`. One reason I chose the qualified import method for the `Graphics.Gloss` and `Vis` modules is that they define several identical names, such as `simulate`. I want to use both definitions of `simulate` in the code I write, and I need a way of telling the compiler which definition I mean in each use.

Having completed our introductory code, let's look at Newton's second law in three dimensions.

Newton's Second Law in Three Dimensions

Equation 14.1 gave Newton's second law in one dimension. In three dimensions, position, velocity, acceleration, and force are described by vectors rather than numbers. In three dimensions, the net force on an object is the vector sum of the forces acting on that object:

Newton's second law in three dimensions

$$\mathbf{F}_{\text{net}}(t, \mathbf{r}(t), \mathbf{v}(t)) = m\mathbf{a}(t) \tag{16.1}$$

Here is Newton's second law in three dimensions as a differential equation:

$$\mathbf{F}_{\text{net}}(t, \mathbf{r}(t), D\mathbf{r}(t)) = mD^2\mathbf{r}(t) \tag{16.2}$$

Figure 16-1 shows a schematic diagram for Newton's second law in three dimensions. Since acceleration, velocity, and position are being treated as vectors now, two of the integrators have vector inputs, vector outputs, and vector state.

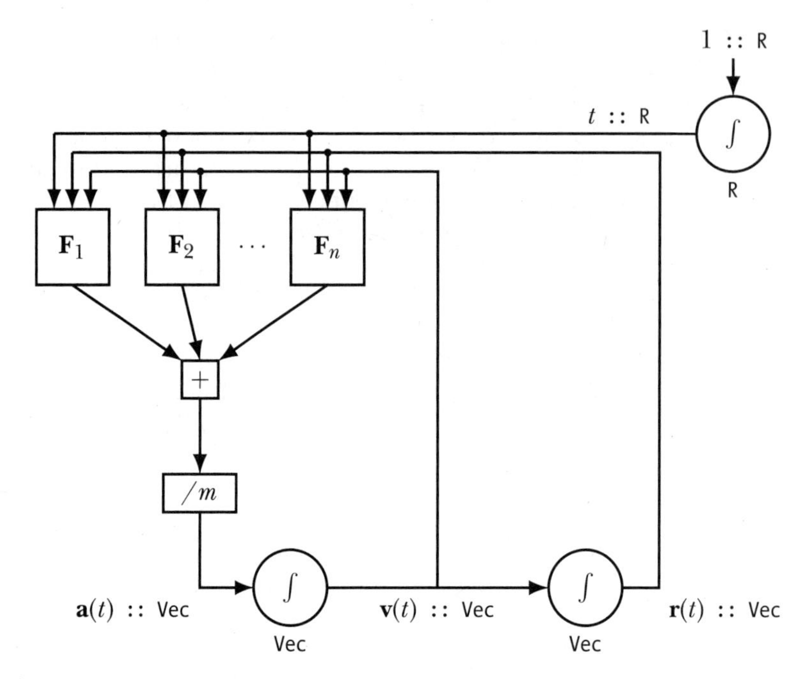

Figure 16-1: Schematic representation of Newton's second law

Here the forces depend on time, position, and velocity. Acceleration depends on net force. Velocity is the integration of acceleration, and position is the integration of velocity. The type below each integrator indicates the type of quantity that the integrator holds as state. The integrator that outputs time holds a real number as state. The integrator that outputs position holds a vector as state. The integrator that outputs velocity also holds a vector as state.

The two integrators in loops in the diagram hold velocity and position as state, so at a minimum, we need velocity and position to be state variables. We included time as a state variable for convenience in Chapters 14 and 15, and we will continue to do so here. Writing Newton's second law as a set of coupled first-order differential equations gives the following equations:

$$\frac{d}{dt}\left[t\right] = 1 \tag{16.3}$$

$$\frac{d}{dt}\left[\mathbf{r}(t)\right] = \mathbf{v}(t) \tag{16.4}$$

$$\frac{d}{dt}\left[\mathbf{v}(t)\right] = \frac{1}{m}\sum_j \mathbf{F}_j(t, \mathbf{r}(t), \mathbf{v}(t)) \tag{16.5}$$

Equations 16.3, 16.4, and 16.5 contain information equivalent to the diagram in Figure 16-1.

Having introduced Newton's second law in three dimensions, let's now turn to the question of how we will describe the state of a particle in three dimensions.

The State of One Particle

The state of a particle plays five roles. First, the state specifies the information necessary for prediction; it's the current information about the system that allows future prediction without historical information about the system (information from the past). Second, the state gives a template for the first-order differential Equations 16.3, 16.4, and 16.5 that form the starting point for our numerical approximation methods; each first-order differential equation expresses the time rate of change of one of the state variables in terms of the state variables themselves. Third, the state describes the information that needs to be updated by a state-update function. Fourth, the state encompasses the information that the forces depend on. Finally, knowing the state at all times counts as a solution to a Newton's second law problem because anything we want to know about a particle is some function of its state.

Since the forces that act on a particle can depend on the particle's mass (such as the force of gravity) or charge (such as the Lorentz force law, which we'll discuss later in the chapter), it's convenient to include mass and charge in the state. However, its certainly not necessary; mass and charge remain constant throughout most physical situations, so we could treat them as global values independent of state. But including mass and charge as state variables will simplify some of our code and allow some forces to be expressed as functions of state only instead of as functions of state and one or more parameters.

The definition of the `ParticleState` data type we'll use for the state of one particle includes mass, charge, time, position, and velocity as state variables.

```
data ParticleState = ParticleState { mass     :: R
                                   , charge   :: R
                                   , time     :: R
                                   , posVec   :: Vec
                                   , velocity :: Vec }
                     deriving Show
```

We use record syntax to give each field of the new data type its own extraction function (`mass`, `charge`, and so on). An extraction function is also called an eliminator or selector. We decided to make a new data type (using the `data` keyword) rather than a type synonym so that this type has no chance of being confused with any other type. We want to be able to display values of this data type, so we want `ParticleState` to be an instance of type class `Show`. By including `deriving Show`, we ask that the compiler automatically figure out how to make a `Show` instance.

For convenience, let's define a default `ParticleState`, which can be used to make new particle states.

```
defaultParticleState :: ParticleState
defaultParticleState = ParticleState { mass     = 1
                                     , charge   = 0
                                     , time     = 0
```

```
                                , posVec   = zeroV
                                , velocity = zeroV }
```

The `defaultParticleState` allows us to define a particle state without needing to explicitly provide all five pieces of information. For example, to specify the state of a 2-kg rock with no net charge, at the origin, moving with velocity $(3\hat{\mathbf{i}} + 4\hat{\mathbf{k}})$ m/s, we can write the following:

```
rockState :: ParticleState
rockState
    = defaultParticleState { mass     = 2                        -- kg
                           , velocity = 3 *^ iHat ^+^ 4 *^ kHat  -- m/s
                           }
```

Because we have the default state, we don't need to explicitly give state variables that are the same as those of the default, such as charge, time, and position. Recall that the operator `*^` is used to scale a vector on the right by a number on the left.

Newton's second law is a recipe for constructing a differential equation from a list of force functions. Throughout Chapters 14 and 15, we saw the usefulness of force functions in which the force depends on the state of the particle. We'll define a *one-body force* to be a force that depends on the current particle state as expressed by `ParticleState`; in other words, a force that could depend on time or the particle's position, velocity, mass, or charge.

```
type OneBodyForce = ParticleState -> Vec
```

We'll see in the following section that many common forces in mechanics are naturally expressed as one-body forces.

The code we write below for Newton's second law will produce a differential equation. In other words, it will produce a function that gives the time derivatives of state variables when presented with the state variables themselves. How should we return these time derivatives of state variables? Since the state variables are bundled together into an object with type `ParticleState`, we will similarly bundle together the time derivatives into an object with type `DParticleState`. Here is our definition of the new data type `DParticleState`:

```
data DParticleState = DParticleState { dmdt :: R
                                     , dqdt :: R
                                     , dtdt :: R
                                     , drdt :: Vec
                                     , dvdt :: Vec }
                      deriving Show
```

As there are five quantities contained in the particle state, so there are five quantities in the state derivative. The real number whose extraction function is `dmdt` (named after the derivative dm/dt) holds the rate at which mass changes. Mass will not change in any of our examples, so this rate will be zero, but the ability to have mass change is useful in some situations, such as rocket motion (in which a rocket expends fuel). Each of the other extraction functions has a name designed to indicate that the quantity represents the rate of change of a state variable. The names `dqdt` and `dtdt` label the real numbers for the rates at which charge and time change with time, respectively. The rate at which time changes with time is 1, so it's a little silly to keep track of this rate. An alternative is to write a data type that omits this quantity; I have chosen a data type whose structure parallels that of the state, even if some slots hold information that seems obvious. The names `drdt` and `dvdt` label the rates at which position and velocity change. These quantities are vectors, as the data type definition shows.

The function `newtonSecondPS` below is the Haskell representation of Newton's second law, which is equivalent to differential Equations 16.3, 16.4, and 16.5.

```
newtonSecondPS :: [OneBodyForce]
               -> ParticleState -> DParticleState  -- a differential equation
newtonSecondPS fs st
    = let fNet = sumV [f st | f <- fs]
          m = mass st
          v = velocity st
          acc = fNet ^/ m
      in DParticleState { dmdt = 0    -- dm/dt
                        , dqdt = 0    -- dq/dt
                        , dtdt = 1    -- dt/dt
                        , drdt = v    -- dr/dt
                        , dvdt = acc  -- dv/dt
                        }
```

The function `newtonSecondPS` is a recipe for converting a list of one-body forces into a differential equation. The PS in the name indicates that the function works with the `ParticleState` data type. The differential equation produced by `newtonSecondPS` expresses the time rate of change of each of the state variables in terms of the state variables themselves. Given values for each of the five state variables, the function `newtonSecondPS` returns values for the time rate of change of each of these five.

The function `newtonSecondPS` consists of a `let` expression, in which we first find the `Vec` representing the net force on the particle in the current state, name it `fNet`, then name the mass and velocity of the particle `m` and `v`, respectively, and finally compute the acceleration of the particle by dividing the net force by the mass. The body of the `let` expression returns a state derivative with type `DParticleState`. The derivatives of mass and charge with respect to time are 0 because mass and charge do not change. The derivative of time with respect to time is 1. Finally, the derivative of position is the velocity in the current state, and the derivative of velocity is the acceleration calculated in the `let` expression.

We regard the function `newtonSecondPS fs`, where `fs` is the list of one-body forces that describes the physical situation, as the Haskell version of the differential equation that expresses Newton's second law. The schematic diagram in Figure 16-1, the differential Equations 16.3, 16.4, and 16.5, and the Haskell function `newtonSecondPS` are different ways of expressing Newton's second law for a single object in three dimensions.

Solving Newton's Second Law

Our strategy for constructing and solving Newton's second law for one particle consists of transforming information about the physical situation through a sequence of four different forms:

1. One-body forces
2. Differential equation
3. State-update function
4. List of states

Figure 16-2 shows a functional diagram of the data representations, shown as vertical arrows, and the functions that transform the data from one representation to another, shown as boxes.

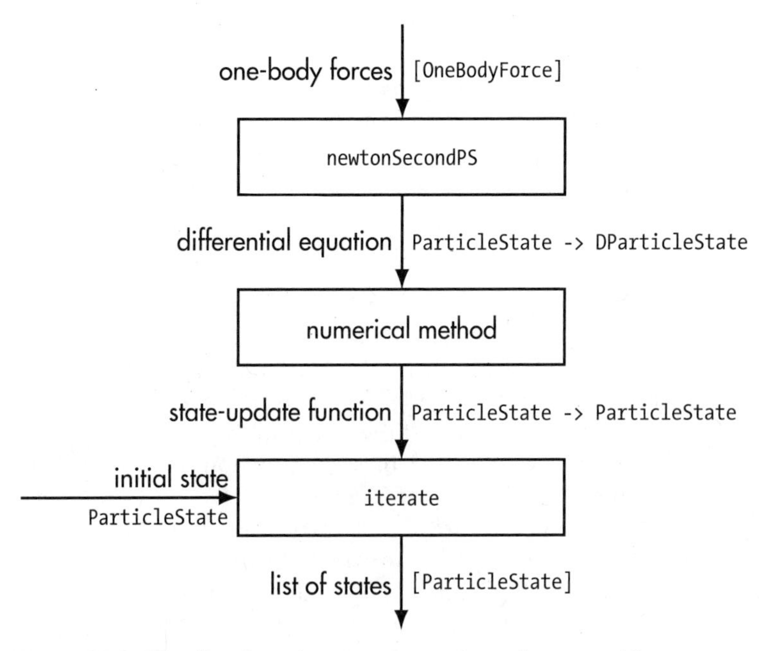

Figure 16-2: Data flow for solving a one-particle mechanics problem

A list of one-body forces is the first in a sequence of four information representations we use to describe a physical situation, each getting closer to a solution. The list of one-body forces characterizes the physical setting or situation in which a particle finds itself, serving as an algebraic analogue to the visual and geometric free-body diagrams that introductory physics classes use to show all of the forces acting on an object.

Newton's second law provides the means to transform the one-body forces into a differential equation, our second information representation. The function `newtonSecondPS` expresses Newton's second law as it applies to the `ParticleState` data type. The Haskell representation of a differential equation is a function `ParticleState -> DParticleState` that gives the time rates of change of the state variables in terms of the state variables themselves.

A state-update function is our third information representation; it describes how to take a small step forward in time, producing a new state from an old state. The two- and three-dimensional animation tools we use take a state-update function as input; inasmuch as an animated visualization of a particle's motion counts as a solution to a mechanics problem, the state-update function at the heart of that animation can also be regarded as a solution. To obtain a state-update function from a differential equation, we need a numerical method. By using a numerical method, we admit that we're looking only for an approximate solution to our mechanics problem rather than an exact solution as can sometimes be found by working with a differential equation analytically. We have a choice of numerical methods; `euler 0.01`, `eulerCromerPS 0.1`, and `rungeKutta4 0.1` will be examples of numerical methods that can be used to produce a state-update function. We'll write the function `eulerCromerPS` later in this chapter, and we'll also show how to make use of the general-purpose functions `euler` and `rungeKutta4` that we wrote in the last chapter. Having chosen a numerical method, we apply it to the differential equation to obtain a state-update function.

The fourth information representation we use is a list of states. The list gives the state of the particle at each time calculated by the numerical method; in other words, each list element is the state at a particular time that is one time step advanced from that of the previous list element. This is pretty much all the information we could hope to know about the particle. From it, we can graph any state variable as a function of time or some other state variable.

Other quantities we might care about, but that are not contained in the state, such as energy or momentum, are functions of the state variables. If we wish, we can write a higher-order function that produces a position function or a velocity function for the particle from the information contained in the list of states. To obtain the list of states from the state-update function, we simply iterate the state-update function using the Prelude function `iterate`, which applies the state-update function to a given initial state and then to the updated state over and over again to make a list.

Figure 16-2 should be regarded as an overview of the process of solving a one-particle mechanics problem in three dimensions. The figure is similar to Figure 15-2; the two main differences are (1) we're using the new `ParticleState` data type, which contains mass, and (2) the new figure allows a choice of numerical method where the previous figure insisted on the Euler method.

In summary, the broad outline of our process is to translate our problem (given by one-body forces) into something mathematical (a differential equation), solve the mathematics problem (by using a numerical method to produce a state-update function and iterating to produce a list of states), and then return to physics to interpret the results.

Having given an overview of the process we'll employ to solve Newton's second law, which begins with a list of one-body forces, let's take a look at some examples of one-body forces.

One-Body Forces

We introduced the definition of a one-body force earlier in this chapter, but we did not give any examples. Many of the common forces that we may want to include in Newton's second law are naturally expressed as one-body forces.

Earth Surface Gravity

An object near Earth's surface feels a gravitational force from the earth. (This is theory 2 from the list of gravity theories on page 148.) If **g** is the acceleration of gravity that points toward the center of the earth, the gravitational force exerted by Earth on a particle or object with mass m that is near Earth's surface is

$$\mathbf{F}_g = m\mathbf{g}$$

If we agree to let the z-direction of our coordinate system point away from Earth's center, and to use SI units, then a one-body force for Earth surface gravity can be written as follows:

```
-- z direction is toward the sky
-- assumes SI units
earthSurfaceGravity :: OneBodyForce
earthSurfaceGravity st
    = let g = 9.80665  -- m/s^2
      in (-mass st * g) *^ kHat
```

Recall that a one-body force is a function from particle state to force vector. The local variable `st` holds the particle state, and `mass st` extracts the mass from the particle state using the extraction function `mass` that is automatically generated because we used record syntax when we defined `ParticleState`.

If Earth's surface gravity is a force that acts on our particle, all we need to do is include `earthSurfaceGravity` in the list of one-body forces that forms the input to `newtonSecondPS`. The appropriate mass will be taken from the state, and the force of gravity will be included in Newton's second law.

Gravity Produced by the Sun

Any object with mass exerts a gravitational force on any other object with mass. (This is theory 3 from the list of gravity theories on page 148.) If the objects are spherical in shape, the force exerted by one object on another is directly proportional to the mass of each object and inversely proportional to the square of the distance between their centers. This is the content of Newton's law of universal gravity, which we will discuss in more detail in Chapter 19.

In our solar system, there are many examples of pairs of objects with one much more massive than the other, such as sun/Earth, Earth/moon, and Earth/telecommunications satellite. If we want to understand Earth's motion as it moves through the solar system, it is a good approximation to suppose two things: one, that the gravitational attraction of other planets such as Mars, Venus, and Jupiter have a very small effect on Earth, and can therefore be ignored; and two, that the sun is so massive compared to Earth that its position can be taken to be fixed. Under these approximations, the universal gravity produced by the sun can be regarded as a one-body force acting on Earth (or on Mars, Venus, Halley's comet, and so on).

The gravitational force exerted by the sun on an object or particle with mass m is

$$\mathbf{F}_{\text{sun gravity}} = -G\frac{M_s m}{r^2}\hat{\mathbf{r}}$$

where G is Newton's gravitational constant (in SI units, $G = 6.67408 \times 10^{-11}$ N m^2/kg^2), M_s is the mass of the sun ($M_s = 1.98848 \times 10^{30}$ kg), r is the distance between the center of the sun and the center of the object, and

$$\hat{\mathbf{r}} = \frac{\mathbf{r}}{|\mathbf{r}|}$$

is a unit vector pointing from the sun toward the object. The minus sign implies that the force on the object is toward the sun. A one-body force for sun gravity can be written as follows:

```
-- origin is at center of sun
-- assumes SI units
sunGravity :: OneBodyForce
sunGravity (ParticleState m _q _t r _v)
    = let bigG = 6.67408e-11  -- N m^2/kg^2
          sunMass = 1.98848e30  -- kg
      in (-bigG * sunMass * m) *^ r ^/ magnitude r ** 3
```

Here we use pattern matching on the input to extract the state variables instead of the extraction function we used for the previous one-body force of Earth surface gravity. Instead of naming the particle state with a local variable like `st`, we match the pattern of a particle state using the `ParticleState` constructor. We assign the five local variables that follow the constructor the values mass, charge, time, position, and velocity, respectively. We don't need charge, time, or velocity to compute the gravitational force that the sun exerts, so they are preceded by underscores. (We could have used *only* the underscore for any or all of the unused variables, but providing a name after the underscore reminds us of exactly what is being ignored.) The choice of whether to use extraction functions or pattern matching to get the state variables out of the state is a matter of style, and you can use whatever you like best.

If we are interested in the motion of the moon around the earth, we can express Earth's universal gravity as a one-body force that acts on the moon. Exercise 16.4 asks you to write a one-body force for the universal gravity produced by Earth. If, on the other hand, we are interested in the motion of the moon through the solar system, then gravity from both the sun and Earth is important, and it's better to use the techniques of Chapter 19.

Air Resistance

Air resistance is a one-body force that depends on the velocity of the object moving through the air. We assume that the air is still with respect to our coordinate system. In Chapter 14, we developed an expression for air resistance that we applied in one-dimensional situations. In three-dimensional situations, velocity is a vector, and the force of air resistance appears as

$$F_{\text{air}}(\mathbf{v}) = -\frac{1}{2} C\rho A \, |\mathbf{v}| \, \mathbf{v} \tag{16.6}$$

with the parameters C, ρ, and A still representing the drag coefficient, the density of the air, and the cross-sectional area of the object, respectively.

Here is the Haskell code for the one-body force that corresponds to Equation 16.6:

```
airResistance :: R  -- drag coefficient
              -> R  -- air density
              -> R  -- cross-sectional area of object
              -> OneBodyForce
airResistance drag rho area (ParticleState _m _q _t _r v)
    = (-0.5 * drag * rho * area * magnitude v) *^ v
```

We are using pattern matching on the input when we name the incoming particle state `ParticleState _m _q _t _r v`. The force depends only on the velocity, so the velocity is the only state variable that needs to be named.

For any situation in which we want to include air resistance, we need to estimate a drag coefficient, determine the cross-sectional area of our object, and determine an appropriate value for the density of air. At a reasonable temperature and pressure near Earth's surface, the density of air is about 1.225 kg/m^3. If, for example, 0.8 was our drag coefficient and 0.003 m^2 was the cross-sectional area of our object, then including

```
airResistance 0.8 1.225 0.003
```

in the list of one-body forces for `newtonSecondPS` would include the force of air resistance in Newton's second law.

If we were going to do a lot of air resistance problems, we might put the cross-sectional area of the object into the state since it's clearly a property of the object. We might even consider putting the drag coefficient into the state, if that can be regarded as a property of the object rather than a property of the interaction between the object and the air. We won't make these modifications to our state data type; instead, we'll stick with our `ParticleState`, and when forces depend on parameters that are not included in the state, we'll just handle it on a case-by-case basis, as we did here.

Wind Force

The one-body force of air resistance just considered assumes that the air is motionless with respect to our coordinate system. The one-body wind force we consider in this section is a generalization of air resistance in that the air moves at some constant velocity with respect to our coordinate system. We can use our air resistance formula to find the wind force, but the appropriate velocity to use is the *relative velocity* between the object and the wind. If $\mathbf{v}$ is the velocity of the object with respect to our coordinate system, and $\mathbf{v}_{\text{wind}}$ is the velocity of the air with respect to our coordinate system, then $\mathbf{v} - \mathbf{v}_{\text{wind}}$ is the velocity of the object with respect to the air. The wind force can be expressed as follows:

$$F_{\text{wind}}(\mathbf{v}) = -\frac{1}{2} C\rho A \left|\mathbf{v} - \mathbf{v}_{\text{wind}}\right| (\mathbf{v} - \mathbf{v}_{\text{wind}}) \tag{16.7}$$

Here is the corresponding Haskell code:

```
windForce :: Vec  -- wind velocity
          -> R    -- drag coefficient
          -> R    -- air density
          -> R    -- cross-sectional area of object
          -> OneBodyForce
windForce vWind drag rho area (ParticleState _m _q _t _r v)
    = let vRel = v ^-^ vWind
      in (-0.5 * drag * rho * area * magnitude vRel) *^ vRel
```

The code for wind force is similar to the code for air resistance. Exercise 17.5 gives an opportunity to try out this force. Notice that if the wind velocity is chosen to be 0, the wind force becomes the force of air resistance we treated in the previous section. Air resistance is the force that stationary air exerts on an object, while wind force is the force that moving air exerts on an object. If the force of air is important in a situation, you'll want air resistance or the wind force, but not both.

Force from Uniform Electric and Magnetic Fields

We haven't talked about electric or magnetic fields yet, but we will in Part III of the book. For now, the important things to know are that these fields are produced by electric charge and that a particle experiences a force in the presence of electric and/or magnetic fields. When these fields are *uniform*, meaning the same at different places in space, a single vector describes the electric field and a single vector describes the magnetic field.

Suppose that $\mathbf{E}$ is a uniform electric field vector and $\mathbf{B}$ is a uniform magnetic field vector. These fields exert a force on a charged particle traveling through them, given by

$$\mathbf{F}_{\text{Lorentz}} = q[\mathbf{E} + \mathbf{v}(t) \times \mathbf{B}] \tag{16.8}$$

where q is the electric charge of the particle and $\mathbf{v}(t)$ is the velocity of the particle. This equation is called the *Lorentz force law*, and we will study it in more detail when we turn to electromagnetic theory, including the more general situation in which the fields need not be uniform. Here is the corresponding Haskell code for the one-body force:

```
uniformLorentzForce :: Vec  -- E
                    -> Vec  -- B
                    -> OneBodyForce
uniformLorentzForce vE vB (ParticleState _m q _t _r v)
    = q *^ (vE ^+^ v >< vB)
```

The function `uniformLorentzForce` has type `Vec -> Vec -> OneBodyForce`, which is the same as `Vec -> Vec -> ParticleState -> Vec`. Given a vector `vE :: Vec` for the electric field, a vector `vB :: Vec` for the magnetic field, and a particle state `ParticleState _m q _t _r v :: ParticleState` using pattern matching on the input, the function returns a force vector by applying the Lorentz force law (Equation 16.8). The charge and velocity of the particle are the state variables needed to compute this electromagnetic force.

Having seen several examples of one-body forces, let's continue along Figure 16-2 and explore the state-update process.

State Update for One Particle

A numerical method transforms a differential equation into a state-update function. The Euler-Cromer method is a numerical method, and because it's not a general-purpose numerical method, we need to write a new function for it for each state data type we want to use. Here is the Euler-Cromer function for the `ParticleState` data type:

```
eulerCromerPS :: TimeStep         -- dt for stepping
              -> NumericalMethod ParticleState DParticleState
eulerCromerPS dt deriv st
    = let t   = time     st
          r   = posVec   st
          v   = velocity st
          dst = deriv st
          acc = dvdt dst
          v'  = v ^+^ acc ^* dt
      in st { time     = t  +        dt
            , posVec   = r ^+^ v'  ^* dt
            , velocity = v ^+^ acc ^* dt
            }
```

As we saw with the Euler-Cromer method in the last chapter, the key difference compared with the Euler method is that it uses an updated velocity to update position. The update equations in `eulerCromerPS` are almost identical to those in `eulerCromer1D` of the previous chapter, with the one difference being that we are now working with vectors.

The Euler and fourth-order Runge-Kutta methods are general-purpose methods for solving any differential equation. In Chapter 15, we wrote the `euler` and `rungeKutta4` functions, which can work with any differential equation and any state type. To use them with the `ParticleState` data type, we must write a `RealVectorSpace` instance for `DParticleState` and a `Diff` instance relating the `ParticleState` and `DParticleState` types.

Here is the `RealVectorSpace` instance:

```
instance RealVectorSpace DParticleState where
    dst1 +++ dst2
        = DParticleState { dmdt = dmdt dst1  +  dmdt dst2
                         , dqdt = dqdt dst1  +  dqdt dst2
                         , dtdt = dtdt dst1  +  dtdt dst2
                         , drdt = drdt dst1 ^+^ drdt dst2
                         , dvdt = dvdt dst1 ^+^ dvdt dst2
                         }
    scale w dst
        = DParticleState { dmdt = w *  dmdt dst
                         , dqdt = w *  dqdt dst
                         , dtdt = w *  dtdt dst
                         , drdt = w *^ drdt dst
                         , dvdt = w *^ dvdt dst
                         }
```

In this instance declaration, we define addition to be item-wise addition for each item, and we define scalar multiplication to be item-wise scaling for each item.

Here is the `Diff` instance:

```
instance Diff ParticleState DParticleState where
    shift dt dps (ParticleState m q t r v)
        = ParticleState (m  +  dmdt dps  * dt)
                        (q  +  dqdt dps  * dt)
                        (t  +  dtdt dps  * dt)
                        (r ^+^ drdt dps ^* dt)
                        (v ^+^ dvdt dps ^* dt)
```

Each item in the state is shifted by the product of its derivative with the time step.

Having made these instance declarations, we now have access to the functions `euler` and `rungeKutta4` we wrote in the previous chapter. We can use any of the three numerical methods, Euler, Euler-Cromer, or fourth-order Runge-Kutta, to produce a state-update function from a differential equation.

Figure 16-2 showed the four data representations we use to solve a mechanics problem and three functions that transform from one data representation to another. Compositions of these three functions are important enough to name and are shown as arrows on the sides of Figure 16-3. We wrote `solver` in the last chapter and will write `updatePS` and `statesPS` shortly.

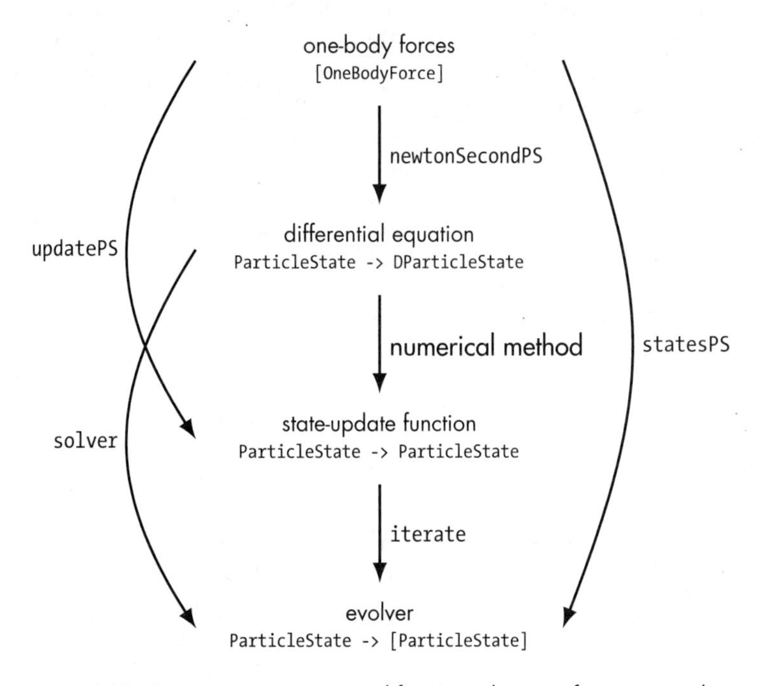

Figure 16-3: Data representations and functions that transform among them

Figure 16-3 again shows the four data representations, with one small change. Where Figure 16-2 has a list of states as the final representation, Figure 16-3 has a function from initial state to list of states, which we call an *evolver*. The reason for this change is that we want to view each representation in Figure 16-3 as the result of applying a single function to the single input consisting of the previous representation. In other words, where the initial state appears as an input in Figure 16-2, it is part of the type of the evolver in Figure 16-3. To transform between adjacent representations in Figure 16-3, we apply Newton's second law, then a numerical method, and then iteration.

Let's write a function `statesPS` that produces all three transformations of Figure 16-3 by producing the differential equation from the forces, using a numerical method to turn the differential equation into a state-update function, and iterating the state-update function to produce an evolver. The inputs to this function will be a numerical method and a list of one-body forces. The output will be an evolver that can act on an initial state to produce an infinite list of states. We call this function `statesPS` because it produces a list of states when supplied with an initial state and is for the `ParticleState` data type.

```
statesPS :: NumericalMethod ParticleState DParticleState
         -> [OneBodyForce]  -- list of force funcs
         -> ParticleState -> [ParticleState]  --evolver
statesPS method = iterate . method . newtonSecondPS
```

The local variable `method` stands for the numerical method we provide when we use `statesPS`. We see from the definition that this function is a composition of three functions, as suggested by Figure 16-3. Recall that numerical methods are things like `euler 0.01`, `eulerCromerPS 0.1`, and `rungeKutta4 0.1`. Notice that the function `newtonSecondPS`, which transforms forces into a differential equation, can be used with any numerical method.

It is similarly useful, especially for animation, to name the function that transforms from a list of forces to a state-update function. We'll call this function `updatePS`, and we see from its definition that it is simply the composition of Newton's second law with a numerical method.

```
updatePS :: NumericalMethod ParticleState DParticleState
         -> [OneBodyForce]
         -> ParticleState -> ParticleState
updatePS method = method . newtonSecondPS
```

Figure 16-3 demonstrates how this function fits into the sequence of data representations.

The final composition of transformations we might want to make is expressed by the function `solver` we wrote in the previous chapter. Unlike `statesPS` and `updatePS`, which require the `ParticleState` data type, the function `solver` works with any data type (any state space). If you look back on its definition, you will see that it is just the composition of a numerical method with iteration.

We are now in a wonderful position. All we need to do to solve any one-body problem in mechanics is give the computer:

- A numerical method
- A list of one-body forces
- The body's initial state

The computer will then calculate a list of states, which we can use to find quantities like position and velocity at arbitrary times.

Putting everything together, we can write a function `positionPS`, similar to `positionFtxv` and others we have written before, that takes the three pieces of information listed above and produces a function that gives the position of the object at any time.

```
positionPS :: NumericalMethod ParticleState DParticleState
           -> [OneBodyForce]  -- list of force funcs
           -> ParticleState   -- initial state
           -> Time -> PosVec  -- position function
positionPS method fs st t
```

```
  = let states = statesPS method fs st
        dt = time (states !! 1) - time (states !! 0)
        numSteps = abs $ round (t / dt)
        st1 = solver method (newtonSecondPS fs) st !! numSteps
    in posVec st1
```

The function begins by naming the incoming numerical method method, the list of one-body forces fs, the initial particle state st, and the time t. The first line in the let clause uses statesPS to create an infinite list of particle states based on the numerical method, forces, and initial particle state given. The second line calculates the time step by subtracting the times of the first and second states in the list. The third line finds the number of time steps necessary to get as close as possible to the desired time t. The fourth line picks out the state that is closest to the desired time, and the body of the let construction, after the in keyword, uses the extraction function posVec to pick out the position from the state.

Having written functions that allow us to solve any one-body mechanics problem using the numerical method of our choice, let's turn our attention to some last details about animation.

Preparing for Animation

In Chapter 13, we discussed how to make 2D and 3D animations with the Graphics.Gloss and Vis modules. Remember that each module has a simulate function, but the two functions are not parallel in terms of the inputs they require. In this section, we'll ease future strain on our brains by creating two new functions, simulateGloss and simulateVis, that take very similar inputs to each other, so we can switch from 2D animation to 3D and back without having to remember all of the details about how the gloss simulate function differs from the not-gloss simulate function.

Two Helpful Animation Functions

Each of the functions simulateGloss and simulateVis calls its own version of simulate to do the actual work. The intention is that we use these new functions instead of either version of simulate. We'll briefly explain how simulateGloss and simulateVis work; however, unlike many of the Haskell functions in this book, the point of writing these functions is not to illustrate important or beautiful ideas about physics or programming. Instead, the point is to make animation easier. We are willing to pay the cost of writing these functions once because we'll then enjoy the benefit of being able to use them again and again to make animations in a more convenient way.

It's more important to understand how to use these functions than it is to understand how they work. If you want to skip the definitions and the explanation of how the functions work, it will not cause you trouble later. However, do pay attention to the types of these two new functions and the inputs that must be provided for them to do their job.

Here are the type signatures and function definitions for simulateGloss and simulateVis:

```
simulateGloss :: R    -- time-scale factor
              -> Int  -- animation rate
              -> s    -- initial state
              -> (s -> G.Picture)
              -> (TimeStep -> s -> s)
              -> IO ()
simulateGloss tsFactor rate initialState picFunc updateFunc
    = G.simulate (G.InWindow "" (1000, 750) (10, 10)) G.black rate
      initialState picFunc
          (\_ -> updateFunc . (* tsFactor) . realToFrac)

simulateVis :: HasTime s => R  -- time-scale factor
            -> Int             -- animation rate
            -> s               -- initial state
            -> (s -> V.VisObject R)
            -> (TimeStep -> s -> s)
            -> IO ()
simulateVis tsFactor rate initialState picFunc updateFunc
    = let visUpdateFunc ta st
              = let dtp = tsFactor * realToFrac ta - timeOf st
                in updateFunc dtp st
      in V.simulate V.defaultOpts (1/fromIntegral rate)
      initialState (orient . picFunc) visUpdateFunc
```

The simulateGloss function makes a 2D animation, while simulateVis makes a 3D animation. Each function takes five inputs. The meaning and order of the inputs are the same in both functions, even though one of the five inputs has a different type in simulateGloss than it has in simulateVis. Let's discuss the meaning and purpose of each of the five inputs.

Time-Scale Factor

The first input to simulateGloss and simulateVis, tsFactor, says how fast we want the animation to run compared with physical evolution. There will be times when we want the animation to run faster or slower than the corresponding physical situation evolves. For example, it takes the moon about a month to orbit Earth, but we might want an animation in which the moon makes a full cycle in six seconds. We're almost always going to want an animation that happens in seconds or minutes. Smaller than that and it's too rapid to see; larger than that and we lose patience.

We can distinguish between two forms of time. *Physical time* is the time for some process to occur in the physical world. The physical time for one period of the moon orbiting Earth is one month. *Animation time* is the time for some process to occur in a computer animation. The animation time for one period of the moon orbiting Earth is six seconds in our example.

To allow a difference between physical time and animation time, our `simulateGloss` and `simulateVis` functions take a *time-scale factor* as their first input, hence the name `tsFactor`. The time-scale factor is the ratio of physical time to animation time. For the moon orbit example, where the physical time is much larger than the animation time, the time-scale factor is a number (much) bigger than 1. A time-scale factor smaller than 1 should be used for a process that occurs very quickly in the physical world that we want to view in "slow motion" so we see interesting changes as they occur. By specifying a time-scale factor as the first input to `simulateGloss` or `simulateVis`, we are declaring how fast we want the animation to run compared with physical evolution.

Animation Rate

The *animation rate*, called `rate` in the code shown earlier, is the number of picture frames displayed per second of animation time, and it's the second input to `simulateGloss` and `simulateVis`. Because a new picture frame is generated every time the state-update function is called to produce a new state, the animation rate is also the number of state updates per second of animation time.

There is a relationship between the time-scale factor, the animation rate, and the time step. If we let α denote the time-scale factor, r the animation rate, and Δt_{p} the time step (which is a physical time, hence subscript p), the relationship is

$$\Delta t_{\mathrm{p}} = \frac{\alpha}{r}$$

Only two of these can be chosen independently. Since the time step is a physical time, and since we are likely to be interested in situations with physical time scales ranging from nanoseconds to years, it's convenient to tell `simulateGloss` or `simulateVis` the time-scale factor and the animation rate and let them calculate the time step to be used for state update. This way, if we choose a sensible time-scale factor, we can pick something like 20 frames/second for the animation rate and we'll have a decent chance of using a reasonable time step. If we find we need a smaller time step, we can increase the animation rate (as we do in the next chapter for Halley's comet).

Initial State

The third input is the initial state of the particle or system we wish to animate, `initialState`. In this chapter, the state of a particle has the type `ParticleState`. In the previous chapter, the state of a particle in one dimension had type `State1D`. In Chapter 19, the state of a system of particles will have type `MultiParticleState`. Our two animation functions can work with any of these, as indicated by the type variable `s` used for the initial state.

Display Function

The fourth input, `picFunc`, is a display function that must explain what 2D or 3D picture to make for a given state. Since `gloss` uses the `Picture` type for a picture and not-gloss uses the VisObject R type, this fourth input has a different type in simulateGloss than in simulateVis. When we want to produce an animation for a specific physical situation, we'll need to write a display function for that situation. The not-gloss package has its own type for three-dimensional vectors, which differs from the Vec type we have been using. Since there is a three-dimensional translation function called Trans that takes a not-gloss vector as input, it will be useful to have a conversion function around when we write the display function for a 3D animation. The function v3FromVec produces a not-gloss vector from a Vec.

```
v3FromVec :: Vec -> V3 R
v3FromVec v = V3 x y z
    where
      x = xComp v
      y = yComp v
      z = zComp v
```

We'll use this function in the 3D animations of projectile motion and of a proton in a magnetic field in the next chapter.

State-Update Function

The fifth and final input is a state-update function, updateFunc. State-update functions have been central to our method for solving Newton's second law, even in the absence of animation. Notice that the type of the state-update function is TimeStep -> s -> s. This function must explain how to create a new state from an old state when given a time step. We are not choosing the time step here. Instead, we are specifying a function that takes a time step as input, along with an old state, and returns a new state. To obtain a state-update function, we can either apply a numerical method to a differential equation that comes from Newton's second law or use the updatePS function we defined earlier with a numerical method and a list of one-body forces.

The five inputs just discussed—the time-scale factor, animation rate, initial state, display function, and state-update function—contain everything about the physical situation we're modeling as well as all the information about how to produce a picture as a function of time.

Having discussed the inputs to the functions simulateGloss and simulateVis to make 2D and 3D animations, let's look at how the functions work.

How the Functions Work

It is easier to understand how the simulateGloss function works, so we'll start with that one. The simulateGloss function names the five inputs, tsFactor for the time-scale factor, and so on. It calls gloss's simulate function to do the actual work, passing six arguments to that function. The first argument to simulate specifies an empty window name, a window size in pixels, and a window location. Since these parameters are not so interesting, and since we are unlikely to want to change these values from animation to animation, we have picked some values that we hope will work once and for all. The second argument to `simulate` is a background color, which we choose to be black. The third, fourth, and fifth inputs are the animation rate, initial state, and display function. These are all inputs to `simulateGloss`, so they can simply be passed along to `gloss`'s `simulate`.

The final input required by `gloss`'s `simulate` is an update function, but it differs from the state-update function we have been working with in three ways. First, `gloss`'s `simulate` expects an update function whose first argument is a `ViewPort`, which we do not intend to use. To give a place for the viewport, we write an anonymous function that throws away its first argument. Second, `gloss`'s `simulate` expects an update function that works with animation time rather than physical time. Since our update functions work with physical time, we need to do a conversion here using the time-scale factor. Third, we need to convert `R` to `Float` using `realToFrac`. In summary, our `simulateGloss` function works by providing inputs to `gloss`'s `simulate` function from the inputs given to `simulateGloss`. Two of the inputs to `gloss`'s `simulate` are simply specified, three are passed unchanged from inputs to `simulateGloss`, and one is a modification of a `simulateGloss` input.

The `simulateVis` function gives the same names to its five inputs that `simulateGloss` uses because the inputs have the same meanings. It calls `not-gloss`'s `simulate` function to do the actual work, passing five arguments to that function. The first argument to `simulate` specifies some options, which we set, once and for all, to the default options. The second argument to `simulate` is the number of seconds per frame at which the animation should run. Since this is just the inverse of the animation rate, we can invert the `rate` after suitably changing its type from an integer to a real number. The third input is the initial state, which we pass along unchanged.

The fourth input is the display function, which we could pass along unchanged, but we don't because I want to take this opportunity to use the `orient` function, originally written in Chapter 13 and repeated below, to rotate the axes so that the y-axis points to the right, the z-axis points up the screen, and the x-axis points to the left and appears to extend out of the screen. In other words, I'm using `orient` so our animations will automatically use my favorite coordinate system.

```
orient :: V.VisObject R -> V.VisObject R
orient pict = V.RotEulerDeg (Euler 270 180 0) $ pict
```

The fifth and final input required by `not-gloss`'s `simulate` is an update function; however, it differs substantially from the state-update function `updateFunc` that we have to work with. Because the difference is so substantial, we use a `let` construction to define a local function `visUpdateFunc` that we will pass as the final input to `not-gloss`'s `simulate` function. We name the first `visUpdateFunc` input `ta` as a reminder that it represents animation time. We name the second `visUpdateFunc` input `st` for state. Our strategy is to use the `updateFunc` passed to `simulateVis` to calculate the value of `visUpdateFunc ta st`, which has type `s`.

The difference between `visUpdateFunc` and `updateFunc` is all in how they interpret their first argument. The first input to `visUpdateFunc`, named `ta`, is the animation time elapsed since the beginning of the animation. In contrast, the first input to `updateFunc` is the physical time *step* since the previous state was calculated. Inside the local definition of `visUpdateFunc`, we have access to the animation time `ta` that we must use to calculate the physical time step `dtp` we will send to `updateFunc`. This conversion is more complicated than for `gloss` because we are really doing two conversions: one from animation time to physical time and one from physical time since the animation's start to the physical time step. We use a nested `let` construction to define the local variable `dtp`, the physical time step we will send to `updateFunc` to produce the new state `visUpdateFunc ta st`. We calculate `dtp`, a real number with type `R`, by first converting the type of `ta` from `Float` to `R`, scaling this animation time by the time-scale factor to produce the physical time since the simulation began, and then subtracting the (physical) time of the old state. The physical time associated with state `st` is `timeOf st`. I explain how that works next.

We need to know the time (which is one of the state variables in `ParticleState`) of a state. This would be no problem if `simulateVis` was intended to work only with the `ParticleState` data type. But we want `simulateVis` to be able to work with any state space `s`, or at least any state space `s` that contains time as a state variable. To solve this problem, it seems we must invent a new type class, called `HasTime`, for types that represent states from which a particular value of time can be extracted. The type class owns just one function, `timeOf`, which pulls the time out of the state. Here is the definition of the type class `HasTime`:

```
class HasTime s where
    timeOf :: s -> Time
```

Each type that aspires to be an instance of `HasTime` must express how to implement `timeOf` through an instance declaration. Here is the instance declaration for `ParticleState`:

```
instance HasTime ParticleState where
    timeOf = time
```

In summary, our `simulateVis` function works by providing inputs to `not-gloss`'s `simulate` function from the inputs given to `simulateVis`. One of the inputs to `not-gloss`'s `simulate` is simply specified, two are passed unchanged from inputs to `simulateVis`, and two are modifications of `simulateVis` inputs.

Summary

This chapter applies Newtonian mechanics to a single object moving in three dimensions. Solving a mechanics problem is a process of transforming information through a sequence of four representations, starting with one-body forces, then a differential equation, then a state-update function, and finally a list of states. Newton's second law appears in this process as the means to transform a list of forces acting on an object into a differential equation.

A numerical method transforms a differential equation into a state-update function. We used the Euler-Cromer method and fourth-order Runge-Kutta method with appropriately chosen time steps as numerical methods in this chapter. A state-update function is an essential ingredient in an animation of motion.

The state-based paradigm remained central in this chapter, where we defined a new data type to hold the state of a particle. This new data type includes the particle's mass and charge, as well as its position and velocity and the time. We introduced the notion of a one-body force, which became the main way we talked about forces in this chapter. In the next chapter, we apply these ideas to specific examples and animate many of our results.

Exercises

Exercise 16.1. Apply the function `newtonSecondPS` to a very simple list of forces, say the list of just a constant force, and a very simple state, say the `defaultParticleState`, and find the type of the resulting expression.

Exercise 16.2. Write a function

```
constantForce :: Vec -> OneBodyForce
constantForce f = undefined f
```

that takes a force vector as input and returns a `OneBodyForce` that will return the given constant force regardless of what state it is provided with. For example, if we use `constantForce` to make a one-body force that always produces $10\hat{\mathbf{i}}$ N,

```
tenNewtoniHatForce :: OneBodyForce
tenNewtoniHatForce = constantForce (10 *^ iHat)
```

then `tenNewtoniHatForce defaultParticleState` should produce `vec 10.0 0.0 0.0`.

Exercise 16.3. Write a function

```
moonSurfaceGravity :: OneBodyForce
moonSurfaceGravity = undefined
```

that returns the gravitational force exerted by Earth's moon on an object near the moon's surface.

Exercise 16.4. Write a one-body force for the universal gravity produced by Earth.

```
earthGravity :: OneBodyForce
earthGravity = undefined
```

Exercise 16.5. Use the function `uniformLorentzForce` to find the direction of the force on a proton moving in the positive x-direction in a uniform magnetic field in the positive z-direction. There is no electric field. You can choose the speed of the proton and the magnitude of the magnetic field to be whatever you like. Given the direction of the force, what do you expect the subsequent motion of the proton to be?

Exercise 16.6. The tools we've developed in this chapter solve a mechanics problem by producing an infinite list of particle states. To make sense of that solution, we often want to extract some of the data and graph it. Suppose we wanted to plot the y-component of velocity as a function of time. We would want a list of pairs of (t, v_y) values. Write a function

```
tvyPair :: ParticleState -> (R,R)
tvyPair st = undefined st
```

that produces the desired pair of numbers from the state of one particle. Then write a function

```
tvyPairs :: [ParticleState] -> [(R,R)]
tvyPairs sts = undefined sts
```

that produces a list of numerical pairs from a list of states of one particle. You can use your `tvyPair` function in the body of this second function.

Exercise 16.7. Write a predicate

```
tle1yr :: ParticleState -> Bool
tle1yr st = undefined st
```

that returns `True` if the time contained in the state (assumed to be a number of seconds) is less than or equal to one year and returns `False` otherwise. This predicate could be used with `takeWhile` to transform an infinite list of states into a finite list in preparation for making a graph.

Exercise 16.8. Write a function

```
stateFunc :: [ParticleState]
          -> Time -> ParticleState
stateFunc sts t
    = let t0 = undefined sts
          t1 = undefined sts
          dt = undefined t0 t1
          numSteps = undefined t dt
      in undefined sts numSteps
```

that produces a function from time to particle state when given a list of states. Assume that the times between adjacent states are all the same.

Exercise 16.9. In all of our work with air resistance so far, we have assumed the density of air to be a constant. However, the density of air near Earth's surface actually decreases with altitude. A useful approximation for how air density changes with altitude is given by

$$\rho = \rho_0 e^{-h/h_0}$$

where ρ_0 is the air density at sea level, h is the distance above sea level, ρ is the air density at height h above sea level, and h_0 is a constant.

Taking h_0 = 8,500 meters and using the z-component of position for height above sea level, write a one-body force

```
airResAtAltitude :: R  -- drag coefficient
                 -> R  -- air density at sea level
                 -> R  -- cross-sectional area of object
                 -> OneBodyForce
airResAtAltitude drag rho0 area (ParticleState _m _q _t r v)
    = undefined drag rho0 area r v
```

that can be used instead of airResistance for situations in which an object finds itself at high altitude. To test this new function, the following function compares the range of a lead ball fired from sea level with some initial velocity at some angle under three different conditions: (a) no air resistance, (b) uniform air resistance, and (c) air resistance that varies with altitude. The diameter of the lead ball is 10 cm. Supply the initial state and the final line of code (the two places marked undefined), and then use the code to see the ranges for a ball fired at 45°. Try initial speeds of 10 m/s, 100 m/s, and 300 m/s.

```
projectileRangeComparison :: R -> R -> (R,R,R)
projectileRangeComparison v0 thetaDeg
    = let vx0 = v0 * cos (thetaDeg / 180 * pi)
          vz0 = v0 * sin (thetaDeg / 180 * pi)
          drag = 1
          ballRadius = 0.05    -- meters
          area = pi * ballRadius**2
          airDensity  =     1.225  -- kg/m^3 @ sea level
          leadDensity = 11342      -- kg/m^3
          m = leadDensity * 4 * pi * ballRadius**3 / 3
          stateInitial = undefined m vx0 vz0
          aboveSeaLevel :: ParticleState -> Bool
          aboveSeaLevel st = zComp (posVec st) >= 0
          range :: [ParticleState] -> R
          range = xComp . posVec . last . takeWhile aboveSeaLevel
          method = rungeKutta4 0.01
          forcesNoAir
              = [earthSurfaceGravity]
          forcesConstAir
              = [earthSurfaceGravity, airResistance    drag airDensity area]
          forcesVarAir
              = [earthSurfaceGravity, airResAtAltitude drag airDensity area]
          rangeNoAir    = range $ statesPS method forcesNoAir    stateInitial
          rangeConstAir = range $ statesPS method forcesConstAir stateInitial
          rangeVarAir   = range $ statesPS method forcesVarAir   stateInitial
      in undefined rangeNoAir rangeConstAir rangeVarAir
```

Exercise 16.10. Consider dropping a ball from a height of 10 meters near Earth's surface. Use the functions we defined in this chapter to write a function that produces a list of particle states for this motion. Extra credit if you can use the `takeWhile` function to extract the particle states with $z \geq 0$ (those in which the ball is still at or above Earth's surface) from the infinite list.

17

SATELLITE, PROJECTILE, AND PROTON MOTION

This chapter discusses three extended examples that use the ideas and code from Chapter 16 to express and solve Newtonian mechanics problems that involve one object. The examples are satellite motion, projectile motion with air resistance, and a proton in a uniform magnetic field. We'll show how to make plots and animations for each of these examples. Note that we will not begin a new module in this chapter; instead, we'll add to the `Mechanics3D` module we started in the last chapter.

Satellite Motion

As an initial example of satellite motion, consider the following: Earth orbits the sun because of the attractive force of gravity between them. Strictly speaking, Earth and the sun each orbit a point located between the two bodies. This point, called the *center of mass*, is much closer to the more massive sun than it is to the less massive Earth, so it's a decent approximation to say that Earth orbits the sun. In Chapter 19, we'll treat universal gravity as a two-body force: both the sun and Earth will accelerate in response to it, and both bodies will orbit about the center of mass. In this chapter, however, we're interested in the motion of a single object, say Earth, and we'll treat the force of gravity on Earth by the sun as a one-body force. This means we'll regard the sun as merely a piece of furniture whose role is to produce a force of gravity on Earth, but it will not partake in the full dance by feeling a force and exhibiting changed motion as a result.

For the satellite motion in this chapter, we'll only pay attention to the satellite. The larger planet or star around which the satellite orbits is imagined to be fixed in place; its only job is creating a force of gravity on the satellite.

Halley's comet orbits the sun, making one orbit every 75 years or so. The orbit is quite elliptical, with the comet traveling quickly when it's close to the sun and slowly when it's far away. In 1986, Halley's comet was close to the sun, and consequently close enough to Earth to be seen without a telescope. It is expected again in our neighborhood in 2061.

Let's animate the orbit of Halley's comet around the sun. In Chapter 16, we described how to produce an animation. We need five pieces of information for the `simulateGloss` or `simulateVis` function: a time-scale factor, an animation rate, an initial state, a display function, and a state-update function. We'll describe these now, starting with the pieces of information that have the most physical content.

State-Update Function

The state-update function `halleyUpdate` can be written using `updatePS`, shown in Figure 16-3 and defined in Chapter 16, which requires a numerical method and a list of one-body forces. The function `halleyUpdate`, and all of the code in this chapter and the next that is not part of a stand-alone program, is part of the `Mechanics3D` module we began in Chapter 16 and should be in the same source code file.

```
halleyUpdate :: TimeStep
             -> ParticleState -> ParticleState
halleyUpdate dt
    = updatePS (eulerCromerPS dt) [sunGravity]
```

For our numerical method, we'll choose the Euler-Cromer method. Recall that for animation, we won't choose the time step of the numerical method directly but rather through the time-scale factor and animation rate we choose later. The time step `dt` appears as an input to `halleyUpdate`, and we pass `dt` along to `eulerCromerPS` to form the numerical method. The list of one-body forces contains only the sun's force of gravity.

Initial State

The initial state `halleyInitial` determines whether we get a circular orbit, an elliptical orbit, or a satellite moving so fast that it escapes the gravitational pull of the sun. The mass of Halley's comet is 2.2×10^{14} kg. The net charge of the comet is zero, and we start our clock at zero as well. It's the initial position and velocity that will determine the subsequent orbit. I've chosen the initial position to be on the positive x-axis at the closest distance that the comet comes to the sun, which is 8.766×10^{10} m. When Halley's comet is closest to the sun, it's moving the fastest it will move in the course of its orbit, 54,569 m/s, in a direction perpendicular to the line connecting the comet with the sun. We call this direction the y-direction. Placing all of this information into the `ParticleState` data type, we arrive at the following expression for the initial state `halleyInitial`:

```
halleyInitial :: ParticleState
halleyInitial = ParticleState { mass     = 2.2e14            -- kg
                              , charge   = 0
                              , time     = 0
                              , posVec   = 8.766e10 *^ iHat  -- m
                              , velocity = 54569 *^ jHat }   -- m/s
```

Time-Scale Factor

Listing 17-1 shows the time-scale factor, animation rate, and display function and gives a stand-alone program for a 2D animation of satellite motion using `gloss`.

```
{-# OPTIONS -Wall #-}

import SimpleVec
    ( xComp, yComp )
import Mechanics3D
    ( ParticleState(..), simulateGloss, disk, halleyInitial, halleyUpdate )
import Graphics.Gloss
    ( Picture(..), pictures, translate, red, yellow )

diskComet :: Picture
diskComet = Color red (disk 10)

diskSun :: Picture
diskSun = Color yellow (disk 20)

halleyPicture :: ParticleState -> Picture
```

```
halleyPicture (ParticleState _m _q _t r _v)
    = pictures [diskSun, translate xPixels yPixels diskComet]
          where
            pixelsPerMeter = 1e-10
            xPixels = pixelsPerMeter * realToFrac (xComp r)
            yPixels = pixelsPerMeter * realToFrac (yComp r)

main :: IO ()
main = simulateGloss (365.25 * 24 * 60 * 60) 400
       halleyInitial halleyPicture halleyUpdate
```

Listing 17-1: Stand-alone program for a 2D animation of Halley's comet in orbit around the sun

We begin by turning on warnings. Then we import the functions we need from the `SimpleVec` module of Chapter 10, the `Mechanics3D` module we began in Chapter 16 and continue to add to in this chapter and the next, and the `Graphics.Gloss` module. The pictures `diskComet` and `diskSun` are markers for Halley's comet and the sun, respectively. The display function `halleyPicture` is one of the five ingredients needed for an animation, and it uses the comet's state to translate the comet marker to the appropriate position. The sun is displayed at the origin and does not move. In the `main` function, we choose a time-scale factor of `365.25 * 24 * 60 * 60` so that one year of physical time is one second of animation time. Since the period of Halley's comet is about 75 years, it will take about a minute and 15 seconds for the animation to show a complete orbit.

Animation Rate

For animations in general, I recommend starting with an animation rate of about 20 frames/second. For Halley's comet, that gives a time step of 1/20 of a year, much smaller than 75 years, which appears to be the important time scale of the situation. If you use 20 frames/second instead of the 400 frames/second shown in Listing 17-1, you'll notice that the orbit has some funny properties. Halley's comet just wanders off the screen and doesn't come back to go around the sun, at least not in anything close to 75 seconds. The trouble is that the comet moves very quickly when it is close to the sun, and relatively slowly when it is far away.

An accurate calculation requires a relatively small time step when the comet is close to the sun, moving rapidly, and changing direction quickly. The time step for the rest of the orbit could be substantially larger without doing any damage. There are some numerical methods that use a variable time step, but they are beyond the scope of this book. We need to increase the animation rate or decrease the time-scale factor in order to use a time step small enough to maintain accuracy during the short period of closest approach. Trying out different animation rates suggests that 400 frames/second is probably sufficient to give reasonably accurate results.

Display Function

Figure 17-1 shows one frame of the animation of Halley's comet around the sun. In this snapshot, the sun is shown in gray at the right of the figure, while Halley's comet moves to the left, away from the sun. The animation we are writing produces a yellow sun and and a red comet.

Figure 17-1: Halley's comet moving away from the sun

The display function `halleyPicture` needs to describe how to produce a picture from a state. The main thing we want to show is the position of the comet. The comet moves in the $z = 0$ plane, so we just need to deal with the x- and y-components of position in this function. The `halleyPicture` function in Listing 17-1 uses pattern matching on the input to assign the local variable `r` to the position of the current state of the comet. Position is the only state variable that the display function cares about; velocity or mass play no role in determining how the picture looks. We use the `xComp` and `yComp` functions from the `SimpleVec` module of Chapter 10 to extract the x- and y-components of the position.

The `where` keyword is similar to the `let` keyword, allowing the code writer to define local variables and functions; however, `where` and its local names come after the body of the principal function rather than before it.

The `realToFrac` function converts real numbers with type `R` into real numbers with type `Float` because `gloss`'s `translate` function requires `Float`s as input. The final picture produced contains a yellow disk to represent the sun and a red disk, translated to the appropriate place, to represent the comet. The `pictures` function in `gloss` produces a single picture from a list of pictures.

Spatial scaling needs to take place in the display function. Physical sizes are expressed in meters, while `gloss` sizes are expressed in pixels. Thus, we need to specify how this conversion is to be done. One natural scaling strategy is to show everything to scale, using a single overall scaling factor to convert meters to pixels. The `scale` function in `gloss` is excellent for this purpose because it can take as input a picture with all lengths in meters and produce another picture scaled by the number of pixels per meter that we want. But in the Halley animation, if we try to show everything to scale, using accurate values in meters for the two radii and the position of the comet and one overall scaling factor from meters to pixels, we will not be able to see the comet or the sun because the distance over which the comet roams is so vast.

Since we can't show the sizes of the sun and comet to scale, the yellow and red disks act only as markers for the location of the sun and comet; the sizes of these disks are not to scale with the orbital motion or with each other. It is easiest to specify the radii of the sun and comet in pixels rather than in meters, which would need to be scaled to pixels and scaled by a different factor than the comet position. The pictures `diskComet` and `diskSun` specify the radii for these two disks to be 10 pixels and 20 pixels, respectively. We won't scale these radii any more. These two pictures make use of the `disk` function, which we defined in Chapter 13 and repeat here:

```
disk :: Float -> G.Picture
disk radius = G.ThickCircle (radius/2) radius
```

Another reason I'd rather specify the radii in pixels is that spatial scaling is often determined by trial and error, reducing or expanding the size of an animation that's working. If this trial-and-error scaling is done on the entire picture, involving both the orbit size and the radii, it's easy to shrink the radii too much so that the disks can't be seen, or to expand the radii too much so that they fill the entire screen. In either of these cases, it's sometimes difficult to know what the problem is.

In the Halley animation, there is only one thing that needs to be scaled, and that's the position of the comet. We use a factor of 10^{-10} pixels/meter to scale the x- and y-components of the position.

Projectile Motion with Air Resistance

For our next example, let's look at a batted baseball. This is an example of projectile motion with air resistance. We'll consider two forces that act on the baseball: Earth's surface gravity and air resistance. We'll use a 145-g baseball with a diameter of 74 mm and a drag coefficient of 0.3. The list `baseballForces` contains the two one-body forces that act on the baseball. The list `baseballForces`, and all of the code in this chapter and the next that is not part of a stand-alone program, is part of the `Mechanics3D` module.

```
baseballForces :: [OneBodyForce]
baseballForces
    = let area = pi * (0.074 / 2) ** 2
      in [earthSurfaceGravity
         ,airResistance 0.3 1.225 area]
```

The first force is Earth surface gravity and the second force is air resistance. We define a local variable `area` to hold the cross-sectional area of the baseball. The number 0.074 is the diameter of the ball in meters, 0.3 is the drag coefficient, and 1.225 is the density of air in kg/m^3.

For situations that take place on or near Earth's surface, I like to use a coordinate system in which x and y are the horizontal coordinates and z is the vertical coordinate, with positive z pointing away from Earth's center. Projectile motion with air resistance takes place in a plane. It would be reasonable to choose the xz-plane or the yz-plane for this motion. We'll choose the yz-plane because the default coordinate system for the `simulateVis` function, should we choose to use it, has y to the right and z up the screen.

Calculating a Trajectory

The function `baseballTrajectory`, defined below, produces a list of (y, z) pairs, where y and z are the horizontal and vertical components of position, respectively. We give this function a time step, an initial speed, and an angle in degrees. The angle is the angle above the horizontal at which the ball leaves the bat.

```
baseballTrajectory :: R  -- time step
                   -> R  -- initial speed
                   -> R  -- launch angle in degrees
                   -> [(R,R)]  -- (y,z) pairs
baseballTrajectory dt v0 thetaDeg
    = let thetaRad = thetaDeg * pi / 180
          vy0 = v0 * cos thetaRad
          vz0 = v0 * sin thetaRad
          initialState
              = ParticleState { mass     = 0.145
                              , charge   = 0
                              , time     = 0
                              , posVec   = zeroV
                              , velocity = vec 0 vy0 vz0 }
      in trajectory $ zGE0 $
         statesPS (eulerCromerPS dt) baseballForces initialState
```

We've defined several local variables to hold the angle in radians, the horizontal and vertical components of initial velocity, and the initial state of the ball. We use `statesPS` to make an infinite list of states, using the Euler-Cromer method with the given step size, the list of forces (`baseballForces`), and the initial state. The function `zGE0`, defined below, truncates the infinite list to a finite list consisting only of states with a vertical position greater than or equal to zero. The function `trajectory`, also defined below, transforms a list of states into a list of (y, z) pairs suitable for plotting.

The infinite list produced by `statesPS` is truncated to a finite list with the function `zGE0`, which takes elements of the infinite list with a vertical position component greater than or equal to zero. As soon as it finds a vertical component less than zero, it stops checking list items and returns the finite list.

```
zGE0 :: [ParticleState] -> [ParticleState]
zGE0 = takeWhile (\(ParticleState _ _ _ r _) -> zComp r >= 0)
```

By returning a finite list of states, we are one step closer to plotting the trajectory since we can't plot an infinite list.

The trajectory function returns the horizontal and vertical components of position for each state in the input list. This would be a natural thing to plot, so we are again one step closer to plotting the trajectory.

```
trajectory :: [ParticleState] -> [(R,R)]
trajectory sts = [(yComp r,zComp r) | (ParticleState _ _ _ r _) <- sts]
```

Finding the Angle for Maximum Range

Let's go a little further with our baseball analysis. The function baseballRange computes the horizontal range of the ball for a given initial speed and angle.

```
baseballRange :: R  -- time step
              -> R  -- initial speed
              -> R  -- launch angle in degrees
              -> R  -- range
baseballRange dt v0 thetaDeg
    = let (y,_) = last $ baseballTrajectory dt v0 thetaDeg
      in y
```

To accomplish this, we use the baseballTrajectory function from earlier, take the last pair with a nonnegative vertical position component, and return the horizontal position component of that pair.

Now let's make a graph of baseball range as a function of the angle at which the ball is hit. In the absence of air resistance, the maximum range is achieved at an angle of 45°. Perhaps the presence of air resistance, which we are including, will produce different results. The function baseballRangeGraph in Listing 17-2 makes such a graph for a baseball batted at 45 m/s (101 mph).

```
baseballRangeGraph :: IO ()
baseballRangeGraph
    = plotFunc [Title "Range for baseball hit at 45 m/s"
               ,XLabel "Angle above horizontal (degrees)"
               ,YLabel "Horizontal range (m)"
               ,PNG "baseballrange.png"
               ,Key Nothing
               ] [10,11..80] $ baseballRange 0.01 45
```

Listing 17-2: Code to produce the graph "Range for baseball hit at 45 m/s"

Figure 17-2 shows the horizontal range of a batted baseball as a function of the angle at which it leaves the bat. We assume an initial speed of 45 m/s (101 mph) at every angle. Notice that the longest range occurs at an angle less than 45° above the horizontal.

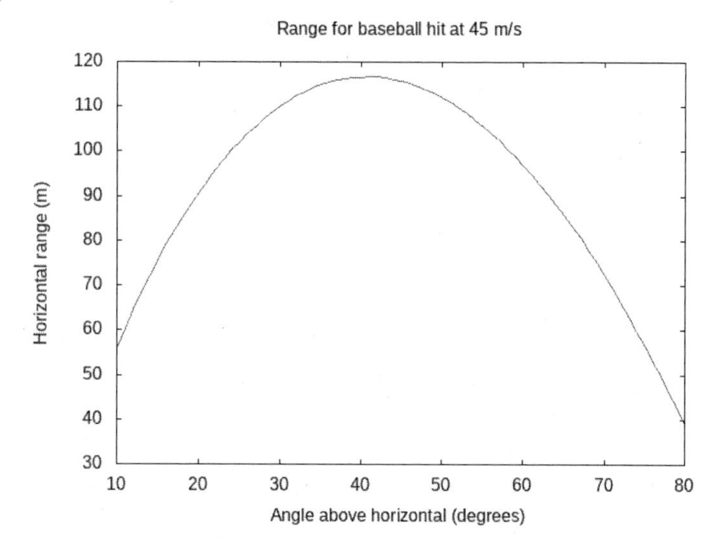

Figure 17-2: Range for a batted baseball. Because of air resistance, the longest range does not occur at an angle of 45° leaving the bat.

We can search for the angle that produces the longest range. The value `bestAngle` looks through all of the angles from 30° to 60° in 1° steps to find the angle that produces the longest range.

```
bestAngle :: (R,R)
bestAngle
    = maximum [(baseballRange 0.01 45 thetaDeg,thetaDeg) |
               thetaDeg <- [30,31..60]]
```

To find the longest range, we want to compare `baseballRange 0.01 45 thetaDeg`, the range for an initial speed of 45 m/s at an angle of `thetaDeg`, for different values of the angle. But we want the `bestAngle` function to return the angle at which we achieve the longest range, so we cannot merely ask for the maximum of `baseballRange 0.01 45 thetaDeg` because that would return only the range and not the angle that achieved that range.

We can get what we want, the longest range and the angle at which it is achieved, by comparing pairs and asking for the maximum pair. The `maximum` function uses dictionary order when comparing pairs, so the maximum pair is the one with the largest first element. If there is a tie in comparing first elements, the function compares second elements to break the tie. By choosing the first element of the pair to be the range, the comparison will be done on the range, and by choosing the second element of the pair to be the angle, the function will return the angle as well. Here is the value of `bestAngle` reported by GHCi:

```
Prelude> :l Mechanics3D
[1 of 4] Compiling Newton2          ( Newton2.hs, interpreted )
[2 of 4] Compiling Mechanics1D      ( Mechanics1D.hs, interpreted )
[3 of 4] Compiling SimpleVec        ( SimpleVec.hs, interpreted )
```

```
[4 of 4] Compiling Mechanics3D      ( Mechanics3D.hs, interpreted )
Ok, four modules loaded.
*Mechanics3D> bestAngle
(116.77499158246208,41.0)
```

We see that, to within an accuracy of 1°, the angle that produces the longest range is 41° above the horizontal.

2D Animation

Let's turn now to making an animation of the motion of the baseball. The motion takes place in a plane, so we'll use the two-dimensional gloss package. We'll make a stand-alone program to do the animation, and then we'll show how the program can take arguments that specify the initial speed and angle.

Main Program

Listing 17-3 gives a stand-alone program for a 2D animation of projectile motion using gloss.

```
{-# OPTIONS -Wall #-}

import SimpleVec
    ( yComp, zComp )
import Mechanics3D
    ( ParticleState(..), simulateGloss, disk
    , projectileInitial, projectileUpdate )
import Graphics.Gloss
    ( Picture(..), red, scale, translate )
import System.Environment
    ( getArgs )

projectilePicture :: ParticleState -> Picture
projectilePicture (ParticleState _m _q _t r _v)
    = scale 0.2 0.2 $ translate yFloat zFloat redDisk
      where
        yFloat = realToFrac (yComp r)
        zFloat = realToFrac (zComp r)
        redDisk :: Picture
        redDisk = Color red (disk 50)

mainWithArgs :: [String] -> IO ()
mainWithArgs args
    = simulateGloss 3 20
      (projectileInitial args) projectilePicture projectileUpdate

main :: IO ()
main = getArgs >>= mainWithArgs
```

Listing 17-3: Stand-alone program for a 2D animation of projectile motion. Initial speed and angle can be specified on the command line when the program is run.

As usual, the first line asks for warnings and the next several lines import functions and types we want to use in the program.

A new feature of this program is that we pass information to the program using command line arguments. A *command line argument* is a piece of information given after the name of the program you execute on the command line. For example, for a stand-alone program called *GlossProjectile*, obtained by compiling a source code file called *GlossProjectile.hs*, we might run the program by entering the following instruction at the command line:

```
$ ./GlossProjectile 30 40
```

We give the name of the executable program we want to run (preceded by a dot-slash to indicate its location in the current directory) followed by some command line arguments that allow us to pass information to the program. We want to pass the initial speed and angle in degrees.

How does the program receive and use this information? The standard module `System.Environment`, which is included when you initially install the GHC compiler, provides a function `getArgs` that returns the command line arguments as a list of strings. For example, the `getArgs` function, if used in the program *GlossProjectile* executed with the command just shown, would return the list `["30","40"]`. We can then use these strings to determine what the program does. The function `getArgs` is simple and sufficient for our purposes, but if you find yourself getting more serious about using command line arguments in your Haskell programs, you may want to look at the standard module `System.Console.GetOpt`, also included with the GHC compiler, as it provides functions to handle command line arguments in a more sophisticated way.

Knowing that we will have access to a list of strings containing the command line arguments, we write a function that does the work our `main` function did in previous animations, namely calling `simulateGloss`, but takes a list of strings as input. The function `mainWithArgs` in Listing 17-3 does exactly that. As before, `simulateGloss` requires five pieces of information: a time-scale factor, an animation rate, an initial state, a display function, and a state-update function.

We chose a time-scale factor of 3 (so the animation takes place more rapidly than the physical evolution) and an animation rate of 20 frames per second. We named the list of strings from the command line `args` in `mainWithArgs` and fed it to the function `projectileInitial`, which creates the initial state based on these strings. We will write the function `projectileInitial` shortly.

The display function `projectilePicture` in Listing 17-3 describes the picture we want to produce for a given particle state. In this display function, we create a picture and then scale the entire picture by a factor of 0.2 pixels/meter. The `scale 0.2 0.2` function shrinks the picture by a factor of five in both the horizontal and vertical dimensions.

The `main` program uses the function `getArgs` to get any command line arguments specified when the program is run. The function `getArgs` is not a pure function; it is an *effectful* function. An effect is a computation that is not purely functional (that is, it depends on or changes the world in some way). Computations that depend on program input, randomness, or the current time are effects. Sending information to a display or writing to the hard drive are also effects. An effect produced by a function is sometimes called a side effect, presumably to distinguish it from the main purpose of a function, which is to produce an output. A pure function is one that involves no effects; its output depends only on its inputs and unchanging global values. An effectful function is one whose output depends on something other than its inputs (such as user input, command line arguments, or randomness) or that has a side effect in addition to its output. In Haskell, an effectful function must have a type that involves the `IO` type constructor.

To see this, let's load the module `System.Environment` prefixed by a plus sign so that the `Mechanics3D` module will not be unloaded, which is the default behavior.

```
*Mechanics3D> :m +System.Environment
*Mechanics3D System.Environment> :t getArgs
getArgs :: IO [String]
```

The fact that `getArgs` is an effectful function is indicated by the `IO` type constructor. While a pure function's output can depend only on its inputs and unchanging global values, the output of `getArgs` depends on command line arguments, which are not function inputs or global values. For this reason, `getArgs` must have type `IO [String]` rather than `[String]`. The `IO` tag on data means that the data may have been acquired through some effect. The absence of an `IO` tag on data is a promise that the data has not been acquired through any effect.

The `IO` type constructor that labels an effectful function is one example of a collection of type constructors known as *monads*. The idea of a monad has been around in the mathematics of category theory for several decades; its use in functional programming is more recent, where it represents a computational abstraction. Haskell has a type class `Monad` for type constructors, such as `IO`, that can implement certain functions. Since the `IO` type constructor is an instance of type class `Monad`, it is also referred to as the `IO` monad. It is not the intent of this book to say much about monads. Monads are an interesting abstraction, but I don't think physics has a strong need for them. The books *Real World Haskell* [2] and *Learn You a Haskell for Great Good* [1] have nice discussions of monads. Stephen Diehl's "What I Wish I Knew When Learning Haskell" at *http://dev.stephendiehl.com/hask* also has a nice discussion of monads.

The operator `>>=`, called "bind," is the most important of the monad functions. In the context of the `IO` monad we have here, it provides a way to use information arising from an effectful function. To understand its use, let's take a look at its type.

```
*Mechanics3D System.Environment> :t (>>=)
(>>=) :: Monad m => m a -> (a -> m b) -> m b
```

The type variables a and b stand for types, while the type variable m stands for a type constructor. The type class Monad has type constructors as instances. Haskell's notion of *kind*, introduced in Chapter 9, helps to classify the possibilities of what a type variable can stand for.

For our purpose, the type variable m can be replaced with IO, which is an instance of type class Monad. Specializing the bind operator to the IO type constructor, bind has the following type:

```
IO a -> (a -> IO b) -> IO b
```

We see that bind takes two things: a value with type a "tagged" by the IO type constructor and an effectful function whose input has type a, which produces a value of type b "tagged" by the IO type constructor. We can think of this IO type constructor as a tag on the value that indicates its origin and/or effect. The bind operator allows an IO-tagged value to be used as a regular value in a function that promises to return an IO-tagged result. Since IO acts as a tag for effects, it is important that the IO tag is not removable once applied. However, if a function returns a tagged output, the bind operator provides a way for a tagged input to be temporarily untagged and used, knowing that the function will retag the output.

We are using the bind operator (>>=) in a setting where type variable a is [String] and type variable b is unit, so the concrete type of bind in our program is the following:

```
IO [String] -> ([String] -> IO()) -> IO ()
```

The bind operator is just what we need to connect the output of getArgs with the input of mainWithArgs. In fact, the main program does nothing other than pass the output of getArgs to the function mainWithArgs.

State-Update Function and Initial State

The state-update function projectileUpdate is written using updatePS, just as in satellite motion.

```
projectileUpdate :: TimeStep
                 -> ParticleState  -- old state
                 -> ParticleState  -- new state
projectileUpdate dt
    = updatePS (eulerCromerPS dt) baseballForces
```

Again, for our numerical method, we chose the Euler-Cromer method. We used the same list of one-body forces, baseballForces, that we used to make the graph shown earlier.

Knowing that we want to pass the initial speed and angle to our program as command line arguments, and that these will be available as a list of strings, we'll write the function projectileInitial to take a list of strings, which will come from the command line when we run the program, and use these strings to determine the initial velocity. We want the list of strings to have two elements: the first string gives the initial speed and the second gives the initial angle in degrees.

```
projectileInitial :: [String] -> ParticleState
projectileInitial []        = error "Please supply initial speed and angle."
projectileInitial [_]       = error "Please supply initial speed and angle."
projectileInitial (_:_:_:_)
    = error "First argument is speed.  Second is angle in degrees."
projectileInitial (arg1:arg2:_)
    = let v0       = read arg1 :: R        -- initial speed, m/s
          angleDeg = read arg2 :: R        -- initial angle, degrees
          theta    = angleDeg * pi / 180  -- in radians
      in defaultParticleState
             { mass     = 0.145  -- kg
             , posVec   = zeroV
             , velocity = vec 0 (v0 * cos theta) (v0 * sin theta)
             }
```

We use pattern matching on the input to give a helpful error message if exactly two arguments are not provided. The first line responds to the empty list, the case in which no command line arguments are given. The second line responds to the case in which only one command line argument is given. The third line responds to the case in which three or more command line arguments are given. Finally, the fourth line treats the case of exactly two command line arguments, which is what we want.

We use the `read` function to convert the strings to real numbers. The `read` function takes a string as input and produces an output that is one of many types. We need to give a type annotation to specify what type we want the string converted to. Here is an example of what `read` does:

```
*Mechanics3D System.Environment> :t read
read :: Read a => String -> a
*Mechanics3D System.Environment> read "56" :: R
56.0
```

3D Animation

Several of the items needed to create a 3D animation are the same as those needed for a 2D animation, but one is different. To compare the process of making a 3D animation with that of making a 2D animation, let's animate the projectile motion using our 3D animation tools. Listing 17-4 gives a stand-alone program for a 3D animation of projectile motion using `not-gloss`.

```
{-# OPTIONS -Wall #-}

import SimpleVec ( R, (*^) )
import Mechanics3D
    ( ParticleState(..), simulateVis
```

```
    , projectileInitial, projectileUpdate, v3FromVec )
import Vis
    ( VisObject(..), Flavour(..), red )
import System.Environment
    ( getArgs )

projectileVisObject :: ParticleState -> VisObject R
projectileVisObject st
    = let r = posVec st
      in Trans (v3FromVec (0.01 *^ r)) (Sphere 0.1 Solid red)

mainWithArgs :: [String] -> IO ()
mainWithArgs args
    = simulateVis 3 20
      (projectileInitial args) projectileVisObject projectileUpdate

main :: IO ()
main = getArgs >>= mainWithArgs
```

Listing 17-4: Stand-alone program for a 3D animation of projectile motion

The main function is exactly the same as in the 2D animation. The mainWithArgs function uses simulateVis instead of simulateGloss, but it uses the same time-scale factor, animation rate, initial-state function, and state-update function we used for the 2D animation.

The only new piece we need is a display function, projectileVisObject. In this display function, the state is named st, and we define a local variable r for the position of the object. We use a solid red sphere with radius 0.1 to represent the projectile. The not-gloss package does not measure distances in pixels; instead, a length of 1 is initially about 20 percent of the height of the screen. You can then zoom in or out by pressing E or Q, or by using the mouse. Before we translate the red sphere to its appropriate position, we need to scale the position r from meters to Vis units, and we need to convert the position to Vis's vector type. We multiply the position by a factor of 0.01 Vis units per meter so the range of the animation is not too big and not too small. We use the v3FromVec function, defined in Chapter 16, to convert to Vis's vector type before using the Trans function.

Having seen an example of projectile motion with air resistance, the technique of using command line arguments to pass information into a program, and a comparison of 2D and 3D animation, let's turn to an example that really requires 3D animation.

Proton in a Magnetic Field

Magnetic fields are used in particle accelerators to get protons or electrons to move in a circular ring to bring them to high speed and slam them into each other. This lets experimentalists look at the particles created in such high-energy collisions and learn things about the nature of particles and their interactions.

A charged particle in a uniform magnetic field will move in a circle or a helix. This is not obvious, but it is a consequence of the Lorentz force law, given as Equation 16.8 for uniform fields. Luckily for us, a helix is a nice motion for showing off our 3D animation tools.

According to the Lorentz force law, the magnetic force on a particle is proportional to the cross product $\mathbf{v}(t) \times \mathbf{B}$, which means that the force is perpendicular to both the particle's velocity and the magnetic field. Since the magnetic force is always perpendicular to the velocity of the particle, it can't make the particle speed up or slow down; it can only make the particle turn (change direction).

To produce an animation of this, we'll need a state-update function, an initial state, and a display function. Here is a state-update function for a particle in a uniform magnetic field of strength 3×10^{-8} Tesla.

```
protonUpdate :: TimeStep -> ParticleState -> ParticleState
protonUpdate dt
    = updatePS (rungeKutta4 dt) [uniformLorentzForce zeroV (3e-8 *^ kHat)]
```

We're using fourth-order Runge-Kutta as our numerical method because it produces good results with a fairly large step size, while Euler-Cromer requires a rather small step size to produce good results. There's no way to know this in advance. It's always a good idea to check that results are stable with changes in step size. The list of forces has only one item, the Lorentz force of a uniform magnetic field. The `zeroV` is for zero electric field.

Here is the initial state of the proton on which the magnetic field acts:

```
protonInitial :: ParticleState
protonInitial
    = defaultParticleState { mass     = 1.672621898e-27  -- kg
                           , charge   = 1.602176621e-19  -- C
                           , posVec   = zeroV
                           , velocity = 1.5*^jHat ^+^ 0.3*^kHat  -- m/s
                           }
```

By giving the proton an initial component of velocity in both the y- and z-directions, we'll get a helix for the motion. If either of these components is set to 0, a different kind of motion will ensue. Play around and see what happens.

Here is the display function for the proton in the magnetic field:

```
protonPicture :: ParticleState -> V.VisObject R
protonPicture st
    = let r0 = v3FromVec (posVec st)
      in V.Trans r0 (V.Sphere 0.1 V.Solid V.red)
```

A red ball is used to mark the location of the proton.

Listing 17-5 shows a stand-alone Haskell program to animate a proton in a magnetic field. The time-scale factor is set to one, so this is a real-time animation. Note that the magnetic field is very small in this example and that a larger magnetic field would cause the proton to complete a turn of the helix in much less time. The animation rate is set to 60 frames/second.

```
{-# OPTIONS -Wall #-}

import Mechanics3D (simulateVis, protonInitial, protonPicture, protonUpdate)

main :: IO ()
main = simulateVis 1 60 protonInitial protonPicture protonUpdate
```

Listing 17-5: Stand-alone Haskell program for 3D animation of a proton in a uniform magnetic field

Summary

In this chapter, we used the ideas and code from Chapter 16 to study three examples of the motion of a single particle in three dimensions subject to different forces. We gave examples of 2D or 3D animation for satellite motion, projectile motion with air resistance, and a particle in a magnetic field. In the next chapter, we'll show how the ideas and code from Chapter 16 can be used or modified slightly to treat single-particle mechanics problems with the theory of relativity instead of Newton's second law.

Exercises

Exercise 17.1. Modify the `halleyPicture` function in Listing 17-1 to include x- and y-axes in the animation. You will be able to see that the comet's aphelion (point of greatest distance from the sun) is not quite aligned with the x-axis. This is an indication of inaccuracy in the numerical method, which can be reduced by decreasing the time-scale factor and reducing the time step of the numerical method.

Exercise 17.2. Let's treat Earth as being fixed at the origin of our coordinate system. Consider the gravitational force on a satellite of mass m, initial position $\mathbf{r}_0$, and initial velocity $\mathbf{v}_0$. Since the motion of the satellite will take place in a plane, we can use vectors that lie in the xy-plane. Plot trajectories of orbits resulting from various initial conditions. Choose some values for initial conditions that give nearly circular orbits and some others that give elliptical orbits. You will find that the Euler method produces orbits that don't close on themselves. Make one plot comparing the Euler and Euler-Cromer methods for one orbit that you like (elliptical or circular). Indicate the step size you used for the Euler and Euler-Cromer methods as well as your choice of initial conditions.

Exercise 17.3. The Lorentz force law, Equation 16.8, describes the force exerted on a particle with charge q and velocity $\mathbf{v}(t)$ by an electric field $\mathbf{E}$ and a magnetic field **B**. Consider a uniform magnetic field in the z-direction. You may already know that a charged particle with initial velocity in the x-direction will go in circles in this magnetic field. Choose some values for the strength of the magnetic field, the charge of the particle, the mass of the particle, and the initial velocity. Confirm, using the Euler-Cromer method, that the particle does indeed go in circles. Plot y versus x for different time steps. Even the Euler-Cromer method will not produce circles that close on themselves if the time step is too big. One of your time steps should be small enough that the orbit appears to close.

Exercise 17.4. Return to the satellite orbiting Earth. Write a Haskell program to animate your satellite's motion around Earth. Show that by using different initial conditions, you can achieve circular orbits and elliptical orbits.

Exercise 17.5. Suppose the wind is blowing horizontally at 10 m/s and you launch a Ping-Pong ball straight up into the air with an initial speed of 5 m/s. How far from the launch point will it hit the ground? You can come up with pretty good estimates for the density of air and the mass and cross-sectional area of the ball, but an estimate for the drag coefficient is more of a guess. Try the calculation for drag coefficients of 0.5, 1.0, and 2.0 to see how they compare. Repeat the calculation for a golf ball launched upward with the same speed.

Exercise 17.6. Make the necessary modifications to the code presented in this chapter so that the initial position of the baseball is 1 meter above the ground. Make a plot of the trajectory for a line drive with a speed of 40 m/s at an angle 5° above horizontal.

Exercise 17.7. Investigate the effect of a 30-mph crosswind on a batted baseball. Assuming the wind is perpendicular to the plane in which the ball would otherwise travel, how far does the wind displace the ball from the place where it would land without the wind? Choose some reasonable values for initial speed and angle of the ball.

Exercise 17.8. Given an initial speed and a drag coefficient, the optimal angle is the angle that produces the longest range for the baseball. Make a graph of the optimal angle as a function of drag coefficient for an initial speed of 45 m/s.

Exercise 17.9. If you can produce uniform electric and magnetic fields, you can make a device called a *velocity selector*. The purpose of a velocity selector is to allow charged particles that are traveling at a specific velocity to proceed in a straight line, while similar particles traveling faster or slower get deflected. From a beam of charged particles moving with a range of speeds, the velocity selector can produce a beam of particles that all have very close to the same speed. In this way, the experimentalist has access to a beam of charged particles with a known velocity that can be used for some experiment.

Let's model a velocity selector using a uniform electric field of 300 N/C in the positive z-direction and a uniform magnetic field of 0.3 T in the positive x-direction. We're interested in the motion of a singly ionized particle with mass 1.00×10^{-22} kg. (Singly ionized means that one electron has been removed so the particle has the charge of a proton.) We'll give this particle an initial velocity in the positive y-direction. If the particle is moving too fast, it will deflect one way; if it's moving too slowly, it will deflect another way.

Using the `Vis` module, make a stand-alone program that takes the initial speed of the particle as input (similar to how our projectile motion program took initial speed and angle as input) and produces an animation for the particle in the velocity selector. To judge whether and how much the particle is deflecting, include a coordinate system in your picture (similar to the coordinate system we displayed in Chapter 13), so you can tell when the particle departs from the y-axis. Use a time-scale factor of 5×10^{-4} and an animation rate of 60 frames/second. Run this program with different initial speeds between 0 and 5,000 m/s.

(a) Confirm that the particle deflects one way for slow speeds.

(b) Confirm that the particle deflects another way for fast speeds.

(c) Extend your program to include a circular aperture at $y = 1$ m that allows particles to pass through. Start with an aperture radius of 4 cm.

```
apR :: R
apR = 0.04  -- meters
```

Particles outside this radius will be blocked by a wall and not allowed to pass through. Modify your state-update function to include a wall force that blocks particles outside the aperture radius when they get to $y = 1$ m. You can use the following wall force:

```
wallForce :: OneBodyForce
wallForce ps
    = let m = mass ps
          r = posVec ps
          x = xComp r
          y = yComp r
          z = zComp r
          v = velocity ps
          timeStep = 5e-4 / 60
      in if y >= 1 && y < 1.1 && sqrt (x**2 + z**2) > apR
         then (-m) *^ (v ^/ timeStep)
         else zeroV
```

This wall force applies a dissipative force that will slow the particle to a crawl in just a few time steps if its y-value is between 100 cm and 110 cm and it is outside the aperture radius. You could think of the wall as 10 cm of lead, but the real reason for the 10 cm is to catch particles that are moving very quickly; the wall thickness could be reduced if the time step was reduced. (You may notice that particles crawl along the wall at a slow speed or move through the 10 cm–long aperture "tube" because the electric field still acts on them.) Modify your display function to include a circle for the aperture. Find the range of velocities allowed through the 4-cm aperture. What range of velocities are allowed through a 1 cm–radius aperture? How about a 1 mm–radius aperture?

Try to guess how the target velocity (velocity of particle when undeflected) is related to the numeric values of electric and magnetic fields.

Exercise 17.10. Use the `Vis` module to animate the motion of Halley's comet. Instead of Euler-Cromer, use fourth-order Runge-Kutta because the `Vis` module can't achieve an animation rate of 400 frames/second (although it won't tell you this and it will do the best it can). Try animation rates of 20 frames/second (too small, the orbits spiral inward toward the sun), 60 frames/second (pretty good), and 400 frames/second (really good, even though 400 frames/second is not being achieved). You may want to use the following function as the last transformation of your picture before handing it off to `simulateVis`:

```
zOut :: V.VisObject R -> V.VisObject R
zOut = V.RotEulerDeg (Euler 90 0 90)
```

The function `zOut` orients the display so that the xy-plane is more or less the plane of the screen, and z points out from the screen. The default orientation has x pointing out of the screen, y to the right, and z up the screen.

Exercise 17.11. Use `gnuplot` to make plots of the Halley's comet orbit using different numerical methods. Use Euler, Euler-Cromer, and fourth-order Runge-Kutta, each with time steps of 1/20 year, 1/60 year, and 1/400 year. The results should look Figure 17-3, where the left column is Euler, the middle column is Euler-Cromer, and the right column is fourth-order Runge-Kutta. The top row uses a time step of 1/20 year, the middle row 1/60 year, and the bottom row 1/400 year.

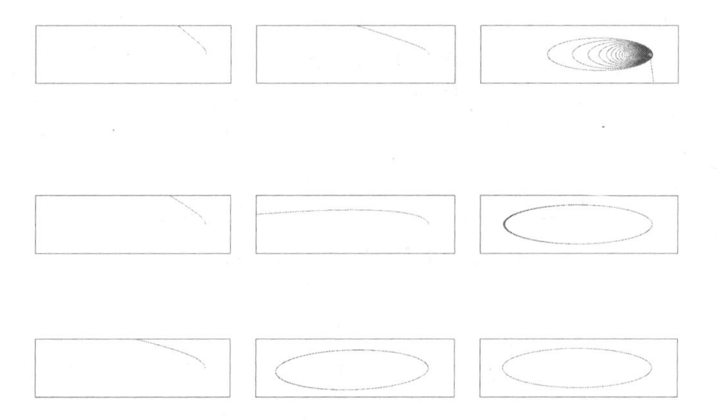

Figure 17-3: Halley's comet orbit using Euler, Euler-Cromer, and fourth-order Runge-Kutta, with time steps of 1/20 year, 1/60 year, and 1/400 year, respectively.

When calculating orbits, we can take advantage of energy conservation to check the numerical accuracy of the integration scheme. The particle state structure has the position of the comet, from which we can find the potential energy, and the velocity, from which we can compute the kinetic energy. Write a function

```
energy :: ParticleState -> R
energy ps = undefined ps
```

that computes the total energy of Halley's comet in a given `ParticleState`. We know that the total energy of Halley's comet is conserved, so any change in the energy we calculate is due to inaccuracies in the numerical method we are using. We can use the fractional change in energy over a period of one orbit as a measure of how good our numerical method is.

For numerical methods that are reasonably good, such as the three in the lower-right corner of Figure 17-3, we can use the following predicate with `takeWhile` to truncate the list of states after one orbit has occurred:

```
firstOrbit :: ParticleState -> Bool
firstOrbit st
    = let year = 365.25 * 24 * 60 * 60
      in time st < 50 * year || yComp (posVec st) <= 0
```

This predicate works by accepting the first 50 years of orbit data (recall the period is about 75 years), at which point the y-component of position is negative, and continuing to accept data until the y-component becomes positive, which indicates the beginning of the second orbit.

Calculate the fractional change in energy over a period of one orbit for (a) the Euler-Cromer method with step size 1/400 year (you should get about one percent), (b) the fourth-order Runge-Kutta method with step size 1/60 year, and (c) the fourth-order Runge-Kutta method with step size 1/400 year. For an extra challenge, label each `gnuplot` graph with the fractional change in energy over a single orbit.

18

A VERY SHORT PRIMER ON RELATIVITY

Albert Einstein was fascinated with the electromagnetic theory that we'll discuss in Part III of this book. An effort to understand it led him to new ideas about space and time, collectively called *special relativity*, which he published in 1905, and which modified the ideas of Newtonian mechanics that had been in place for more than 200 years.

Special relativity departs from Newtonian physics in several ways, the most significant conceptual departure being the non-universality of time—that is, the idea that clocks in different patterns of motion evolve at different rates. The subject of special relativity deserves an entire course for one to develop insight and intuition into relativistic kinematics and dynamics, but here we'll only scratch the surface.

In this chapter, we'll go beyond Newtonian mechanics to show how the theory of special relativity makes different predictions for the motion of a particle, especially when the particle is moving very fast. The framework of Figure 16-2, in which we transform from forces to a differential equation to a state-update function and finally to a list of states, still works well. It's just that Newton's second law needs to be replaced with a relativistic law of motion to compute special relativity's prediction for the motion of a particle experiencing forces. The relativistic law will transform the forces into a differential equation different from the one that Newton's second law produces. The remaining steps of solving the differential equation are the same in relativity as they are in Newtonian mechanics. At the end of the chapter, we'll show some examples where Newtonian mechanics and relativity make different predictions. Let's begin by getting specific about how special relativity departs from Newtonian mechanics.

A Little Theory

In special relativity, the net force acting on a particle is still the vector sum of all of the forces acting on the particle.

$$\mathbf{F}_{\text{net}}(t, \mathbf{r}(t), \mathbf{v}(t)) = \sum_j \mathbf{F}_j(t, \mathbf{r}(t), \mathbf{v}(t))$$

There are no new forces in relativity. The theory of special relativity claims that net force is close to, but not quite equal to, mass times acceleration, as Newton's second law says. The difference is more noticeable as objects move faster and closer to the speed of light. However, there is a version of Newton's second law that continues to hold in relativity. Net force is still the rate of change of momentum with respect to time. Equation 16.1 needs to be replaced by

$$\mathbf{F}_{\text{net}}(t, \mathbf{r}(t), \mathbf{v}(t)) = \frac{d\mathbf{p}(t)}{dt}$$

where $\mathbf{p}(t)$ is the momentum of the object under consideration.

The relationship between velocity and momentum is different in special relativity than in Newtonian mechanics. In Newtonian mechanics, the momentum of a particle is its mass times its velocity, $\mathbf{p}(t) = m\mathbf{v}(t)$. In relativistic physics, the momentum of a particle is

$$\mathbf{p}(t) = \frac{m\mathbf{v}(t)}{\sqrt{1 - \frac{\mathbf{v}(t) \cdot \mathbf{v}(t)}{c^2}}}$$

where $c = 299{,}792{,}458$ m/s, the speed of light in vacuum. We can algebraically invert this equation to give an expression for velocity in terms of momentum.

$$\mathbf{v}(t) = \frac{\mathbf{p}(t)}{\sqrt{m^2 + \frac{\mathbf{p}(t) \cdot \mathbf{p}(t)}{c^2}}}$$

Acceleration is still the rate of change of velocity with respect to time, so by taking a time derivative of the previous equation and substituting net force for the time derivative of momentum, we arrive at a relativistic expression for acceleration in terms of net force.

$$\frac{d\mathbf{v}(t)}{dt} = \sqrt{1 - \frac{\mathbf{v}(t) \cdot \mathbf{v}(t)}{c^2}} \left[\frac{\mathbf{F}_{\text{net}}(t, \mathbf{r}(t), \mathbf{v}(t))}{m} - \left(\frac{\mathbf{F}_{\text{net}}(t, \mathbf{r}(t), \mathbf{v}(t))}{m} \cdot \frac{\mathbf{v}(t)}{c} \right) \frac{\mathbf{v}(t)}{c} \right] \tag{18.1}$$

Equation 18.1 is the relativistic replacement for Equation 16.5. You can see that if the ratio of particle velocity to light velocity is much smaller than 1, the right side of this equation reduces to net force divided by mass, and we regain the original form of Newton's second law. This means that for something like a baseball traveling through the air, Newton's second law is basically sound. But if it were hit close to the speed of light, we'd need relativity.

If Equation 18.1 seems too complicated and ugly to be correct, you should know that special relativity has its own notation that makes equations like Equation 18.1 appear much nicer. Relativity's notation uses 4-vectors because spacetime has four dimensions. The vectors of Newtonian mechanics we are using in this book are called 3-vectors because space has three dimensions. From the perspective of relativity, 3-vectors are based on an arbitrary division of space-time into a particular three-dimensional space and a particular one-dimensional time. Some quantities that we think of as distinct, like momentum and energy, come together in relativity to form the 4-vector that Taylor and Wheeler call *momenergy*.[1] It's only when we cast the newer ideas of relativity in the older notation of Newtonian mechanics that they appear so complicated. Nevertheless, although there is a different notation that can be used for special relativity, it gives the same results as the notation we use in this book.

A Replacement for Newton's Second Law

In Chapters 16 and 17, we used the function `newtonSecondPS` to produce a differential equation that expresses Newton's second law. The function `relativityPS`, which we'll write next, produces the differential equation that special relativistic dynamics prescribes, and therefore it serves as a replacement for `newtonSecondPS`. Fortunately, we can use the same data type for particle state, namely `ParticleState`, that we have been using throughout the previous chapters.

The key difference between `newtonSecondPS` and `relativityPS` is in the expression we return for acceleration. We want to use Equation 18.1 instead of net force divided by mass. The function `relativityPS` assumes SI units, so velocity is expressed in meters per second. Relativity is more elegantly expressed in natural, or geometrized, units in which $c = 1$, implying that one second is interchangeable with 299,792,458 meters. Exercise 18.2 asks you to write a similar function that does not assume SI units.

Here is `relativityPS`, which we include in the `Mechanics3D` module that includes all of the code in Chapters 16, 17, and 18 that is not part of a standalone program.

```
relativityPS :: [OneBodyForce]
             -> ParticleState -> DParticleState  -- a differential equation
```

1. Taylor and Wheeler's *Spacetime Physics*, available at *https://www.eftaylor.com/spacetimephysics*, is a good introduction to special relativity. What Taylor and Wheeler call momenergy is called 4-momentum by most other books.

```
relativityPS fs st
    = let fNet = sumV [f st | f <- fs]
          c = 299792458  -- m / s
          m = mass st
          v = velocity st
          u = v ^/ c
          acc = sqrt (1 - u <.> u) *^ (fNet ^-^ (fNet <.> u) *^ u) ^/ m
      in DParticleState { dmdt = 0     -- dm/dt
                        , dqdt = 0     -- dq/dt
                        , dtdt = 1     -- dt/dt
                        , drdt = v     -- dr/dt
                        , dvdt = acc   -- dv/vt
                        }
```

The let clause introduces local variables for the net force, the speed of light, the mass and velocity contained in the state, the velocity u expressed in units of the speed of light, and the acceleration determined by Equation 18.1. The time derivative of the state is then prepared and returned in the body of the let construction.

Let's now look at the first of two examples comparing Newtonian mechanics to special relativity theory.

Response to a Constant Force

Let's contrast the predictions of special relativity with those of Newtonian mechanics. The first situation we'll explore is the motion of a particle, initially at rest, that experiences a constant force for some extended time period.

Figure 18-1 shows a graph of velocity as a function of time for a 1-kg object experiencing a 10-N force. This is close to the gravitational force acting on a 1-kg object at the surface of the earth (a 1-*g* acceleration).

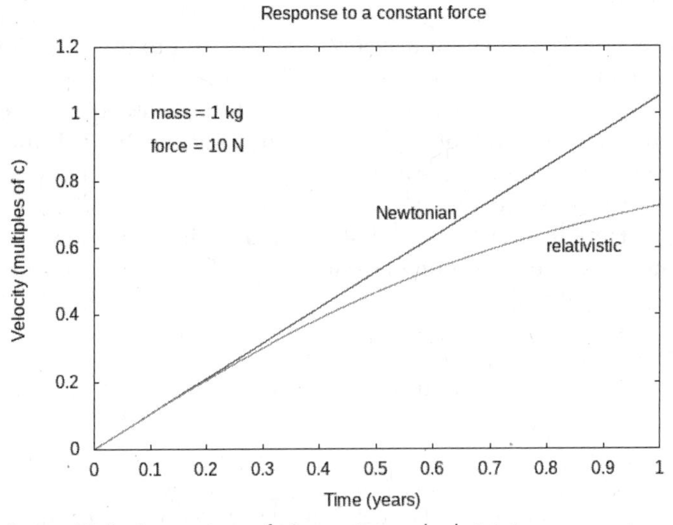

Figure 18-1: Comparison of Newtonian and relativistic response to a constant force. A mass of 1 kg experiences a constant force of 10 N.

For the first few months, there is little difference in velocity between the relativistic prediction and that of Newtonian mechanics. However, as the velocity gets closer to that of light, we do see a difference, with the relativistic curve predicting a velocity asymptotically approaching, but never reaching, that of light, while the Newtonian velocity increases linearly and eventually exceeds that of light. As there is very strong experimental evidence that objects with mass cannot travel faster than light, the Newtonian prediction is clearly incorrect.

Listing 18-1 shows the code that produced the graph.

```
constantForcePlot :: IO ()
constantForcePlot
    = let year = 365.25 * 24 * 60 * 60  -- seconds
          c = 299792458                 -- m/s
          method = rungeKutta4 1000
          forces = [const (10 *^ iHat)]
          initialState = defaultParticleState { mass = 1 }
          newtonStates = solver method (newtonSecondPS forces) initialState
          relativityStates = solver method (relativityPS forces) initialState
          newtonTVs = [(time st / year, xComp (velocity st) / c)
                          | st <- takeWhile tle1yr newtonStates]
          relativityTVs = [(time st / year, xComp (velocity st) / c)
                              | st <- takeWhile tle1yr relativityStates]
      in plotPaths [Key Nothing
                   ,Title "Response to a constant force"
                   ,XLabel "Time (years)"
                   ,YLabel "Velocity (multiples of c)"
                   ,PNG "constantForceComp.png"
                   ,customLabel (0.1,1) "mass = 1 kg"
                   ,customLabel (0.1,0.9) "force = 10 N"
                   ,customLabel (0.5,0.7) "Newtonian"
                   ,customLabel (0.8,0.6) "relativistic"
                   ] [newtonTVs,relativityTVs]
```

Listing 18-1: Code to produce the graph "Response to a constant force"

Several local variables are defined in the beginning of the code, such as the number of seconds in a year, the speed of light in meters per second, a numerical method, an initial state, and so on. The first five local variables are used in both the Newtonian and relativistic calculations. The lists `newtonStates` and `relativityStates` are infinite lists of states for the Newtonian and relativistic theories, respectively. In comparing their definitions, we see that they use the same numerical method, the same forces (a single 10-N force in the x-direction), and the same initial state. The only difference is that we replaced `newtonSecondPS` with `relativityPS` as the function that produces the differential equation we're solving.

Finally, the lists `newtonTVs` and `relativityTVs` are lists of time-velocity pairs suitable for plotting. The definitions of these two lists are almost identical. In each case, the code uses a list comprehension along with the `takeWhile` function to produce a finite list. The predicate `tle1yr` asks whether the time associated with a state is less than or equal to one year. You were asked to write this function in Exercise 16.7.

The code places several labels on the graph using the `customLabel` function, which I first introduced in Chapter 11 and repeat here for convenience.

```
customLabel :: (R,R) -> String -> Attribute
customLabel (x,y) label
    = Custom "label"
      ["\"" ++ label ++ "\"" ++ " at " ++ show x ++ "," ++ show y]
```

Proton in a Magnetic Field

As a second example contrasting the predictions of special relativity with those of Newtonian mechanics, let's look at the motion of a charged particle in a magnetic field. While our first example of special relativity took place in one spatial dimension, this example takes place in two. Figure 18-2 shows trajectories of a proton in a 1-Tesla magnetic field pointing in the z-direction (perpendicular to the plane of the circular trajectories).

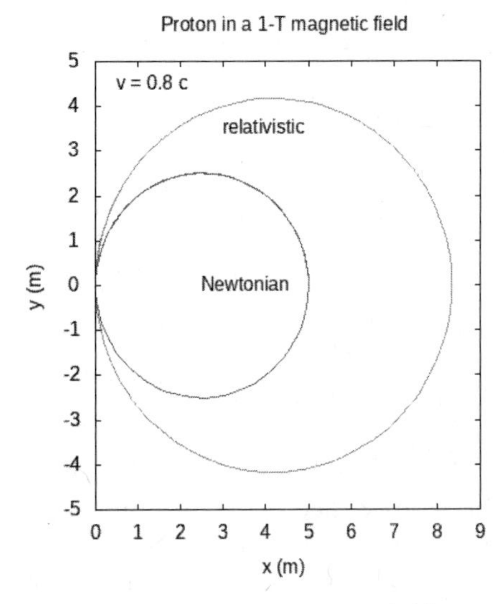

Figure 18-2: Proton in a magnetic field

The proton has a speed $4/5$ that of light. Both the Newtonian theory and the relativistic theory predict circular motion, but the radii of the circles are different. We see from the graphs that relativity predicts a larger radius compared with the Newtonian theory. The relativistic radius turns out to be bigger by a factor of $1/\sqrt{1-v^2/c^2}$, a factor that shows up in many places in relativity. In this case, the relativistic radius is $5/3$ times that of the Newtonian radius.

Listing 18-2 shows the code that produced the trajectories.

```
circularPlot :: IO ()
circularPlot
    = let c = 299792458  -- m/s
          method = rungeKutta4 1e-9
          forces = [uniformLorentzForce zeroV kHat]    -- 1 T
          initialState = defaultParticleState
                         { mass     = 1.672621898e-27  -- kg
                         , charge   = 1.602176621e-19  -- C
                         , velocity = 0.8 *^ c *^ jHat
                         }
          newtonStates = solver method (newtonSecondPS forces) initialState
          relativityStates = solver method (relativityPS forces) initialState
          newtonXYs = [(xComp (posVec st), yComp (posVec st))
                          | st <- take 100 newtonStates]
          relativityXYs = [(xComp (posVec st), yComp (posVec st))
                              | st <- take 120 relativityStates]
      in plotPaths [Key Nothing
                   ,Aspect (Ratio 1)
                   ,Title "Proton in a 1-T magnetic field"
                   ,XLabel "x (m)"
                   ,YLabel "y (m)"
                   ,PNG "circularComp.png"
                   ,customLabel (0.5,4.5) "v = 0.8 c"
                   ,customLabel (2.5,0.0) "Newtonian"
                   ,customLabel (3.0,3.5) "relativistic"
                   ] [newtonXYs,relativityXYs]
```

Listing 18-2: Code to produce the graph "Proton in a 1-T magnetic field"

The first four local variables, which are used in the predictions of both theories, define the speed of light, a numerical method, a list of forces, and an initial state.

How can we choose an appropriate time step for the numerical method? Guessing can work, but a time step that is much too large usually gives unintelligible results, and a time step that is much too small may result in very little motion, or a calculation that takes a very long time, depending on what you ask of the computer. The key is that we want the time step to be small compared with the important time scales of the situation. The important time scales of a situation can be found from dimensional analysis. Using dimensional analysis, multiplying or dividing relevant parameters to produce a quantity with dimensions of time, we can find one or more characteristic time scales. The relevant parameters in this situation are the proton charge, the proton mass, the 1-T magnetic field, and the initial velocity of $4c/5$.

The only way to form a quantity with dimensions of time from these parameters is to divide the proton mass by the product of the proton charge with the magnetic field. This combination of parameters produces a time of $m_p/(q_pB) = 1.04 \times 10^{-8}$ s. To make the time step small compared with the relevant time scale of the problem, we should divide this time by 100 or 1,000. Therefore, a time step of 10^{-10} s would be a good first guess.

The lists newtonStates and relativityStates are, as in the constant force example from earlier, infinite lists of states for the Newtonian and relativistic cases, respectively. The lists newtonXYs and relativityXYs are lists of (x, y) pairs suitable for plotting. Since we end up with circular motion, it's aesthetically pleasing to use the same scale for the x-axis that we use for the y-axis. This can be achieved with the Aspect (Ratio 1) option in the list of options.

The speeds in the relativistic and Newtonian calculations are the same, but since the relativistic circle is bigger, the period of the proton's motion (the time to go around the circle once) is larger in the relativistic theory. However, this fact is not apparent in the graph. Because of this, and also to show a technique for animating two separate motions, we'll make an animation for these protons.

Since the motion takes place in two dimensions, we'll use gloss for the animation. What we're animating here is not the interaction between two protons, which would be one physical problem consisting of multiple particles, and the kind of thing we'll discuss in the next chapter. Rather, we're interested in an animation that shows the independent motion of the two protons at the same time. Up to now, the state space for every animation we have written has been the same as the state space for the underlying physical situation. For a single particle in three dimensions, that state space is ParticleState. Now we want to animate two particles, each of which uses the state space ParticleState. This means that the state space for the animation needs to be (ParticleState,ParticleState) so the animation can keep track of both particles.

The following state-update function for animation combines two state-update functions: one for the Newtonian theory and one for the relativistic.

```
twoProtUpdate :: TimeStep
              -> (ParticleState,ParticleState)
              -> (ParticleState,ParticleState)
twoProtUpdate dt (stN,stR)
    = let forces = [uniformLorentzForce zeroV kHat]
      in (rungeKutta4 dt (newtonSecondPS forces) stN
         ,rungeKutta4 dt (relativityPS   forces) stR)
```

The local variable stN represents the incoming (not-yet-updated) state for the Newtonian calculation, while stR is the analogous state for the relativistic calculation.

The initial state for the animation combines the initial states for the two situations (Newtonian and relativistic), which are the same.

```
twoProtInitial :: (ParticleState,ParticleState)
twoProtInitial
    = let c = 299792458  -- m/s
          pInit = protonInitial { velocity = 0.8 *^ c *^ jHat }
      in (pInit,pInit)
```

The display function produces a blue disk for the Newtonian calculation and a red disk for the relativistic calculation.

```
twoProtPicture :: (ParticleState,ParticleState) -> G.Picture
twoProtPicture (stN,stR)
    = G.scale 50 50 $ G.pictures [G.translate xN yN protonNewtonian
                                 ,G.translate xR yR protonRelativistic]
      where
        xN = realToFrac $ xComp $ posVec stN
        yN = realToFrac $ yComp $ posVec stN
        xR = realToFrac $ xComp $ posVec stR
        yR = realToFrac $ yComp $ posVec stR
        protonNewtonian = G.Color G.blue (disk 0.1)
        protonRelativistic = G.Color G.red (disk 0.1)
```

Listing 18-3 shows the main program for the animation. This, and all of the other stand-alone programs, are not part of the Mechanics3D module. It uses a time-scale factor of 10^{-8}, an animation rate of 20 frames/second, and the three functions we just defined.

```
{-# OPTIONS -Wall #-}

import Mechanics3D
    ( simulateGloss
    , twoProtInitial, twoProtPicture, twoProtUpdate
    )

main :: IO ()
main = simulateGloss 1e-8 20
       twoProtInitial twoProtPicture twoProtUpdate
```

Listing 18-3: Stand-alone program for 2D animation of proton motion in a magnetic field

Summary

In this chapter, we introduced special relativity as a different, more modern theory of mechanics, which our methods are capable of treating if we replace Newton's second law with the appropriate relativistic recipe for creating a differential equation from a list of forces. Solving a mechanics problem using relativity is still a process of transforming information through a sequence of four representations, starting with one-body forces, then a differential equation, then a state-update function, and finally a list of states. The relativity law, Equation 18.1, appears in this process as the means to transform a list of forces acting on an object into a differential equation. A numerical method still transforms a differential equation into a state-update function, and we still use iteration to produce a list of states as the solution to a mechanics problem. We were able to use the same ParticleState data type that we used for Newtonian mechanics. We developed ideas and tools to solve any single-particle mechanics problem using the laws of special relativity. This is the last chapter to focus on a single particle. In the next chapter, we'll discuss multiple interacting particles.

Exercises

Exercise 18.1. What time step is being used in the calculations of Figure 18-1?

Exercise 18.2. The function `relativityPS` we wrote to do the dynamics for relativity assumed that velocities would be given in SI units. However, this may not always be convenient. We may want to use natural units instead, in which $c = 1$. Let's write a function that takes a value of c as input, thereby allowing us to use SI units, natural units, or whatever other units we might want.

Use Equation 18.1 to write the function.

```
relativityPS' :: R  -- c
              -> [OneBodyForce]
              -> ParticleState -> DParticleState
relativityPS' c fs st = undefined c fs st
```

Exercise 18.3. Explore the relativistic harmonic oscillator by comparing it to a Newtonian harmonic oscillator. The only force is a linear restoring force chosen with a spring constant that will give a Newtonian period of 1 second. Use a mass of 1 kg, an initial position of 0, and an initial velocity of $4/5c$ in whatever direction you like. (The motion will be one dimensional.) Use one of the examples in this chapter as a template for your code. Plot velocity versus time for the Newtonian result and the relativistic result. Your results should look something like Figure 18-3.

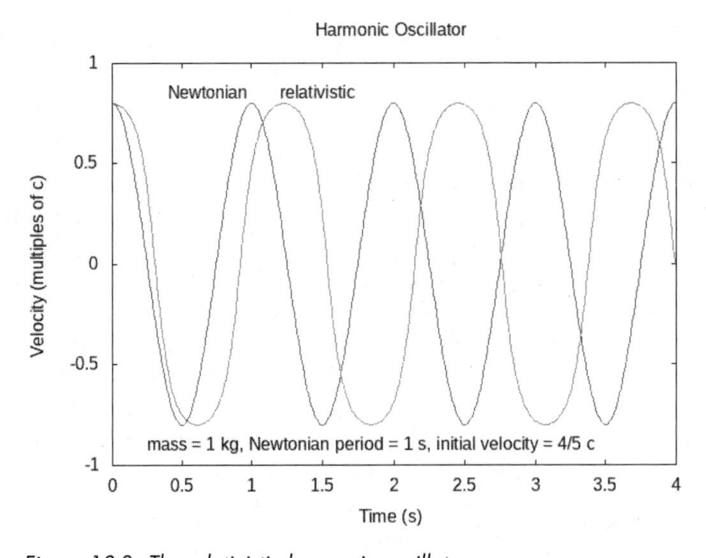

Figure 18-3: The relativistic harmonic oscillator

19

INTERACTING PARTICLES

In Newtonian mechanics, forces that act on particles are produced by other particles. The goal of this chapter is to develop the key ideas that will allow us to predict the motion of multiple interacting particles: Newton's third law, the two-body force, and the distinction between internal and external forces. As usual, we'll express these ideas in code.

We'll begin the chapter by discussing Newton's third law. We'll then develop the idea of a two-body force to express the notion that an interaction between two particles leads to one force produced by one particle on a second particle and simultaneously leads to another force produced by the second particle on the first particle. Two-body force is a significant enough idea in multi-particle situations that we'll define a data type for two-body force. We'll decide which particles are going to be in our system and then distinguish between an internal force and an external force. We'll finish the chapter by considering the state of a multi-particle system and writing a state-update rule that automatically applies Newton's third law, so we need not apply it manually. In the next chapter, we'll apply the ideas discussed here to specific example situations.

Newton's Third Law

If we stand on slippery ice and push our friend, we might find ourselves accelerating in the direction opposite from the way we pushed. We produced a force on our friend, but our friend also produced a force on us, whether or not that was their intention. Newton's third law claims that these two forces are equal and opposite.

> **Newton's third law, Newton's words [15]**
> *To any action there is always an opposite and equal reaction; in other words, the actions of two bodies upon each other are always equal and always opposite in direction.*
>
> **Newton's third law, modern version**
> *If object A exerts a force on object B, then object B exerts a force on object A. This second force is equal in magnitude to the first force, but opposite in direction.*

When we say the *second* force, we're not implying an order in time. The forces arise together, from the same process, whatever the interaction between the objects is.

When dealing with Newton's second law, as we did in the previous five chapters, we're only concerned with the forces that act *on* the object we're applying Newton's second law to. If an object also produces forces on other objects, that's fine, but those forces only get counted when we apply Newton's second law to one of those other objects. Newton's second law cares about the forces *on* an object, not the forces produced *by* an object.

Newton's third law, on the other hand, cares about both and asserts a relationship between them for each interaction. Newton's second law applies to an object; Newton's third law applies to an interaction between two objects.

Listing 19-1 shows the first lines of code for the `MultipleObjects` module we'll develop in this chapter.

```
{-# OPTIONS -Wall #-}
{-# LANGUAGE MultiParamTypeClasses #-}

module MultipleObjects where

import SimpleVec
    ( Vec, R, (^+^), (^-^), (*^), (^*), (^/), zeroV, magnitude )
import Mechanics1D
    ( RealVectorSpace(..), Diff(..), NumericalMethod, Mass, TimeStep, euler )
import Mechanics3D
    ( OneBodyForce, ParticleState(..), DParticleState(..), HasTime(..)
    , defaultParticleState, newtonSecondPS )
```

Listing 19-1: Opening lines of code for the `MultipleObjects` module

You should be familiar with the first line by now: it turns on warnings. The second line turns on a language option that allows us to use multi-parameter type classes; we'll explain this later in the chapter. We name the module `MultipleObjects`. We import the data types `Vec` and `R` from `SimpleVec` so we can refer to them in our type signatures. We also import the zero vector, the `magnitude` function, and the basic vector operators from `SimpleVec`. We use the type classes `RealVectorSpace` and `HasTime` to extend the general-purpose numerical methods `euler` and `rungeKutta4` to the multi-particle setting of this chapter. By making the new data type for multi-particle state an instance of these two type classes, we'll be able to use the two general-purpose numerical methods. We import `newtonSecondPS`, the function that applies Newton's second law to a single particle, to use in the function we write that applies Newton's second law to a collection of particles. We import `euler` to use as a basis for writing an Euler-Cromer method for multiple particles. We'll fill out the rest of the module as we move through the chapter.

NOTE *We'll be using the terms* body, object, *and* particle *interchangeably. A particle connotes a small thing; the words* body *and* object *sometimes connote things that are larger and have an orientation in space. Change in orientation is called* rotation, *and the study of objects with orientations that can rotate as well as move through space is called* rigid-body mechanics. *We won't be getting into rigid-body mechanics here, although this chapter contains important prerequisite material for the subject. What we mean when using the word* body, object, *or* particle *is something that can experience forces, move, and accelerate through space but that either has no orientation or allows us to ignore its orientation. Sometimes the term point particle is used to emphasize that orientation is irrelevant.*

Two-Body Forces

In Chapter 16, we defined a `OneBodyForce` to be a function taking the state of a particle as input and producing a (vector) force on the particle as output. The one-body force is appropriate when the force acting on the one body depends only on that body's state of affairs: namely, its position, velocity, mass, charge, or current time.

Many forces in mechanics are fundamentally *two-body* in nature, meaning the force vector depends on the states of both the particle producing the force and the particle experiencing the force. A *two-body force* is a force that depends on the states of two particles.

```
type TwoBodyForce
    = ParticleState  -- force is produced BY particle with this state
    -> ParticleState  -- force acts ON particle with this state
    -> ForceVector
```

The type `ForceVector` is a type synonym for `Vec`, the name suggesting that the particular vector we have in mind represents a force.

```
type ForceVector = Vec
```

The force vector returned by a two-body force is produced *by* the particle whose state is given first in `TwoBodyForce`, and it acts *on* the particle whose state is given second. Let's call this the *by-on convention*. The comments in the code just shown remind us of this convention.

Every two-body force we write should obey Newton's third law. If the two particle states are exchanged, the force vector produced should be negated from what it was. That is, if `f` is a two-body force, the vector `f st2 st1` should be the negative of `f st1 st2`. Since the force acting on one particle is the negative of the force acting on the other, a convention like the by-on convention is important so that, for example, gravity acts as an attractive force and not as a repulsive force.

There is a relationship between a two-body force and a one-body force. Given the state of a particle that is producing a force, we can turn a two-body force into a one-body force by providing the two-body force with its first input and nothing else. This creates a function. The function takes the state of a particle on which the force acts and returns a force vector, making the function a one-body force. Here is the Haskell code to express this idea:

```
oneFromTwo :: ParticleState  -- state of particle PRODUCING the force
           -> TwoBodyForce
           -> OneBodyForce
oneFromTwo stBy f = f stBy
```

This code is deceptively simple. By applying the two-body force `f` to the state of the particle that is producing the force, we obtain a one-body force. The local variable `stBy` holds the state of the particle producing the force; equivalently, the force is produced *by* the particle with the state `stBy`.

If our brains worked more like the Haskell compiler, we might not bother to make this definition of `oneFromTwo` because in any place we use this function, we could achieve equivalent behavior with fewer keystrokes by omitting the name `oneFromTwo` and reversing the order of its arguments. However, my brain does not work enough like the Haskell compiler for this to be an easy or natural thing to do. I believe that this function, silly as it is for the Haskell compiler, offers value to the human reader and writer of code because it engages the ideas and terminology of mechanics, namely one- and two-body forces. With experience in programming, you will come across more ways of writing code that the compiler sees as equivalent but that sit differently on your brain. Use the flexibility of the language to write in a way that is easy for you, and perhaps others, to read and understand. Your code is not just for the computer; it's for you and other people too.

We'll use the function `oneFromTwo` when we talk about springs later in the chapter. We'll also use it when we want to collect all the forces that act on one particle.

Let's look at some examples of two-body forces. We need to be careful of two issues as we write a two-body force. First, a two-body force needs to respect Newton's third law. To achieve this, as we will see in the examples that follow, the states of the two bodies need to be used in a symmetric way. (More precisely, the two states need to be used in an *anti-symmetric* way so that interchanging them produces a minus sign.) Second, we need to respect the by-on convention so we have a clear understanding of which body the force acts on. Haskell's type system will not prevent us from mistakenly writing a `TwoBodyForce` that violates Newton's third law, or one that returns the wrong force, so we need to be careful.

Universal Gravity

Newton was the first to give a quantitative relationship describing the gravitational force between two massive spherical objects. He showed that the force exerted by one object on another is directly proportional to the mass of each object and inversely proportional to the square of the distance between their centers. As an equation, Newton's law of universal gravity can be written as

$$F = G\frac{m_1 m_2}{r^2} \tag{19.1}$$

where m_1 is the mass of object 1, m_2 is the mass of object 2, and r is the distance between the centers of the objects. This equation gives the magnitude of the force produced by object 1 on object 2 (which, by Newton's third law, is the same as the magnitude of the force produced by object 2 on object 1). In SI units, the constant $G = 6.67408 \times 10^{-11}$ N m^2/kg^2. Equation 19.1 can be translated into Haskell as follows:

```
gravityMagnitude :: Mass -> Mass -> R -> R
gravityMagnitude m1 m2 r = let gg = 6.67408e-11  -- N m^2 / kg^2
                           in gg * m1 * m2 / r**2
```

We can use vector notation to give a more comprehensive version of Newton's law of universal gravity, which includes the direction of the force in the equation. Define the displacement vector $\mathbf{r}_{21}$ to be the vector that points from particle 1 to particle 2, as in Figure 19-1.

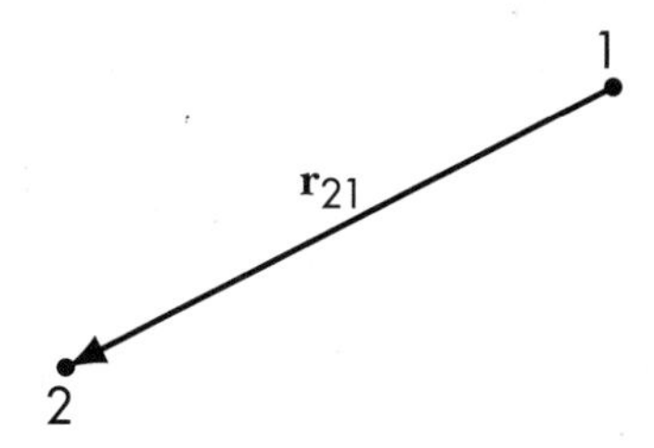

Figure 19-1: The displacement vector $\mathbf{r}_{21}$ points from particle 1 to particle 2.

We'll also define a unit vector $\hat{\mathbf{r}}_{21}$ to point in the direction from particle 1 to particle 2.

$$\hat{\mathbf{r}}_{21} = \frac{\mathbf{r}_{21}}{|\mathbf{r}_{21}|} \tag{19.2}$$

The force $\mathbf{F}_{21}$ exerted on particle 2 produced by particle 1 is given in vector notation by replacing r in Equation 19.1 with $|\mathbf{r}_{21}|$ and indicating the direction of the force with $-\hat{\mathbf{r}}_{21}$ because the force on particle 2 points toward particle 1.

$$\mathbf{F}_{21} = G\frac{m_1 m_2}{|\mathbf{r}_{21}|^2}\left(-\hat{\mathbf{r}}_{21}\right) = -Gm_1 m_2 \frac{\mathbf{r}_{21}}{|\mathbf{r}_{21}|^3} \tag{19.3}$$

Notice that the force $\mathbf{F}_{21}$ on object 2 points opposite to the displacement vector $\mathbf{r}_{21}$; that is, it points toward object 1. This makes sense because gravity is an attractive force.

Comparing Equations 19.1 and 19.3, we see that Equation 19.1 is simpler, while Equation 19.3 is more powerful since the direction of the force is encoded in the equation.

If $\mathbf{r}_1$ is the position vector for particle 1 and $\mathbf{r}_2$ is the position vector for particle 2, then $\mathbf{r}_{21} = \mathbf{r}_2 - \mathbf{r}_1$, and we can write the force on particle 2 as:

$$\mathbf{F}_{21} = -Gm_1 m_2 \frac{\mathbf{r}_2 - \mathbf{r}_1}{|\mathbf{r}_2 - \mathbf{r}_1|^3} \tag{19.4}$$

Here is the Haskell definition for the two-body force of universal gravity:

```
universalGravity :: TwoBodyForce
universalGravity st1 st2
    = let gg = 6.67408e-11  -- N m^2 / kg^2
          m1 = mass st1
          m2 = mass st2
          r1 = posVec st1
          r2 = posVec st2
          r21 = r2 ^-^ r1
      in (-gg) *^ m1 *^ m2 *^ r21 ^/ magnitude r21 ** 3
```

We use the extraction (also called eliminator or selector) functions `mass` and `posVec` from the `ParticleState` data type to extract the masses and position vectors of the two particles and we give local names to these values. The final expression we return comes from Equation 19.4.

Notice that universal gravity obeys Newton's third law. We calculate the force $\mathbf{F}_{12}$ produced by particle 2 on particle 1 by exchanging the roles of m_1 with m_2 and of $\mathbf{r}_1$ with $\mathbf{r}_2$.

$$\mathbf{F}_{12} = -Gm_2 m_1 \frac{\mathbf{r}_1 - \mathbf{r}_2}{|\mathbf{r}_1 - \mathbf{r}_2|^3} = Gm_1 m_2 \frac{\mathbf{r}_2 - \mathbf{r}_1}{|\mathbf{r}_2 - \mathbf{r}_1|^3} = -\mathbf{F}_{21}$$

Constant Repulsive Force

Let's try to write a constant repulsive force between two objects; in other words, a force that doesn't depend on the distance between the objects.

Here's a wrong way to do it:

```
constantRepulsiveForceWrong :: ForceVector -> TwoBodyForce
constantRepulsiveForceWrong force = \_ _ -> force
```

The intent of the code is clear: we intend to ignore the states of the particles and return whatever force was given. This code satisfies the Haskell type checker and compiles, but it does not satisfy Newton's third law. Since the particle states are completely ignored, there is no chance that exchanging the particle states will reverse the direction of the force, as required by Newton's third law.

Here is a constant repulsive force that does obey Newton's third law:

```
constantRepulsiveForce :: R -> TwoBodyForce
constantRepulsiveForce force st1 st2
    = let r1 = posVec st1
          r2 = posVec st2
          r21 = r2 ^-^ r1
      in force *^ r21 ^/ magnitude r21
```

Instead of passing a force vector to our function, we now pass only a force magnitude. We use the positions of the two bodies to determine what direction "repulsive" is. When we exchange the two particle states in this two-body force, the direction of the force is properly reversed.

Linear Spring

A spring, typically made of coiled metal or plastic, can be extended by pulling its ends apart and compressed by pushing them toward each other. The spring has some equilibrium length, which is the distance between the two ends when the spring is detached and allowed to assume its natural shape.

If objects are connected to the ends of the spring, the spring can exert forces on these objects. If the distance between the two objects becomes less than the equilibrium length of the spring, the spring compresses and exerts repulsive forces on the objects in an effort to restore its equilibrium length. Similarly, an extended spring seeks restoration by exerting attractive forces on the objects at its ends. If the mass of the spring is negligible compared with the masses of the objects at its ends, then the force exerted by one end of the spring on one object will be equal and opposite to the force exerted by the other end of the spring on the other object. The spring acts as a Newton's third law–respecting two-body force between the two objects. Since the magnitudes of the forces at each end are equal, we sometimes speak of the force of the spring as if it were one single value.

We'll assume a spring that does not bend but only compresses or extends along a line connecting the ends. If r_{21} is the distance between the ends of the spring and r_e is the equilibrium length of the spring, the spring is in extension when $r_{21} > r_e$, in compression when $r_{21} < r_e$, and in equilibrium when $r_{21} = r_e$.

The size of the force exerted by a spring depends on how much the spring has been extended or compressed from its equilibrium length. A larger extension or compression produces a larger force. The force exerted by the spring depends on the difference $r_{21} - r_e$. A *linear spring* is one in which the force is proportional to this difference. The constant of proportionality k is called the *spring constant*.

Let's call $\mathbf{r}_{21}$ the displacement vector from object 1 at one end of the spring to object 2 at the other, as in Figure 19-1. Then $r_{21} = |\mathbf{r}_{21}|$ is the distance from one end to the other, and $\hat{\mathbf{r}}_{21} = \mathbf{r}_{21}/r_{21}$ is a unit vector from end 1 toward end 2. Table 19-1 shows the force on each end of the spring.

Table 19-1: Forces on Each End of a Linear Spring

	Spring state	Force at end 1	Force at end 2
$r_{21} > r_e$	Extension	$k(r_{21} - r_e)\hat{\mathbf{r}}_{21}$	$-k(r_{21} - r_e)\hat{\mathbf{r}}_{21}$
$r_{21} = r_e$	Equilibrium	0	0
$r_{21} < r_e$	Compression	$k(r_{21} - r_e)\hat{\mathbf{r}}_{21}$	$-k(r_{21} - r_e)\hat{\mathbf{r}}_{21}$

The force $\mathbf{F}_{21}$ exerted on object 2 produced by the spring connected to object 1 is as follows:

$$\mathbf{F}_{21} = -k\left(r_{21} - r_e\right)\frac{\mathbf{r}_{21}}{r_{21}} \tag{19.5}$$

This equation holds whether the spring is in extension, compression, or equilibrium. If $\mathbf{r}_1$ is the position vector for object 1 and $\mathbf{r}_2$ is the position vector for object 2, then $\mathbf{r}_{21} = \mathbf{r}_2 - \mathbf{r}_1$, and we can write the force on object 2 as follows:

$$\mathbf{F}_{21} = -k\left(\left|\mathbf{r}_2 - \mathbf{r}_1\right| - r_e\right)\frac{\mathbf{r}_2 - \mathbf{r}_1}{\left|\mathbf{r}_2 - \mathbf{r}_1\right|} \tag{19.6}$$

Here is Haskell code for the two-body force of a linear spring with spring constant `k` and equilibrium length `re`:

```
linearSpring :: R  -- spring constant
             -> R  -- equilibrium length
             -> TwoBodyForce
linearSpring k re st1 st2
    = let r1 = posVec st1
          r2 = posVec st2
          r21 = r2 ^-^ r1
          r21mag = magnitude r21
      in (-k) *^ (r21mag - re) *^ r21 ^/ r21mag
```

There may be times when we want to attach one end of a spring to a fixed wall or ceiling. In that case, the spring is better represented by a one-body force. This is a good opportunity to use the function `oneFromTwo` that we wrote earlier in this chapter to produce a one-body force from a two-body force. Given a spring constant, an equilibrium length, and a fixed position for one end of the spring, the function `fixedLinearSpring` produces a one-body force for the object attached to the other end of the spring.

```
fixedLinearSpring :: R -> R -> Vec -> OneBodyForce
fixedLinearSpring k re r1
    = oneFromTwo (defaultParticleState { posVec = r1 }) (linearSpring k re)
```

The function `fixedLinearSpring` works by creating a fake particle state at the fixed end of the spring. The particle state is fake in that its only use is to supply a position; we do not intend to allow this particle state to evolve as we would for a real particle. When this fake particle state is given to the function `oneFromTwo` along with the two-body force `linearSpring k re`, we obtain a one-body force that describes the force the spring exerts on the movable mass.

Central Force

The three two-body forces we have considered so far—universal gravity, a constant repulsive force, and a linear spring—are all examples of a *central force*, which is a force between two particles that depends only on the distance between them and acts along the line that joins them. A central force can be either attractive or repulsive. A general expression for the force on object 2 produced by a central force from object 1 is

$$\mathbf{F}_{21} = f(r_{21})\frac{\mathbf{r}_{21}}{r_{21}} \tag{19.7}$$

or

$$\mathbf{F}_{21} = f\left(\left|\mathbf{r}_2 - \mathbf{r}_1\right|\right) \frac{\mathbf{r}_2 - \mathbf{r}_1}{\left|\mathbf{r}_2 - \mathbf{r}_1\right|} \tag{19.8}$$

if $\mathbf{r}_1$ is the position vector for object 1 and $\mathbf{r}_2$ is the position vector for object 2. Here is the central force in Haskell:

```
centralForce :: (R -> R) -> TwoBodyForce
centralForce f st1 st2
    = let r1 = posVec st1
          r2 = posVec st2
          r21 = r2 ^-^ r1
          r21mag = magnitude r21
      in f r21mag *^ r21 ^/ r21mag
```

We provide `centralForce` with a scalar function `f` that describes how the force depends on the distance between the two objects.

The linear spring force of the previous section can be alternatively defined using this `centralForce` function as follows:

```
linearSpringCentral :: R  -- spring constant
                    -> R  -- equilibrium length
                    -> TwoBodyForce
linearSpringCentral k re = centralForce (\r -> -k * (r - re))
```

Here we send the scalar function

$$f(r) = -k(r - r_e)$$

to `centralForce`, where the negative sign indicates that the force is attractive when $r > r_e$. Exercise 19.3 asks you to write universal gravity as a central force.

Elastic Billiard Interaction

In an elastic collision between two objects, the objects compress a bit and store energy, much like a spring, before bouncing apart. In an introductory physics course, collisions are typically treated as "black box" events, where we don't get involved in the particular forces that act between the colliding objects but rather use conservation of momentum to figure out how the objects will move after the collision instead. Here, we'll view the force between objects as a two-body force that is 0 when the objects are apart from one another and that acts as a spring when the objects come into contact.

The crucial thing to know is whether the objects are touching or not touching. We only keep track of the position of the center of each object in the state, so the question becomes whether the centers of the two objects are closer than some threshold distance between the centers that we'll call r_e. If the distance between the objects is greater than r_e, there is no force. If the distance is less than r_e, we'll model the force as a compressed linear spring with spring constant k. The following equation gives the force on object 2 produced by object 1:

$$\mathbf{F}_{21} = \begin{cases} 0 & , \quad r_{21} \geq r_e \\ -k(r_{21} - r_e)\frac{\mathbf{r}_{21}}{r_{21}} & , \quad r_{21} < r_e \end{cases} \tag{19.9}$$

This force is like half of a linear spring. It behaves like a linear spring under compression, when the centers are closer than the threshold distance r_e, but there is no force when the spring would exhibit extension. Here is the Haskell code:

```
billiardForce :: R  -- spring constant
              -> R  -- threshold center separation
              -> TwoBodyForce
billiardForce k re
    = centralForce $ \r -> if r >= re
                             then 0
                             else (-k * (r - re))
```

When the distance between the particles is greater than or equal to the threshold separation, the particles feel no force. When the distance between the particles is less than the threshold separation, the objects are touching and compressing a bit, feeling a force of repulsion. We'll use this two-body force to animate a collision in the next chapter.

Internal and External Forces

When we have multiple interacting particles, the forces that act on any one particle can be classified into two sorts. On the one hand, there are the forces produced by other particles in the collection of particles we are paying attention to. These are the forces to which Newton's third law applies. If we care about particles A and B, and A feels a force from B, then Newton's third law reminds us that somewhere in our calculation we need to account for the fact that B feels a force from A.

On the other hand, there are the forces produced by things outside of the collection of particles we are paying attention to. We may want Earth's surface gravity as a force without having to include Earth as one of the particles we care about. We may want to include forces produced by electric or magnetic fields without including the sources of these fields in our calculation. For this second sort of force, Newton's third law is irrelevant; it tells us about a force that acts on something we don't care about and don't need to account for in our calculation. The distinction between these two sorts of forces motivates the following definitions.

A *system of particles* is simply a choice of which particles to pay attention to. We decide which particles to include in our system, and these are the particles whose motion we calculate by applying Newton's second law.

For a system of particles, it's useful to distinguish between an *internal force* (a force produced by a particle in our system) and an *external force* (a force produced by something outside of our system). By making this distinction between forces, we will be able to write a state-update rule that automatically applies Newton's third law for us. An external force does not require Newton's third law; it's treated the same as last chapter because we're not concerned with the motion of the object creating the force. For an internal force, both particles are in our system, and we can treat them symmetrically, making sure that each particle experiences the appropriate force.

Let's make a new data type for force that demands every force be either an external force or an internal force.

```
data Force = ExternalForce Int OneBodyForce
           | InternalForce Int Int TwoBodyForce
```

The `Int`s in this data type definition are particle numbers. We are going to number the particles in our system, starting with 0. A particular external force is specified by giving the particle number that experiences the force along with the one-body force that describes it. For example, the `Force`

```
ExternalForce 98 (fixedLinearSpring 1 0.5 (vec 100 0 0))
```

indicates that particle 98 experiences the force of a linear spring with spring constant 1 and equilibrium length 0.5, and whose other end is fixed at position $100\hat{\mathbf{i}}$.

A particular internal force is specified by giving numbers for the two particles that participate in the interaction, followed by the two-body force that describes the interaction. For example, the `Force`

```
InternalForce 0 1 universalGravity
```

indicates that particles 0 and 1 interact via universal gravity. The `Force`

```
InternalForce 1 0 universalGravity
```

means the same thing. To indicate that particles 0 and 1 interact by universal gravity, we include one, but not both, of these forces in the list of forces that describes the setting for our system of particles.

The State of a Multi-Particle System

The state of a system of particles consists of the information contained in the states of each particle. A list of particle states is an appropriate type for the state of a system of multiple particles.

There are a few ways we could approach this. We could use the data type `[ParticleState]` to describe the state of a system of particles. We could also write a type synonym to give an alternate name to a list of single-particle states. However, we won't follow either of these two paths because we already use a list of single-particle states to represent a solution to a one-particle mechanics problem. In this solution, each single-particle state describes the same particle at a different time. The state of a multi-particle system wants a list of particles for a different purpose; each single-particle state describes a different particle at the same time.

Since we do not want to confuse a list of single-particle states used as a solution to a one-particle mechanics problem with a list used to describe a multi-particle system, we create a new data type using the `data` keyword so that the compiler regards the two types as different. Different purposes suggest different types.

```
data MultiParticleState
    = MPS { particleStates :: [ParticleState] } deriving Show
```

We build a value with type `MultiParticleState` with the data constructor `MPS`, which is short for multi-particle state. We could have used `MultiParticleState` as the name of the data constructor; however, I chose `MPS` because it's shorter and appears less awkward to use in code to me. Underneath the data constructor is a plain old list of single-particle states. The `data` constructor makes the type `MultiParticleState` distinct from the type `[ParticleState]`. We use record syntax to obtain the extraction function `particleStates` without having to explicitly define it.

Underneath the data constructor, the single-particle states reside in a list, meaning we can refer to the particles by number, starting with `0`. Each particle is labeled by a number with type `Int`.

We note in passing that using a list for the states of each particle is not the most efficient way to handle data. In this book, we are primarily concerned with the clarity, beauty, and simplicity of the code we write, and we are less concerned with its efficiency. A good rule in Haskell programming is not to worry about efficiency until your code runs slower than you would like. At that point, it makes sense to ask what can be done to make it faster. In the old days, functional programming was saddled with a reputation for slowness. However, that is no longer true. Haskell, in particular, provides data structures such as arrays that can be more efficient than the list structure we are using. The list-based method we use works well for tens of particles, but it may be too slow for hundreds of particles. If you get to the point where you want to run simulations with hundreds or thousands of particles, I recommend looking into array types. For simplicity, we stick to the list data type in this book.

To do animation with the `Vis` module using the function `simulateVis` we wrote in Chapter 16, a data type that represents the state of something needs to be an instance of type class `HasTime`, which means that a state needs to have a time associated with it. Each single-particle state has a time; in fact, every single-particle state that makes up a multi-particle state has the same time. So, we'll just take the time from particle number 0. Here is the instance declaration:

```
instance HasTime MultiParticleState where
    timeOf (MPS sts) = time (sts !! 0)
```

We use pattern matching on the input to define the function `timeOf`. By giving the data constructor `MPS` followed by a list, we have access to that list in the body of the function and can use the list element operator `(!!)`.

In the next chapter, we'll animate a system of two masses and two springs. The animation uses `simulateVis`, which uses `timeOf`. Since the state space for that animation is `MultiParticleState`, `simulateVis` needs the `timeOf` that goes with `MultiParticleState`, which is exactly what the instance declaration provides.

Recall that in Chapter 16 we introduced the data type `DParticleState` to hold the time derivatives of the state variables in `ParticleState`. Here, in the many-particle setting, we do something similar and define a new data type `DMultiParticleState` to hold the time derivatives of the state variables in `MultiParticleState`. Here is the data type definition:

```
data DMultiParticleState = DMPS [DParticleState] deriving Show
```

You can see from this definition that we are just packaging a list of `DParticleStates`, analogous to the way we packaged a list of `ParticleStates` in the data type definition above for `MultiParticleState`.

With a new data type in hand to represent the state of a multi-particle system, let us turn to the question of how that state evolves in time—in other words, how the state gets updated.

State Update for Multiple Particles

For a system of multiple particles, Figure 19-2 gives an overview of the data representations and functions that transform among them, much as Figure 16-3 did for a single particle.

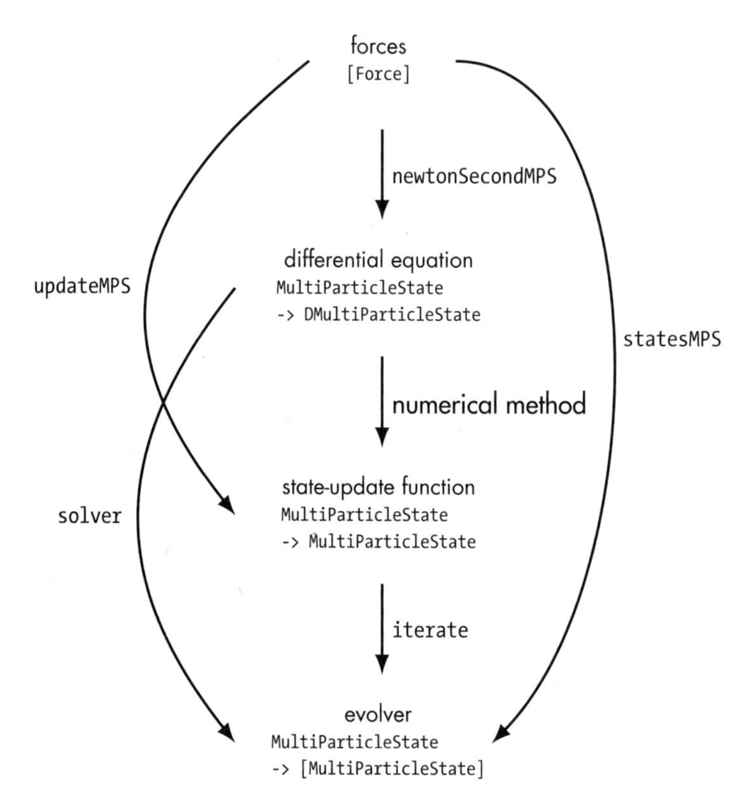

Figure 19-2: Data representations and functions that transform among them

The four representations are forces, differential equation, state-update function, and evolver. In the next section, we'll discuss Newton's second law and the function `newtonSecondMPS` that implements it in the multi-particle setting. We'll see how to use our numerical methods in this setting, and we'll define the composite functions `updateMPS` and `statesMPS` that are shown in Figure 19-2.

Implementing Newton's Second Law

For a system of interacting particles, both Newton's second law and Newton's third law are involved in producing a differential equation. Figure 19-3 shows a schematic diagram for a two-body mechanics problem.

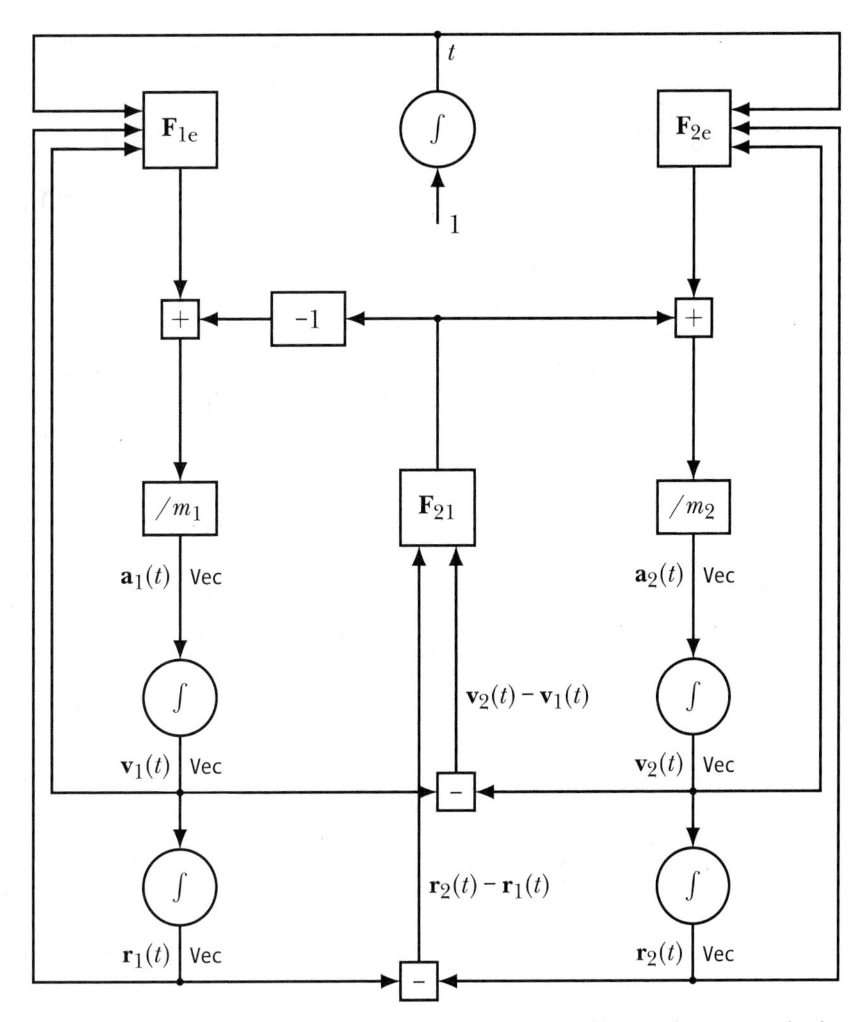

Figure 19-3: Schematic representation of Newton's second law and Newton's third law working together in a two-body situation. The two bodies interact with each other, and each body has external forces acting on it as well.

The function $\mathbf{F}_{1e}$, shown in the upper left of Figure 19-3, produces the net external force on particle 1. The external forces on particle 1 can depend on time, the position of particle 1, and the velocity of particle 1. These three quantities are shown as inputs to $\mathbf{F}_{1e}$. We find the net external force on particle 1 by supplying the inputs to the function to form $\mathbf{F}_{1e}(t, \mathbf{r}_1(t), \mathbf{v}_1(t))$. If there are multiple external forces acting on particle 1, they need to be added together (as vectors). The function $\mathbf{F}_{1e}$ needs to return this sum.

Moving down the diagram from $\mathbf{F}_{1e}$, we find a summation that adds together the net external force and the net internal force to produce the net force on particle 1 that appears in Newton's second law. We'll talk about the net internal force shortly; for now let's keep moving down the left column of the diagram. Newton's second law says that dividing the net force on particle 1 by the mass of particle 1 will give the acceleration of particle 1. Integrating the acceleration produces the velocity, and integrating the velocity produces the position. The position and velocity of particle 1 feed back as inputs to the net external force function $\mathbf{F}_{1e}$.

The position and velocity of particle 1 also get combined with those of particle 2, governed by the right column of Figure 19-3, to produce a relative position and a relative velocity that serve as inputs to $\mathbf{F}_{21}$, the internal force of particle 1 on particle 2. If particle 1 exerts multiple forces on particle 2, perhaps a spring force and an electrical force, these must be added as vectors to form $\mathbf{F}_{21}(\mathbf{r}_2(t) - \mathbf{r}_1(t), \mathbf{v}_2(t) - \mathbf{v}_1(t))$, the net internal force of particle 1 on particle 2. An internal force between two particles shouldn't depend explicitly on time, and it should depend on the positions and velocities of the two particles only through their relative values. All of the two-body forces we introduced earlier in the chapter have this property.

Since $\mathbf{F}_{21}(\mathbf{r}_2(t)-\mathbf{r}_1(t), \mathbf{v}_2(t)-\mathbf{v}_1(t))$ is the force on particle 2, it goes straight into the summation on the right to be added to the net external force on particle 2. By Newton's third law, the force on particle 1 produced by particle 2 is equal and opposite to this, hence the multiplication by -1 to produce the net internal force on particle 1 before it gets added to the net external force on particle 1.

Time is produced by integrating the constant 1, as in our previous schematic diagrams. Time is an input to each of the net external force functions, but not to the net internal force function.

To summarize Figure 19-3, the internal and external forces on each particle must be added together to form the net force on the particle, which, per Newton's second law, gets divided by its mass to calculate the acceleration of the particle. Newton's third law is carried out in the center column, where the force of interaction is calculated, sent unchanged to particle 2, and sent negated to particle 1. The internal forces express the interaction between the two particles, while the external forces represent interactions with things outside of the system. All of the feedback means that a set of coupled differential equations will be the mathematical expression of a multi-particle mechanics problem.

We can write Newton's second law as a set of coupled differential equations.

$$\frac{d}{dt}\left[t\right] = 1 \tag{19.10}$$

$$\frac{d}{dt}\left[\mathbf{r}_1(t)\right] = \mathbf{v}_1(t) \tag{19.11}$$

$$\frac{d}{dt}\left[\mathbf{v}_1(t)\right] = \frac{1}{m_1}\left(\mathbf{F}_{1e}(t, \mathbf{r}_1(t), \mathbf{v}_1(t)) + \sum_n \mathbf{F}_{1n}(\mathbf{r}_1(t) - \mathbf{r}_n(t), \mathbf{v}_1(t) - \mathbf{v}_n(t))\right) \tag{19.12}$$

$$\frac{d}{dt}\left[\mathbf{r}_2(t)\right] = \mathbf{v}_2(t) \tag{19.13}$$

$$\frac{d}{dt}\left[\mathbf{v}_2(t)\right] = \frac{1}{m_2}\left(\mathbf{F}_{2e}(t, \mathbf{r}_2(t), \mathbf{v}_2(t)) + \sum_n \mathbf{F}_{2n}(\mathbf{r}_2(t) - \mathbf{r}_n(t), \mathbf{v}_2(t) - \mathbf{v}_n(t))\right) \tag{19.14}$$

The time derivative of time is 1, as in single-particle mechanics. The time derivative of position is velocity, and this holds for each particle. The time derivative of velocity is acceleration, which is found by dividing net force by mass for each particle. The net force on a particle is the sum of the net external force and the net internal force. The function $\mathbf{F}_{2e}$ produces the net external force on particle 2 when given the time, the position of particle 2, and the velocity of particle 2. The function $\mathbf{F}_{mn}$ produces the internal force of particle n on particle m when given the relative position and relative velocity of particles m and n. The net internal force on particle 2, for example, is

$$\sum_n \mathbf{F}_{2n}(\mathbf{r}_2(t) - \mathbf{r}_n(t), \mathbf{v}_2(t) - \mathbf{v}_n(t))$$

where we add the internal forces produced by all of the other particles in the system. We have only given equations for the first two particles in a system, but there is a similar pair of equations for every particle in the system. Hopefully the pattern is clear.

In Chapter 16, which was concerned with the motion of a single particle, we used the function `newtonSecondPS` to transform a list of forces into a differential equation. We would now like an analogous function for multiple particles. We would like a function with the following type signature:

```
newtonSecondMPS :: [Force]
                -> MultiParticleState -> DMultiParticleState  -- a diff eqn
```

The name of this function has *MPS* on the end to remind us that it works with the `MultiParticleState` state space.

Our plan is to apply Newton's second law to each particle in the system. For each particle, we identify all of the external and internal forces that act on it, and we transform each of these into a one-body force. Once we have the list of one-body forces that act on a particle, we can use the function `newtonSecondPS` to calculate the time derivatives of all the state variables for that particle. When we have the time derivatives of each state variable for each particle, we'll bundle it all up and ask `newtonSecondMPS` to return that bundle. Here's the code:

```
newtonSecondMPS fs mpst@(MPS sts)
    = let deriv (n,st) = newtonSecondPS (forcesOn n mpst fs) st
      in DMPS $ map deriv (zip [0..] sts)
```

The first line in the code names the incoming list of forces `fs` and gives *two* names to the incoming multi-particle state. When placed between two names on the left side of a definition, the `@` ("at symbol") allows the code writer to give a simple identifier to the incoming value *and* simultaneously do pattern matching on the input. The simple identifier `mpst` stands for the incoming multi-particle state with type `MultiParticleState`. Since it appears underneath the data constructor `MPS`, the name `sts` stands for the list of single-particle states (type `[ParticleState]`). We want to use both `mpst` and `sts` in our definition.

In the second line of code, we define a local function deriv to calculate the time derivatives of single-particle state variables. Its argument (n,st) is a pair of particle number and single-particle state. Its return value is a collection of derivatives (type DParticleState). This local function uses newtonSecondPS to calculate the derivatives. The expression forcesOn n mpst fs is the list of one-body forces that act on particle number n when the multi-particle state is mpst and the list of all system forces (external and internal) is fs. We will write the function forcesOn next.

In the final line, we zip together particle numbers with associated particle states to make a list of number-state pairs. We then map the local function deriv onto that list, producing a list of time derivatives of single-particle states (type [DParticleState]). Finally, we apply the DMPS data constructor to form the time multi-particle state derivative (type DMultiParticleState).

From a physics standpoint, all we are doing is applying Newton's second law to each particle in the system. The apparent complexity comes partly from our need to number the particles and partly from our desire to use clear types to represent the things we care about (such as internal and external forces). The clear types benefit us in two ways. First, types represent the important ideas in physics and help us think about them. Second, types help the compiler find our mistakes. By doing some of the heavy lifting in functions like this, we will end up with a collection of powerful functions that let us solve multi-particle problems with relative ease.

How do we find the net force on each particle? In other words, how do we write the function forcesOn we used earlier? What type should forcesOn have? The function forcesOn needs to take a particle number, a multi-particle state, and a list of forces as inputs, and it needs to produce a list of one-body forces as output. Here is the type signature and the definition for forcesOn:

```
forcesOn :: Int -> MultiParticleState -> [Force] -> [OneBodyForce]
forcesOn n mpst = map (forceOn n mpst)
```

The incoming particle number is named n, and the incoming multi-particle state is named mpst. We don't bother to name the list of forces, which means that forcesOn n mpst has type [Force] -> [OneBodyForce]. You can see from the definition that forcesOn passes most of the work on to another as-yet-undefined function called forceOn (note the dropped *s*). The idea is that the function forceOn n mpst has type Force -> OneBodyForce and transforms an external or internal force in the list of system forces into a one-body force acting on particle n. If we can transform a single force, we can use map to transform a list of forces.

The function forceOn needs to deal with both external and internal forces. External forces are easier. We just have to check whether the external force acts on particle n. If so, we return the one-body force held inside the external force. If not, we make up a one-body force that is zero.

For an internal force, we need to check whether either of the particle numbers specified in the internal force are particle number n that we care about. If so, we use the oneFromTwo function written earlier in the chapter that constructs a one-body force from a two-body force. If not, we again make up a zero force. Here is the code for forceOn:

```
forceOn :: Int -> MultiParticleState -> Force -> OneBodyForce
forceOn n _          (ExternalForce n0 fOneBody)
    | n == n0     = fOneBody
    | otherwise   = const zeroV
forceOn n (MPS sts) (InternalForce n0 n1 fTwoBody)
    | n == n0     = oneFromTwo (sts !! n1) fTwoBody  -- n1 acts on n0
    | n == n1     = oneFromTwo (sts !! n0) fTwoBody  -- n0 acts on n1
    | otherwise   = const zeroV
```

The function uses pattern matching on the Force input, so there is one part of the definition for each of the two data constructors of Force (that is, one part for an external force and a second part for an internal force). This definition uses a Haskell feature called *guards*. A guard is the vertical bar on the left of several of the lines of code. The guard is a convenient alternative to an if-then-else construction, especially when there are more than two possibilities, as in the part of the definition that works with internal forces.

Each line in the guard construction consists of four items: a vertical bar, a Boolean condition, an equal sign, and a result. At each vertical bar, the condition is checked. If the condition is true, the corresponding result is returned; if the condition is false, we go to the next vertical bar and repeat the process. It is good practice for the condition in the last guard line to be otherwise, which is just another name for True. By using otherwise, we are guaranteed that one of the guard conditions will hold and consequently that one of the results will be returned.

The first part of the definition for forceOn is for an external force. We check whether particle number n, the particle we are currently interested in, matches particle number n0, the particle on which the current external force acts. If they match, we return the one-body force fOneBody contained in the external force. If they don't match, we return the one-body force const zeroV, a one-body force that ignores the state of the particle and simply returns the zero vector.

The second part of the definition for forceOn is for an internal force. An internal force involves two particles; if the particle we care about is either of these, the function needs to return the appropriate one-body force. We first check whether particle number n, the particle we are currently interested in, matches particle number n0, the first of the two particles involved in the internal force. If n equals n0, we are interested in the force that particle n1 exerts on particle n0. We provide oneFromTwo, the function we wrote earlier in the chapter to convert a two-body force to a one-body force, with the single-particle state for n1, called sts !! n1, and the two-body force contained in the internal force, called fTwoBody. The function oneFromTwo returns the one-body force that acts on particle n.

If n does not equal n0, we check whether n equals n1, the other particle involved in the internal force under consideration. If so, we are interested in the force that particle n0 exerts on particle n1. We provide oneFromTwo with the single-particle state for n0, called sts !! n0, and with the two-body force contained in the internal force. The function oneFromTwo again returns the one-body force that acts on particle n. Finally, if particle n matches neither n0 nor n1, we return a zero force.

The part of the definition of forceOn that deals with an internal force is how we apply Newton's third law. This is the part that guarantees the forces will be equal and opposite because they come from the same internal force. For each of the two particles involved, we use the same two-body force to produce the one-body force that acts on each particle; only the single-particle states are interchanged. Since a two-body force has the property that particle interchange negates the force, Newton's third law is automatically applied. We can't make the mistake of remembering that n1 exerts a force on n0 but forgetting that n0 exerts a force on n1. Our language of internal forces and our code to deal with them ensure that Newton's third law holds without further attention from the code writer. In particular, each time we investigate a new multi-particle system, all we give is a list of external and internal forces. We don't have to ensure that the force n1 exerts on n0 is equal and opposite to the force n0 exerts on n1. Haskell deals with this automatically.

Numerical Methods for Multiple Particles

Recall that the Euler and fourth-order Runge-Kutta methods are general methods for solving any differential equation. At the end of Chapter 15, we wrote functions euler and rungeKutta4 that work with any state space s that is an instance of type class RealVectorSpace. To make these two functions usable with the MultiParticleState state space, we need to write two instance declarations. Here they are:

```
instance RealVectorSpace DMultiParticleState where
    DMPS dsts1 +++ DMPS dsts2 = DMPS $ zipWith (+++) dsts1 dsts2
    scale w (DMPS dsts) = DMPS $ map (scale w) dsts
```

The content of this instance declaration is that the sum of two multi-particle states is just the particle-wise sum.

```
instance Diff MultiParticleState DMultiParticleState where
    shift dt (DMPS dsts) (MPS sts) = MPS $ zipWith (shift dt) dsts sts
```

This instance declaration says that "shifting" a multi-particle state by a state derivative is just shifting each single-particle state by the associated single-particle state derivative.

The Euler-Cromer method is not a general method for any differential equation, so we need to write an explicit numerical method for Euler-Cromer as it applies to the MultiParticleState data type. Here it is:

```
eulerCromerMPS :: TimeStep        -- dt for stepping
               -> NumericalMethod MultiParticleState DMultiParticleState
eulerCromerMPS dt deriv mpst0
    = let mpst1 = euler dt deriv mpst0
```

```
        sts0 = particleStates mpst0
        sts1 = particleStates mpst1
        -- now update positions
        in MPS $ [ st1 { posVec = posVec st0 ^+^ velocity st1 ^* dt }
                   | (st0,st1) <- zip sts0 sts1 ]
```

We choose to calculate the Euler-Cromer derivatives by first taking an Euler step, which is easy to do and correctly updates the mass, charge, time, and velocity for each particle. However, the position needs to be fixed because it needs to be based on the updated velocity. The local variable `mpst0` stands for the incoming multi-particle state, while `mpst1` stands for the Euler-updated multi-particle state. The variables `sts0` and `sts1` are the lists of single-particle states underneath the `MPS` data constructor for the incoming and Euler-updated multi-particle states.

To form the Euler-Cromer-updated multi-particle state, we use a list comprehension to go through all the particles and apply the Euler-Cromer update Equation 15.10, which updates the incoming position with the Euler-updated velocity.

Composite Functions

As in the single-particle situation, it's convenient to have composite functions that take two or three steps in Figure 19-2. The function `updateMPS` is the composition of Newton's second law with a numerical method and is useful for animation. You can see from its definition that it's just this composition:

```
updateMPS :: NumericalMethod MultiParticleState DMultiParticleState
          -> [Force]
          -> MultiParticleState -> MultiParticleState
updateMPS method = method . newtonSecondMPS
```

The function `solver` in Figure 19-2 also takes two steps in data representations, solving a differential equation by producing an evolver that can then generate a list of states from an initial state. We wrote `solver` in Chapter 15, and that code works just fine in the multi-particle setting.

The function `statesMPS` takes all three steps in Figure 19-2, transforming a list of forces into an evolver. Its definition is what you would expect: a composition of Newton's second law, a numerical method, and iteration.

```
statesMPS :: NumericalMethod MultiParticleState DMultiParticleState
          -> [Force]
          -> MultiParticleState -> [MultiParticleState]
statesMPS method = iterate . method . newtonSecondMPS
```

Summary

In this chapter, we applied Newtonian mechanics to multiple interacting objects moving in three dimensions. Newton's third law governs the interaction between particles. A two-body force is a force that depends on the states of two particles. We classify the forces that act on particles in our system into internal forces, which are produced by other particles in our system, and external forces, which are produced by something outside of our system. The state of a system of particles is described by giving the single-particle states for each of the particles in the system. Our `MultiParticleState` data type does exactly that. Our state-update procedure is still based on Newton's second law, but now it automatically applies Newton's third law to all internal forces.

As in the single-particle situation, solving a mechanics problem is still a process of transforming information through a sequence of four representations. For the multi-particle case, we start with a list of internal and external forces, produce a differential equation, produce a state-update function, and finally produce a list of multi-particle states.

The process of producing the differential equation from the forces has evolved over the course of Part II of this book. Newton's second law is always present, but the state has contained more information as we moved from one dimension to three dimensions to multiple particles. Table 19-2 shows the functions we have used to carry out Newton's second law by producing a differential equation. The function `newtonSecondV` is for one particle in one dimension when the forces depend only on velocity. The function `newtonSecondTV` is for one particle in one dimension when the forces depend only on time and velocity. The function `newtonSecond1D` is for one particle in one dimension when the forces could depend on any combination of time, position, or velocity. The function `newtonSecondPS` is for one particle in three dimensions when the forces could depend on time, position, or velocity. Finally, the function `newtonSecondMPS` is for multiple particles in three dimensions where the forces could depend on time, position, or velocity.

Table 19-2: Functions for Newton's Second Law

Function	Type
`newtonSecondV`	`Mass -> [Velocity -> Force] -> Velocity -> R`
`newtonSecondTV`	`Mass -> [(Time, Velocity) -> Force] -> (Time, Velocity) -> (R, R)`
`newtonSecond1D`	`Mass -> [State1D -> Force] -> State1D -> (R, R, R)`
`newtonSecondPS`	`[OneBodyForce] -> ParticleState -> DParticleState`
`newtonSecondMPS`	`[Force] -> MultiParticleState -> DMultiParticleState`

We have used the type `Force` in two different ways over the course of Part II. In a one-dimensional setting, `Force` is simply a type synonym for a real number. In the three-dimensional, multi-particle setting of this chapter, the definition of `Force` is much more complex, describing a data type that could be an internal force or an external force and including the dependence of force on state. We have come a long way. In the next chapter, we'll apply these ideas to specific examples of interacting particles and animate our results.

Exercises

Exercise 19.1. Write a function speed

```
speed :: ParticleState -> R
speed st = undefined st
```

that returns the speed of a particle from its state.

Exercise 19.2. We could use pattern matching on the input rather than extraction functions to write the two-body force for universal gravity. The resulting definition has the benefit of being a bit shorter. Complete the following definition:

```
universalGravity' :: TwoBodyForce
universalGravity' (ParticleState m1 _ _ r1 _) (ParticleState m2 _ _ r2 _)
    = undefined m1 r1 m2 r2
```

Exercise 19.3. Universal gravity is a central force. Use the function centralForce to write the function

```
universalGravityCentral :: TwoBodyForce
universalGravityCentral = undefined
```

that expresses the same two-body force as universalGravity.

Exercise 19.4. Our constant repulsive force is a central force. Rewrite the constant repulsive force using centralForce.

Exercise 19.5. No real spring is completely linear across its entire range. In Exercise 15.10, we introduced the Lennard-Jones spring as an example of a nonlinear spring.

The force on end 2 of the spring is given by the following expression, in which r_e is the equilibrium length and D_e is the dissociation energy (that is, the energy required to extend the spring so that the ends are very far apart):

$$\mathbf{F}_{21} = \frac{12D_e}{r_e}\left[\left(\frac{r_{21}}{r_e}\right)^{-13} - \left(\frac{r_{21}}{r_e}\right)^{-7}\right]\frac{\mathbf{r}_{21}}{r_{21}}$$

If $r_{21} < r_e$, the force on particle 2 will be in the direction of $\mathbf{r}_{21}$, which is repulsive. If $r_{21} > r_e$, the force on particle 2 will be in the direction of $-\mathbf{r}_{21}$, which is attractive.

Write the function lennardJones, which takes a dissociation energy and an equilibrium length and returns a two-body force for the Lennard-Jones spring.

```
lennardJones :: R  -- dissociation energy
             -> R  -- equilibrium length
             -> TwoBodyForce
lennardJones de re = centralForce $ \r -> undefined de re r
```

Exercise 19.6. Write a function systemKE

```
systemKE :: MultiParticleState -> R
systemKE mpst = undefined mpst
```

that returns the kinetic energy of a system of particles by adding up the kinetic energy of each particle.

Exercise 19.7. Each wire in a schematic diagram can be labeled with a type. Label each wire in Figure 19-3 with a type.

Exercise 19.8. An alternative way to write the function forcesOn is to form the list of one-body forces by appending a list of one-body forces that come from external forces to a list of one-body forces that come from internal forces. This method has the advantage of not requiring us to make up any fake zero forces and not needing the function forceOn.

```
forcesOn' :: Int -> MultiParticleState -> [Force] -> [OneBodyForce]
forcesOn' n mpst fs = externalForcesOn n fs ++ internalForcesOn n mpst fs

externalForcesOn :: Int -> [Force] -> [OneBodyForce]
externalForcesOn n fs = undefined n fs

internalForcesOn :: Int -> MultiParticleState -> [Force] -> [OneBodyForce]
internalForcesOn n (MPS sts) fs
    = [oneFromTwo (sts !! n1) f | InternalForce n0 n1 f <- fs, n == n0] ++
      [oneFromTwo (sts !! n0) f | InternalForce n0 n1 f <- fs, n == n1]
```

In defining the function internalForcesOn, we use pattern matching inside a list comprehension. We exclude any force that does not match the pattern. Following the model of internalForcesOn, write the function externalForcesOn.

20

SPRINGS, BILLIARD BALLS, AND A GUITAR STRING

This chapter applies the ideas and theory of the previous chapter to three specific examples. The first is a system of two masses and two springs hanging from a fixed ceiling, the second is a billiard ball collision, and the third is a guitar string modeled as a long line of particles connected to their neighbors with springs.

A few things will be different in this chapter. We'll take a closer look at the approximate numerical calculation we have been doing, treating numerical issues in more detail than we have in previous chapters. We'll examine momentum and energy conservation in the context of approximate numerical calculation. We'll also introduce a method for asynchronous animation where the calculations are done first and then made into a movie that can be watched afterward. Asynchronous animation is appropriate when the calculations get too intense to be done on the time scale of human impatience to see the results.

Introductory Code

Listing 20-1 shows the introductory code for the `MOExamples` module we'll develop in this chapter (*MO* for *Multiple Objects*). As usual, we import the functions and types we want to use in this module.

```
{-# OPTIONS -Wall #-}

module MOExamples where

import SimpleVec
    ( R, Vec, (^+^), (^-^), (*^), vec, zeroV, magnitude
    , sumV, iHat, jHat, kHat, xComp, yComp, zComp )
import Mechanics1D ( TimeStep, NumericalMethod, euler, rungeKutta4 )
import Mechanics3D
    ( ParticleState(..), HasTime(..), defaultParticleState
    , earthSurfaceGravity, customLabel, orient, disk )
import MultipleObjects
    ( MultiParticleState(..), DMultiParticleState, Force(..), TwoBodyForce
    , newtonSecondMPS, updateMPS, statesMPS, eulerCromerMPS
    , linearSpring, fixedLinearSpring, billiardForce )
import Graphics.Gnuplot.Simple
import qualified Graphics.Gloss as G
import qualified Vis as V
```

Listing 20-1: Opening lines of code for the `MOExamples` module

The types `ParticleState` and `MultiParticleState` are the state-description types we use for one particle and multiple particles, respectively. The function `newtonSecondMPS` creates a differential equation from a list of internal and external forces. The functions `euler`, `eulerCromerMPS`, and `rungeKutta4` are used to solve differential equations. We import the type class `HasTime` so we can use the `timeOf` function that belongs to it, and because we make explicit reference to `HasTime` in one of the exercises.

Two Masses and Two Springs

As a first example of a system with multiple objects, let's analyze the situation in Figure 20-1.

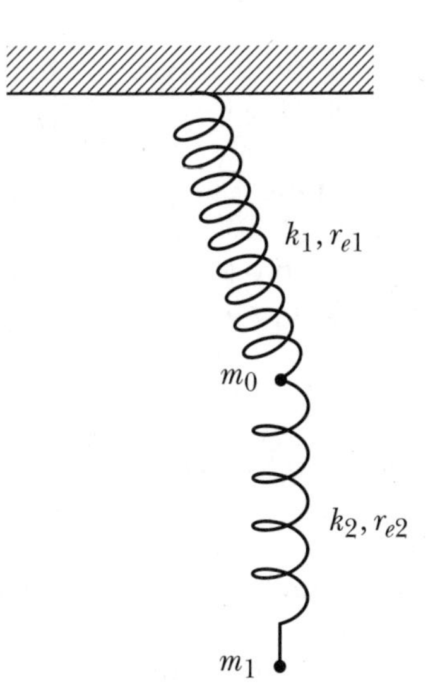

Figure 20-1: A system of two masses and two springs

In Figure 20-1 we have two masses and two springs. The upper spring is attached to a fixed ceiling and has spring constant k_1 and equilibrium length r_{e1}. The lower spring connects the two objects and has spring constant k_2 and equilibrium length r_{e2}. The upper object has mass m_0, and the lower object has mass m_1. Earth surface gravity acts on each of the objects.

Forces

There are a total of four forces acting in this situation: three external forces and one internal force. The list `twoSpringsForces` contains the four forces, which we will describe in the order they are listed.

```
twoSpringsForces :: [Force]
twoSpringsForces
    = [ExternalForce 0 (fixedLinearSpring 100 0.5 zeroV)
      ,InternalForce 0 1 (linearSpring 100 0.5)
      ,ExternalForce 0 earthSurfaceGravity
      ,ExternalForce 1 earthSurfaceGravity
      ]
```

Since one end is fixed, the upper spring acts as an external force on object 0. The relevant one-body force is `fixedLinearSpring 100 0.5 zeroV`, the force from a linear spring with spring constant 100 N/m and equilibrium length 0.5 m attached to the ceiling at the origin. The lower spring acts as an internal force between objects 0 and 1. It also has spring constant 100 N/m and equilibrium length 0.5 m. The last two forces describe the gravitational force Earth exerts on each of the two objects.

Animation Functions

We'd like to animate the oscillation of the objects using `not-gloss`. We need to specify the five ingredients that serve as inputs to the `simulateVis` function: a time-scale factor, an animation rate, an initial state, a display function, and a state-update function. We choose a time-scale factor of 1 and an animation rate of 20 frames/second.

In the initial state, we must give the mass, initial position, and velocity of each object. We let $m_0 = 2$ kg, $m_1 = 3$ kg, $\mathbf{r}_0(0) = (0.4\hat{\mathbf{j}} - 0.3\hat{\mathbf{k}})$ m, $\mathbf{r}_1(0) = (0.4\hat{\mathbf{j}} - 0.8\hat{\mathbf{k}})$ m, $\mathbf{v}_0(0) = 0$ m/s, and $\mathbf{v}_1(0) = 0$ m/s, where $\mathbf{r}_j$ and $\mathbf{v}_j$ are the position and velocity functions of particle j. The code for the initial state is as follows:

```
twoSpringsInitial :: MultiParticleState
twoSpringsInitial
    = MPS [defaultParticleState
           { mass   = 2
           , posVec = 0.4 *^ jHat ^-^ 0.3 *^ kHat }
          ,defaultParticleState
           { mass   = 3
           , posVec = 0.4 *^ jHat ^-^ 0.8 *^ kHat }
          ]
```

We use the `MPS` constructor, the one constructor for the `MultiParticleState` data type, to transform a list of single-particle states into a multi-particle state. We set the charge, time, and velocity of each particle to 0 since that is the value in `defaultParticleState`.

The function that involves most of the physics is `twoSpringsUpdate`, the state-update function for the system. Here is the code:

```
twoSpringsUpdate :: TimeStep
                 -> MultiParticleState  -- old state
                 -> MultiParticleState  -- new state
twoSpringsUpdate dt = updateMPS (eulerCromerMPS dt) twoSpringsForces
```

The time step gets the local name `dt`. Recall that, in animation, we do not choose the time step directly. We choose it indirectly through the time-scale factor and animation rate, and then the animation packages do their best to adhere to that rate, although they make no promises. Since the old state is not named on the left of the equal sign of the definition, the return type is `MultiParticleState -> MultiParticleState`. We use `updateMPS` to create the state-update function. It returns a state-update function when given a numerical method and a list of forces. We choose Euler-Cromer with time step dt as our numerical method, and we give the list of forces we wrote earlier.

We have specified the time-scale factor, the animation rate, the initial state, and the state-update function we will use for our animation. We will discuss the display function next when we look at the stand-alone program for the animation.

Stand-Alone Animation Program

Listing 20-2 shows a stand-alone program for 3D animation of the two masses and two springs.

```
❶ {-# OPTIONS -Wall #-}

import SimpleVec ( R, zeroV )
import Mechanics3D ( posVec, simulateVis, v3FromVec )
import MultipleObjects ( MultiParticleState(..) )
import MOExamples ( twoSpringsInitial, twoSpringsUpdate )
import Vis ( VisObject(..), Flavour(..), red, green, blue )

main :: IO ()
main = simulateVis 1 20 twoSpringsInitial twoSpringsVisObject twoSpringsUpdate

twoSpringsVisObject :: MultiParticleState -> VisObject R
❷ twoSpringsVisObject (MPS sts)
  ❸ = let r0 = posVec (sts !! 0)
        ❹ r1 = posVec (sts !! 1)
        ❺ springsObj = Line Nothing [v3FromVec zeroV
                                    ,v3FromVec r0
                                    ,v3FromVec r1] ❻ blue
        ❼ objs = [Trans (v3FromVec r0) (Sphere 0.1 Solid red)
                ,Trans (v3FromVec r1) (Sphere 0.1 Solid green)
                ,springsObj
                ]
        ❽ vpm = 1  -- Vis units per meter
    ❾ in Scale (vpm,vpm,vpm) $ VisObjects objs
```

Listing 20-2: Stand-alone program for a 3D animation of two masses and two springs

The program begins by turning on warnings ❶ and then imports the needed types and functions. The program makes imports from the SimpleVec module of Chapter 10, the Mechanics3D module of Chapter 16, the Multiple Objects module of Chapter 19, and the MOExamples module of the current chapter. It imports zeroV from the SimpleVec module to be able to refer to the origin, where a spring is attached. The program imports R because it is used in the type VisObject R.

Let's discuss the imports from the Mechanics3D module. The program imports the posVec function, the extraction function of the ParticleState data type that returns the position vector of a particle state. The only state variable needed in the display function is position. None of the other state variables have anything to contribute to what the picture should look like for a given state. The program imports `simulateVis`, the principal function that produces the animation. The program also imports `v3FromVec` to convert a vector with type `Vec` into a vector with type `V3`, `not-gloss`'s vector type.

From the `MultipleObjects` module, we import the `MultiParticleState` type with its constructor so we can refer to the type in the type signature for the display function and use the constructor `MPS` to pattern match on the input in the definition of the display function. From the `MOExamples` module of the current chapter, we import the initial state and the state-update function, two of the five ingredients needed for an animation.

From the module `Vis`, the program imports the `VisObject` type with its constructors, which include `Line`, `Sphere`, `Trans`, `Scale`, and `VisObjects`. We import the `Flavour` type with its constructors because a sphere is required to be either solid or wire-frame, and the `Solid` data constructor we use is a constructor of `Flavour`. The program also imports the colors `red`, `green`, and `blue`.

The main program is named `main` and has type `IO ()`. It calls `simulateVis` with the five ingredients needed to make an animation, which includes the display function `twoSpringsVisObject` defined in the stand-alone program.

The display function `twoSpringsVisObject` produces a picture from the state of the system. The definition of the display function begins by pattern matching on the input so that the body of the function has access to the two-element list of single-particle states `sts` ❷. We give local names `r0` and `r1` to the positions of the two objects ❸ ❹. The local variable `springsObj` is a picture of two lines that represent the two springs ❺. To construct a picture of two lines, we use the `Line` data constructor.

Let's take a look at the type of `Line` in `not-gloss`.

```
Prelude Graphics.Gloss> :m Vis Linear.V3
Prelude Vis Linear.V3> :t Line
Line :: Maybe a -> [V3 a] -> Color -> VisObject a
```

This `Line` has three inputs. The first has to do with line width. To get the default line width, in Listing 20-2 we supply `Nothing` for the first input ❺. The second input is a list of vectors, each vector having `not-gloss`'s native `V3 a` type. In our use, the type variable `a` stands for the type `R`. We convert from `Vec` to `V3 R` with `v3FromVec` ❺. The third input is a color, for which the program supplies `blue` ❻.

We define `objs` to be a list of pictures of the two masses and the two springs ❼. The pictures of the masses are spheres, translated to the positions contained in the state.

We then define a spatial-scale factor `vpm` to be 1 Vis-unit/meter ❽. Of course, scaling by 1 is unnecessary, but the code makes it easy to change the value to something else. Finally, `VisObjects` combines multiple pictures into a single picture and `Scale` scales the entire result ❾.

Using Mechanical Energy as a Guide to Numerical Accuracy

The system of two masses and two springs that we've been exploring should conserve mechanical energy. In this section, we'll discuss the types of energy a system of particles can have, and we'll see how to use energy as a tool to assess the accuracy of our numerical methods.

Kinetic Energy

Each moving particle has an energy of motion called *kinetic energy*. The kinetic energy of a single particle is one half the particle's mass multiplied by the square of its speed. We'll use small capitals KE for the kinetic energy of a single particle. Kinetic energy is a scalar whose SI units are Joules (J).

$$\mathrm{KE} = \frac{1}{2} m v^2$$

The speed $v = |\mathbf{v}|$ is the magnitude of the velocity $\mathbf{v}$. Here is a Haskell function that returns the kinetic energy of one particle:

```
kineticEnergy :: ParticleState -> R
kineticEnergy st = let m = mass st
                       v = magnitude (velocity st)
                   in (1/2) * m * v**2
```

The kinetic energy of a system of particles is the sum of the kinetic energies for each particle in the system. We use uppercase KE to denote system kinetic energy. In a system of particles, the kinetic energy of particle n, which has mass m_n and velocity $\mathbf{v}_n$, is given by

$$\mathrm{KE}_n = \frac{1}{2} m_n v_n^2$$

and the system kinetic energy is

$$\mathrm{KE} = \sum_n \mathrm{KE}_n = \sum_n \frac{1}{2} m_n v_n^2$$

Here is a Haskell function that returns the kinetic energy of a system of particles:

```
systemKE :: MultiParticleState -> R
systemKE (MPS sts) = sum [kineticEnergy st | st <- sts]
```

Potential Energy

Some forces are distinguished in that they can be associated with a *potential energy*. Such forces are called *conservative* and include the elastic force of a spring and the force of gravity as examples.

A spring acquires potential energy by being compressed or extended from its equilibrium position. The spring can be used to store energy in this way. A linear spring with spring constant k that is displaced (compressed or extended) a distance x from its equilibrium position has a potential energy

$$\mathrm{PE}_{\mathrm{spring}} = \frac{1}{2}kx^2 \tag{20.1}$$

The type of potential energy associated with a spring is called *elastic potential energy*. The function `linearSpringPE` computes the elastic potential energy of a spring given its spring constant, equilibrium length, and particle states at each end.

```
linearSpringPE :: R              -- spring constant
               -> R              -- equilibrium length
               -> ParticleState  -- state of particle at one end of spring
               -> ParticleState  -- state of particle at other end of spring
               -> R              -- potential energy of the spring
linearSpringPE k re st1 st2
    = let r1 = posVec st1
          r2 = posVec st2
          r21 = r2 ^-^ r1
          r21mag = magnitude r21
      in k * (r21mag - re)**2 / 2
```

This function is similar to the function `linearSpring` we wrote in Chapter 19, except that instead of calculating a force, it calculates a potential energy. The displacement from equilibrium x in Equation 20.1 is the difference between the distance `r21mag` from one end of the spring to the other and the spring's equilibrium length `re`.

An object near Earth's surface has a *gravitational potential energy* that depends on its height. An object with mass m has potential energy

$$\mathrm{PE}_{\mathrm{g}} = mgh$$

where g is Earth's gravitational acceleration constant and h is the object's height above some reference level, such as Earth's surface. The function `earthSurfaceGravityPE` computes the gravitational potential energy of an object near Earth's surface given its particle state.

```
-- z direction is toward the sky
-- assumes SI units
earthSurfaceGravityPE :: ParticleState -> R
earthSurfaceGravityPE st
    = let g = 9.80665  -- m/s^2
          m = mass st
          z = zComp (posVec st)
      in m * g * z
```

This function is similar to the function `earthSurfaceGravity` we wrote in Chapter 16, except that instead of calculating a force, it calculates a potential energy.

Returning now to the example of two masses and two springs, the total potential energy is the elastic potential energy of each spring plus the gravitational potential energy of each mass.

```
twoSpringsPE :: MultiParticleState -> R
twoSpringsPE (MPS sts)
    = linearSpringPE 100 0.5 defaultParticleState (sts !! 0)
      + linearSpringPE 100 0.5 (sts !! 0) (sts !! 1)
      + earthSurfaceGravityPE (sts !! 0)
      + earthSurfaceGravityPE (sts !! 1)
```

Since the top spring is connected to a fixed ceiling, we use the default particle state for one end of the top spring to indicate that one end of the spring is fixed at the origin.

Mechanical Energy

The *mechanical energy* of a system of particles is the sum of its kinetic energy and its potential energy. Systems with no non-conservative forces conserve mechanical energy. The mechanical energy of such a system at a later time is the same as it was earlier. Since we are doing approximate calculation, we cannot expect that our calculation of mechanical energy will stay exactly the same over time. Since we know that it would stay the same if we could do exact calculation, we can take the deviation that occurs in our calculation as a guide to the level of inaccuracy our numerical method is producing. For a system that should conserve mechanical energy, how well it is conserved in our calculations is an indication of our numerical method's accuracy.

The function `twoSpringsME` computes the mechanical energy for the system of two masses and two springs.

```
twoSpringsME :: MultiParticleState -> R
twoSpringsME mpst = systemKE mpst + twoSpringsPE mpst
```

Mechanical energy is conserved for this system because all of the forces involved are conservative. For the system of two masses and two springs, Figure 20-2 shows mechanical energy as a function of time for different numerical methods. The first column shows the Euler method, the second the Euler-Cromer, and the third the fourth-order Runge-Kutta. The first row uses a time step of 0.1 s, the second row 0.01 s, the third row 10^{-3} s, and the fourth row 10^{-4} s.

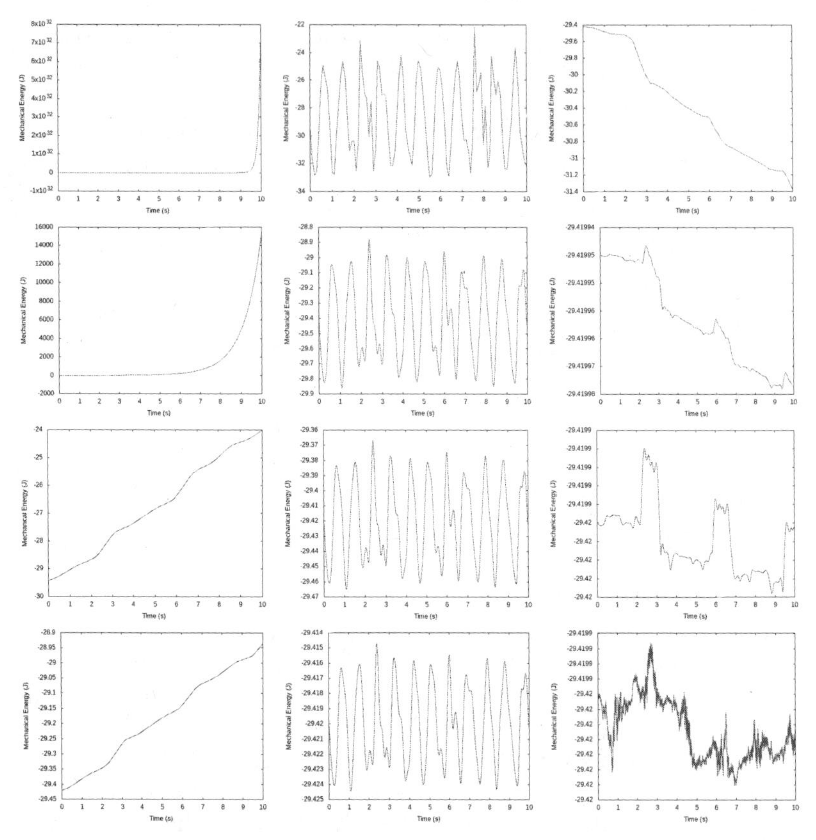

Figure 20-2: Mechanical energy as a function of time for different numerical methods. The change in mechanical energy is a measure of numerical inaccuracy.

Mechanical energy tends to increase in the Euler method, oscillate in the Euler-Cromer method, and perhaps decrease in the fourth-order Runge-Kutta method. The horizontal axis shows a period of 10 seconds in each graph. The vertical axes show vastly different scales. If we take the difference between maximum mechanical energy and minimum mechanical energy over a 10-second time period as our measure of inaccuracy, we can make a table comparing the numerical methods. Table 20-1 shows such a comparison.

Table 20-1: Change in Calculated Mechanical Energy for the Two-Mass and Two-Spring System over a 10-Second Interval for Different Numerical Methods

Time step	Euler	Euler-Cromer	4th order RK
10^{-1} s	Way off	40%	7%
10^{-2} s	Way off	4%	10^{-4}%
10^{-3} s	20%	0.4%	10^{-8}%
10^{-4} s	2%	0.04%	10^{-11}%

Table 20-1 shows maximum mechanical energy minus minimum mechanical energy over a 10-second period, expressed as a percentage of initial mechanical energy. The Euler method is not even close for time steps less than a millisecond. You can see this in the animation if you change the numerical method to the Euler method. The springs extend and droop in a horrible display of numerical inaccuracy. The table also shows how each 10-fold reduction in time step produces approximately a 10-fold improvement in accuracy for the two first-order methods, Euler and Euler-Cromer. A 10-fold decrease in time step for the fourth-order Runge-Kutta method produces a roughly 10^4-fold improvement in accuracy. In this way, the fourth-order method shows why it is considered a fourth-order method.

Having seen a first example of multiple interacting objects, let's look at a second example of interacting objects, that of a collision.

A Collision

Our second example of multiple objects interacting is a collision between two billiard balls. For this example, we'll go through the four data representations of Figure 19-2, discuss the choice of parameters (including time step), look at momentum and energy conservation, discuss some numerical issues, and finally produce some animated results.

Data Representations

The four data representations are a list of forces, a differential equation, a state-update function, and a list of states. We'll discuss each of these in turn.

Forces

We suppose that the elastic billiard interaction described in Chapter 19 is the only force that acts on either particle. This elastic billiard interaction is an internal force that acts between the two billiard balls. Here is the list of one force:

```
billiardForces :: R -> [Force]
billiardForces k = [InternalForce 0 1 (billiardForce k (2*ballRadius))]
```

The elastic billiard interaction requires that we specify a spring constant for the elastic repulsion and a threshold distance at which the repulsive force kicks in. We make the spring constant `k` an argument to the function `billiardForces` so we can delay committing to a particular value and so it will be easy to try out different values.

Each ball has a diameter of 6 cm. The threshold distance occurs when the centers of the balls are two radii apart. Because it's always the same, we specify a particular value, `2*ballRadius`, for the threshold distance rather than making it an argument like we did for the spring constant. We name the ball radius because it's used in two places: the list of forces we just wrote and the display function we write later in the chapter.

```
ballRadius :: R
ballRadius = 0.03  -- 6cm diameter = 0.03m radius
```

Differential Equation

Newton's second law transforms a list of forces into a differential equation. For multi-particle systems, `newtonSecondMPS` is the function that does this transformation. We give the name `billiardDiffEq k` to the differential equation that expresses Newton's second law for two billiard balls acted on by `billiardForces k`.

```
billiardDiffEq :: R -> MultiParticleState -> DMultiParticleState
billiardDiffEq k = newtonSecondMPS $ billiardForces k
```

We continue to parameterize the differential equation by the spring constant `k` that we have not yet specified.

State-Update Function

Next, we need a state-update function. The simplest way to write a state-update function uses the `updateMPS` function of Figure 19-2.

```
billiardUpdate
    :: (TimeStep -> NumericalMethod MultiParticleState DMultiParticleState)
    -> R         -- k
    -> TimeStep  -- dt
    -> MultiParticleState -> MultiParticleState
billiardUpdate nMethod k dt = updateMPS (nMethod dt) (billiardForces k)
```

This state-update function has the same form as others we have written, except that we included a numerical method and a spring constant as parameters so we can specify these items later. There are only three things that have type `TimeStep -> NumericalMethod MultiParticleState DMultiParticleState` and can serve for the input `nMethod`: `euler`, `eulerCromerMPS`, and `rungeKutta4`. We need to specify one of these when we use this function, but for now we'll delay that decision.

List of States

The fourth data representation of Figure 19-2 is an evolver, which is a function that will produce a list of states when given an initial state. The simplest way to write an evolver uses the `statesMPS` function of Figure 19-2.

```
billiardEvolver
    :: (TimeStep -> NumericalMethod MultiParticleState DMultiParticleState)
    -> R         -- k
    -> TimeStep  -- dt
    -> MultiParticleState -> [MultiParticleState]
billiardEvolver nMethod k dt = statesMPS (nMethod dt) (billiardForces k)
```

To get a list of states from an evolver requires an initial state. In the initial state, we give the masses of each object as well as their initial positions and velocities. We'll let each billiard ball have a mass of 160 grams. The first ball starts at the origin with an initial velocity of 0.2 m/s in the x-direction. The second ball starts at rest at coordinates (1 m, 0.02 m) in the xy-plane. The small y-component is present so that the collision will be slightly oblique rather than one dimensional. The code for the initial state is as follows:

```
billiardInitial :: MultiParticleState
billiardInitial
    = let ballMass = 0.160  -- 160g
      in MPS [defaultParticleState { mass     = ballMass
                                   , posVec   = zeroV
                                   , velocity = 0.2 *^ iHat }
             ,defaultParticleState { mass     = ballMass
                                   , posVec   = iHat ^+^ 0.02 *^ jHat
                                   , velocity = zeroV }
             ]
```

Let's now name a list of states based on this initial state. The list

```
billiardStates nMethod k dt
```

is an infinite list of states for the billiard collision when the calculation is done with numerical method `nMethod` (`euler`, `eulerCromerMPS`, or `rungeKutta4`), spring constant `k`, and time step `dt`.

```
billiardStates
    :: (TimeStep -> NumericalMethod MultiParticleState DMultiParticleState)
    -> R          -- k
    -> TimeStep   -- dt
    -> [MultiParticleState]
billiardStates nMethod k dt
    = statesMPS (nMethod dt) (billiardForces k) billiardInitial
```

Next we will want a finite list of states that we can use to make a graph, or to compare the value of some physical quantity, like momentum or kinetic energy, before the collision with the corresponding value after. The list `billiardStatesFinite nMethod k dt` is a finite list of states for the billiard collision when the calculation is done with numerical method `nMethod`, spring constant `k`, and time step `dt`.

```
billiardStatesFinite
    :: (TimeStep -> NumericalMethod MultiParticleState DMultiParticleState)
    -> R          -- k
    -> TimeStep   -- dt
    -> [MultiParticleState]
billiardStatesFinite nMethod k dt
    = takeWhile (\st -> timeOf st <= 10) (billiardStates nMethod k dt)
```

To form the finite list, we use `takeWhile` to select all of the states before 10 seconds elapse. As we will see soon, the collision occurs about 5 seconds into the simulation.

So far, we have delayed making any choices about numerical method, spring constant, or time step. Let's turn to that issue next.

Spring Constant and Time Step

In an introductory physics course, collisions are usually dealt with using conservation of momentum rather than by giving an explicit force for the interaction between two particles. Using conservation of momentum is elegant because we don't need to know the force that acts between the particles; as long as it is short-lived, the system momentum before the collision must equal the system momentum after the collision. However, there are downsides to relying solely on conservation of momentum to analyze collisions. For example, for collisions in two dimensions, one usually needs information that is not present in the initial state of the system, such as the velocity of one of the particles *after* the collision, to find the velocities of both particles after the collision. If we know the nature of the force between the particles, on the other hand, then initial conditions suffice to determine the future motion of the particles.

On the practical side of things, our method of analyzing a collision by specifying an explicit force between the particles means that we need to make some choices. The elastic billiard collision requires that we specify a spring constant. Values that are too small or too large can introduce trouble for the numerical analysis of the collision with a finite time step. If the spring constant is too small, the objects will squish together during the collision, making their centers very close, and there is a risk that one object will pass through the other object instead of bouncing off of it. If the spring constant is too large, the spring will apply a very large force the first time the objects are within their separation threshold. The force may be so large that the objects may be outside their separation threshold by the very next time step. This would be a poor sampling of the force and could lead to poor numerical results.

If the time step is too large, one particle may move too far during one time step and entirely miss any state in which the particles are within their threshold separation. Even if the collision is not completely missed, a time step that is too large may give inaccurate results. Our general advice is to choose a time step that is small compared with the characteristic time scales of the situation.

We need to choose a spring constant for the interaction force in addition to the usual time step for numerical analysis. How should we choose these two things?

One way to choose a spring constant for the elastic interaction is to suppose that all of the initial kinetic energy of motion is transformed into potential energy in the spring. If this were true, we could write the following equation:

$$\frac{1}{2}mv^2 = \frac{1}{2}kx^2$$

The separation threshold for the two billiard balls we are considering is 6 cm; perhaps we would like the ball centers to get no closer than 5 cm to each other. Then the displacement from equilibrium should be no more than 1 cm. Using 1 cm for x in the equation above and solving for k gives us the following:

$$k = \frac{mv^2}{x^2} = \frac{(0.16\ \text{kg})(0.2\ \text{m/s})^2}{(0.01\ \text{m})^2} \approx 60\ \text{N/m}$$

This calculation is a rough way of making a reasonable guess for a spring constant. The initial kinetic energy in this collision is only partially transformed into elastic potential energy.

What are the relevant time scales for the collision? One time scale is the time it takes the moving billiard ball to traverse a distance equal to the threshold separation. This time is given by the threshold separation divided by the initial velocity of the moving ball.

$$\frac{0.06\ \text{m}}{0.2\ \text{m/s}} = 0.3\ \text{s}$$

A second time scale comes from the spring constant and mass in the problem. If this were a problem in which the mass could oscillate on the spring, the period of oscillation would be proportional to $\sqrt{m/k}$. Oscillation will not occur in this situation, but the collision can be viewed as taking place over the course of half a period of the oscillation that would occur if this were a full-fledged spring. The half period consists of the compression of the spring from equilibrium to closest approach, followed by the expansion of the spring back to equilibrium. This second spring-based time scale is

$$\sqrt{\frac{m}{k}} = \sqrt{\frac{0.16\ \text{kg}}{60\ \text{N/m}}} \approx 0.05\ \text{s}$$

The time step we choose for numerical analysis needs to be small compared with the smaller of the two time scales, namely 0.05 s. At this point, we have rough estimates for a spring constant and a time step. We will sharpen these estimates shortly, after we identify several desirable properties we want of our calculation and explore how these properties depend on the spring constant and time step. Two of the desirable properties are conservation of momentum and conservation of energy, which we look at next.

Momentum and Energy Conservation

The basic wisdom about collisions that comes from an introductory physics course is that momentum is conserved in all collisions; however, energy is conserved only in *elastic collisions*. Our collision is elastic, so we expect both momentum and energy to be conserved.

Momentum Conservation

The momentum of a single particle is the particle's mass multiplied by its velocity. The symbol **p** is conventionally used for momentum, which is a vector whose SI units are kg m/s.

$$\mathbf{p} = m\mathbf{v}$$

(We saw in Chapter 18 that relativity theory uses a different definition of momentum, but in Chapter 19 and the current chapter, we're focusing again on Newtonian mechanics.) Here is a Haskell function that returns the momentum of one particle:

```
momentum :: ParticleState -> Vec
momentum st = let m = mass st
                  v = velocity st
              in m *^ v
```

The momentum of a system of particles is the vector sum of the momenta for each particle in the system. We use an uppercase **P** to denote system momentum. In a system of particles, the momentum of particle n, which has mass m_n and velocity $\mathbf{v}_n$, is given by

$$\mathbf{p}_n = m_n \mathbf{v}_n$$

and the system momentum is

$$\mathbf{P} = \sum_n \mathbf{p}_n = \sum_n m_n \mathbf{v}_n$$

Here is a Haskell function that returns the momentum of a system of particles.

```
systemP :: MultiParticleState -> Vec
systemP (MPS sts) = sumV [momentum st | st <- sts]
```

In any system with only internal forces, the momentum of the system is conserved, meaning it stays the same over time. Our numerical calculations involve a finite time step, which must be small compared to characteristic time scales of the physical situation to give accurate results. Most quantities of physical interest become less and less accurate as the time step is increased. An exception is system momentum in the case where only internal forces are present. Our practice of categorizing forces as external or internal, along with our automatic application of Newton's third law, guarantees that system momentum will be conserved in any situation without external forces, regardless of numerical method and regardless of time step. This is because each internal force, acting over the course of one time step, will change the momentum vector of one particle by some amount and the momentum vector of another particle by the opposite amount. System momentum does not change from time step to time step, even if the time step is so large that the results of the calculation are poor.

Even in collisions where external forces are present, system momentum is usually approximately conserved because the internal forces of collision are usually strong compared with any external forces. Since collisions are typically short-lived, the effect of external forces over the short duration of the collision is usually quite small.

To confirm this conservation of system momentum, let's write a function that computes percent change in system momentum. Since a list of states is a common information representation, we'll use a finite list of multi-particle states for the input to this function, but we'll only be comparing the first and last states in the list. Here is the function:

```
percentChangePMag :: [MultiParticleState] -> R
percentChangePMag mpsts
    = let p0 = systemP (head mpsts)
          p1 = systemP (last mpsts)
      in 100 * magnitude (p1 ^-^ p0) / magnitude p0
```

We name the incoming list of multi-particle states `mpsts`, and we use the Prelude functions `head` and `last` to pick out the first and last states in the list, giving their system momenta the local names `p0` and `p1`, respectively. We proceed to take the difference between the final system momentum `p1` and the initial system momentum `p0`, form the magnitude of that momentum change vector, divide by the magnitude of the initial system momentum, and multiply by 100 to make a percentage.

Creating Tables

To see how well system momentum is conserved in our calculations, let's make a small table showing the percent change in system momentum for a few different time steps and a few different spring constants. The table will appear unwieldy if I allow the computer to display all 15 digits that it keeps around for double precision floating-point numbers (numbers of type `R` that we have been calling real numbers). The following `sigFigs` function rounds a number to a specified number of significant figures:

```
sigFigs :: Int -> R -> Float
sigFigs n x = let expon :: Int
                  expon = floor (logBase 10 x) - n + 1
                  toInt :: R -> Int
                  toInt = round
              in (10^^expon *) $ fromIntegral $ toInt (10^^(-expon) * x)
```

This function works by dividing the input number `x` by 10^m for some integer m, rounding the number, and then re-multiplying the number by 10^m. The integer m is called `expon` in the code; its value depends on the number `n` of significant figures requested. The Prelude function `round` has a quite general type; I specialize it to my needs by defining a local function `toInt` with a simple, concrete type.

The final tool we need to make cute little tables, which we will use for momentum, energy, and a few other things, is a data type for tables with a

Show instance that makes them appear in a nicely formatted way. First, we define a new data type Table a, which is a table of items of type a.

```
data Justification = LJ | RJ deriving Show

data Table a = Table Justification [[a]]
```

The data type Justification is to specify whether we want a left-justified or right-justified table. A Table a contains a Justification along with a list of lists of items with type a.

We write an explicit show instance for the new data type that formats the output in a nice way.

```
instance Show a => Show (Table a) where
    show (Table j xss)
        = let pairWithLength x = let str = show x in (str, length str)
              pairss = map (map pairWithLength) xss
              maxLength = maximum (map maximum (map (map snd) pairss))
              showPair (str,len)
                  = case j of
                      LJ -> str ++ replicate (maxLength + 1 - len) ' '
                      RJ -> replicate (maxLength + 1 - len) ' ' ++ str
              showLine pairs = concatMap showPair pairs ++ "\n"
          in init $ concatMap showLine pairss
```

In the first line, we see a type class constraint; type a must be an instance of type class Show for Table a to be an instance of type class Show. An instance declaration for Show requires only that we define a function show that takes a Table a as input and produces a string as output. We define a local function pairWithLength that pairs the string representation of a value with the length of that string. We care about the length because we want to make columns that line up nicely. The local variable pairss is a list of lists of pairs of strings and lengths. The double *s* at the end of the name suggests a list of lists. We form pairss by mapping map pairWithLength onto the input list of lists, xss. Since each element of xss is a list, we apply map pairWithLength to each list in xss, so pairWithLength gets applied to each item in the list of lists.

The local variable maxLength finds the length of the longest item in the table. We then use this longest length to set the width of all the columns that will be displayed. We write local functions to show an individual item and a line of the table. Finally, we form the table by mapping showLine onto pairss and concatenating the results. If you find this trick for displaying tables interesting, by all means study it; otherwise, let's move on and use it.

Here are tables that show the percent change in momentum for the Euler, Euler-Cromer, and fourth-order Runge-Kutta methods for several values of spring constant and time step:

```
pTable :: (TimeStep -> NumericalMethod MultiParticleState DMultiParticleState)
       -> [R]         -- ks
       -> [TimeStep]  -- dts
```

```
        -> Table Float
pTable nMethod ks dts
    = Table LJ [[sigFigs 2 $
                 percentChangePMag (billiardStatesFinite nMethod k dt)
                     | dt <- dts] | k <- ks]

pTableEu :: [R]          -- ks
         -> [TimeStep]   -- dts
         -> Table Float
pTableEu = pTable euler
```

We can view these tables in GHCi.

```
Prelude Vis> :m
Prelude> :l MOExamples
[1 of 6] Compiling Newton2          ( Newton2.hs, interpreted )
[2 of 6] Compiling Mechanics1D      ( Mechanics1D.hs, interpreted )
[3 of 6] Compiling SimpleVec        ( SimpleVec.hs, interpreted )
[4 of 6] Compiling Mechanics3D      ( Mechanics3D.hs, interpreted )
[5 of 6] Compiling MultipleObjects  ( MultipleObjects.hs, interpreted )
[6 of 6] Compiling MOExamples       ( MOExamples.hs, interpreted )
Ok, six modules loaded.
*MOExamples> pTable euler [10,30,100] [0.003,0.01,0.03,0.1]
4.3e-14 0.0     0.0     0.0
0.0     0.0     0.0     0.0
2.2e-14 0.0     0.0     8.7e-14
*MOExamples> pTable eulerCromerMPS [10,30,100] [0.003,0.01,0.03,0.1]
0.0 0.0 0.0 0.0
0.0 0.0 0.0 0.0
0.0 0.0 0.0 0.0
*MOExamples> pTable rungeKutta4 [10,30,100] [0.003,0.01,0.03,0.1]
4.3e-14 2.2e-14 0.0     0.0
2.2e-14 0.0     2.2e-14 0.0
0.0     0.0     0.0     0.0
```

Regardless of numerical method, spring constant, or time step, the percentage change in momentum is either 0 or something times 10^{-14}. Since this is a percentage, we are really talking about a few parts in 10^{16}, which is the accuracy of double-precision floating-point numbers. This deviation from 0 is not from the finite-step-size calculation we are doing; it's because any calculation at all with floating-point numbers is approximate. The computer can't divide by 10 exactly because it represents the fraction 1/10 with the repeating binary expansion 0.0001100110011 . . . (recall Table 1-4). A few parts in 10^{16} is the deviation from exactness we expect to see for any calculation that involves double-precision floating-point numbers.

With this understanding, these tables are showing us that system momentum is conserved for the billiard collision for any numerical method, any spring constant, and any step size. This is an example of the claim made earlier that for any situation with only internal forces, system momentum is conserved regardless of numerical method, step size, or other parameters that describe the problem. Let's now turn to look at the energy of a multi-particle system, which does not share this desirable property.

Energy Conservation

With the exception of the short time in which the colliding objects are in contact with each other, the only form of energy present is the kinetic energy of the objects.

Let's look at how system kinetic energy evolves in time for the billiard ball collision. During all of the time in which the collision is not occurring, kinetic energy will be conserved just fine because the velocities of the balls are not changing at all. During the short time in which the collision is occurring, some kinetic energy is transformed into elastic potential energy, held by our spring, before it is transformed back into kinetic energy. Figure 20-3 shows a graph of system kinetic energy versus time for the collision of two billiard balls.

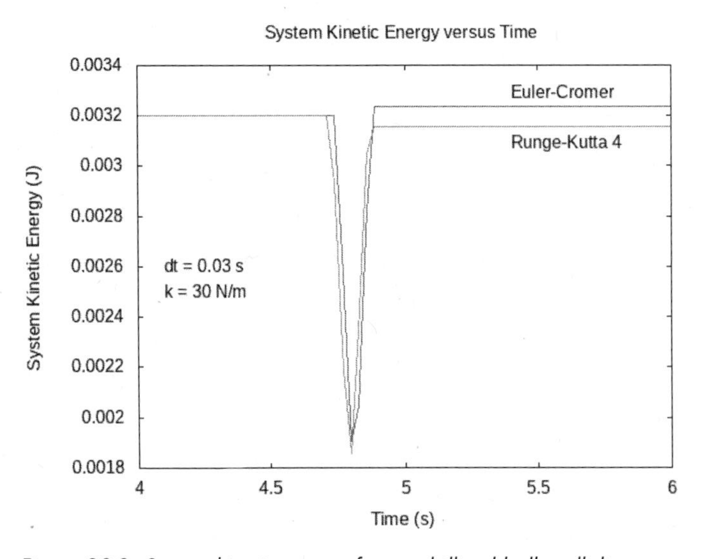

Figure 20-3: System kinetic energy for two billiard balls colliding

We see that before the collision, which begins at a time of about 4.8 s, the system kinetic energy is conserved, as the incoming particle moves with a constant velocity. We see the expected dip in system kinetic energy as it is converted into elastic potential energy. The graph suggests that only about 40 percent of the system kinetic energy is so converted since it drops to about 1.9 mJ from its initial value of 3.2 mJ. As the spring expands from its state of maximal compression, elastic potential energy is converted back into kinetic energy, which remains constant after the collision.

In an elastic collision, the system kinetic energy after the collision should be the same as before the collision. We can see from the graph that system kinetic energy after the collision is close to, but not precisely, the same as before. This difference is a result of the finite step size of our method, and it depends on numerical method, step size, and other parameters of the situation. We see from the graph that the Euler-Cromer and fourth-order Runge-Kutta methods produce slightly different results, including slightly different results for the final system kinetic energy.

The following code produced the graph in Figure 20-3.

```
systemKEWithTime :: IO ()
systemKEWithTime
    = let timeKEPairsEC
              = [(timeOf mpst, systemKE mpst)
                     | mpst <- billiardStatesFinite eulerCromerMPS 30 0.03]
          timeKEPairsRK4
              = [(timeOf mpst, systemKE mpst)
                     | mpst <- billiardStatesFinite rungeKutta4    30 0.03]
      in plotPaths [Key Nothing
                   ,Title "System Kinetic Energy versus Time"
                   ,XLabel "Time (s)"
                   ,YLabel "System Kinetic Energy (J)"
                   ,XRange (4,6)
                   ,PNG "SystemKE.png"
                   ,customLabel (4.1,0.0026) "dt = 0.03 s"
                   ,customLabel (4.1,0.0025) "k = 30 N/m"
                   ,customLabel (5.4,0.00329) "Euler-Cromer"
                   ,customLabel (5.4,0.00309) "Runge-Kutta 4"
                   ] [timeKEPairsEC,timeKEPairsRK4]
```

Local variables `timeKEPairsEC` and `timeKEPairsRK4` hold lists of pairs of time and system kinetic energy for the Euler-Cromer and fourth-order Runge-Kutta methods. We then plot these lists of pairs using `gnuplot`'s `plotPaths` function.

The Effect of the Step Size and Spring Constant on Kinetic Energy

System kinetic energy does not share the nice property of system momentum that guarantees its conservation regardless of step size. An accurate calculation of energy, like most physical quantities, requires a reasonably small step size. In fact, looking at how well system kinetic energy is conserved is a good way to determine whether we are using a step size that is small enough.

To study conservation of system kinetic energy for different step sizes and spring constants, let's write a function that computes percent change in system kinetic energy. Since a list of states is a common information representation, we'll use a finite list of multi-particle states for the input to this function, but we'll only be comparing the first and last states in the list. Here is the function:

```
percentChangeKE :: [MultiParticleState] -> R
percentChangeKE mpsts
    = let ke0 = systemKE (head mpsts)
```

```
    ke1 = systemKE (last mpsts)
  in 100 * (ke1 - ke0) / ke0
```

We name the incoming list of multi-particle states mpsts and use the Prelude functions head and last to pick out the first and last states in the list, giving their system kinetic energies the local names ke0 and ke1, respectively. We proceed to take the difference between the final system kinetic energy ke1 and the initial system kinetic energy ke0, divide by the initial system kinetic energy, and multiply by 100 to make a percentage.

To explore conservation of kinetic energy, we'll make some small tables showing the percent change in system kinetic energy for a few different time steps and a few different spring constants. The following tenths function rounds a number to one digit after the decimal point and helps to make a handsome table.

```
tenths :: R -> Float
tenths = let toInt :: R -> Int
             toInt = round
         in (/ 10) . fromIntegral . toInt . (* 10)
```

This function works by multiplying the input number x by 10, rounding the number, and then dividing the result by 10. The Prelude function round has a quite general type; I specialize it to my needs by defining a local function toInt with a simple, concrete type.

The function keTable produces a table of percent change in system kinetic energy when given a numerical method, a list of spring constants, and a list of time steps.

```
keTable
    :: (TimeStep -> NumericalMethod MultiParticleState DMultiParticleState)
    -> [R]          -- ks
    -> [TimeStep]   -- dts
    -> Table Float
keTable nMethod ks dts
    = Table RJ [[tenths $
                 percentChangeKE (billiardStatesFinite nMethod k dt)
                     | dt <- dts] | k <- ks]
```

We can view these tables in GHCi.

```
*MOExamples> keTable euler [10,30,100] [0.003,0.01,0.03,0.1]
     4.2    15.9    68.7   705.7
     8.3    34.1   185.4  3117.9
    16.9    82.9   642.2 39907.1
*MOExamples> keTable eulerCromerMPS [10,30,100] [0.003,0.01,0.03,0.1]
    0.0    0.0   -0.3    6.2
    0.0    0.1    1.1  154.1
    0.0    0.3   -8.9 3705.2
```

```
*MOExamples> keTable rungeKutta4 [10,30,100] [0.003,0.01,0.03,0.1]
   0.0   0.0   0.0  -2.8
   0.0  -0.1  -1.4 -14.6
   0.0  -0.5  -1.6  90.3
```

Conservation is best in the upper-left corner of each table, where time step and spring constant are both small. It is not surprising that a smaller time step produces better results. A smaller spring constant causes the collision to last longer, taking place over a larger number of time steps. A calculation in which a collision takes place over only a handful of time steps is unlikely to be very accurate. On the other hand, a spring constant that is too small runs the risk of allowing the objects to get too close to each other. We take up this concern in the next section.

Numerical Issues

We have suggested that an accurate calculation for a collision requires that multiple time steps take place during the collision. We have also noted that we don't want the colliding objects to get too close to each other. These two desired properties for our calculation are in tension with each other because the first benefits from a small spring constant while the second benefits from a large spring constant. Let's analyze these two desired properties in more detail.

Time Steps During Collision

As mentioned earlier, if only a handful of time steps, or worse yet, only one or zero, elapse over the course of the collision, we are unlikely to get accurate results. This observation motivates us to ask how many time steps take place during the collision, or equivalently how many time steps occur with the balls within the threshold separation. The answer to this question depends on numerical method, spring constant, and time step. We'd like a large number of time steps (for example, at least 10).

The function `contactSteps` returns the number of time steps during which the balls are within their threshold separation of 6 cm. It takes a finite list of multi-particle states as input.

```
contactSteps :: [MultiParticleState] -> Int
contactSteps = length . takeWhile inContact . dropWhile (not . inContact)
```

The function works by using `dropWhile` to discard the multi-particle states before there is contact between the balls; in other words, to discard the states before the balls come within 6 cm of each other. We use the `inContact` predicate, defined next, to decide whether, in a given multi-particle state, the balls are in contact, meaning that their centers are within 6 cm of each other. We then use `takeWhile` to keep the states in which the balls are in contact. Finally, we calculate the length of this list, which is the number of states, or time steps, in which the balls are in contact.

The predicate inContact works by calculating the distance between the particle centers and comparing it to the threshold separation of twice the ball radius (6 cm).

```
inContact :: MultiParticleState -> Bool
inContact (MPS sts)
    = let r = magnitude $ posVec (sts !! 0) ^-^ posVec (sts !! 1)
      in r < 2 * ballRadius
```

The function contactTable returns this number of time steps in contact for a numerical method, a list of spring constants, and a list of time steps.

```
contactTable
    :: (TimeStep -> NumericalMethod MultiParticleState DMultiParticleState)
    -> [R]          -- ks
    -> [TimeStep]   -- dts
    -> Table Int
contactTable nMethod ks dts
    = Table RJ [[contactSteps (billiardStatesFinite nMethod k dt)
                    | dt <- dts] | k <- ks]
```

We use a list comprehension to form the list of lists that will get displayed as a table. Here are the results from GHCi:

```
*MOExamples> contactTable euler [10,30,100] [0.003,0.01,0.03,0.1]
 89 27  9  3
 53 16  6  2
 29  9  3  2
*MOExamples> contactTable eulerCromerMPS [10,30,100] [0.003,0.01,0.03,0.1]
 89 27  9  2
 53 16  5  1
 29  9  3  1
*MOExamples> contactTable rungeKutta4 [10,30,100] [0.003,0.01,0.03,0.1]
 89 27  9  2
 53 16  5  1
 29  9  3  0
```

There is not much difference among numerical methods. Regardless of numerical method, the best results appear in the upper-left corner of each table, where the spring constant and time step are both smallest. A small spring constant gives a loose (not stiff) spring, which compresses over a greater distance, allowing more time steps to take place during the collision.

Something may strike you as strange in the lower-right corner of the Runge-Kutta table, corresponding to k = 100 N/m and dt = 0.1 s. How is it possible that no times take place within the threshold distance? It is not because the time step is so large that the moving ball completely skips over the stationary ball without colliding. Rather, it is because a fourth-order Runge-Kutta step is made up of four substeps, using derivatives at four different places to compute a final change in values for the time step. When the two balls are close but just outside the threshold separation, the Runge-Kutta step senses the repulsive force from one or more of the substeps. The spring constant is so large that a large repulsive force acts to repel the balls, and by the next actual time step, the balls have already repelled. So, the 0 listed in the table for number of time steps within the threshold separation is not so different from having one time step within the threshold; it's not enough for an accurate calculation.

Closest Separation

We don't want the centers of the balls to get too close to each other. If we use a very small spring constant, the balls may compress until their centers coincide, or even move past each other. This is certainly not how billiard balls work. Billiard balls hardly compress at all, so to model them accurately requires a fairly large spring constant.

It is interesting and important to know how close the ball centers get to each other, so we can avoid choosing a spring constant that is too small. We wish to know the minimum separation of the balls, center to center, that occurs during the collision. The answer to this question depends on numerical method, spring constant, and time step.

The function closest returns the closest separation that the balls achieve during the collision. It takes a finite list of multi-particle states as input.

```
closest :: [MultiParticleState] -> R
closest = minimum . map separation
```

All this function does is apply the separation function below to each multi-particle state in the finite list and calculate the minimum.

The separation function works by finding the displacement between ball centers and computing its magnitude.

```
separation :: MultiParticleState -> R
separation (MPS sts)
    = magnitude $ posVec (sts !! 0) ^-^ posVec (sts !! 1)
```

The function closestTable returns a table of closest separations when given a numerical method, a list of spring constants, and a list of time steps.

```
closestTable
    :: (TimeStep -> NumericalMethod MultiParticleState DMultiParticleState)
    -> [R]          -- ks
    -> [TimeStep]   -- dts
    -> Table Float
closestTable nMethod ks dts
    = Table RJ [[tenths $ (100*) $
                closest (billiardStatesFinite nMethod k dt)
                    | dt <- dts] | k <- ks]
```

We multiply by 100 to convert meters into centimeters, which are shown in the table.

Here are the results from GHCi:

```
*MOExamples> closestTable euler [10,30,100] [0.003,0.01,0.03,0.1]
 4.4 4.3 4.0 2.8
 5.0 4.9 4.6 2.8
 5.4 5.3 5.0 2.8
*MOExamples> closestTable eulerCromerMPS [10,30,100] [0.003,0.01,0.03,0.1]
 4.4 4.4 4.4 4.5
 5.1 5.1 5.0 4.5
 5.5 5.5 5.5 4.5
*MOExamples> closestTable rungeKutta4 [10,30,100] [0.003,0.01,0.03,0.1]
 4.4 4.4 4.4 4.7
 5.1 5.1 5.1 5.2
 5.5 5.5 5.5 6.3
```

If our goal is to have minimal compression, resulting in a large closest separation, the lower-left corner of each table is where we want to be. This suggests a large spring constant.

We see, in the lower-right corner of the Runge-Kutta table, a closest distance of 6.3 cm, which seems impossible. How can the balls repel if they never get within the threshold separation? Again, the answer is that a time step of fourth-order Runge-Kutta is based on four substeps, some of which sample the repulsive force inside the threshold distance.

Suppose we desire parameters (spring constant and time step) that produce at least 10 time steps during the collision, allow a closest separation of no less than 5 cm, and conserve kinetic energy within one percent.

The Euler method is out. It doesn't conserve kinetic energy to within one percent for any of the spring constants and time steps we sampled. Euler-Cromer could be used with k = 30 N/m and dt = 0.003 s or dt = 0.01 s, or with k = 100 N/m and dt = 0.003 s. Fourth-order Runge-Kutta could be used with the same parameters.

Animated Results

We'd like to animate the collision of the billiard balls using `gloss`. We have already written a state-update function and an initial state. What remains is to write a display function, which we'll do now.

```
billiardPicture :: MultiParticleState -> G.Picture
billiardPicture (MPS sts)
    = G.scale ppm ppm $ G.pictures [place st | st <- sts]
      where
        ppm = 300  -- pixels per meter
        place st = G.translate (xSt st) (ySt st) blueBall
        xSt = realToFrac . xComp . posVec
        ySt = realToFrac . yComp . posVec
        blueBall = G.Color G.blue (disk $ realToFrac ballRadius)
```

We use pattern matching on the input to give the name sts to the incoming list of single-particle states. This is a list with length 2 because there are two particles. Our display function uses the "scale the whole picture at the end" paradigm, using the G.scale function with the constant ppm as the number of pixels per meter for our one universal spatial scale factor. The pre-scaled picture is made with the G.pictures function, which combines a list of pictures, one for each ball. The list of pictures is formed using a list comprehension and the place function, to be defined in a few lines. This code could be used for a multi-particle system with any number of particles, as long as we are happy having each particle represented by a blue disk. (See Exercise 20.2 if you are offended by every particle being blue.)

The remainder of the billiardPicture display function consists of local constants and functions defined after the where keyword. Recall that where, like the let-in construction, allows us to define local variables; the local variables are used before the where keyword and defined after. The local variables in a let-in construction are defined before the in keyword and used after. The difference between the let-in construction and the where construction is similar to the difference between bottom-up thinking, where we define the smallest pieces first and build upward to the whole function, and top-down thinking, where we define the whole function first in terms of pieces we have not yet defined. Haskell supports and encourages both kinds of thinking by supplying these two constructions and allowing us to make definitions in terms of as-yet-undefined constants and functions.

The first local variable we define is the spatial scale factor ppm, which we set to 300 pixels per meter. Next, we define the local function place, which we have already used to translate a picture to the xy-coordinates specified in the position of the state. The function place uses the as-yet-undefined functions xSt and ySt to pull the coordinates out of a state, and the as-yet-undefined picture blueBall for a blue disk. The function place uses the G.translate function to translate the picture by the coordinates.

The local function xSt picks out the x-coordinate of position from the state and uses realToFrac to return a Float, which is the type that G.translate expects. The definition of xSt is written in point-free style, as a composition of three functions: posVec, which extracts the position from the state; xComp, which extracts the x-coordinate from the position; and realToFrac, which converts an R into a Float. The function ySt is just like xSt but for the y-coordinate. Finally, we define the local constant blueBall as a blue disk with radius ballRadius, which must get converted to a Float to match the expected input type of the disk function from Chapter 17.

Listing 20-3 shows a stand-alone program that uses the simulateGloss function we wrote in Chapter 16. The stand-alone program consists of one definition: a main function called main.

```
{-# OPTIONS -Wall #-}

import Mechanics3D ( simulateGloss )
import MultipleObjects (eulerCromerMPS )
```

```
import MOExamples ( billiardInitial, billiardPicture, billiardUpdate )

main :: IO ()
main = simulateGloss 1 100 billiardInitial billiardPicture
       (billiardUpdate eulerCromerMPS 30)
```

Listing 20-3: Stand-alone program for a 2D animation of two billiard balls colliding

The `main` function uses the imported `simulateGloss` function to perform the animation. We choose a time-scale factor of 1 and an animate rate of 100 frames/second, giving a time step of 0.01 s. We choose a spring constant of 30 N/m. We import the initial state `billiardInitial`, the state-update function `billiardUpdate`, and the display function `billiardPicture` from the `MOExamples` module we have been writing in this chapter. When you run the animation, you'll see one blue billiard ball moving to the right collide with a stationary blue billiard ball. The originally moving ball moves downward after the collision, while the originally stationary ball moves up and to the right.

Wave on a Guitar String

In this section, we'll model a wave on a guitar string. In particular, we'll focus on the G string of a guitar. A typical G string has a mass of 0.8293 grams per meter. The distance from the neck of the guitar to the bottom, the two places where the string will stay fixed, is 65 cm. The fundamental vibration of the string, when played in an open position to make the note G, has a wavelength of 130 cm since the displacement of the string from equilibrium is a sine function that starts at 0 at the neck and returns to 0 at the bottom, completing only half a wavelength. Any open string on the guitar has 130 cm as the wavelength for its fundamental vibration.

We want a frequency of 196 Hz to make the note G. This is because convention dictates that frequencies of 55 Hz, 110 Hz, 220 Hz, 440 Hz, and 880 Hz each give some version of the note A. Doubling the frequency takes you to the same note, up one octave. The chromatic scale that people use has 12 "half steps" in one octave, or one doubling of the frequency. The note G is two half steps below A, so we must multiply the frequency of an A by $2^{-2/12}$ to get the frequency of a G. For the guitar, we multiply $2^{-2/12}$ by 220 Hz to get 196 Hz.

For any wave, wavelength λ and period T are related to wave speed v by

$$v = \frac{\lambda}{T}$$

This equation is easiest to understand for a traveling wave (that is, one in which a wave crest simply travels along at speed v). For a traveling wave, the crest passes through one wavelength of distance during each period of time, so its speed is its wavelength divided by its period, which is what the equation claims. Frequency f and period T are related by

$$f = \frac{1}{T}$$

leading to the equation

$$v = \lambda f$$

relating wave speed, wavelength, and frequency.

NOTE *Some people like to use the Greek letter ν (nu) for frequency, which allows them to respond to a common greeting with the following physics joke:*

Friend: *What's new? (nu?)*

Jokester: *v over λ!*

Most people are unable to laugh outwardly at this joke, but surely they are laughing on the inside.

Since the two ends of a guitar string are fixed, the guitar string exhibits standing waves rather than traveling waves, but the equations just shown are still useful because the wave speed is related to the tension F in the string and the mass per unit length μ by

$$v = \sqrt{\frac{F}{\mu}}$$

When we tune the guitar, we change the tension, which changes the wave speed, which changes the frequency. Let's calculate the tension we need to achieve a frequency of 196 Hz.

To achieve 196 Hz, we need a wave speed,

$$v = \lambda f = (130 \text{ cm})(196 \text{ Hz}) = 254.8 \text{ m/s}$$

which requires a tension:

$$F = \mu v^2 = (0.8293 \text{ g/m})(254.8 \text{ m/s})^2 = 53.84 \text{ N}$$

Thus, we need a tension of 53.84 N in the G string of our guitar.

Forces

We will model the guitar string as 64 little point masses, spaced 1 cm from each other along the 65-cm length. Each mass is linked by springs to its two nearest neighbors. If we give each spring an equilibrium length of zero and a spring constant of 5384 N/m, each spring will produce a force of 53.84 N when it is extended by 1 cm, which is the distance between masses when the string is at rest. So, there will be 64 masses, 63 internal springs, and 2 external springs to connect to the two fixed ends at 0 cm and 65 cm. Here is the list of forces:

```
-- 64 masses (0 to 63)
-- There are 63 internal springs, 2 external springs
forcesString :: [Force]
forcesString
    = [ExternalForce  0 (fixedLinearSpring 5384 0 (vec    0 0 0))
      ,ExternalForce 63 (fixedLinearSpring 5384 0 (vec 0.65 0 0))] ++
      [InternalForce n (n+1) (linearSpring 5384 0) | n <- [0..62]]
```

State-Update Function

To make an animation, we need a state-update function, and for that we must choose a numerical method. Either Euler-Cromer or fourth-order Runge-Kutta would do just fine. Here we choose Runge-Kutta because it's slightly more accurate.

```
stringUpdate :: TimeStep
             -> MultiParticleState  -- old state
             -> MultiParticleState  -- new state
stringUpdate dt = updateMPS (rungeKutta4 dt) forcesString
```

Initial State

We'll need an initial state. In fact, it is interesting to explore several different initial states of the string. The function `stringInitialOvertone` produces an initial state in which the string lies in the xy-plane, not initially moving, with a sinusoidal pattern.

```
stringInitialOvertone :: Int -> MultiParticleState
stringInitialOvertone n
    = MPS [defaultParticleState
          { mass     = 0.8293e-3 * 0.65 / 64
          , posVec   = x *^ iHat ^+^ y *^ jHat
          , velocity = zeroV
          } | x <- [0.01, 0.02 .. 0.64],
          let y = 0.005 * sin (fromIntegral n * pi * x / 0.65)]
```

Using this function with an input of 1 produces the fundamental vibration we discussed earlier. Figure 20-4 demonstrates what `stringInitialOvertone 1` looks like.

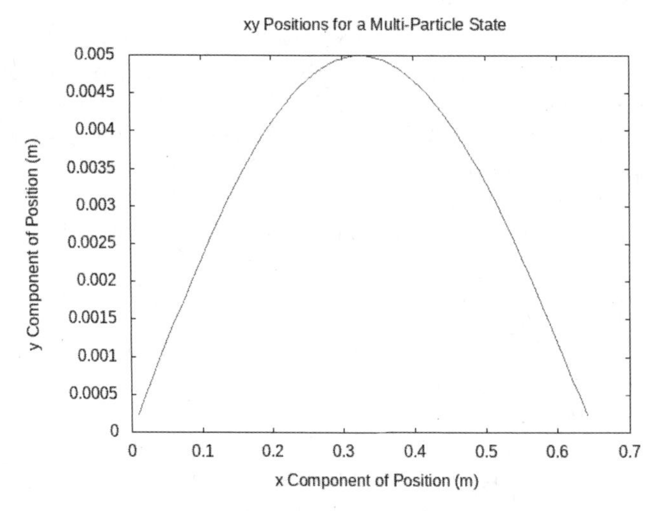

Figure 20-4: Initial state for the fundamental mode of vibration for a guitar string, given by `stringInitialOvertone 1`

Higher numbers produce overtones that vibrate at higher frequencies. Using 2 will produce an overtone that vibrates at 392 Hz, while 3 will produce one that vibrates at 588 Hz. Figure 20-5 shows what `stringInitialOvertone 3` looks like.

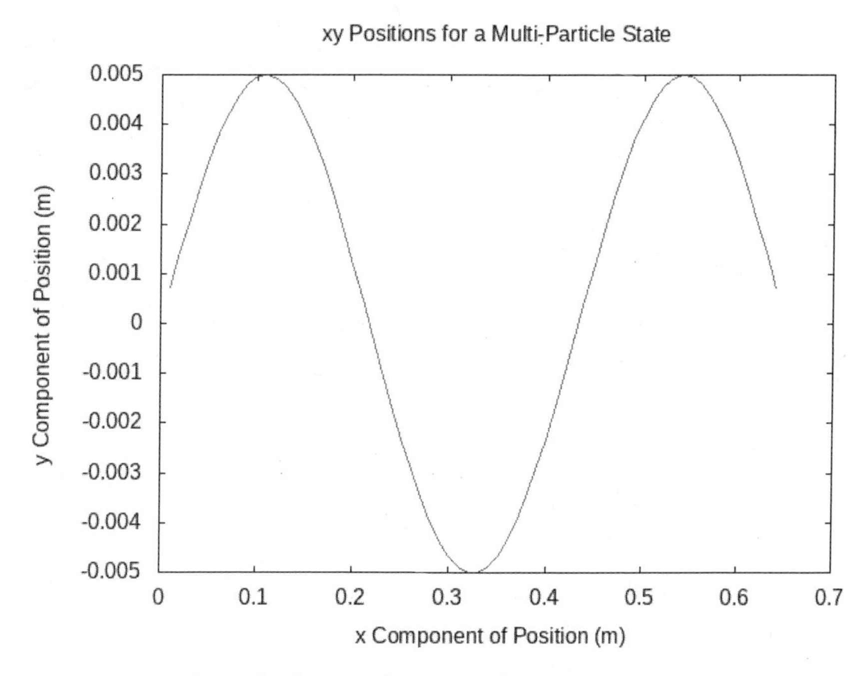

Figure 20-5: Initial state for the second overtone of a guitar string, given by `stringInitialOvertone 3`

The sound you hear from a guitar string is a mixture of the fundamental frequency along with overtones. The initial state `stringInitialPluck` is intended to approximate a pluck of the guitar string.

```
stringInitialPluck :: MultiParticleState
stringInitialPluck = MPS [defaultParticleState
             { mass     = 0.8293e-3 * 0.65 / 64
             , posVec   = x *^ iHat ^+^ y *^ jHat
             , velocity = zeroV
             } | x <- [0.01, 0.02 .. 0.64], let y = pluckEq x]
    where
      pluckEq :: R -> R
      pluckEq x
          | x <= 0.51  = 0.005 / (0.51 - 0.00) * (x - 0.00)
          | otherwise  = 0.005 / (0.51 - 0.65) * (x - 0.65)
```

Suppose the pick touches the string 51 cm from the neck, in front of the hole in the guitar body. If the pick moves the string 5 mm at that point, the resulting string shape is given by `stringInitialPluck`, as shown in Figure 20-6.

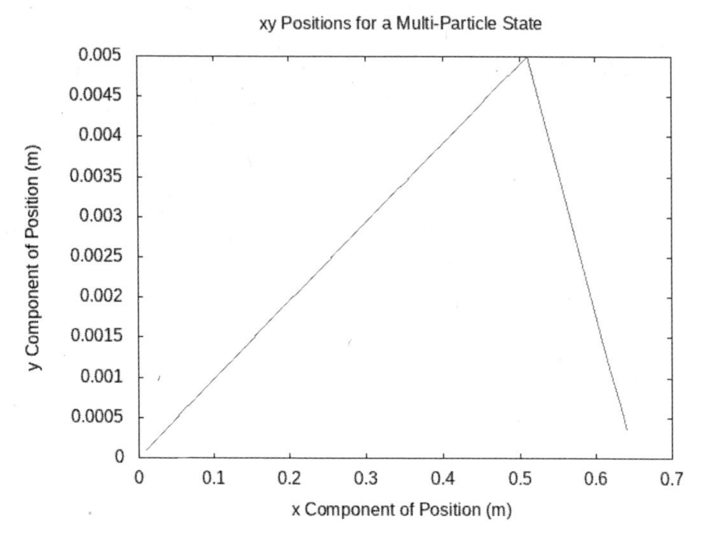

Figure 20-6: Initial state for a pluck of the guitar string, given by `stringInitialPluck`

Exercise 20.10 asks you to write a function that produces an *xy* picture like this from a `MultiParticleState`.

Stand-Alone Program

Listing 20-4 shows a stand-alone program to make a 2D animation of the wave on a guitar string.

```
{-# OPTIONS -Wall #-}

import SimpleVec ( zeroV, iHat, (*^), xComp, yComp )
import Mechanics3D ( ParticleState(..), simulateGloss )
import MultipleObjects ( MultiParticleState(..) )
import MOExamples
import Graphics.Gloss ( Picture(..), scale, blue )

stringPicture :: MultiParticleState -> Picture
stringPicture (MPS sts)
    = let rs = [zeroV] ++ [posVec st | st <- sts] ++ [0.65 *^ iHat]
          xy r = (realToFrac $ xComp r, realToFrac $ yComp r)
          xys = map xy rs
          ppm = 400  -- pixels per meter
      in scale ppm (20*ppm) $ Color blue $ Line xys

main :: IO ()
main = let initialState = stringInitialOvertone 3
       in simulateGloss 0.001 40 initialState stringPicture stringUpdate
```

Listing 20-4: Stand-alone program for a 2D animation of a guitar string

We use a time-scale factor of 0.001, meaning that 1 ms of physical time corresponds to 1 s of animation time. The code in Listing 20-4 uses the initial state `stringInitialOvertone 3`, but we could replace that initial state with `stringInitialOvertone 1` to animate the fundamental vibration or with `stringInitialPluck` to animate the vibration resulting from a pluck. The fundamental frequency of 196 Hz represents a period of roughly 5 ms, so the fundamental vibration will take about 5 seconds of animation time to complete a period, while the overtone number 3 will take only 1.7 seconds of animation time to complete a period.

We use an animation rate of 40 frames/second, giving a time step of 25μs. This choice is based on the important time scales of the problem. First, there is the period of the fundamental vibration, which is about 5 ms. The overtones have progressively shorter periods, namely

$$\frac{5\text{ ms}}{2}, \frac{5\text{ ms}}{3}, \frac{5\text{ ms}}{4}, \frac{5\text{ ms}}{5}, \ldots$$

Evidently, any finite time step we employ is going to lose some information about some of the higher overtones because their periods get arbitrarily small. (Our method of modeling the string with 64 masses also places limits on the number of overtones that can accurately be accounted for. Overtone number 200, for example, has about 100 crests and 100 troughs; we can't possibly account for that if we're only keeping track of the position of 64 masses.)

Besides the period of the fundamental vibration, one other time scale is very important to the vibrating string: the time it takes the wave to travel from one little mass to a neighboring mass. This time is given by the distance between masses divided by the wave speed.

$$\frac{1\text{ cm}}{254.8\text{ m/s}} \approx 40\mu\text{s}$$

This time is about 40μs, which is substantially shorter than the fundamental period.

There is a stability criterion for wave situations or any situation that has a spatial step size Δx in which information travels at a finite velocity. That criterion says that the time step must be smaller than the time it takes information to travel one spatial step.

$$\Delta t < \frac{\Delta x}{v}$$

More about this stability criterion can be found in [18]. Using a time step above this threshold runs the risk of a numerical instability that produces nonsensical results. Figure 20-7 shows an example of numerical instability.

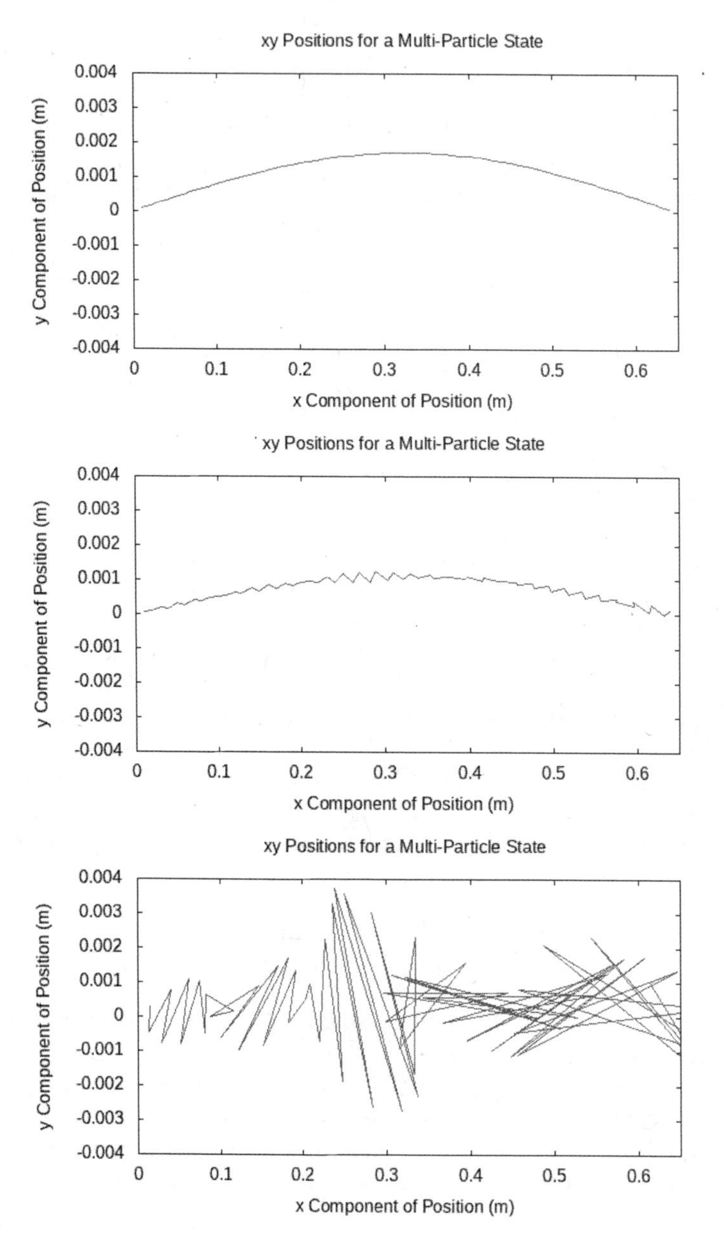

Figure 20-7: Example of numerical instability when the time step exceeds the time it takes information to travel between masses in the string. From top to bottom are three successive time steps.

Figure 20-7 shows the string for three successive time steps, numbers 10, 11, and 12, for a simulation with a time step of 100μs, which is above the stability threshold. In just two time steps, the calculation goes from reasonable to ridiculous. The calculation with a 100-μs time step is unstable. If we used this time step in an animation, the picture would rapidly become chaotic. I created the panels in Figure 20-7 with the commands

```
mpsPos (iterate (stringUpdate 100e-6) (stringInitialOvertone 1) !! 10)
mpsPos (iterate (stringUpdate 100e-6) (stringInitialOvertone 1) !! 11)
mpsPos (iterate (stringUpdate 100e-6) (stringInitialOvertone 1) !! 12)
```

where mpsPos is the function you are asked to write in Exercise 20.10.

The time step must be smaller than $40\mu s$, which is why we chose $25\mu s$ earlier.

Asynchronous Animation

The animations for the guitar string produced by gloss and not-gloss are on the edge of my computer's ability to carry out gracefully. After the animation runs for a little while, the frames seem to update less quickly, indicating that the computer is having trouble doing all of the calculations we are asking of it while at the same time displaying the results. As we ask more and more of the computer, there will come a time when it cannot do all of the calculations we want at a rate that would it allow it to display the results at the same time.

One solution for this situation is to use *asynchronous animation*, in which we do all of the calculations first and then sew the results together into a movie that we can watch later. Listing 20-5 shows a stand-alone program that creates 1,000 PNG files, each showing a picture of the guitar string, spaced at 25-μs intervals. These can be combined into an MP4 movie with an external program such as ffmpeg.

```
{-# OPTIONS -Wall #-}

import SimpleVec ( R, zeroV, iHat, (*^), xComp, yComp )
import Mechanics3D ( ParticleState(..) )
import MultipleObjects ( MultiParticleState(..) )
import MOExamples
import Graphics.Gnuplot.Simple

makePNG :: (Int,MultiParticleState) -> IO ()
makePNG (n,MPS sts)
    = let rs = [zeroV] ++ [posVec st | st <- sts] ++ [0.65 *^ iHat]
          xy r = (xComp r, yComp r)
          xys :: [(R,R)]
          xys = map xy rs
          threeDigitString = reverse $ take 3 $ reverse ("00" ++ show n)
          pngFilePath = "GnuplotWave" ++ threeDigitString ++ ".png"
      in plotPath [Title "Wave"
                  ,XLabel "Position (m)"
                  ,YLabel "Displacement (m)"
                  ,XRange (0,0.65)
                  ,YRange (-0.01,0.01)
                  ,PNG pngFilePath
```

```
                   ,Key Nothing
                   ] xys

main :: IO ()
main = sequence_ $ map makePNG $ zip [0..999] $
       iterate (stringUpdate 25e-6) (stringInitialOvertone 3)
```

Listing 20-5: Stand-alone program for a 2D asynchronous animation of a flexible string

The function makePNG takes as input an integer and a multi-particle state and produces a PNG file showing the position of the string. The purpose of the integer is to label the output file; 0 produces the file *GnuplotWave000.png*, 8 produces the file *GnuplotWave008.png*, and 167 produces the file *GnuplotWave167.png*. Only integers between 0 and 999, inclusive, should be used with this function.

The function begins by naming the incoming integer n and, using pattern matching on the input, the incoming list of single-particle states sts. The function then defines several local variables in the let construction. The local variable rs is a list of 66 position vectors describing the position of the string. The list consists of the positions of the 64 little masses, augmented with the fixed positions of the string at each end. The local function xy produces an (x, y) pair from a position. The list xys, formed by mapping xy onto the list rs of positions, is the list of pairs we will ask gnuplot to plot.

The local string threeDigitString is a three-digit string based on the integer n given as input. The function works by using show to convert n to a string, prepending that string with zeros, and then taking the last three digits. We take the last three digits by reversing the string, taking the first three digits with take 3, and then reversing back. The threeDigitString is then used as part of the filename pngFilePath. In the body of the let construction, we use gnuplot's plotPath function to plot the pairs xys that we defined earlier. Since we intend to animate the frames that gnuplot produces, it's important that we specify the XRange and YRange attributes so that each frame has the same range.

Let's turn our attention to the function main. We want to apply the function makePNG to each of 1,000 pairs of an integer with a multi-particle state. In an imperative language like Python, this would be an opportunity to use a loop. In a functional language like Haskell, this is an opportunity to use a list. The main function consists of several phrases separated by the function application operator $. Since this operator is right associative (recall Table 1-2), it's easiest to read the definition of main from right to left. The rightmost phrase,

```
iterate (stringUpdate 25e-6) (stringInitialOvertone 3)
```

is an infinite list of multi-particle states, starting with overtone number 3, spaced 25μs apart. Applying zip [0..999] to this infinite list produces a finite list, with each element being a pair of an integer and a multi-particle state. Applying map makePNG to this list of pairs produces a length-1,000 list with type [IO ()]. This is not the type we want the function main to have. We want main to have type `IO ()`, which simply means that it does something. Haskell provides a function `sequence_` to convert a list of actions into a single action.

Here is the type of `sequence_`:

```
*MOExamples> :t sequence_
sequence_ :: (Foldable t, Monad m) => t (m a) -> m ()
```

We are using `sequence_` in a context where the `Foldable` thing is a list, the `Monad` is `IO`, and the type variable `a` is unit, so the concrete type for `sequence_` in our use is

```
sequence_ :: [IO ()] -> IO ()
```

which is just what we need to produce the correct type for `main`. The function `sequence_` assembles a list of actions into a single action by sequencing them. The following command asks the external program `ffmpeg` to combine all PNG files named *GnuplotWaveDDD.png*, where the capital Ds are digits. We ask for a frame rate of 40 frames/second. The final movie is called *GnuplotWave.mp4*.

```
$ ffmpeg -framerate 40 -i GnuplotWave%03d.png GnuplotWave.mp4
```

Note that the character after the percent sign is a zero and not the letter O. If you are using a Unix-like system, you can find documentation about `ffmpeg`, after you have installed it, with the command `man ffmpeg`.

When we are doing asynchronous animation, we specify the time step and the animation rate rather than the time-scale factor and the animation rate.

Summary

In this chapter we examined three physical situations involving multiple interacting particles, applying the theory and ideas of Chapter 19. The first situation, with two masses and two springs, contains two particles and both internal and external forces.

Our second situation, a collision, has only internal forces and consequently conserves system momentum. We looked at conservation of momentum and energy in the context of the approximate numerical calculations we have been doing, and we found that momentum is conserved for all time steps, no matter how large. The collision also gave us an opportunity to look more deeply into numerical issues that affect the choice of technical parameters, such as the spring constant and time step.

Our third situation, the guitar string, involves many particles and suggests a transition into fields and waves. All along we have been discretizing time as a method to achieve practical results in mechanics problems; here, in using many particles to model a string, we are coming close to discretizing space as one would do to numerically solve field equations such as the Maxwell equations.

Part II of this book has dealt with Newtonian mechanics. Starting with a single particle moving in one dimension, we have slowly and steadily introduced ideas and code to deal with more and more sophisticated situations in mechanics. We talked about which mechanics problems can be solved with algebra, which require integration from calculus, and which require differential equations. We developed some general-purpose ways of solving systems of differential equations, and we put them to use in the service of mechanics. We took advantage of Haskell's type system to make simple data structures, like 3D vectors, to construct a sequence of information representations that lead from problem specification to problem solution, and to create a modular system in which it is easy to switch out one numerical method for another. We viewed Newton's second law as the rule for constructing a differential equation from a list of forces. We built Newton's third law into the infrastructure for interacting particles so it gets automatically applied to all of the internal forces acting in a multi-particle situation. I hope that I have convinced you that a functional language is a fruitful way to express the ideas of mechanics alongside the ideas required for solving problems in mechanics.

The next part of the book explores electromagnetic theory. It begins with a chapter about Coulomb's law, which fits nicely into the framework of interacting particles we have developed. Coulomb's law, like Newton's law of universal gravity, is expressed as an internal force between two particles.

Exercises

Exercise 20.1. What is the type of local variable `v` in the function `kineticEnergy`? What is the type of local variable `v` in the function `momentum`?

Exercise 20.2. The `billiardPicture` display function we wrote will display any number of balls, but they are all blue. If you want an animation with billiard balls of different colors, you can modify the `billiardPicture` function to cycle through a list of colors.

Make a new function `billiardPictureColors` by starting with a copy of `billiardPicture` and making the following changes. First, replace the local name `blueBall` with a local function `coloredBall` that takes a color as input. Next, modify the local function `place` to take a color as a second argument and use the new `coloredBall` function instead of `blueBall`. Finally, replace the list comprehension `[place st | st <- sts]` with

```
(zipWith place sts (cycle [G.blue, G.red]))
```

The latter will cycle through the colors blue and red. You can change this list to cycle through as many colors as you want.

Modify the main billiard collision program to use your new display function, `billiardPictureColors`, and check that it works.

Exercise 20.3. Using Listing 20-3 as a starting point, animate a two-body gravitational system such as the sun and Earth. In place of `billiardForces`, write a list of forces called `sunEarthForces` that contains the only force present:

the universal gravity between the sun and Earth. In place of `billiardUpdate`, write an update function called `sunEarthUpdate`. In place of `billiardInitial`, write an initial state called `sunEarthInitial`, choosing appropriate values for the initial positions and velocities of the sun and Earth. In place of `billiard Picture`, write a display function called `sunEarthPicture`, making the sun yellow and Earth blue. You will not be able to display the size of the sun or Earth to scale with the orbital motion. Choose any convenient values for the object sizes. Choose a time-scale factor of `365*24*60` so that one year of physical time is one minute of animation time. Choose an animation rate of 60 frames/second. Confirm that the orbital period is approximately one year by observing that it takes one minute for the animated Earth to complete a revolution around the sun.

Exercise 20.4. In this problem, we investigate how Jupiter makes the sun wobble. We say that Jupiter orbits the sun because the sun is much more massive, but in a system consisting only of the sun and Jupiter, both objects orbit a point at the system's center of mass. The center of mass is a weighted average of positions.

$$\mathbf{r}_{CM} = \frac{m_S \mathbf{r}_S + m_J \mathbf{r}_J}{m_S + m_J}$$

The distance between the sun and Jupiter is about 8×10^{11} m. The radius of the sun is 6.96×10^{8} m. The mass of the sun is 1.99×10^{30} kg. The mass of Jupiter is 1.90×10^{27} kg.

Placing the sun at the origin in the equation above, and looking only at the radial component of the vectors, we find that the center of mass

$$r_{CM} = \frac{m_J}{m_S + m_J} r_J \approx 8 \times 10^{8} \text{ m}$$

lies slightly outside the radius of the sun. We should be able to see the sun orbit about this center of mass in an animation.

Make an animation for the sun-Jupiter system using `gloss` or `not-gloss`. The only force is the universal gravity between the sun and Jupiter. Display the size of the sun to scale with the orbital motion, but zoom in on the sun, ignoring the display of Jupiter. If things go well, you should see the sun orbit a point slightly outside of its radius.

You do not need to include any of the center-of-mass calculations above to make this animation. However, for the center of mass to stay fixed, you will need to supply initial conditions that give zero momentum to the sun-Jupiter system. If you take the initial velocity of Jupiter to be in the y-direction, the sun will need to have a small initial velocity in the negative y-direction so that the total momentum is 0.

To obtain an initial velocity estimate for Jupiter, you may assume that Jupiter's orbit is circular and that the radius of its orbit is the same as the sun-Jupiter distance. (This is only off by 0.1 percent.)

Exercise 20.5. Using realistic initial conditions, program an animation for the sun, Earth, and moon mutually interacting though gravity. The actual Earth-sun separation is about 500 times the Earth-moon separation, so you won't be able to resolve the Earth and moon as separate objects on the screen. To be able to see where the moon is relative to Earth, I suggest the following: instead of displaying the moon at the position you calculate, display the moon at a fake position that has the correct orientation but is 50 times as far from Earth as you calculate. An equation to use to calculate a fake moon position is

$$\mathbf{r}_{\text{FM}} = \mathbf{r}_{\text{E}} + A(\mathbf{r}_{\text{M}} - \mathbf{r}_{\text{E}})$$

where $\mathbf{r}_{\text{FM}}$ is the position of the fake moon, $\mathbf{r}_{\text{M}}$ is the position of the (real) moon, $\mathbf{r}_{\text{E}}$ is the position of Earth, and A is a magnification factor that artificially magnifies the vector from Earth to moon for display purposes. Try $A = 50$ and see what happens. Note that the fake moon need only appear in the display function. This situation should involve three internal forces and no external forces.

Exercise 20.6. Using the `fixedLinearSpring` function, investigate a springy pendulum. Choose values for the spring constant k, the equilibrium length of the spring r_e, and the mass m.

(a) Confirm, with an animation or a graph, that placing the mass a distance $r_e + mg/k$ directly below the spot where the spring attaches to the ceiling, giving it no initial velocity, results in an equilibrium situation that allows the mass to hang motionless on the spring.

(b) Choose an initial state in which the mass has no initial velocity and sits a distance r_e directly below the spot where the spring attaches to the ceiling. In this situation, the spring initially exerts no force on the mass, but gravity does exert a force. Confirm, with an animation or a graph, that the mass oscillates with an angular frequency $\omega = \sqrt{k/m}$. This is equivalent to a period of $T = 2\pi\sqrt{m/k}$.

(c) Investigate the horizontal oscillation of a pendulum by placing the mass, with no initial velocity, a distance $r_e + mg/k$ from the ceiling attachment point, but not directly below that point. Allow the mass to evolve in time. Confirm, with an animation or a graph, that the angular frequency of oscillation is close to $\omega = \sqrt{g/l}$, with $l = r_e + mg/k$, or, equivalently, that the period is $T = 2\pi\sqrt{l/g}$.

(d) Find an initial position and velocity for the mass so that it undergoes horizontal circular motion.

(e) If you now change the parameters so that

$$\frac{g}{l} \approx \frac{k}{m}$$

you may be able to find initial conditions in which the mass undergoes circular or elliptical motion in a vertical plane. See if this is possible.

Exercise 20.7. Consider two equal-mass billiard balls traveling toward each other with equal and opposite velocities. If they travel directly toward each other, say along the x-axis, the collision will be one dimensional. If, on the other hand, their initial velocities are in the positive and negative x-direction, but there is a small y-component of displacement between them, the collision will be two dimensional. The angle at which the balls travel after the collision (which is the same for both balls because of the symmetry) depends on this initial y-component of displacement. There is some initial y-displacement that will produce a right angle. Find, by trial and error, this y-displacement for some spring constant and time step. It may not be what you expect. Try a different spring constant and/or time step. Can you explain why the y-displacement required to produce a right angle depends on the spring constant?

Exercise 20.8. Write an animation for the guitar string in which the string is modeled with 9 masses rather than 64. The distance between masses will be 6.5 cm rather than 1 cm. This increases the time-step stability threshold to $255\mu s$, so we can use a larger time step. We're asking less of the computer, so the animation should run smoothly, even on older hardware. You'll need to write new definitions for `forcesString`, `stringUpdate`, and `stringInitialOvertone`. The function `stringPicture` can remain the same. The spring constant for the springs will need to be different.

Exercise 20.9. To see what an animation looks like when a numerical instability occurs, modify the `gloss` animation of the guitar string so that the time step is $100\mu s$. You can do this by increasing the time-scale factor or by decreasing the animation rate.

Exercise 20.10. Write the functions

```
mpsPos :: MultiParticleState -> IO ()
mpsPos = undefined

mpsVel :: MultiParticleState -> IO ()
mpsVel = undefined
```

that use `gnuplot` to graph the x- and y-components of position (for the first function) and velocity (for the second) for a multi-particle state as if it was a guitar string. Functions like these can help with debugging. You can use them to visualize what happens in the first several time steps, giving you clues to what might be wrong when things aren't working.

Exercise 20.11. Write a 3D animation for the guitar string using `simulateVis`.

Exercise 20.12. Make a 3D animation of the guitar string in which the motion looks like that of a jump rope. You should be able to do this by changing the initial state so that the masses that make up the string have some initial velocity.

Exercise 20.13. Modify the function `makePNG` so that it uses four digits rather than three to label the output file. This allows for a longer animation of up to 10,000 frames. Test your function by making an animation of 2,000 frames.

Exercise 20.14. Write code to produce graphs like those in Figure 20-2.

Exercise 20.15. Explore energy conservation for the guitar string. Mechanical energy should be conserved, but how well it is conserved depends on the time step. You will need to write an expression for the mechanical energy of this system.

Exercise 20.16. Most of our animated results up to now have been based on a state-update function and have used the `simulateGloss` or `simulateVis` functions. There is a way to animate a list of states, and we'll explore that now. The functions `animateGloss` and `animateVis` that follow take as input a time-scale factor, a display function, and a list of states, and they produce an animation. The time step for the animation is obtained from the list of states. We do not specify an animation rate; that is calculated from the time-scale factor and time step.

```
animateGloss :: HasTime s => R  -- time-scale factor
             -> (s -> G.Picture)
             -> [s]
             -> IO ()
animateGloss tsFactor displayFunc mpsts
    = let dtp = timeOf (mpsts !! 1) - timeOf (mpsts !! 0)
          n tp = round (tp / dtp)
          picFromAnimTime :: Float -> G.Picture
          picFromAnimTime ta = displayFunc (mpsts !! n (tsFactor * realToFrac ta))
          displayMode = G.InWindow "My Window" (1000, 700) (10, 10)
      in G.animate displayMode G.black picFromAnimTime

animateVis :: HasTime s => R  -- time-scale factor
           -> (s -> V.VisObject R)
           -> [s]
           -> IO ()
animateVis tsFactor displayFunc mpsts
    = let dtp = timeOf (mpsts !! 1) - timeOf (mpsts !! 0)
          n tp = round (tp / dtp)
          picFromAnimTime :: Float -> V.VisObject R
          picFromAnimTime ta = displayFunc (mpsts !! n (tsFactor * realToFrac ta))
      in V.animate V.defaultOpts (orient . picFromAnimTime)
```

Listing 20-6 is a stand-alone program using `animateGloss` to make the animation.

```
{-# OPTIONS -Wall #-}

import MultipleObjects ( eulerCromerMPS )
import MOExamples
    ( animateGloss, billiardPicture, billiardStates )
```

```
main :: IO ()
main = animateGloss 1 billiardPicture (billiardStates eulerCromerMPS 30 0.01)
```

Listing 20-6: Stand-alone program for a 2D animation of two billiard balls colliding

Write a stand-alone program using `animateVis` to make the animation.

Exercise 20.17. The billiard collision we studied in this chapter is an elastic collision. Any collision in which the only force is `billiardForce` must be an elastic collision. How can we produce an inelastic collision? We need some sort of two-body force that can dissipate energy. The following two-body force can provide dissipation in a collision:

```
dissipation :: R  -- damping constant
            -> R  -- threshold center separation
            -> TwoBodyForce
dissipation b re st1 st2
    = let r1 = posVec st1
          r2 = posVec st2
          v1 = velocity st1
          v2 = velocity st2
          r21 = r2 ^-^ r1
          v21 = v2 ^-^ v1
      in if magnitude r21 >= re
         then zeroV
         else (-b) *^ v21
```

When the distance between the particles is greater than or equal to the threshold separation, the particles feel no dissipation force. When the distance between the particles is less than the threshold separation, the particles feel a force proportional to the relative velocity in a direction that helps the relative velocity decrease in magnitude. Including this dissipation force along with `billiardForce` in the list of forces acting in a system results in an inelastic collision.

Revise the stand-alone program in Listing 20-3 to produce an inelastic collision. A damping constant of 4 kg/s should produce a totally inelastic collision.

To confirm that this collision is inelastic, make a graph of mechanical energy as a function of time. Since a non-conservative force is present, mechanical energy will not be conserved. Mechanical energy should be conserved before the collision, drop rapidly during the collision, and then maintain at a lower value after the collision.

PART III

EXPRESSING ELECTROMAGNETIC THEORY AND SOLVING PROBLEMS

21

ELECTRICITY

When we think about electricity, we often think about how it's used, like the electric current flowing through a telephone wire or emerging from a battery. But all electrical technologies start from a single concept: electric charge. Electric charge has something to do with all electrical phenomena, and it's the logical starting place for our discussion.

Accordingly, we'll begin this chapter by describing electric charge. We'll then discuss Coulomb's late 18th century theory of electricity, which is specified by Coulomb's law. Finally, we'll put this theory to use in examining the motion of two charged particles repelling each other.

Electric Charge

Electric charge is a quantity associated with a particle or object that determines whether and how it can participate in electrical phenomena. In the 1700s, people discovered that there were two types of electric charge. Charges of the same type repelled each other and charges of different types attracted each other. Later, when physicists discovered subatomic particles, they decided that the proton was positive and the electron negative, but that was an arbitrary choice that everyone now respects as a convention.

The SI unit of charge is the Coulomb (C), named after Charles-Augustin de Coulomb, the French physicist of the late 18th century who did groundbreaking work on electricity. Table 21-1 gives the charges of the proton, the electron, and the neutron. The charge of a proton is *exactly* $1.602176634 \times 10^{-19}$ C. How can the charge of a proton be known exactly? Since 2019, the SI *defines* the Coulomb to be that quantity of charge such that an *elementary charge* is exactly $1.602176634 \times 10^{-19}$ C. The proton is thought to possess one unit of elementary charge, and the electron negative one unit. The proton and electron are known experimentally to have equal (but opposite) charges to better than one part in 10^{18}. In equations, we use q or Q as a symbol for charge.

Table 21-1: Electric Charge and Mass of Some Common Particles

Particle	Charge	Mass
Proton	1.602×10^{-19} C	1.673×10^{-27} kg
Neutron	0 C	1.675×10^{-27} kg
Electron	-1.602×10^{-19} C	9.109×10^{-31} kg

Listing 21-1 shows the first lines of code for the `Electricity` module we will develop in this chapter. We import `TwoBodyForce` and `MultiParticleState` from the `MultipleObjects` module of Chapter 19 because Coulomb's law, the theory of electricity that we'll describe in this chapter, is a two-body force.

```
{-# OPTIONS -Wall #-}

module Electricity where

import SimpleVec
    ( Vec(..), R, (*^), iHat )
import Mechanics3D
    ( ParticleState(..), defaultParticleState )
import MultipleObjects
    ( TwoBodyForce, MultiParticleState(..), Force(..), statesMPS
    , eulerCromerMPS, centralForce )
import Graphics.Gnuplot.Simple
    ( Attribute(..), plotPaths )
```

Listing 21-1: Opening lines of code for the `Electricity` module

Charge is a scalar, not a vector. Charge is represented by a real number. This suggests that the type for charge should be real numbers.

```
type Charge = R
```

Let's encode the value of the elementary charge.

```
elementaryCharge :: Charge
elementaryCharge = 1.602176634e-19  -- in Coulombs
```

Charge is quantized—that is, it occurs in discrete lumps—but this fact does not play a role in classical electromagnetic theory. In fact, the size of the lumps of charge are so small that we often want to think of charge as more like a fluid. Don't worry if that doesn't make sense quite yet; we'll discuss continuous charge distributions in Chapter 24.

Charge is also conserved. If the charge in any volume changes, it must flow in or out through the boundary surface of the volume.

The most important and interesting questions about charge are not about its intrinsic nature but rather about the relationships and interactions between charged particles. How do charges interact?

Coulomb's Law

Charles-Augustin de Coulomb was the first to give a quantitative relationship describing the interaction of two charged particles. He showed that the force exerted by one point charge on another is directly proportional to each charge and inversely proportional to the square of the distance between them. As an equation, Coulomb's law can be written as

$$F = k\frac{|q_1 q_2|}{r^2} \tag{21.1}$$

where q_1 is the charge of particle 1, q_2 is the charge of particle 2, and r is the distance between the particles. This equation gives the magnitude of the force produced by particle 1 on particle 2, which, by Newton's third law, is the same as the magnitude of the force produced by particle 2 on particle 1. The direction of the force depends on the signs of the charges; the force is repulsive for like charges and attractive for unlike charges. In SI units, the constant k is

$$k = \frac{1}{4\pi\epsilon_0} \approx 9 \times 10^9 \text{ N m}^2/\text{C}^2$$

The constant ϵ_0, called the *vacuum electric permittivity*, *electric constant*, or *permittivity of free space*, serves as a kind of bridge between electrical units, such as the Coulomb, and mechanical units, such as the Newton. The expression $1/(4\pi\epsilon_0)$ is often used instead of k as k is an overused symbol in physics.

Here is a translation of Equation 21.1 into Haskell:

```
coulombMagnitude :: Charge -> Charge -> R -> R
coulombMagnitude q1 q2 r
    = let k = 9e9  -- in N m^2 / C^2
      in k * abs (q1 * q2) / r**2
```

We can use vector notation to give a more comprehensive version of Coulomb's law, which includes the direction of the force in the equation.

We'll define the displacement vector $\mathbf{r}_{21}$ to be the vector that points from particle 1 to particle 2, as in Figure 21-1.

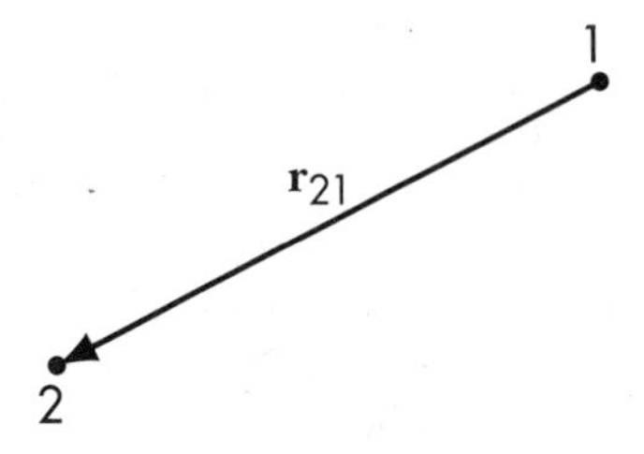

Figure 21-1: The displacement vector $\mathbf{r}_{21}$ points from particle 1 to particle 2.

The force $\mathbf{F}_{21}$ exerted *on* particle 2 produced *by* particle 1 is given in vector notation as follows:

$$\mathbf{F}_{21} = kq_1q_2\frac{\mathbf{r}_{21}}{|\mathbf{r}_{21}|^3} \tag{21.2}$$

Notice that if both charges are positive, the force $\mathbf{F}_{21}$ on particle 2 points in the same direction as the displacement vector $\mathbf{r}_{21}$, away from particle 1, as we expect for like charges. If the charges have unlike signs, the direction of $\mathbf{F}_{21}$ will flip, indicating an attractive force.

If $\mathbf{r}_1$ is the position vector for particle 1 and $\mathbf{r}_2$ is the position vector for particle 2, then $\mathbf{r}_{21}=\mathbf{r}_2-\mathbf{r}_1$ and we can write the force on particle 2 as follows:

$$\mathbf{F}_{21} = kq_1q_2\frac{\mathbf{r}_2-\mathbf{r}_1}{|\mathbf{r}_2-\mathbf{r}_1|^3} \tag{21.3}$$

In summary, Coulomb's law 21.1 is simpler, and Coulomb's laws 21.2 and 21.3 are more powerful since the direction of the force is encoded in the equations.

This Coulomb interaction is a `TwoBodyForce` of the kind we discussed in Chapter 19. Here is Equation 21.3 in Haskell:

```
coulombForce :: TwoBodyForce
coulombForce st1 st2
    = let k = 9e9  -- N m^2 / C^2
          q1 = charge st1
          q2 = charge st2
      in centralForce (\r -> k * q1 * q2 / r**2) st1 st2
```

The Coulomb force is another example of a central force, so here we use the `centralForce` function we defined in Chapter 19.

Having encoded Coulomb's law as a two-body force, let's apply it to a situation of two protons repelling one another.

Two Charges Interacting

Suppose we release two protons with an initial separation of 1 cm. How far will they travel in five milliseconds? This is a good problem for the tools we developed in the previous chapters, especially Chapter 19. The problem cannot be solved using only algebra because the force diminishes as the particles move farther apart. The particles start from rest and accelerate away from each other, and this acceleration decreases as the repulsive force decreases. When the two protons are far apart, the force decreases to a negligible magnitude and the protons approach a terminal velocity.

Looking at Extremes

Before we apply the multi-particle tools of Chapter 19, let's see how much we can learn about this situation by thinking about two extremes: what happens in the first few moments and what happens after a long time.

For very short times, before the particles move much, we can approximate the initial acceleration as constant. We can obtain the initial acceleration of one of the protons by dividing the net force

$$k\frac{q_p^2}{d^2}$$

where q_p is the charge of a proton and d is 1 cm, by the mass m_p of a proton:

$$a(0) = \frac{kq_p^2}{m_p d^2} = 1379 \text{ m/s}^2$$

Treating this acceleration as constant, the velocity and position of one proton are

$$v(t) = a(0)t \tag{21.4}$$

$$x(t) = x(0) + \frac{1}{2}a(0)t^2 \tag{21.5}$$

from the constant acceleration Equations 4.14 and 4.15. These equations are a good approximation for a short time, but extending them for too long is overly ambitious and gives poor results. We'll call this approximation the "constant acceleration approximation."

After the protons have been moving for some time, the particles will approach a terminal velocity. We can find this terminal velocity using conservation of energy. The potential energy of two charges q_1 and q_2 separated by a distance d is kq_1q_2/d, so the potential energy of two protons a distance d from each other is kq_p^2/d. The initial electric potential energy of the two protons is converted into kinetic energy. The kinetic energy of a particle with mass m moving with speed v is $mv^2/2$. The two protons will approach the same terminal speed v_T, so conservation of energy leads to the following equation:

$$\frac{kq_p^2}{d} = \frac{1}{2}m_p v_T^2 + \frac{1}{2}m_p v_T^2$$

The terminal speed of each proton is given through conservation of energy by

$$v_T = \sqrt{\frac{kq_p^2}{m_p d}} \approx 3.71 \text{ m/s}$$

For very long times, we can treat the terminal velocity as constant, so the velocity and position of one proton are

$$v(t) = v_T \quad (21.6)$$

$$x(t) = x_1 + v_T t \quad (21.7)$$

where x_1 is some as-yet-undetermined distance. These equations are a good approximation when t is very large, but applying them for shorter times gives poor results. We will call this approximation the "terminal velocity approximation."

Let's summarize what we've learned from the short-time extreme and the long-time extreme. When released from rest, each proton experiences an acceleration of 1379 m/s^2 away from the other proton. The acceleration decreases as the protons move farther apart, until the acceleration is negligible and the protons obtain the terminal speed of 3.71 m/s. If we were to plot proton velocity as a function of time, the velocity would start at 0 and increase with a slope of 1379 m/s^2. As time increases, velocity increases, asymptotically approaching the terminal speed of 3.71 m/s.

Modeling the Situation in Haskell

Now that we have a basic idea of what to expect, let's apply the tools we developed in Chapter 19 for multiple-particle situations. The only force we need to include is the internal force of the Coulomb interaction between the protons.

By using `statesMPS` from Chapter 19, we can form an infinite list of multiparticle states.

```
twoProtonStates :: R                    -- time step
                -> MultiParticleState   -- initial 2-particle state
                -> [MultiParticleState] -- infinite list of states
twoProtonStates dt
    = statesMPS (eulerCromerMPS dt) [InternalForce 1 0 coulombForce]
```

We supply this function with a time step and an initial two-particle state, and it will give back an infinite list of two-particle states that we can mine for any information we want.

Here is a function that sets an initial state with both protons at rest and with an initial separation given as a parameter to the function. The origin in this function is midway between the two protons. The proton mass is from Table 21-1.

```
-- protons are released from rest
initialTwoProtonState :: R  -- initial separation
                      -> MultiParticleState
initialTwoProtonState d
    = let protonMass = 1.673e-27  -- in kg
      in MPS [defaultParticleState { mass   = protonMass
                                   , charge = elementaryCharge
```

```
                              , posVec = (-d/2) *^ iHat
                              }
          ,defaultParticleState { mass   = protonMass
                              , charge = elementaryCharge
                              , posVec = ( d/2) *^ iHat
                              }
          ]
```

Let's start by making a graph of proton velocity as a function of time. The function `oneProtonVelocity` returns an infinite list of time-velocity pairs.

```
oneProtonVelocity :: R          -- dt
                  -> R          -- starting separation
                  -> [(R,R)]    -- (time,velocity) pairs
oneProtonVelocity dt d
    = let state0 = initialTwoProtonState d
      in [(time st2, xComp $ velocity st2)
              | MPS [_,st2] <- twoProtonStates dt state0]
```

We construct the list using a list comprehension, and we use pattern matching in the list comprehension to give the name `st2` to the state of the second proton. We choose the second proton rather than the first because, based on our initial state, the second proton will have a positive velocity component, while the first proton will have a negative velocity component. Finally, we use the functions `time`, `velocity`, and `xComp` to pick out the values we want to plot.

It's not so obvious what time step to use. Let's try to use dimensional analysis with the parameters of this problem to estimate a characteristic time scale. The relevant parameters for this situation are the proton charge q_p, the proton mass m_p, the electrical constant k, and the distance d of 1 cm. Can we combine these parameters to get a quantity with dimensions of time? We can. The characteristic time scale of the problem is given by

$$\sqrt{\frac{m_p d^3}{k q_p^2}} \approx 2.7 \times 10^{-3} \text{ s}$$

We'll use a time step of 10^{-5} s, which is small compared with the characteristic time scale we just found.

The following list of time-velocity pairs is a finite list of the results we will plot.

```
tvPairs :: [(R,R)]
tvPairs = takeWhile (\(t,_) -> t <= 2e-2) $
          oneProtonVelocity 1e-5 1e-2
```

We pass the time step `1e-5` (10^{-5} s) and the initial proton separation `1e-2` (1 cm) to the function `oneProtonVelocity` to get an infinite list of states. We then truncate this infinite list to a finite list of states occurring up to 20 ms.

Figure 21-2 shows the velocity of a proton as a function of time. The straight lines on the graph are the constant acceleration approximation and the terminal velocity approximation. The calculated velocity transitions smoothly from linearly increasing at the initial acceleration at very early times to approaching the terminal velocity at later times.

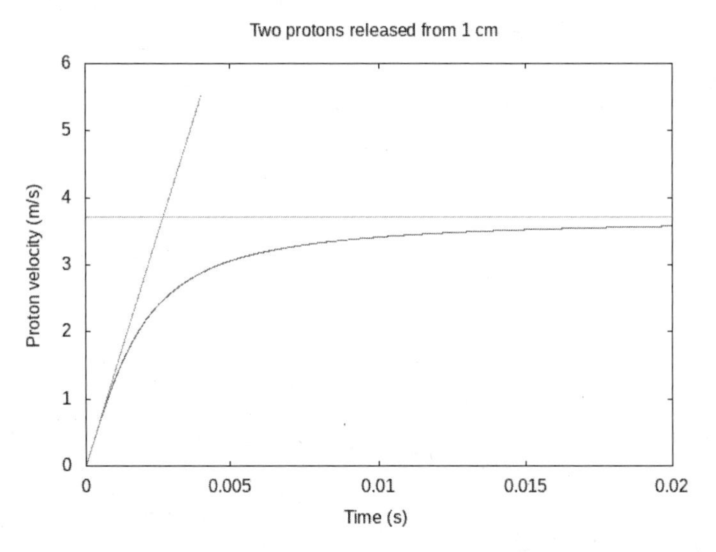

Figure 21-2: Two protons repelling one another. The curve shows one proton's velocity as a function of time. The horizontal line is the terminal velocity. The sloped line indicates the initial acceleration of the proton.

Here is the code that produced the graph in Figure 21-2.

```
velocityPlot :: IO ()
velocityPlot
    = plotPaths [Title "Two protons released from 1 cm"
                ,XLabel "Time (s)"
                ,YLabel "Proton velocity (m/s)"
                ,PNG "protons.png"
                ,Key Nothing
                ] $ [tvPairs
                    ,[(t,1379*t) | t <- [0,1e-5..4e-3]]
                    ,[(t,3.71)   | t <- [0,1e-3..2e-2]]]
```

Since we created the finite concrete list `tvPairs` to hold the data, the plotting code consists mostly in using the `plotPaths` function. The two approximations are graphed by constructing time-velocity pairs with list comprehensions. The `1379` is the initial acceleration of a proton in m/s^2, and the `3.71` is the terminal speed in m/s.

Our original question asked how far a proton would travel in 5 ms. Let's produce a graph of position versus time for a proton and then answer the original question directly. Figure 21-3 shows the position of a proton as a function of time. It also shows the constant acceleration approximation, which is the parabola on the left that appears to give good results for about 2 ms.

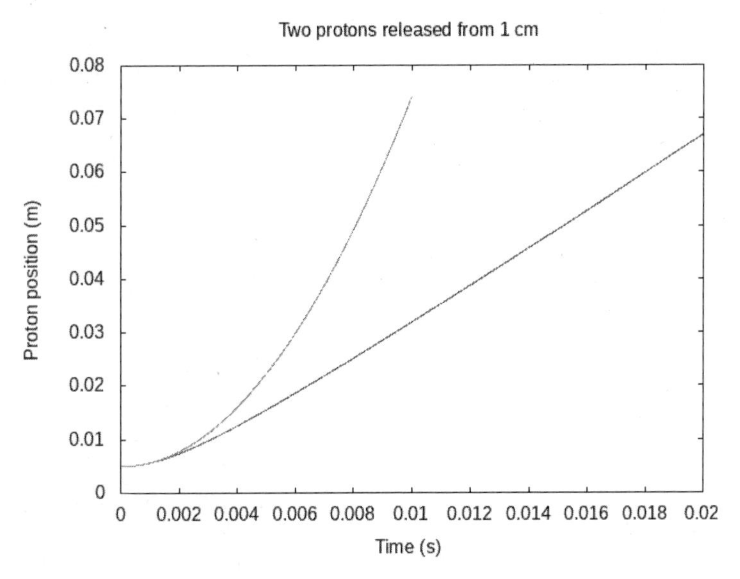

Figure 21-3: Two protons repelling one another. One proton's position as a function of time is shown by the curve that becomes linear over time. The parabola, shown for comparison, gives position if the acceleration initially experienced by the proton were maintained at a constant value.

Finally, we ask GHCi to give us the position of a proton at 5 ms. Since `initialTwoProtonState 0.01` is an initial two-particle state with the protons separated by 1 cm, and `twoProtonStates 1e-5 (initialTwoProtonState 0.01)` is an infinite list of two-proton states, formed with a time step of 10^{-5} s, a time of 5 ms occurs 500 time steps into the list. We can ask for the information we want as follows:

```
Prelude> :l Electricity
[1 of 6] Compiling Newton2          ( Newton2.hs, interpreted )
[2 of 6] Compiling Mechanics1D      ( Mechanics1D.hs, interpreted )
[3 of 6] Compiling SimpleVec        ( SimpleVec.hs, interpreted )
[4 of 6] Compiling Mechanics3D      ( Mechanics3D.hs, interpreted )
[5 of 6] Compiling MultipleObjects  ( MultipleObjects.hs, interpreted )
[6 of 6] Compiling Electricity      ( Electricity.hs, interpreted )
Ok, six modules loaded.
*Electricity> twoProtonStates 1e-5 (initialTwoProtonState 0.01) !! 500
MPS {particleStates =
    [ParticleState {mass = 1.673e-27,
                    charge = 1.602176634e-19,
                    time = 4.9999999999999645e-3,
                    posVec = vec (-1.550866906307423e-2) 0.0 0.0,
                    velocity = vec (-3.0582222353914252) 0.0 0.0},
```

```
ParticleState {mass = 1.673e-27,
               charge = 1.602176634e-19,
               time = 4.9999999999999645e-3,
               posVec = vec 1.550866906307423e-2 0.0 0.0,
               velocity = vec 3.0582222353914252 0.0 0.0}]}
```

GHCi returns the two-particle state at 5 ms. I have formatted the output for easy readability. The protons are located at −1.55 cm and 1.55 cm along the x-axis, so they are 3.1 cm apart at 5 ms.

Summary

We have given an overview of 18th century electrical theory, which works well when particles move slowly compared to the speed of light and don't experience extreme acceleration. Coulomb's 18th century theory is still a good theory of static electricity, also called *electrostatics*. Coulomb's law is a two-body force, just like Newton's law of universal gravity. Coulomb's law is intended to be used in the context of the multi-particle Newtonian mechanics we studied in Part II. An example of this is two protons repelling each other, a problem that, simple as it is to state, is not solvable by simple algebraic methods, but rather requires the ideas and tools we have developed.

In the 19th century, Michael Faraday discovered an electrical phenomenon that was not (directly) caused by electric charge. This led to the concepts of electric and magnetic fields, which, in the modern Faraday-Maxwell theory of electricity and magnetism, are the mediators of electric charge. This newer theory is a theory of *electrodynamics*, and it makes good predictions even when charges are moving rapidly and accelerating strongly. Since this newer theory is a field theory, meaning the players are fields rather than particles, and since a field in physics is a function of three-dimensional space or spacetime, we'll spend the next two chapters studying coordinate systems and geometry for three-dimensional space.

Exercises

Exercise 21.1. Make a plot similar to Figure 21-2 for two electrons released from rest with a separation of 1 cm. What are the terminal velocity and characteristic time scale in this case?

Exercise 21.2. Coulomb's theory of electricity predicts that an electron could orbit a proton in much the same way that the Earth orbits the sun. We might call this "classical hydrogen." (The modern Faraday-Maxwell theory of electricity and magnetism that we will touch on later in this part of the book presents problems for this picture because an accelerating charged particle radiates, making classical hydrogen unstable.) Write an animation for classical hydrogen in which the Coulomb force is the only internal force between the proton and the electron and there are no external forces. You will need to choose some initial conditions for the proton and the electron.

Exercise 21.3. Consider a proton and an electron released from rest. Write a function to calculate the time until collision given the initial separation. How far apart should they be initially so that it will take one second for them to collide?

Exercise 21.4. Animate the two-proton repulsion using `gloss` or `not-gloss`.

Exercise 21.5. Write code to produce the graph in Figure 21-3. Here is some starting code you may use if you wish:

```
oneProtonPosition :: R         -- dt
                  -> R         -- starting separation
                  -> [(R,R)]  -- (time,position) pairs
oneProtonPosition dt d
    = undefined dt d

positionPlot :: IO ()
positionPlot = plotPaths [Title "Two protons released from 1 cm"
                         ,XLabel "Time (s)"
                         ,YLabel "Proton position (m)"
                         ,PNG "ProtonPosition.png"
                         ,Key Nothing
                         ] $ [undefined $ oneProtonPosition 1e-5 1e-2
                             ,undefined :: [(R,R)]]
```

Exercise 21.6. By trial and error, find a value for x_1 in Equation 21.7 so that the position-time curve in Figure 21-3 appears asymptotic to the straight line of Equation 21.7 for large times.

22

COORDINATE SYSTEMS AND FIELDS

In this chapter we'll begin exploring Faraday and Maxwell's electromagnetic theory, which broke from Coulomb's particle-based ideas by introducing the notion of a field. The Faraday-Maxwell theory is the best theory we have for explaining electrical, magnetic, and optical phenomena. As a field theory, the Faraday-Maxwell theory supported the locality ideas of relativity 40 years before Einstein wrote about it, served as an inspiration for other field theories like general relativity, and became the prototype for contemporary gauge field theories of particle physics. The field idea now plays an important role in many areas of physics, such as continuum mechanics, fluid dynamics, and quantum field theory. That a field is a function is one reason why functional programming serves physics so well.

It is possible, and elegant, to give an exposition of electromagnetic theory in four-dimensional spacetime (and functional languages like Haskell are especially well suited to the task), but we'll follow the more common practice of using three-dimensional notation since physical insight and geometric insight into four-dimensional relativistic spacetime language take some time to acquire. Accordingly, this chapter describes coordinate systems for three-dimensional space, defines a data type for position in three-dimensional space, and introduces the idea of a *field*, which is a function whose input is a position in three-dimensional space.

We'll gain a bit of insight by looking first at polar coordinates, the most common coordinates for a two-dimensional plane after Cartesian. Then, we'll look at cylindrical and spherical coordinates, the two most common coordinate systems for three-dimensional space after Cartesian coordinates. We'll make a new data type for positions in three-dimensional space that accommodates Cartesian, cylindrical, and spherical coordinates as well as any other coordinate system we might want to use. We'll introduce scalar and vector fields, and data types for them, so that we have the basic mathematical framework to talk about things like charge density (a scalar field) and electric field (a vector field).

Polar Coordinates

Polar coordinates are a way of assigning two numbers to each point in the plane so that one of the numbers is the distance from the origin to the point. Polar coordinates are a natural choice for situations with rotational symmetry about a point in the plane, although their use need not be confined to such situations. We'll use the variables s and ϕ for polar coordinates. The names s and ϕ are from Griffiths' electrodynamics text [19].

The Cartesian coordinates x and y are related to the polar coordinates s and ϕ by the following equations:

$$x = s \cos \phi$$
$$y = s \sin \phi$$

The coordinate s is the distance from the origin to a point in the plane, and the coordinate ϕ is the angle between the x-axis and a line joining the origin to a point (see Figure 22-1).

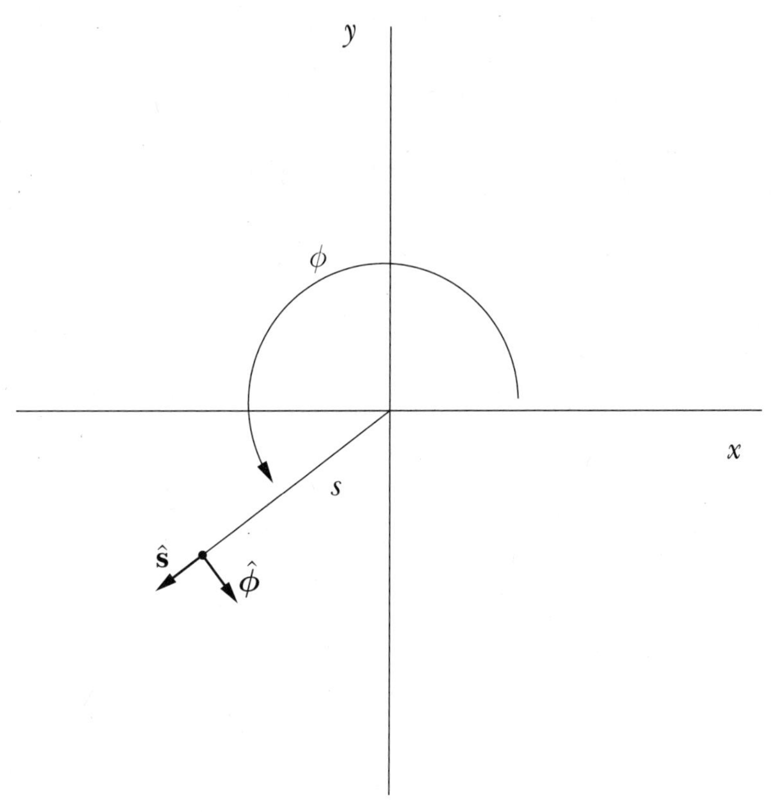

Figure 22-1: Polar coordinates

In Figure 22-1, we've also introduced polar coordinate unit vectors. The unit vector $\hat{\mathbf{s}}$ points away from the origin. (This is a well-defined direction at every point in the plane except for the origin itself.) Equivalently, the unit vector $\hat{\mathbf{s}}$ points in the direction for which ϕ stays constant and s increases. Analogously, the unit vector $\hat{\phi}$ points in the direction for which s stays constant and ϕ increases. We can write the polar coordinate unit vectors $\hat{\mathbf{s}}$ and $\hat{\phi}$ in terms of the Cartesian coordinate unit vectors $\hat{\mathbf{x}}$ and $\hat{\mathbf{y}}$ as follows:

$$\hat{\mathbf{s}} = \frac{x\hat{\mathbf{x}} + y\hat{\mathbf{y}}}{\sqrt{x^2 + y^2}} = \cos\phi\hat{\mathbf{x}} + \sin\phi\hat{\mathbf{y}}$$

$$\hat{\phi} = -\sin\phi\hat{\mathbf{x}} + \cos\phi\hat{\mathbf{y}}$$

Unlike the Cartesian unit vectors $\hat{\mathbf{x}}$ and $\hat{\mathbf{y}}$, the polar unit vectors $\hat{\mathbf{s}}$ and $\hat{\phi}$ point in different directions at different points in the plane. The picture on the right of Figure 22-5 later in this chapter shows the unit vector $\hat{\phi}$ at different points in the xy-plane, and you can see how its direction changes.

The definition of polar coordinates in two dimensions makes it easy to define cylindrical coordinates in three dimensions, to which we now turn.

Cylindrical Coordinates

Cylindrical coordinates are an extension of polar coordinates into three dimensions and are a natural choice of coordinates for situations with rotational and translational symmetry about some axis. We can use the cylindrical coordinates s, ϕ, and z to represent the location of a point in three-dimensional space, as shown in Figure 22-2. The coordinate s is the distance from the z-axis to the point in space, the coordinate ϕ is the angle between the xz-plane and the plane containing the z-axis and the point, and the coordinate z means the same thing as in Cartesian coordinates: the distance from the xy-plane. Cylindrical coordinates are closely related to polar coordinates in that cylindrical coordinates describe the xy-plane in a polar fashion but continue to use the Cartesian z-coordinate.

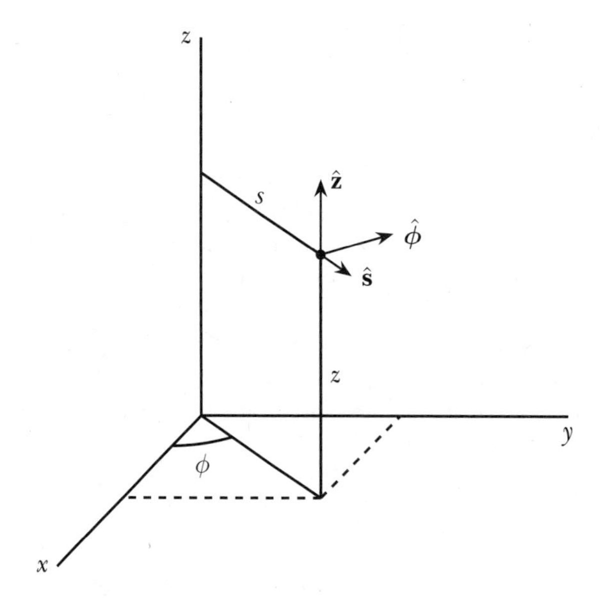

Figure 22-2: Cylindrical coordinates

The Cartesian coordinates x, y, and z are related to the cylindrical coordinates s, ϕ, and z by the following equations:

$$x = s\cos\phi \tag{22.1}$$

$$y = s\sin\phi \tag{22.2}$$

$$z = z \tag{22.3}$$

Also shown in Figure 22-2 are the cylindrical coordinate unit vectors. The unit vector $\hat{\mathbf{s}}$ points away from the z-axis. (This is a well-defined direction at every point in space except for points on the z-axis.) Equivalently, the unit vector $\hat{\mathbf{s}}$ points in the direction for which ϕ and z stay constant and s increases. The unit vector $\hat{\boldsymbol{\phi}}$ points in the direction for which s and z stay constant and ϕ increases. Finally, the unit vector $\hat{\mathbf{z}}$ points in the direction for which s and ϕ stay constant and z increases.

We can write the cylindrical coordinate unit vectors $\hat{\mathbf{s}}$, $\hat{\boldsymbol{\phi}}$, and $\hat{\mathbf{z}}$ in terms of the Cartesian coordinate unit vectors $\hat{\mathbf{x}}$, $\hat{\mathbf{y}}$, and $\hat{\mathbf{z}}$ as follows:

$$\hat{\mathbf{s}} = \cos\phi\hat{\mathbf{x}} + \sin\phi\hat{\mathbf{y}} \tag{22.4}$$

$$\hat{\boldsymbol{\phi}} = -\sin\phi\hat{\mathbf{x}} + \cos\phi\hat{\mathbf{y}} \tag{22.5}$$

$$\hat{\mathbf{z}} = \hat{\mathbf{z}} \tag{22.6}$$

Now that we've talked about the system of cylindrical coordinates and shown how it's an alternative to Cartesian coordinates for describing points in three-dimensional space, let's discuss one more three-dimensional coordinate system.

Spherical Coordinates

Spherical coordinates are a natural choice in situations with rotational symmetry about a point in space, but, like all of the three-dimensional coordinate systems described in this chapter, they are also a general system of coordinates capable of describing arbitrary positions in 3D space. We can use the spherical coordinates r, θ, and ϕ to represent the location of a point in space, as shown in Figure 22-3.

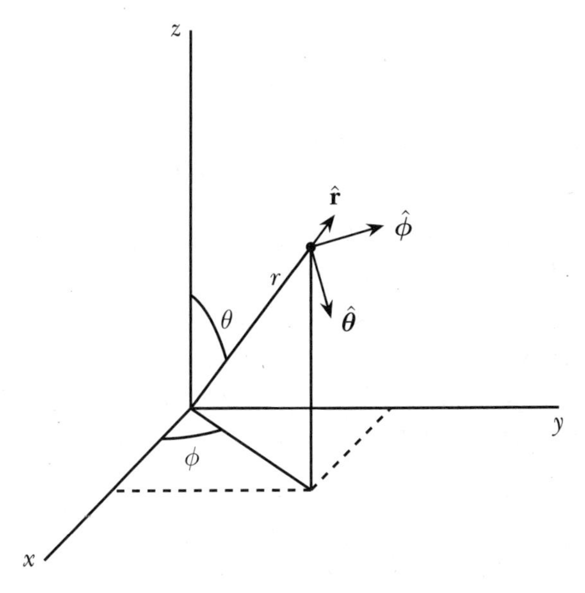

Figure 22-3: Spherical coordinates

The coordinate r is the distance from the origin to the point in space, the coordinate θ is the angle between the z-axis and a line from the origin to the point, and the coordinate ϕ is the angle between the xz-plane and the plane containing the z-axis and the point. (The coordinate ϕ has the same meaning in spherical coordinates that it has in cylindrical coordinates.)

The Cartesian coordinates x, y, and z are related to the spherical coordinates r, θ, and ϕ by the following equations:

$$x = r \sin\theta \cos\phi \tag{22.7}$$

$$y = r \sin\theta \sin\phi \tag{22.8}$$

$$z = r \cos\theta \tag{22.9}$$

Also shown in Figure 22-3 are the spherical coordinate unit vectors. The unit vector $\hat{\mathbf{r}}$ points away from the origin. (This is a well-defined direction at every point in space except for the origin itself.) Equivalently, the unit vector $\hat{\mathbf{r}}$ points in the direction for which θ and ϕ stay constant and r increases. The unit vector $\hat{\boldsymbol{\theta}}$ points in the direction for which r and ϕ stay constant and θ increases. Finally, the unit vector $\hat{\boldsymbol{\phi}}$ points in the direction for which r and θ stay constant and ϕ increases.

To write $\hat{\mathbf{r}}$ in terms of the Cartesian unit vectors, we divide the position vector $\mathbf{r} = x\hat{\mathbf{x}} + y\hat{\mathbf{y}} + z\hat{\mathbf{z}}$ by its magnitude $r = \sqrt{x^2 + y^2 + z^2}$. The expression for $\hat{\boldsymbol{\phi}}$ is the same as it was for cylindrical coordinates. An expression for $\hat{\boldsymbol{\theta}}$ can be found from $\hat{\boldsymbol{\theta}} = \hat{\boldsymbol{\phi}} \times \hat{\mathbf{r}}$. We can write the spherical coordinate unit vectors $\hat{\mathbf{r}}$, $\hat{\boldsymbol{\theta}}$, and $\hat{\boldsymbol{\phi}}$ in terms of the Cartesian coordinate unit vectors $\hat{\mathbf{x}}$, $\hat{\mathbf{y}}$, and $\hat{\mathbf{z}}$ as follows:

$$\hat{\mathbf{r}} = \frac{x\hat{\mathbf{x}} + y\hat{\mathbf{y}} + z\hat{\mathbf{z}}}{\sqrt{x^2 + y^2 + z^2}} = \sin\theta \cos\phi \hat{\mathbf{x}} + \sin\theta \sin\phi \hat{\mathbf{y}} + \cos\theta \hat{\mathbf{z}} \tag{22.10}$$

$$\hat{\boldsymbol{\theta}} = \cos\theta \cos\phi \hat{\mathbf{x}} + \cos\theta \sin\phi \hat{\mathbf{y}} - \sin\theta \hat{\mathbf{z}} \tag{22.11}$$

$$\hat{\boldsymbol{\phi}} = -\sin\phi \hat{\mathbf{x}} + \cos\phi \hat{\mathbf{y}} \tag{22.12}$$

Now that we've completed our introduction to spherical coordinates, and to all of the coordinate systems we intend to use, our next mission is to define a new type for positions in three-dimensional space that will work well with all of our three-dimensional coordinate systems. However, before we do that, let's lay down some introductory code for this chapter.

Introductory Code

Listing 22-1 shows the first lines of code for the `CoordinateSystems` module we'll develop in this chapter.

```
{-# OPTIONS -Wall #-}

module CoordinateSystems where
```

```
import SimpleVec
    ( R, Vec, (^/), vec, xComp, yComp, zComp, iHat, jHat, kHat
    , magnitude, sumV, zeroV )
import Mechanics3D ( orient, v3FromVec )
import MOExamples ( Table(..), Justification(..) )
import qualified Vis as V
import SpatialMath ( V3(..) )
import Diagrams.Prelude
    ( Diagram, V2(..), PolyType(..), PolyOrientation(..), PolygonOpts(..)
    , (#), , dims, p2, r2, arrowAt, position, fc, black, white
    , blend, none, lw, rotate, deg, rad, scale, polygon, sinA )
import Diagrams.Backend.Cairo ( B, renderCairo )
```

Listing 22-1: Opening lines of code for the `CoordinateSystems` module

Here, we import functions and types that we've previously written from the `SimpleVec`, `Mechanics3D`, and `MOExamples` modules. We'll use the `Vis` module for visualizing scalar and vector fields, and we'll use the `V3` type from `SpatialMath` since it's the native vector type for the `Vis` module. The `Diagrams.Prelude` and `Diagrams.Backend.Cairo` modules are part of the `diagrams` package, which we'll use for vector field visualization. The appendix contains information on installing the `diagrams` package.

A Type for Position

We'd like to have a Haskell type to describe the position of a point in space. We'd also like to be able to specify points in three-dimensional space in Cartesian, cylindrical, or spherical coordinates, and to access previously defined positions in any of the coordinate systems, including a system different from the one used to define it.

Defining the New Type

How can we use Haskell to describe a point in space? We have three options. Option A is to use a triple `(R,R,R)` of Cartesian coordinates. This is fine for many purposes. It has the advantage of simplicity, but it has the disadvantage that we already know we're interested in using cylindrical and spherical coordinates, which are also triples of numbers. This puts us in the dangerous position of mistaking a Cartesian (x, y, z) triple for a spherical (r, θ, ϕ) triple. The compiler can help us avoid this mistake, but only if we make intelligent use of the type system. Option A is workable but dangerous. We can make better use of the computer to help us avoid mistakes.

Option B is to use the `Vec` type for position, as we did in mechanics. The `Vec` type clearly has Cartesian components, so it's harder to get confused compared with Option A. If we run into a triple `(R,R,R)` somewhere in code we've previously written, the type does not tell us whether it's a Cartesian triple or a spherical triple. On the other hand, if we run into a `Vec`, we know it is a Cartesian triple under the hood. Option B is workable. One downside of Option B is that position is not really a vector because vectors are, by definition, things that can be added, and it doesn't make sense to add positions. If we think of position as a vector, it is a vector from some fixed origin. But adding vectors means putting them tip-to-tail, and this isn't really allowed for position "vectors" whose tails are fixed at the origin. The other disadvantage of using `Vec` for position (Option B) is that the Haskell type system cannot help us to distinguish position from any other `Vec` (such as velocity, acceleration, or momentum).

Option C is to use Haskell's facilities to make a brand-new data type ourselves, which can't be confused with any other data type. This is not the simplest option, but it will give us the power of working with the three coordinate systems we're interested in, and it will give us the advantage that the compiler will not allow us to confuse position with velocity. We'll pursue Option C.

We'll construct a new type in Haskell with the `data` keyword.

```
data Position = Cart R R R
                deriving (Show)
```

The `Position` that appears immediately to the right of the `data` keyword is the name we give to the new type. The `Cart` that appears to the right of the equal sign is the type's one data constructor, so named to remind us that we are storing the position information in Cartesian coordinates, regardless of the coordinate system in which any particular `Position` is defined or used.

With the new `Position` data type, we have a way to store three numbers that the compiler will not confuse with any other way of storing three numbers (like a `Vec`). But the real usefulness of `Position` is that we can now define three ways of *making* a `Position` (one for each coordinate system) and three ways of *using* a `Position` (again, one for each coordinate system).

Making a Position

At the beginning of this chapter, we showed how Cartesian, cylindrical, and spherical coordinates can be used to describe a position in space. Each coordinate system uses three numbers to specify a position. A coordinate system is a function from three real numbers to space.

```
type CoordinateSystem = (R,R,R) -> Position
```

Here are the definitions for the three coordinate systems. For Cartesian coordinates, we just stick the coordinates behind the data constructor `Cart`. For cylindrical coordinates (s, ϕ, z), we convert to Cartesian using Equations 22.1 and 22.2 and then apply the `Cart` constructor to the Cartesian values. For spherical coordinates (r, θ, ϕ), we again apply the data constructor to the converted Cartesian values using Equations 22.7, 22.8, and 22.9.

```
cartesian   :: CoordinateSystem
cartesian (x,y,z)
    = Cart x y z

cylindrical :: CoordinateSystem
cylindrical (s,phi,z)
    = Cart (s * cos phi) (s * sin phi) z

spherical   :: CoordinateSystem
spherical (r,theta,phi)
    = Cart (r * sin theta * cos phi)
           (r * sin theta * sin phi)
           (r * cos theta)
```

The functions `cartesian`, `cylindrical`, and `spherical` are our three ways of making a `Position`. Before we turn to the three ways of using a `Position`, we'll define three helper functions that are almost the same as `cartesian`, `cylindrical`, and `spherical`. These three functions have the shortened names `cart`, `cyl`, and `sph`, and the only difference is that they take their arguments in a curried style, one right after the other, rather than as a triple. They are convenient helping functions.

```
cart :: R  -- x coordinate
     -> R  -- y coordinate
     -> R  -- z coordinate
     -> Position
cart = Cart

cyl  :: R  -- s   coordinate
     -> R  -- phi coordinate
     -> R  -- z   coordinate
     -> Position
cyl s phi z = cylindrical (s,phi,z)

sph  :: R  -- r     coordinate
     -> R  -- theta coordinate
     -> R  -- phi   coordinate
     -> Position
sph r theta phi = spherical (r,theta,phi)
```

The function cart is a helping function to take three numbers (x, y, z) and form the appropriate position using Cartesian coordinates. The definition of cart is given in point-free style, meaning we omitted the parameters because they are identical on both sides of the equation.

The function `cyl` is a helping function to take three numbers (s, ϕ, z) and form the appropriate position using cylindrical coordinates. We just call the function `cylindrical` to do the real work. The function `sph` is a helping function to take three numbers (r, θ, ϕ) and form the appropriate position using spherical coordinates.

Let's use the `cart` function to define the `origin`, which is the position where all three Cartesian coordinates are 0.

```
origin :: Position
origin = cart 0 0 0
```

Using a Position

We said earlier that we would like to be able to look at an existing `Position` in Cartesian, cylindrical, or spherical coordinates, regardless of the coordinate system used to define the position. The following three functions show how to *use* a position to obtain a triple in the desired coordinate system:

```
cartesianCoordinates   :: Position -> (R,R,R)
cartesianCoordinates   (Cart x y z) = (x,y,z)

cylindricalCoordinates :: Position -> (R,R,R)
cylindricalCoordinates (Cart x y z) = (s,phi,z)
    where
      s = sqrt(x**2 + y**2)
      phi = atan2 y x

sphericalCoordinates   :: Position -> (R,R,R)
sphericalCoordinates   (Cart x y z) = (r,theta,phi)
    where
      r = sqrt(x**2 + y**2 + z**2)
      theta = atan2 s z
      s = sqrt(x**2 + y**2)
      phi = atan2 y x
```

The mathematical content of these three functions is merely to convert Cartesian coordinates to any of the three systems. However, the worth of these functions lies in their type. They allow us to express a `Position` in any of the three coordinate systems, giving the numerical values of the coordinates so they can be used for something. The value of the `Position` data type is that it abstracts away from a specific coordinate system, allowing us to use any coordinate system without getting confused about what a set of three numbers might mean. In practice, then, we'll keep our `Position`s for as long as we can, converting to a particular coordinate system only when we need access to particular coordinate values.

In physics language, both position and displacement have the dimension of length and the SI unit of meter. The next section endeavors to clarify the relationship between position and displacement.

Displacement

A *displacement* is a vector that points from a source position to a target position. We have argued earlier that position in physics in not really a vector. Physicists use the term *displacement* when they want to refer to a vector with the dimension of length.

It is useful and natural to want a type `Displacement` for these vectors with the dimension of length. As usual, we have the choice of whether to make a brand-new type using the `data` keyword or to merely make a type synonym using the `type` keyword. The former option protects us from confusing displacement with any other vector, but at the cost of introducing a new data constructor, while the latter option is convenient but provides no such protection. We choose the latter option and make `Displacement` a type synonym for `Vec`.

```
type Displacement = Vec
```

The `displacement` function allows us to "subtract" positions (recall we cannot add positions) to get a vector.

```
displacement :: Position  -- source position
             -> Position  -- target position
             -> Displacement
displacement (Cart x' y' z') (Cart x y z)
    = vec (x-x') (y-y') (z-z')
```

Since a displacement vector points from the source position to the target position, we subtract the Cartesian source coordinates from the target coordinates.

The `shiftPosition` function allows us to add a displacement to a position to get a new position.

```
shiftPosition :: Displacement -> Position -> Position
shiftPosition v (Cart x y z)
  = Cart (x + xComp v) (y + yComp v) (z + zComp v)
```

We'll use the `shiftPosition` function in the next chapter to define some geometric objects.

Having introduced coordinate systems, a type for position, and the distinction between position and displacement, we now turn to the last major idea of the chapter—that of a field.

The Scalar Field

Some physical quantities, like volume charge density and electric potential, are best described by giving a number for each point in space. These physical quantities are called *scalar fields*. The word *field* in physics means a function of physical space or spacetime; in other words, something that can take a different value at each point in space. (The word *field* in mathematics means something else.) A scalar field is a field in which the value assigned at each point in space is a scalar (that is, a number). Temperature is another example of a scalar field. The temperature in one place (Annville, Pennsylvania, for example) is usually different from the temperature at another place (Vero Beach, Florida, say).

Since a scalar field associates a number with each position in space, it makes sense to define a scalar field type to be a function from space to numbers.

```
type ScalarField = Position -> R
```

When we're using a coordinate system, we can define scalar fields for each of the coordinates. For example, we can have a scalar field that associates each position in space with the value of its x-coordinate.

```
xSF :: ScalarField
xSF p = x
    where
      (x,_,_) = cartesianCoordinates p
```

Here is the coordinate scalar field that is associated with the spherical coordinate r:

```
rSF :: ScalarField
rSF p = r
    where
      (r,_,_) = sphericalCoordinates p
```

In Chapter 9, we defined functions that extract components from a triple:

```
fst3 :: (a,b,c) -> a
fst3 (u,_,_) = u

snd3 :: (a,b,c) -> b
snd3 (_,u,_) = u

thd3 :: (a,b,c) -> c
thd3 (_,_,u) = u
```

We can use these functions to express the y-coordinate scalar field as the scalar field associated with the second Cartesian coordinate.

```
ySF :: ScalarField
ySF = snd3 . cartesianCoordinates
```

We can define any of the coordinate scalar fields in this way.

Figure 22-4 shows a visualization of the scalar field `ySF` using a coordinate system in which x comes out of the page, y increases to the right, and z

increases upward. Associated with each position in space is its y-value, so the numbers increase to the right but do not change moving upward or out of the page. Later in the chapter, we will show how to make scalar field visualizations like that in Figure 22-4.

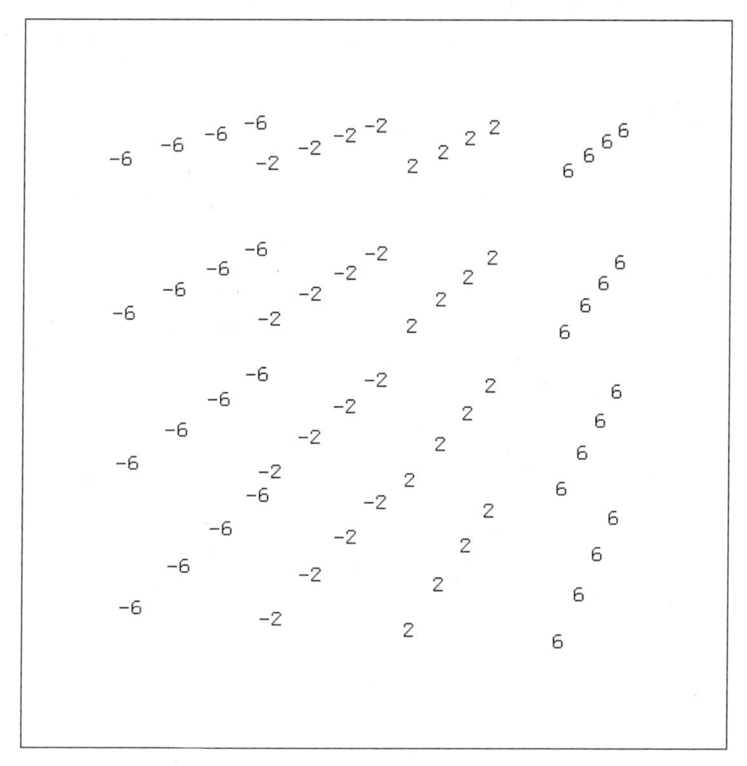

Figure 22-4: A screenshot of the y-coordinate scalar field ySF produced with the program ySF3D. The mouse and keyboard can be used to zoom in or out and rotate the visualization, a standard feature of the `Vis` module.

Because charge density is a scalar field, the scalar field will play an important role in Chapter 24 when we define charge distributions.

The second type of field used in physics, and possibly the more important, is the vector field, to which we turn next.

The Vector Field

A *vector field* associates a vector with each point in space.

```
type VectorField = Position -> Vec
```

In Chapters 25 and 27, we'll discuss electric fields and magnetic fields, respectively, which are vector fields.

When we're using a coordinate system, we can define vector fields that come from coordinates. The unit vectors used with cylindrical and spherical coordinates, such as $\hat{\mathbf{s}}$, $\hat{\phi}$, $\hat{\mathbf{r}}$, and $\hat{\theta}$, are really *unit vector fields* because their directions change depending on their location in space.

The vector fields $\hat{\mathbf{s}}$ and $\hat{\phi}$ are defined using Equations 22.4 and 22.5.

```
sHat   :: VectorField
sHat   r = vec ( cos phi) (sin phi) 0
    where
      (_,phi,_) = cylindricalCoordinates r

phiHat :: VectorField
phiHat r = vec (-sin phi) (cos phi) 0
    where
      (_,phi,_) = cylindricalCoordinates r
```

Figure 22-5 shows visualizations of the vector field `phiHat`. Associated with each position in space is a vector, whose tail is located at the point in space, and whose magnitude and direction show the value of the vector at that point. The picture on the left shows the vector field in three dimensions, where *x* comes out of the page, *y* increases to the right, and *z* increases upward. The z-axis is the central axis of symmetry for the `phiHat` vector field. The picture on the right shows the vector field in the xy-plane. Since `phiHat` is a unit vector field, all of the vectors in these pictures have the same length. The pictures make clear how the unit vector field $\hat{\phi}$ points in different directions at different points in space. Later in the chapter I will show how to produce visualizations like this.

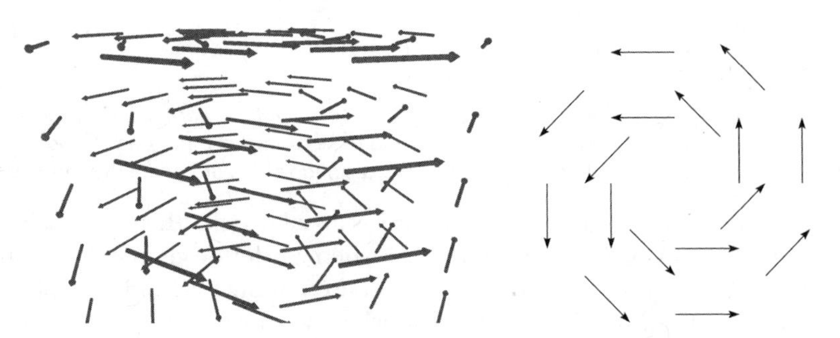

Figure 22-5: Two ways of visualizing the vector field $\hat{\phi}$, or `phiHat`. The left shows a screenshot of the image produced by `phiHat3D`. The right shows an image in the xy-plane produced by `phiHatPNG`.

Here are definitions for the unit vector fields $\hat{\mathbf{r}}$ and $\hat{\theta}$, using Equations 22.10 and 22.11:

```
rHat :: VectorField
rHat rv = let d = displacement origin rv
          in if d == zeroV
             then zeroV
             else d ^/ magnitude d
```

```
thetaHat :: VectorField
thetaHat r = vec ( cos theta * cos phi)
                 ( cos theta * sin phi)
                 (-sin theta          )
    where
      (_,theta,phi) = sphericalCoordinates r
```

We regard $\hat{\mathbf{i}}$, $\hat{\mathbf{j}}$, and $\hat{\mathbf{k}}$ as simple unit vectors (Vecs), but we define $\hat{\mathbf{x}}$, $\hat{\mathbf{y}}$, and $\hat{\mathbf{z}}$ as unit vector fields (VectorFields), analogous to $\hat{\mathbf{s}}$, $\hat{\boldsymbol{\phi}}$, $\hat{\mathbf{r}}$, and $\hat{\boldsymbol{\theta}}$.

```
xHat :: VectorField
xHat = const iHat

yHat :: VectorField
yHat = const jHat

zHat :: VectorField
zHat = const kHat
```

One important vector field that is not a unit vector field is the vector field $\mathbf{r}$, which associates each position with the displacement vector from the origin to that position. We'll give the name rVF to this vector field.

```
rVF :: VectorField
rVF = displacement origin
```

The function displacement takes a source Position and a target Position and returns the displacement vector from the source to the target. By omitting the target position in the definition, the function rVF takes a target position as input and produces a displacement vector as output, which is just the VectorField we want.

Figure 22-6 shows visualizations of the vector field rVF. Both pictures show the vector field in the xy-plane. The picture on the left places the tail of each vector at the position it is associated with, and it displays vectors with greater magnitude as arrows with longer length. The picture on the right places the center of each vector at the position it is associated with, and it displays vectors with greater magnitude as darker arrows.

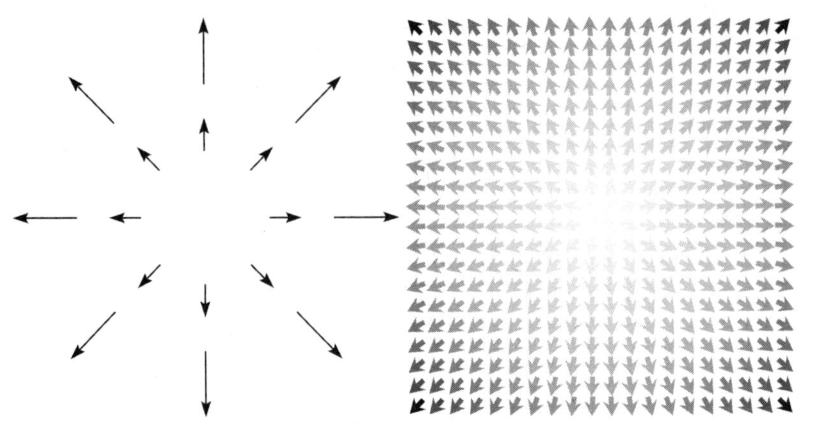

Figure 22-6: Two ways of visualizing the vector field **r***, or* `rVF`*, in the xy-plane. The left image is produced by* `rVFpng`*; the right image is produced by* `rVFGrad`*.*

Later in the chapter, we'll introduce functions that produce pictures, like those in Figure 22-6, when given a vector field as input; however, producing pictures is only one of several things we can do with vector fields. Vector fields admit two kinds of derivatives, called *divergence* and *curl*, that express how the vectors of a vector field change in space. There are several integrals of vector fields over lines, surfaces, and volumes that are used to extract information and assert relationships among physical quantities. It is useful in physics to think of the vector field as a single mathematical entity. One of the advantages of a functional language for physics is the ease with which a vector field can be handled and written about as a single entity. This book aspires to make vector fields more accessible by presenting functions that allow you to play with vector fields.

Electric and magnetic fields are the most important vector fields in electromagnetic theory, although current density is also a vector field that appears in the famous Maxwell equations. We'll discuss the electric field in Chapter 25, current density in Chapter 26, and the magnetic field in Chapter 27.

Scalar and vector fields can be added. The following are some functions to do that:

```
addScalarFields :: [ScalarField] -> ScalarField
addScalarFields flds r = sum  [fld r | fld <- flds]

addVectorFields :: [VectorField] -> VectorField
addVectorFields flds r = sumV [fld r | fld <- flds]
```

We'll use these functions in Chapter 25 to add electric potential and electric field produced by multiple sources. For now, let's turn to the question of how to visualize scalar and vector fields.

Functions for Visualizing Scalar Fields

A scalar field associates a number with each point in space. There are lots of ways to visualize a scalar field. We'll develop two: one using `Vis` and one using text.

3D Visualization

One simple way to visualize a scalar field is to ask `Vis` to display numbers for the values of the scalar field at a list of positions. The function `sf3D` takes a list of positions and a scalar field as input and returns the action of displaying a 3D picture on the screen.

```
sf3D :: [Position]   -- positions to use
     -> ScalarField  -- to display
     -> IO ()
sf3D ps sf
    = V.display whiteBackground $ orient $
      V.VisObjects [V.Text3d (show (round $ sf p :: Int))
                    (v3FromPos p) V.Fixed9By15 V.black
                        | p <- ps]
```

We name the incoming list of positions `ps` and the incoming scalar field `sf`. We use a list comprehension to make a list of pictures, one for each position `p` in `ps`. Each picture is a piece of text showing the value of the scalar field at that position. The value `sf p` is the `R` expressing the value of the scalar field `sf` at position `p`. The value `round $ sf p :: Int` is the `Int` obtained by rounding the scalar field value. We round so that the numbers take up only a small amount of space and don't overlap each other in the final picture. The value `show (round $ sf p :: Int)` is the `String` we give to `Vis`'s constructor `V.Text3d` to be shown on the screen. The value `v3FromPos p` is the `V3` (`Vis`'s native vector type) giving the position where the text should be displayed. The definition of `v3FromPos` is similar to that of `v3FromVec` from Chapter 16.

```
v3FromPos :: Position -> V3 R
v3FromPos p = V3 x y z
    where
      (x,y,z) = cartesianCoordinates p
```

The `V.VisObjects` constructor sews the list of pictures together into a single picture, which we orient to use my favorite coordinate system with `orient` and display with the `V.display` function using a set of options called `whiteBackground`, which we'll define next.

The option set `whiteBackground` differs from the option set `V.defaultOpts` only in that the background color has been set to white.

```
whiteBackground :: V.Options
whiteBackground = V.defaultOpts {V.optBackgroundColor = Just V.white}
```

This definition uses record syntax to specify that all fields of the `V.Options` data type should have the same values as those in `V.defaultOpts`, except for `V.optBackgroundColor`, which is set to white.

If you find you'd like to control the camera position from code, you can add options to do that. For example, the option set `whiteBackground'` sets the viewpoint to be a distance of 40 Vis units from the center.

```
whiteBackground' :: V.Options
whiteBackground'
    = V.defaultOpts {V.optBackgroundColor = Just V.white,
                     V.optInitialCamera   = Just V.Camera0 {V.rho0   = 40.0,
                                                            V.theta0 = 45.0,
                                                            V.phi0   = 20.0}}
```

Here is an example of how to use this scalar field visualization function for the *y* scalar field `ySF`:

```
ySF3D :: IO ()
ySF3D = sf3D [cart x y z | x <- [-6,-2..6]
                         , y <- [-6,-2..6]
                         , z <- [-6,-2..6]] ySF
```

Figure 22-4 from earlier in this chapter shows the resulting picture. Perhaps the most useful feature of a 3D scalar field visualization like that in Figure 22-4 is that it helps us to develop a visual and geometric understanding of what a scalar field is by imagining a number at each point in space. Once we have this geometric idea under our belts, and we wish to look in detail at a specific scalar field, it's often simpler and more convenient to use a 2D visualization, which we'll describe next.

2D Visualization

A 3D visualization of a scalar field can become unwieldy and hard to read, so it's useful to have tools to view the scalar values on a 2D plane or surface. Our 2D visualization functions will allow the user to specify any plane or surface to focus on. We can do this by specifying how two numbers, representing horizontal and vertical position on the 2D visualization, map into three-dimensional space—in other words, by giving a function `(R,R) -> Position`. The functions that follow refer locally to this function as `toPos`. The function `sfTable` allows the user to visualize a scalar field by specifying a surface on which to view the values of the scalar field.

```
sfTable :: ((R,R) -> Position)
        -> [R]  -- horizontal
        -> [R]  -- vertical
        -> ScalarField
        -> Table Int
sfTable toPos ss ts sf
    = Table RJ [[round $ sf $ toPos (s,t) | s <- ss] | t <- reverse ts]
```

The first input to sfTable, locally called toPos, specifies the surface of interest. If we wanted to specify the xz-plane, for example, we'd send the function \(x,z) -> cart x 0 z in for toPos.

The second and third inputs to sfTable, locally called ss and ts, give the horizontal and vertical two-dimensional coordinates at which scalar values will be displayed. For a visualization of the xz-plane, the horizontal values could be x-values and vertical values could be z-values. The fourth input is the scalar field to be visualized.

The function works by sampling and displaying the values of the scalar field at the given points. We use the function toPos to produce a Position from an (s,t) pair of horizontal and vertical two-dimensional coordinates. We then apply the scalar field sf to this position, which is rounded so as not to take up too much space on the screen. The list of vertical coordinates is reversed so that the vertical values start at the bottom of the table and proceed to the top. We use the Table data type from Chapter 20.

Here is an example using sfTable to visualize the y-coordinate scalar field:

```
Prelude> :l CoordinateSystems
[1 of 7] Compiling Newton2          ( Newton2.hs, interpreted )
[2 of 7] Compiling Mechanics1D      ( Mechanics1D.hs, interpreted )
[3 of 7] Compiling SimpleVec        ( SimpleVec.hs, interpreted )
[4 of 7] Compiling Mechanics3D      ( Mechanics3D.hs, interpreted )
[5 of 7] Compiling MultipleObjects  ( MultipleObjects.hs, interpreted )
[6 of 7] Compiling MOExamples       ( MOExamples.hs, interpreted )
[7 of 7] Compiling CoordinateSystems ( CoordinateSystems.hs, interpreted )
Ok, 7 modules loaded.
*CoordinateSystems> sfTable (\(x,y) -> cart x y 0) [-6,-2..6] [-6,-2..6] ySF
  6  6  6  6
  2  2  2  2
 -2 -2 -2 -2
 -6 -6 -6 -6
```

We can use a 2D scalar field visualization to show the temperature in a room or the electric potential in a capacitor, for example.

Functions for Visualizing Vector Fields

A vector field associates a vector with each position in space. In this section, we'll write three functions for visualizing a vector field: vf3D, vfPNG, and vfGrad.

These functions essentially have type VectorField -> IO (), meaning they take a vector field as input and do something, either displaying a picture on the screen or producing a graphics file on the hard drive.

3D Visualization

The Vis module can produce a 3D visualization of a vector field. The basic idea is to choose a list of positions at which vectors will be displayed.

We use the vector field to compute the vector at each listed position and then display that vector with its tail at that position. The vector field will often have units that are different from the units of position (meters), so we need a scale factor to specify the number of vector field units that should be displayed per meter of space. Here is the code for the function vf3D that does this.

```
vf3D :: R             -- scale factor, vector field units per meter
     -> [Position]    -- positions to show the field
     -> VectorField   -- vector field to display
     -> IO ()
vf3D unitsPerMeter ps vf
    = V.display whiteBackground $ orient $
      V.VisObjects [V.Trans (v3FromPos p) $
                    visVec V.black (vf p ^/ unitsPerMeter)
                        | p <- ps]
```

The function vf3D takes a scale factor, a list of positions, and a vector field as input, and it produces a picture on the screen that can be enlarged and rotated with the mouse. As in sf3D, this function uses a list comprehension to make a list of pictures, one for each position p in ps. Each picture is a black arrow, produced by the visVec function defined below, representing the vector at position p, appropriately scaled and translated to the correct location. The V.VisObjects constructor sews the list of pictures together into a single picture, which is oriented to use my favorite coordinate system with orient and displayed with the V.display function using the option set whiteBackground, defined earlier in the chapter.

The visVec function takes a color and a vector as input and produces a picture of an arrow as output. Here is the code:

```
visVec :: V.Color -> Vec -> V.VisObject R
visVec color v = let vmag = magnitude v
                 in V.Arrow (vmag,20*vmag) (v3FromVec v) color
```

This function uses Vis's V.Arrow constructor to make a picture of a vector. The first argument to V.Arrow is a pair of numbers. The first number is the requested length of the arrow, for which we choose vmag, the magnitude of the input vector. The second number is an aspect ratio for the desired ratio of arrow length to arrow shaft diameter. I chose 20*vmag because I want the arrows to have a uniform shaft diameter. The shaft diameter is the arrow length vmag divided by the aspect ratio 20*vmag, which is $1/20$, independent of the arrow length.

The second argument to V.Arrow is a vector in Vis's native V3 type, which specifies the direction of the arrow. We send v3FromVec v, our input vector converted to type V3. The third and final argument to V.Arrow is a color, and we simply pass on the input color visVec is given.

The following program uses the vf3D function to produce a visualization of the $\hat{\phi}$ unit vector field, defined as phiHat earlier in the chapter:

```
phiHat3D :: IO ()
phiHat3D = vf3D 1 [cyl r ph z | r  <- [1,2,3]
                              , ph <- [0,pi/4..2*pi]
                              , z  <- [-2..2]] phiHat
```

The left side of Figure 22-5 from earlier in the chapter shows a screenshot of the image `phiHat3D` produces. The image on the screen is interactive and can be rotated and zoomed with the mouse.

Sometimes a 3D visualization of a vector field can appear cluttered, so we want tools to show a slice of a vector field in two dimensions. The right side of Figure 22-5 shows such a 2D visualization, and we'll turn next to how to make this kind of picture.

2D Visualization

How can we hope to visualize a 3D vector field in two dimensions? In general, we can't. Even if we limit our attention to a plane in three-dimensional space, say the xy-plane, the vectors could have a z-component so that they can't be represented in the xy-plane. Nevertheless, there are enough examples of vector fields that have planes in which the vectors point *in the plane* that 2D visualization is a worthwhile endeavor.

As with 2D scalar field visualization, the function we write will take an argument, locally called `toPos`, with type `(R,R) -> Position`, that maps the two-dimensional coordinates we supply to 3D `Positions`. After we gather a collection of vectors at positions in the plane, we need a second function to specify how these 3D vectors are to be regarded as 2D vectors in the plane. We can do this with a function `Vec -> (R,R)` that we will name with the local variable `fromVec`.

We could use `gloss` for our 2D vector field visualization, but because we may want a platform for the asynchronous animation we first explored in Chapter 20, we'll instead pursue a graphics library called `diagrams` that produces PNG files that could be sewn together into an asynchronous animation. The function `vfPNG` we are about to write takes a `VectorField` as input, along with some other parameters, and produces a PNG file.

```
vfPNG :: ((R,R) -> Position)
      -> (Vec -> (R,R))
      -> FilePath      -- file name
      -> R             -- scale factor in units per meter
      -> [(R,R)]       -- positions to use
      -> VectorField
      -> IO ()
vfPNG toPos fromVec fileName unitsPerMeter pts vf
    = let vf2d = r2 . fromVec . (^/ unitsPerMeter) . vf . toPos
          pic  = mconcat [arrowAt (p2 pt) (vf2d pt) | pt <- pts]
      in renderCairo fileName (dims (V2 1024 1024)) pic
```

The function takes five items as input before the vector field we want to display. The first two items are the functions, locally named `toPos` and `fromVec`, that manage the connection between 2D and 3D vector fields. The third item is a filename for the PNG file. The fourth is a scale factor, in (vector field) units per meter, to control the length of displayed vectors. The fifth item is a list of 2D points at which we want vectors displayed. Finally, the sixth item is the vector field itself.

The local function `vf2d` is a composition of five functions. It takes a 2D point as input and produces a 2D vector as output, with a type that `diagrams` likes for the positioning of arrows. Starting with a 2D position `(R,R)`, the function `vf2d` begins by applying `toPos`, the function the user of `vfPNG` provided to transform a 2D position into a `Position`. The vector field `vf` is then applied to produce a `Vec`. This vector is scaled by the scale factor `unitsPerMeter`, after which the function `fromVec` transforms the `Vec` into a pair of real numbers representing a 2D vector. Finally, `diagrams`'s function `r2` transforms a pair of real numbers `(R,R)` into `diagrams`'s 2D vector type.

The local variable `pic` is for the picture to be displayed, which is made by combining a list of arrow pictures formed with a list comprehension. Each arrow picture is made with `diagrams`'s `arrowAt` function, which places the tail of the 2D vector in its second argument at the 2D position in its first argument. The `diagrams` package makes a distinction between a 2D position, formed from a pair of numbers with its `p2` function, and a 2D vector, formed from a pair of numbers with its `r2` function.

The last line in `vfPNG` produces the PNG file with `diagrams`'s `renderCairo` function, which takes a filename, a pixel size, and a picture as input.

If the xy-plane happens to be our plane of interest, we can write a helping function by supplying `vfPNG` with its first two arguments. The function `vfPNGxy` has these first two arguments supplied:

```
vfPNGxy :: FilePath       -- file name
        -> R              -- scale factor
        -> [(R,R)]        -- positions to use
        -> VectorField
        -> IO ()
vfPNGxy = vfPNG (\(x,y) -> cart x y 0) (\v -> (xComp v, yComp v))
```

The function that `vfPNG` locally calls `toPos` is specified here as the function that maps the pair `(x,y)` into the xy-plane. The function that `vfPNG` locally calls `fromVec` projects the 3D vector into the xy-plane.

The following program produces a PNG file for the vector field $\hat{\phi}$, or `phiHat`, the unit vector field in cylindrical and spherical coordinates corresponding to the coordinate ϕ:

```
phiHatPNG :: IO ()
phiHatPNG
    = vfPNGxy "phiHatPNG.png" 1
      [(r * cos ph, r * sin ph) | r  <- [1,2]
                                , ph <- [0,pi/4..2*pi]] phiHat
```

The right side of Figure 22-5 earlier in the chapter shows the vector field $\hat{\phi}$ produced by `phiHatPNG`.

Here is code to produce a PNG picture of the vector field `rVF` introduced earlier in the chapter:

```
rVFpng :: IO ()
rVFpng
    = vfPNGxy "rVFpng.png" 2
      [(r * cos ph, r * sin ph) | r  <- [1,2]
                                , ph <- [0,pi/4..2*pi]] rVF
```

The left side of Figure 22-6 earlier in the chapter shows the resulting picture.

Physicists use the notation **r** in at least three ways. It can stand for a single position vector, which we would call a `Vec`. It can stand for a position function, like what we worked with in Part II, returning a position when given a time. In Part II, this position function would have type `R -> Vec` because position was regarded as a vector back then. Now that we have a data type for position, such a function has type `R -> Position`. A third use of the symbol **r** is for the vector field we just introduced. This has type `VectorField`, which is a type synonym for `Position -> Vec`. The type system helps clarify that these three uses of the symbol **r** are distinct.

Before we leave the topic of vector field visualization, we need to look at one more visualization method.

Gradient Visualization

When we visualize electric and magnetic fields, which we will do a few chapters from now, the magnitudes of the vectors can change enormously over short distances. Thus, displaying the magnitude of the vector as the length of an arrow can produce a burdensome picture. An alternative is to use shading to indicate magnitude, with short fat arrows to indicate direction. I call this style of vector field visualization *gradient visualization*.

The function `vfGrad` we define below takes a vector field, along with some other parameters, and produces a PNG file.

```
vfGrad :: (R -> R)
       -> ((R,R) -> Position)
       -> (Vec -> (R,R))
       -> FilePath
       -> Int     -- n for n x n
       -> VectorField
       -> IO ()
vfGrad curve toPos fromVec fileName n vf
 ❶ = let step = 2 / fromIntegral n
       ❷ xs = [-1+step/2, -1+3*step/2 .. 1-step/2]
       ❸ pts = [(x, y) | x <- xs, y <- xs]
       ❹ array = [(pt,magRad $ fromVec $ vf $ toPos pt) | pt <- pts]
```

```
  ❺ maxMag = maximum (map (fst . snd) array)
  ❻ scaledArrow m th = scale step $ arrowMagRad (curve (m/maxMag)) th
  ❼ pic = position [(p2 pt, scaledArrow m th) | (pt,(m,th)) <- array]
❽ in renderCairo fileName (dims (V2 1024 1024)) pic
```

The first argument to vfGrad is a monotonic function curve that maps the unit interval [0, 1] onto itself. The purpose of this argument is to make some accommodation for the possibility that a vector field may have very large magnitudes at some positions and rather small magnitudes elsewhere. The largest magnitude vectors will be colored black and those closest to zero will be colored white. Sometimes a linear scaling results in a picture in which there are black vectors close to a source and white vectors everywhere else. In those cases, a power law such as cube root or fifth root can boost the smaller magnitudes so that a continuous transition from black to white becomes evident. We can achieve a linear scaling with the identity function id and a fifth root scaling with the section `(**0.2)`.

The next arguments, with local names toPos and fromVec, are the same as in the function vfPNG. However, toPos plays a double role in this function because vfGrad does not ask for a list of positions at which to show vectors. Instead, vfGrad displays the square from (−1, −1) to (1, 1). This square must be mapped to some square in three-dimensional space, the vectors at which will be displayed. If we wanted to see the square in the xy-plane with corners at Cartesian coordinates (−10, −10, 0) and (10, 10, 0), we would send the function `\(x,y) -> cart (10*x) (10*y) 0` in for toPos.

The argument fileName is a filename for the PNG file. The argument n is an integer specifying the number of arrows to use in each direction. Sending in 20, for example, will produce an image of 20 arrows by 20 arrows. The last input vf is the vector field itself.

The function vfGrad consists of several local definitions that build a picture pic ❼, followed by the same renderCairo line ❽ used in vfPNG to make the PNG file. The first three lines ❶ ❷ ❸ in the let clause serve to choose the points pts at which the vector field will be sampled and displayed. The next line ❹ defines array (type `[((R,R),(R,R))]`) as a list of pairs of points and 2D vectors. We calculate the 2D vector at pt by applying toPos to convert pt to a 3D Position, applying the vector field vf, then using fromVec to convert the 3D vector to a 2D vector, and finally applying magRad, defined next, to express the 2D vector in magnitude-angle form.

The local variable maxMag ❺ searches the list array to find the maximum magnitude of all of the vectors in the list. Vectors with this magnitude will be colored black. The local function scaledArrow ❻ describes how to make a picture of a single arrow from a magnitude m and angle th. It normalizes the magnitude m by dividing it by the maximum magnitude maxMag, resulting in a normalized magnitude between 0 and 1. This normalized magnitude is then scaled, or curved, by the function curve, a monotonic function mapping the unit interval [0, 1] to itself. The normalized and scaled magnitude then passes with the angle to the arrowMagRad function, defined next, to get a picture of the arrow. Finally, the code scales the size of the arrow based on the number of arrows requested. We form the final picture pic ❼ with a list comprehension by placing each of the arrows at the appropriate position.

The function magRad converts a pair of Cartesian coordinates to polar coordinates, with the angle in radians.

```
magRad :: (R,R) -> (R,R)
magRad (x,y) = (sqrt (x*x + y*y), atan2 y x)
```

The function arrowMagRad produces a picture of an arrow based on a normalized magnitude in the range 0 to 1 and an angle in radians.

```
-- magnitude from 0 to 1
arrowMagRad :: R  -- magnitude
            -> R  -- angle in radians, counterclockwise from x axis
            -> Diagram B
arrowMagRad mag th
    = let r      = sinA (15 @@ deg) / sinA (60 @@ deg)
          myType = PolyPolar [120 @@ deg, 0 @@ deg, 45 @@ deg, 30 @@ deg,
                              45 @@ deg, 0 @@ deg, 120 @@ deg]
                   [1,1,r,1,1,r,1,1]
          myOpts = PolygonOpts myType NoOrient (p2 (0,0))
      in scale 0.5 $ polygon myOpts # lw none # fc (blend mag black white) #
         rotate (th @@ rad)
```

The function defines the shape of the arrow as a polygon and chooses the color based on the normalized magnitude. A normalized magnitude of 1 results in a black arrow, 0 results in white, and numbers in between result in some shade of gray.

Here is an example of gradient visualization for the vector field **r**, or rVF:

```
rVFGrad :: IO ()
rVFGrad = vfGrad id
          (\(x,y) -> cart x y 0)
          (\v -> (xComp v,yComp v))
          "rVFGrad.png" 20
          rVF
```

The right side of Figure 22-6 earlier in the chapter shows the vector field **r**, or rVF produced by rVFGrad.

Summary

This chapter introduced the idea of a field, which is a function from a position in three-dimensional space. Scalar fields and vector fields are the two most important types of field for electromagnetic theory. We introduced several ways of visualizing scalar and vector fields as well as coordinate systems for three-dimensional space, in particular for cylindrical and spherical coordinates. We then wrote a new data type for position.

Since electromagnetic theory is geometric, the next chapter introduces data types for geometric objects like curves, surfaces, and volumes.

Exercises

Exercise 22.1. Show that the polar coordinate unit vectors form an orthonormal system. Orthonormal means both orthogonal (different vectors are perpendicular to each other) and normalized (each vector has length one). In other words, show that

$$\hat{\mathbf{s}} \cdot \hat{\mathbf{s}} = 1$$
$$\hat{\boldsymbol{\phi}} \cdot \hat{\boldsymbol{\phi}} = 1$$
$$\hat{\mathbf{s}} \cdot \hat{\boldsymbol{\phi}} = 0$$

Exercise 22.2. Write $\hat{\mathbf{x}}$ and $\hat{\mathbf{y}}$ in terms of $\hat{\mathbf{s}}$ and $\hat{\boldsymbol{\phi}}$. Your results should contain s, ϕ, $\hat{\mathbf{s}}$, and $\hat{\boldsymbol{\phi}}$ but not x or y.

Exercise 22.3. Show that the spherical coordinate unit vectors form an orthonormal system. In other words, show that

$$\hat{\mathbf{r}} \cdot \hat{\mathbf{r}} = 1$$
$$\hat{\boldsymbol{\theta}} \cdot \hat{\boldsymbol{\theta}} = 1$$
$$\hat{\boldsymbol{\phi}} \cdot \hat{\boldsymbol{\phi}} = 1$$
$$\hat{\mathbf{r}} \cdot \hat{\boldsymbol{\theta}} = 0$$
$$\hat{\mathbf{r}} \cdot \hat{\boldsymbol{\phi}} = 0$$
$$\hat{\boldsymbol{\theta}} \cdot \hat{\boldsymbol{\phi}} = 0$$

Exercise 22.4. Write $\hat{\mathbf{x}}$, $\hat{\mathbf{y}}$, and $\hat{\mathbf{z}}$ in terms of $\hat{\mathbf{r}}$, $\hat{\boldsymbol{\theta}}$, and $\hat{\boldsymbol{\phi}}$. Your results can contain r, θ, ϕ, $\hat{\mathbf{r}}$, $\hat{\boldsymbol{\theta}}$, and $\hat{\boldsymbol{\phi}}$ but not x, y, or z.

Exercise 22.5. Define a coordinate scalar field

```
thetaSF :: ScalarField
thetaSF = undefined
```

for the θ coordinate in spherical coordinates.

Exercise 22.6. Use 3D visualization to make an image of the vector field $\hat{\boldsymbol{\theta}}$, or `thetaHat`.

```
thetaHat3D :: IO ()
thetaHat3D = undefined
```

Exercise 22.7. Use the `vf3D` function to visualize the vector field $\mathbf{r}$, or `rVF`. You may need to use a scale factor greater than 1 so the arrows don't overlap each other. A larger scale factor shrinks the arrows because the scale factor is in units per meter.

Exercise 22.8. Use gradient vector field visualization to make an image of the vector field $\hat{\theta}$, or `thetaHat`, in the xz-plane. In the first `undefined` below, you must say how to map a pair of coordinates to a `Position`, knowing that you are interested in the xz-plane. In the second `undefined`, you must say how to map a `Vec` to a pair of numbers describing the two components of the vector to be displayed.

```
thetaHatGrad :: IO ()
thetaHatGrad = vfGrad id undefined undefined "thetaHatGrad.png" 20 thetaHat
```

Exercise 22.9. Use gradient vector field visualization to make an image of the vector field $\hat{\phi}$, or `phiHat`, in the xy-plane.

```
phiHatGrad :: IO ()
phiHatGrad = undefined
```

23

CURVES, SURFACES, AND VOLUMES

Electrodynamics is a geometric subject. Curves, surfaces, and volumes play a dual role in electromagnetic theory. They serve as the places where electric charge and current can reside, and they play an essential role in the formulation of the Maxwell equations, the modern expression of how electric and magnetic fields are created and how they evolve in time.

Before we can explore the Maxwell equations, we'll need data types for curves, surfaces, and volumes—we'll build them in this chapter. A curve can be specified by giving a function from a single real parameter to a position in space. A surface can be specified as a function from a pair of real numbers to a position in space. A volume can be specified as a function from a triple of numbers to a position in space. These mathematical parameterizations lead naturally to data type definitions. We'll package the parameterizations with appropriate boundaries to form the types `Curve`, `Surface`, and `Volume`.

Let's start with some introductory code.

Introductory Code

Listing 23-1 shows the introductory code for the `Geometry` module we'll develop in this chapter.

```
{-# OPTIONS -Wall #-}

module Geometry where

import SimpleVec ( R, Vec, (*^) )
import CoordinateSystems ( Position, cylindrical, spherical, cart, cyl, sph
                         , shiftPosition, displacement )
```

Listing 23-1: Opening lines of code for the `Geometry` module

We'll use the type `Position` and the related functions we defined in the `CoordinateSystems` module in Chapter 22, so we've imported these and a few types and functions from the `SimpleVec` module of Chapter 10.

Our first geometric objects are one-dimensional curves embedded in three-dimensional space.

Curves

Curves have two distinct uses in electromagnetic theory. First, we use them to describe the place that electric charge and current live. Current in a wire can flow along a curve. Static charge can also be placed along a curve.

The second place we use them is in Ampere's law, which asserts a relationship between the magnetic field along a closed curve in space (a loop) and the electric current that flows through a surface with the closed curve as its boundary. This second use of curves is more abstract since the curve doesn't need to be the location of any actual material, but it's also more important for a deep understanding of modern electromagnetic theory.

Parameterizing Curves

How can we describe a curve in space? We can parameterize the curve so there's a real number associated with each point on the curve and then give (by way of a function) the position in space associated with each value of the parameter. For example, a line along the y-axis could be parameterized with the following function:

$$t \mapsto (0, t, 0)$$

A circle with radius 2 in the xy-plane centered at the origin could be parameterized with the following function:

$$t \mapsto (2\cos t, 2\sin t, 0) \tag{23.1}$$

In these functions, t serves only as the name of a parameter (we could have chosen s or any convenient symbol) and has nothing to do with time.

A parameterized curve therefore requires a function with type `R -> Position` sending a parameter `t :: R` along the curve to a point `r :: Position` in space. But we also need starting and ending points for our curve. For example, the circle in the xy-plane with radius 2 centered at the origin can be specified with the function

$$t \mapsto (2 \cos t, 2 \sin t, 0)$$

as well as the starting parameter $t_a = 0$ and the ending parameter $t_b = 2\pi$. If we use the same function and starting parameter but change the ending parameter to $t_b = \pi$, we get a semicircle (the half circle above the x-axis).

The starting and ending points can be specified by a starting parameter `startingCurveParam :: R` (which we called t_a earlier) and an ending parameter `endingCurveParam :: R` (which we called t_b). Thus, we specify a curve with three pieces of data: a function, a starting parameter, and an ending parameter.

A data type can be used to combine pieces of data that really belong together. For the curve, it will be very convenient to have a single type `Curve` that contains the function, the starting point, and the ending point.

```
data Curve = Curve { curveFunc          :: R -> Position
                   , startingCurveParam :: R  -- t_a
                   , endingCurveParam   :: R  -- t_b
                   }
```

The data type `Curve` has a single data constructor that is also called `Curve`.

Examples of Curves

Let's encode the example of the circle with radius 2 in the xy-plane centered at the origin.

```
circle2 :: Curve
circle2 = Curve (\t -> cart (2 * cos t) (2 * sin t) 0) 0 (2*pi)
```

We're naming our curve `circle2` to remind us of the radius 2. The parameterization 23.1 is given as the first argument to the data constructor `Curve`, followed by the starting and ending curve parameters.

A circle in the xy-plane centered at the origin is easier to express in cylindrical coordinates than in Cartesian. In cylindrical coordinates, our circle has the constant values $s = 2$ and $z = 0$. Only the ϕ coordinate changes from 0 to 2π. This suggests that we use the ϕ coordinate as our parameter for the curve.

```
circle2' :: Curve
circle2' = Curve (\phi -> cyl 2 phi 0) 0 (2*pi)
```

We use the `cyl` function to specify the curve in cylindrical coordinates. The curve `circle2'` is the same as the curve `circle2`.

Here's the definition for a unit circle:

```
unitCircle :: Curve
unitCircle = Curve (\t -> cyl 1 t 0) 0 (2 * pi)
```

There are families of curves for which we need to provide additional information before we've defined a specific curve. A straight line segment is such a curve. We need to provide both a starting position and an ending position, which is the perfect job for a higher-order function.

```
straightLine :: Position  -- starting position
             -> Position  -- ending position
             -> Curve     -- straight-line curve
straightLine r1 r2 = let d = displacement r1 r2
                         f t = shiftPosition (t *^ d) r1
                     in Curve f 0 1
```

We define the local name `d` to be the displacement vector pointing from position `r1` to position `r2`. We also define a local function `f` as our curve function by using the `shiftPosition` function to pick out the position that is shifted from `r1` by the displacement vector `t *^ d`. The curve parameter `t` runs from `0` to `1`, so `t *^ d` is a scaled version of the displacement vector `d` that runs from length 0 to the full length of `d`.

We've seen how to talk about one-dimensional curves in Haskell. Now let's move up a dimension and talk about surfaces.

Surfaces

Surfaces have two distinct uses in electromagnetic theory. We use them to describe the place that electric charge and current live. Current can flow along a surface. Static charge can also be placed on a surface.

We also use them in Gauss's law, which asserts a relationship between the electric field on a closed surface in space and the electric charge inside that surface. This second use of surfaces is more abstract since the surface need not be the location of any actual material, but it's also more important for a deep understanding of modern electromagnetic theory.

Parameterizing Surfaces

A surface is a parameterized function from two parameters to space. For example, we can parameterize the unit sphere with two parameters, θ and ϕ, as the function

$$(\theta, \phi) \mapsto (\sin\theta\cos\phi, \sin\theta\sin\phi, \cos\theta)$$

and the ranges $0 \leq \theta \leq \pi$ and $0 \leq \phi \leq 2\pi$.

For a second example, suppose we want to parameterize the surface that lies in the xy-plane, bounded by the parabola $y = x^2$ and the line $y = 4$. This surface is shown in Figure 23-1.

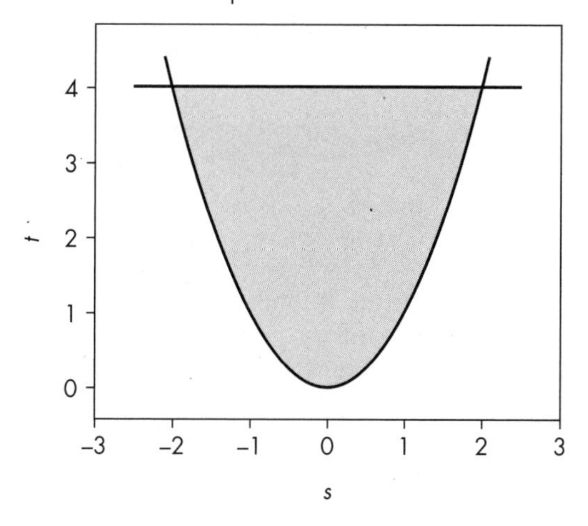

Figure 23-1: A parameterized surface

In this case, it makes sense to use x and y as the parameters. The parameterized function for the surface is not very exciting:

$$(x, y) \mapsto (x, y, 0)$$

The interesting part about this surface is the specification of the boundary. There's a lower curve of $y = x^2$ that gives the bottom boundary, an upper curve of $y = 4$ that gives the top boundary, a lower limit of $x = -2$ that specifies the left boundary, and an upper limit of $x = 2$ that specifies the right boundary.

For a general surface, we will call our two parameters s and t. (This parameter s is unrelated to the s of cylindrical coordinates discussed in Chapter 22.) To specify a general surface, we must give five pieces of data: a parameterizing function of two variables, a lower curve, an upper curve, a lower limit, and an upper limit. Here's the data type definition for a general surface:

```
data Surface = Surface { surfaceFunc :: (R,R) -> Position
                       , lowerLimit  :: R          -- s_l
                       , upperLimit  :: R          -- s_u
                       , lowerCurve  :: R -> R     -- t_l(s)
                       , upperCurve  :: R -> R     -- t_u(s)
                       }
```

The function `surfaceFunc` is the parameterizing function that maps (s, t) into a `Position`. The lower curve is given as a function $t_l(s)$ that gives the lowest value of t on the surface for each value of the parameter s. The upper curve is given as a function $t_u(s)$ that gives the highest value of t on the surface for each value of the parameter s. The lower limit s_l is the lowest value of s on the surface, and the upper limit s_u is the largest value of s on the surface.

Examples of Surfaces

To encode the unit sphere we discussed earlier, we can write the following:

```
unitSphere :: Surface
unitSphere = Surface (\(th,phi) -> cart (sin th * cos phi)
                                        (sin th * sin phi)
                                        (cos th))
                     0 pi (const 0) (const $ 2*pi)
```

In this case, we want constant functions for the lower and upper curves, so we use the `const` function to turn a number into a constant function and the `$` operator to avoid the need for parentheses around `2*pi`.

Unsurprisingly, it's easier to specify a unit sphere in spherical coordinates.

```
unitSphere' :: Surface
unitSphere' = Surface (\(th,phi) -> sph 1 th phi)
                      0 pi (const 0) (const $ 2*pi)
```

In spherical coordinates, we use the same parameters (θ, ϕ), the same lower and upper curves, and the same limits. Only the parameterizing function changes. The surface `unitSphere'` is the same surface as `unitSphere`.

Let's encode our parabolic surface from Figure 23-1.

```
parabolaSurface :: Surface
parabolaSurface = Surface (\(x,y) -> cart x y 0)
                          (-2) 2 (\x -> x*x) (const 4)
```

We use anonymous functions to specify both the surface parameterization and the parabolic lower boundary curve.

What about a sphere centered at an arbitrary position with an arbitrary radius? We could parameterize it by hand, but instead let's define a function that shifts the location of any surface. That seems like a useful function to have around.

```
shiftSurface :: Vec -> Surface -> Surface
shiftSurface d (Surface g sl su tl tu)
    = Surface (shiftPosition d . g) sl su tl tu
```

The `shiftSurface` function doesn't change the limits of the parameters being used. Instead, it shifts the positions that the parameterizing function `g` was providing by the displacement vector `d`.

Next, we define a centered sphere with an arbitrary radius.

```
centeredSphere :: R -> Surface
centeredSphere r = Surface (\(th,phi) -> sph r th phi)
                           0 pi (const 0) (const $ 2*pi)
```

Finally, we define a sphere with an arbitrary center and arbitrary radius.

```
sphere :: R -> Position -> Surface
sphere radius center
    = shiftSurface (displacement (cart 0 0 0) center)
      (centeredSphere radius)
```

Here's the northern hemisphere of the unit sphere:

```
northernHemisphere :: Surface
northernHemisphere = Surface (\(th,phi) -> sph 1 th phi)
                             0 (pi/2) (const 0) (const $ 2*pi)
```

Here's a disk in the xy-plane, centered at the origin:

```
disk :: R -> Surface
disk radius = Surface (\(s,phi) -> cyl s phi 0)
                      0 radius (const 0) (const (2*pi))
```

I don't think the term "unit cone" is standard terminology, but here is a cone in which the circular boundary of the base lies on a unit sphere, with the vertex of the cone at the center of the sphere:

```
unitCone :: R -> Surface
unitCone theta = Surface (\(r,phi) -> sph r theta phi)
                         0 1 (const 0) (const (2*pi))
```

These surfaces, or ones you write, can be used in Chapter 24 to form a charge distribution in which charge is distributed across a surface, or in Chapter 26 to form a current distribution in which current flows across a surface. Closed surfaces, such as spheres, can be used with Gauss's law.

Orientation

Our surfaces are oriented surfaces. An *orientation* is a choice of which direction (perpendicular to the surface) to consider "positive." If $\hat{\mathbf{s}}$ is a unit vector pointing in the direction of increasing s, and $\hat{\mathbf{t}}$ is a unit vector pointing in the direction of increasing t, then the positive direction for orientation is $\hat{\mathbf{s}} \times \hat{\mathbf{t}}$. (The parameter s used in specifying a surface and its associated unit vector $\hat{\mathbf{s}}$ are unrelated to the cylindrical coordinate s and its associated unit vector $\hat{\mathbf{s}}$. Context should make clear which is meant.) The orientation of a surface is important in a flux integral, which is used to calculate electric flux, magnetic flux, and current flowing through a surface.

Let's determine the orientation for the `unitSphere`. We used spherical coordinates to parameterize this surface, with the first parameter (in general called s) being θ for the unit sphere and the second parameter (in general called t) being ϕ for the unit sphere. Therefore, as shown in Figure 23-2, the orientation of the unit sphere is positive in the $\hat{\boldsymbol{\theta}} \times \hat{\boldsymbol{\phi}}$ direction. In spherical coordinates, $\hat{\boldsymbol{\theta}} \times \hat{\boldsymbol{\phi}} = \hat{\mathbf{r}}$, meaning that "outward" is the positive direction of orientation for the unit sphere.

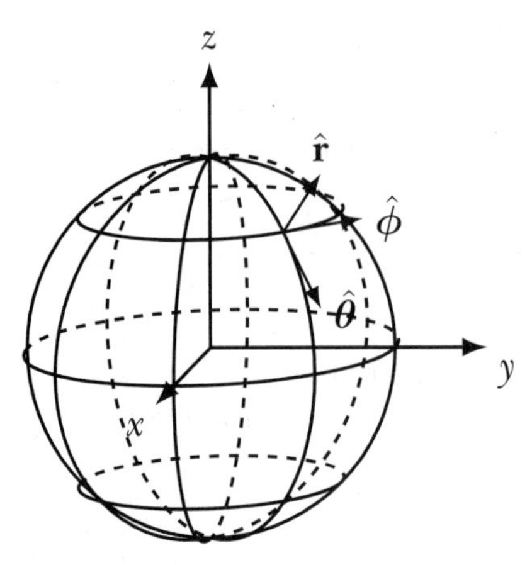

Figure 23-2: When the first parameter `s` *is θ and the second parameter* `t` *is ϕ, the direction of orientation is $\hat{\boldsymbol{\theta}} \times \hat{\boldsymbol{\phi}}$, which is the same as $\hat{\mathbf{r}}$, so the orientation is outward.*

We could make a unit sphere with "inward" as the orientation, but we would need to parameterize it differently from the `unitSphere`. If we take ϕ as the first parameter and θ as the second, the orientation is inward.

Volumes

When we have a charge that's distributed throughout a volume, we'll use a volume charge density to describe it; therefore, we'll need a new data type to describe a volume. We need to specify seven pieces of data to describe a volume:

1. A parameterizing function from three parameters (s, t, u) into space
2. A lower surface $u_l(s, t)$ describing the lowest value of u for each (s, t)
3. An upper surface $u_u(s, t)$ describing the highest value of u for each (s, t)
4. A lower curve $t_l(s)$ describing the lowest value of t for each value of s
5. An upper curve $t_u(s)$ describing the highest value of t for each value of s
6. A lower limit s_l describing the lowest value of s
7. An upper limit s_u describing the highest value of s

Here's the definition of the `Volume` data type:

```
data Volume = Volume { volumeFunc :: (R,R,R) -> Position
                     , loLimit    :: R              -- s_l
                     , upLimit    :: R              -- s_u
                     , loCurve    :: R -> R         -- t_l(s)
                     , upCurve    :: R -> R         -- t_u(s)
                     , loSurf     :: R -> R -> R    -- u_l(s,t)
                     , upSurf     :: R -> R -> R    -- u_u(s,t)
                     }
```

The `volumeFunc` for a given `Volume` has type `(R,R,R) -> Position`. Recall from Chapter 22 that this type is the same as `CoordinateSystem`. We'll often want to use `cartesian`, `cylindrical`, or `spherical` as our `volumeFunc`, although it's possible to invent your own coordinate system.

Here's a unit ball, centered at the origin:

```
unitBall :: Volume
unitBall = Volume spherical 0 1 (const 0) (const pi)
                  (\_ _ -> 0) (\_ _ -> 2*pi)
```

For the `volumeFunc`, we use `spherical`, which means that the parameters (s, t, u) are the spherical coordinates (r, θ, ϕ). We must provide a lower limit r_l, an upper limit r_u, a lower curve $\theta_l(r)$, an upper curve $\theta_u(r)$, a lower surface $\phi_l(r, \theta)$, and an upper surface $\phi_u(r, \theta)$. For a ball, we should pick the following:

$$r_l = 0$$
$$r_u = 1$$
$$\theta_l(r) = 0$$
$$\theta_u(r) = \pi$$
$$\phi_l(r, \theta) = 0$$
$$\phi_u(r, \theta) = 2\pi$$

Notice that θ_l is the function $r \mapsto 0$ (in Haskell notation `\r -> 0` or `\_ -> 0`). This the same as the constant function that returns 0 for any input (in Haskell notation `const 0`). The function ϕ_l takes *two* inputs and returns 0 (in Haskell notation `\_ _ -> 0`).

Here's a cylinder with a circular base centered at the origin and circular top in the plane $z = h$. We give the radius and height of the cylinder as inputs to the function `centeredCylinder`.

```
centeredCylinder :: R       -- radius
                 -> R       -- height
                 -> Volume  -- cylinder
centeredCylinder radius height
  = Volume cylindrical 0 radius (const 0) (const (2*pi))
           (\_ _ -> 0) (\_ _ -> height)
```

These volumes, or ones you write, can be used in Chapter 24 to form a charge distribution in which charge is distributed throughout a volume, or in Chapter 26 to form a current distribution in which current flows throughout a volume.

Summary

In this chapter, we developed the data types `Curve`, `Surface`, and `Volume` for describing geometric objects. We defined some particular geometric objects, such as `unitCircle`, `sphere`, and `unitBall`. These curves, surfaces, and volumes will become objects we integrate over to calculate electric fields, and they will also serve as the abstract settings for Gauss's law and Ampere's law. The next chapter discusses charge distributions in preparation for the following chapter on electric fields.

Exercises

Exercise 23.1. Replace the `undefined r radius` below with a definition that will take a center position and radius and produce a circle parallel to the xy-plane.

```
circle :: Position  -- center position
       -> R         -- radius
       -> Curve
circle r radius = undefined r radius
```

Exercise 23.2. A helix can be parameterized most easily in cylindrical coordinates. In cylindrical coordinates (s, ϕ, z), a helix with radius 1 can be parameterized as

$$\phi \mapsto (1, \phi, \phi/5)$$

Define a `Curve` for this helix. Choose end points so that the helix makes five loops around.

Exercise 23.3. A square has four sides. Let's make a `Curve` to represent a square with vertices $(-1, -1, 0)$, $(1, -1, 0)$, $(1, 1, 0)$, and $(-1, 1, 0)$. Make the orientation of the curve counterclockwise. Fill in the parts that are `undefined`.

```
square :: Curve
square = Curve squareFunc 0 4

squareFunc :: R -> Position
squareFunc t
    |              t < 1  = cart undefined    (-1)   0
    | 1 <= t && t < 2  = cart      1      undefined 0
    | 2 <= t && t < 3  = cart undefined       1    0
    | otherwise         = cart    (-1)     undefined 0
```

Exercise 23.4. Define a `Surface` for a cone with height h and radius r. Do not include the surface of the base of the cone. Position and orient the cone however it's convenient.

Exercise 23.5. Replace the `undefined` that follows with a definition of an upper-half ball ($z \geq 0$) with unit radius, centered at the origin.

```
northernHalfBall :: Volume
northernHalfBall = undefined
```

Exercise 23.6. Replace the `undefined` that follows with a definition of a ball with given a radius centered at the origin. (The `R` is the type of the radius, and you may want to put a variable for the radius on the left of the equal sign.)

```
centeredBall :: R -> Volume
centeredBall = undefined
```

Exercise 23.7. What is the type of `shiftPosition d` in the definition of `shiftSurface` given earlier?

Exercise 23.8. Define a function

```
shiftVolume :: Vec -> Volume -> Volume
shiftVolume = undefined
```

that takes a displacement vector and a volume as input and returns a shifted volume as output.

Exercise 23.9. Define a function

```
quarterDiskBoundary :: R -> Curve
quarterDiskBoundary = undefined
```

that takes a radius as input and gives a `Curve` as output corresponding to Figure 23-3.

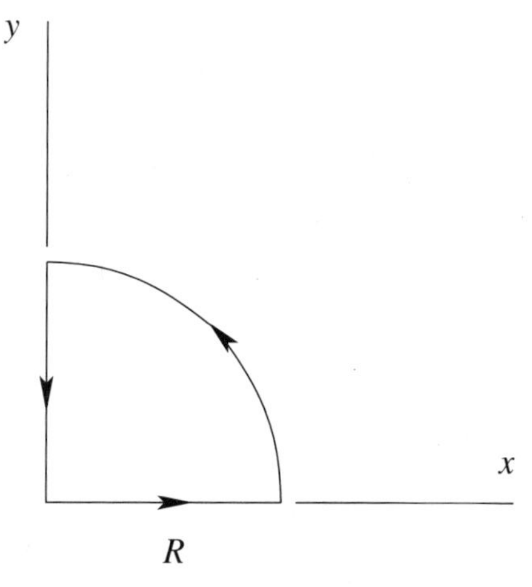

Figure 23-3: A curve representing the boundary of a quarter disk

Exercise 23.10. Define a `Surface` for the rectangular region that is shown in Figure 23-4. Choose your parameterization so that the orientation is in the $\hat{\mathbf{i}}$ direction (the positive x-direction).

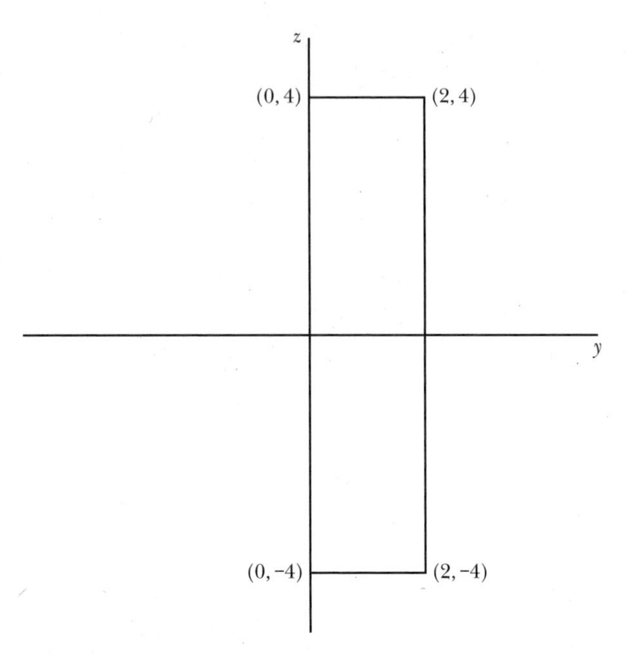

Figure 23-4: A surface in the yz-plane

Exercise 23.11. Define a function

```
quarterCylinder :: R -> R -> Volume
quarterCylinder = undefined
```

that takes a height h and radius R as input and gives a `Volume` as output corresponding to Figure 23-5.

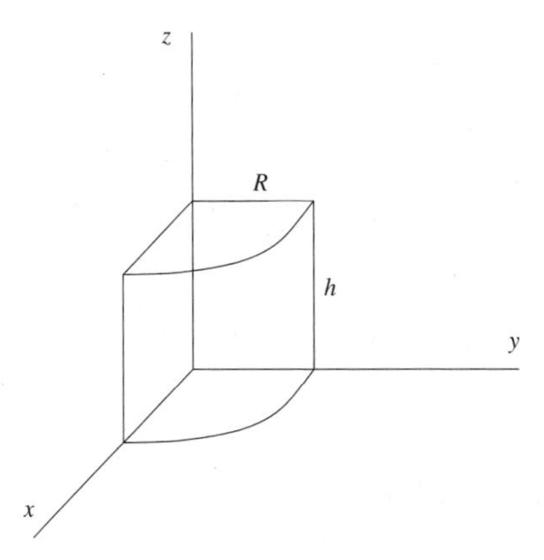

Figure 23-5: A volume representing a quarter cylinder

Exercise 23.12.

(a) Define a `Surface` for a torus with major radius 3 and minor radius 0.5.

(b) Define a `Volume` for the space inside the torus of part (a).

24

ELECTRIC CHARGE

Electric charge is ultimately responsible for all electrical phenomena. It's useful to be able to talk about electric charge that's localized at a particular point in space, but it's also useful to be able to talk about charge that's distributed along a curve, along a surface, or throughout a volume. Such charge distributions are the subject of this short chapter.

We'll first introduce the ideas of linear charge density, surface charge density, and volume charge density. We'll then define a data type for charge distribution capable of representing a point charge, a line charge, a surface charge, a volume charge, or any combination of these. We'll write functions to find the total charge and the electric dipole moment of a charge distribution. Electric charge gives rise to electric fields. Having a good language for charge distributions prepares us for the next chapter, in which we'll find the electric field produced by a charge distribution.

Charge Distributions

Electric charge is the fundamental quantity responsible for electromagnetic effects, and it plays a key role in electromagnetic theory. In classical electromagnetic theory, which we're studying in this book, we'll sometimes think of charge as associated with a particle, in which case we call the charge a *point charge*, imagining that it has a location in space but no spatial extent.

We also sometimes think of charge as a fluid (that is, something that can be continuously distributed throughout a region of space). In fact, we use three types of continuous charge distributions. First, there is charge continuously distributed along a one-dimensional path such as a line or a curve. In this case, we speak of the *linear charge density* λ, meaning the charge per unit length. This use of the Greek letter lambda is separate and independent of its use in Chapter 20 to mean wavelength. The SI unit for linear charge density is the Coulomb per meter (C/m).

Second, there is charge continuously distributed along a two-dimensional surface. In this case, we speak of the *surface charge density* σ (Greek letter sigma), meaning the charge per unit area. The SI unit for surface charge density is the Coulomb per square meter (C/m^2).

Third, there is charge continuously distributed throughout a three-dimensional volume. In this case, we speak of the *volume charge density* ρ (Greek letter rho), meaning the charge per unit volume. The SI unit for volume charge density is the Coulomb per cubic meter (C/m^3). Table 24-1 summarizes these charge distributions.

Table 24-1: Charge Distributions

Charge distribution	Dimensionality	Symbol	SI unit
Point charge	0	q, Q	C
Linear charge density	1	λ	C/m
Surface charge density	2	σ	C/m^2
Volume charge density	3	ρ	C/m^3

Electric charge plays a role in microscopic physics, in that we associate an electric charge with elementary particles such as electrons and quarks. Charge also plays a role in macroscopic physics because electrons can pile up in places, leading to a net negative charge, or they can be absent from atoms, leading to a net positive charge.

Materials can be classified as insulators or conductors based on how easy it is for electrons to move around. Insulators are materials in which it is difficult for electrons to leave their atoms and move about the material, while conductors are materials in which it is easy for electrons to move about.

Charge distributions on insulators can be more or less arbitrary, and charge distributions of things like atoms can be nonuniform—determined, ultimately, by quantum mechanics. But charge distributions on conductors are constrained because the electrons will not stay where you put them. *Fixed* macroscopic charge distributions can be realized only on insulators, on which the charges cannot move. So I can charge the tail of my cat to 10 times the charge density as his nose because the electrons stay put. Macroscopic *conductors* cannot support arbitrary charge densities: the charges move, redistributing themselves very rapidly. So you should imagine as we talk about charge distributions that we are distributing charge on an insulator.

Before we introduce the various charge distributions, some introductory code needs to be at the top of the source code file; we'll look at that next.

Introductory Code

Listing 24-1 shows the first lines of the code in the Charge module we'll write in this chapter.

```
{-# OPTIONS -Wall #-}

module Charge where

import SimpleVec ( R, Vec, vec, sumV, (*^), (^/), (<.>), magnitude, negateV )
import Electricity ( elementaryCharge )
import CoordinateSystems ( Position, ScalarField, origin, cart, sph
                         , rVF, displacement, shiftPosition )
import Geometry ( Curve(..), Surface(..), Volume(..)
                , straightLine, shiftSurface, disk )
import Integrals
    ( scalarLineIntegral, scalarSurfaceIntegral, scalarVolumeIntegral
    , vectorLineIntegral, vectorSurfaceIntegral, vectorVolumeIntegral
    , curveSample, surfaceSample, volumeSample )
```

Listing 24-1: Opening lines of code for the Charge module

We use types and functions from the SimpleVec module of Chapter 10, the Electricity module of Chapter 21, the CoordinateSystems module of Chapter 22, the Geometry module of Chapter 23, and the Integrals module that includes various functions introduced in the next chapter.

Let's define a type synonym for charge:

```
type Charge = R
```

Defining a new type for charge in this way is half good and half silly. It's good in that human readers of the code (including the writer of the code) will know the intent of an expression with type Charge. It is, in this sense, a form of documentation for the code. However, it is silly because the compiler doesn't make any distinction between Charge and R and Double, so it cannot help the writer to avoid using Charge in any place an R or a Double could be used. One of the main purposes of types is separating things that should be separated and letting the computer help enforce that separation. Charge is not at all the same as time, for example, which might also be described by a real number R.

Once again, we must choose between simplicity and power. We could use Haskell's data keyword to define a new type for charge that couldn't be confused with any other type. Defining a new data type would be a reasonable thing to do here, but there is a bit of extra effort and overhead involved, so I have chosen the simplicity of the type method over the power of the data method.

A Type for Charge Distribution

We want to define a new data type ChargeDistribution that can hold a point charge, a line charge, a surface charge, a volume charge, or a combination of these. This is not mandatory; we could settle for separate types for line charge, surface charge, and so on. Introducing a single ChargeDistribution type allows us to write a function

```
eField :: ChargeDistribution -> VectorField
```

in the next chapter to emphasize the central idea that charge is the source of electric fields. Once you understand the options a language allows, you can use them to your advantage to underscore the central ideas of the discipline you're writing about.

What information is required to specify each of these distributions? For a point charge, we need to specify how much charge there is and where the charge is located, so we need to give a Charge and a Position. For a line charge, we need to specify a curve along which the charge sits and the linear charge density at each point on the curve. The linear charge density need not be uniform along the curve. It could be high in some places and low in others, or even positive in some places and negative in others. We will use a scalar field to specify the linear charge density. So, a line charge requires that we give a ScalarField for the linear charge density and a Curve along which the charge lies.

Specification of a surface charge requires that we give a scalar field for the surface charge density, which may vary from place to place, as well as a surface along which the charge lies. A surface charge is specified by giving a ScalarField and a Surface. Similarly, a volume charge is specified by giving a ScalarField and a Volume.

Finally, we can specify a combination of charge distributions by giving a list of charge distributions. Let's take a look at the code defining the data type ChargeDistribution.

```
data ChargeDistribution
    = PointCharge   Charge      Position
    | LineCharge    ScalarField Curve
    | SurfaceCharge ScalarField Surface
    | VolumeCharge  ScalarField Volume
    | MultipleCharges [ChargeDistribution]
```

The type ChargeDistribution has five data constructors, one for each situation we described earlier. To construct a ChargeDistribution, we use one of the five data constructors along with the relevant information for that sort of charge distribution. A curious attribute of this data type is that it is recursive. The information required by the MultipleCharges constructor is a list of ChargeDistributions, the very type we are defining. We can think that there are four basic sorts of charge distribution (point, line, surface, and volume) and one composite sort that combines distributions.

Examples of Charge Distributions

Let's write some examples of charge distributions. We can define the charge distribution of a proton at the origin as follows:

```
protonOrigin :: ChargeDistribution
protonOrigin = PointCharge elementaryCharge origin
```

Here is a uniform line charge with total charge `q` and length `len`, centered at the origin:

```
chargedLine :: Charge -> R -> ChargeDistribution
chargedLine q len
    = LineCharge (const $ q / len) $
      Curve (\z -> cart 0 0 z) (-len/2) (len/2)
```

We pass in the total charge `q` and the length `len`. The uniform linear charge density is `q / len`, which we pass to the `const` function since `LineCharge` requires a scalar field for its first argument.

Here is a uniform ball of charge with total charge `q` and radius `radius`, centered at the origin:

```
chargedBall :: Charge -> R -> ChargeDistribution
chargedBall q radius
    = VolumeCharge (const $ q / (4/3*pi*radius**3)) $
      Volume (\(r,theta,phi) -> sph r theta phi)
             0 radius (const 0) (const pi) (\_ _ -> 0) (\_ _ -> 2*pi)
```

We pass in the total charge `q` and the `radius` of the ball. We can then find the uniform volume charge density by dividing `q` by the volume $4\pi r^3/3$ of a spherical ball.

A parallel-plate capacitor consists of two conducting plates, parallel to each other. It is typically used so that one plate has a positive charge and the other has an equal-but-opposite negative charge. The charge on each plate will not be distributed in a precisely uniform way, but if the plates are close together, it is a good approximation to regard the charge on each plate as uniformly distributed over the surface of the plate.

The following model of a parallel-plate capacitor is formed with two plates, each the shape of a disk with radius `radius`. The plates are parallel to each other and separated by a distance of `plateSep`. The surface charge density on the positive plate is `sigma`, and that on the negative plate is `-sigma`.

```
diskCap :: R -> R -> R -> ChargeDistribution
diskCap radius plateSep sigma
    = MultipleCharges
      [SurfaceCharge (const sigma) $
```

```
 shiftSurface (vec 0 0 (plateSep/2)) (disk radius)
,SurfaceCharge (const $ -sigma) $
 shiftSurface (vec 0 0 (-plateSep/2)) (disk radius)
]
```

Here we see our first use of the `MultipleCharges` constructor. The charge distribution consists of two `disk`s of charge, of the sort we wrote in the last chapter. Since the `disk` we wrote last chapter is centered at the origin, we use `shiftSurface`, also from last chapter, to place the disks above and below the origin.

Having written some examples of charge distributions, let's turn to the question of finding the total electric charge of a charge distribution.

Total Charge

If we distribute 2 C of charge uniformly over a 4-m length, the linear charge density is 0.5 C/m. Here, we're speaking in terms of charge density, but sometimes it's more convenient to speak in terms of total charge. For example, when we're interested in the electric field produced by this line charge at a place several hundred meters away, total charge is more relevant because the line charge looks like a simple point charge from that distance.

Total Charge of a Line Charge

The total charge of a line charge can be found by integrating the linear charge density over the curve on which the charge sits. If P is the path, or curve, along which the charge lies, dl is a length element along the curve, and λ is the scalar field for linear charge density, then the following is the total charge of the line charge:

$$Q = \int_P \lambda(\mathbf{r})\, dl \tag{24.1}$$

In the case where linear charge density is uniform, multiplying it by the length of the curve gives the total charge. In general, linear charge density is not uniform. We perform the integral by dividing the curve into many small segments, multiplying the length of each segment by a representative value of charge density on the segment to determine the charge on the segment, and adding all of these segment charges together. When working analytically, we look at the limit in which the length of each segment goes to zero and the number of segments becomes infinite. When working numerically, we choose some large but finite number of segments.

Total Charge of a Surface Charge

The total charge of a surface charge can be found by integrating the surface charge density over the surface on which the charge sits. If S is the surface, da is an area element for the surface, and σ is the scalar field for surface charge density, then

$$Q = \int_S \sigma(\mathbf{r})\, da \tag{24.2}$$

is the total charge of the surface charge.

In the case where surface charge density is uniform, multiplying it by the area of the surface gives the total charge. In general, surface charge density is not uniform. We perform the integral by dividing the surface into a large number of small patches, multiplying the area of each patch by a representative value of charge density on the patch to determine the charge on the patch, and adding all of these patch charges together. When working analytically, we look at a limit in which the area of each patch goes to zero and the number of patches becomes infinite. When working numerically, we choose some large but finite number of patches.

Total Charge of a Volume Charge

The total charge of a volume charge can be found by integrating the volume charge density over the volume in which the charge sits. If V is the volume, dv is a volume element for the volume, and ρ is the scalar field for volume charge density, then

$$Q = \int_V \rho(\mathbf{r})\, dv \tag{24.3}$$

is the total charge of the volume charge.

In the case where volume charge density is uniform, multiplying it by the volume gives the total charge. In general, volume charge density is not uniform. We perform the integral in much the same way as we did for the line charge and the surface charge.

Calculating Total Charge in Haskell

Here is a function called `totalCharge` that computes the total charge of a charge distribution:

```
totalCharge :: ChargeDistribution -> Charge
totalCharge (PointCharge   q      _)
    = q
totalCharge (LineCharge    lambda c)
    = scalarLineIntegral    (curveSample  1000) lambda c
totalCharge (SurfaceCharge sigma  s)
    = scalarSurfaceIntegral (surfaceSample 200) sigma s
totalCharge (VolumeCharge  rho    v)
    = scalarVolumeIntegral  (volumeSample   50) rho v
totalCharge (MultipleCharges ds    )
    = sum [totalCharge d | d <- ds]
```

The function `totalCharge` uses pattern matching on the input to treat each of the five data constructors separately. In the case of a point charge, the function simply returns the charge of the point charge. For a line charge, the function uses scalarLineIntegral, which we'll write in Chapter 25, to perform the integral in Equation 24.1. The scalarLineIntegral function takes a method to approximate the curve, a scalar field, and a curve as inputs, and it returns an approximation to the line integral of the scalar field over the curve.

For a surface charge, the function uses scalarSurfaceIntegral, which we'll write in Chapter 25, to perform the integral shown in Equation 24.2. The scalarSurfaceIntegral function takes as inputs a method to approximate the surface, a scalar field, and a surface, and it returns an approximation to the surface integral of the scalar field over the surface. As we'll see in the next chapter, the method surfaceSample approximates the surface with a number of patches equal to twice the square of the number given; the value of 200 shown earlier will use 80,000 patches.

For a volume charge, the function uses scalarVolumeIntegral, which we will write in Chapter 25, to perform the integral shown in Equation 24.3. The scalarVolumeIntegral function takes a method to approximate the volume, a scalar field, and a volume as inputs, and it returns an approximation to the volume integral of the scalar field over the volume. As we'll see in the next chapter, the method volumeSample approximates the volume with a number of volume elements equal to five times the cube of the number given; the value of 50 shown earlier will use 625,000 volume elements.

In the case of multiple charges, the function calculates the total charge of each distribution in the list and adds the results together. Because totalCharge uses totalCharge to carry this out, it is a recursive function.

Let's check the total charge of the distributions we defined earlier.

```
Prelude> :l Charge
[ 1 of 11] Compiling Newton2          ( Newton2.hs, interpreted )
[ 2 of 11] Compiling Mechanics1D      ( Mechanics1D.hs, interpreted )
[ 3 of 11] Compiling SimpleVec        ( SimpleVec.hs, interpreted )
[ 4 of 11] Compiling Mechanics3D      ( Mechanics3D.hs, interpreted )
[ 5 of 11] Compiling MultipleObjects  ( MultipleObjects.hs, interpreted )
[ 6 of 11] Compiling MOExamples       ( MOExamples.hs, interpreted )
[ 7 of 11] Compiling Electricity      ( Electricity.hs, interpreted )
[ 8 of 11] Compiling CoordinateSystems ( CoordinateSystems.hs, interpreted )
[ 9 of 11] Compiling Geometry         ( Geometry.hs, interpreted )
[10 of 11] Compiling VectorIntegrals  ( VectorIntegrals.hs, interpreted )
[11 of 11] Compiling Charge           ( Charge.hs, interpreted )
Ok, 11 modules loaded.
*Charge> totalCharge protonOrigin
1.602176634e-19
*Charge> totalCharge $ chargedLine 0.25 2
0.2500000000000002
```

The total charge of a proton is just the charge of a proton. The total charge of a line charge is the value we gave for the total charge of the line charge.

Electric Dipole Moment

An *electric dipole* is a combination of positive and negative electric charges separated in space. The simplest case consists of a point charge q and a point charge $-q$ separated by some distance d. An electric dipole creates an electric field and also responds to an electric field by feeling a force and/or a torque, so it can be thought of as an electrically active entity similar to electric charge itself. An electric charge distribution with a total charge of 0 often looks like an electric dipole. A neutral diatomic molecule such as sodium chloride is an example of an electric dipole.

We characterize an electric dipole by its *electric dipole moment*, a vector that points from the negative charge toward the positive charge. For the case of charges q and $-q$ separated by distance d, the dipole moment is $\mathbf{p} = q\mathbf{d}$, where $\mathbf{d}$ is the displacement vector from the position of $-q$ to the position of q. The $\mathbf{p}$ used for electric dipole moment is not related to the $\mathbf{p}$ used for momentum.

The charge distribution `simpleDipole` describes two charges, q and $-q$, separated by distance d. The dipole is centered at the origin. The function takes an electric dipole moment and the distance between the point charges as input.

```
simpleDipole :: Vec  -- electric dipole moment
             -> R    -- charge separation
             -> ChargeDistribution
simpleDipole p sep
    = let q    = magnitude p / sep
          disp = (sep/2) *^ (p ^/ magnitude p)
      in MultipleCharges
             [PointCharge   q  (shiftPosition          disp  origin)
             ,PointCharge (-q) (shiftPosition (negateV disp) origin)
             ]
```

The function computes the charge `q` of the point charges by dividing the magnitude of the dipole moment by the separation. The displacement vector `disp` points from the origin to the positive point charge. The displacement vector `negateV disp` points from the origin to the negative point charge.

An electric dipole moment can be associated with any charge distribution. A volume charge density ρ has an electric dipole moment given by

$$\mathbf{p} = \int \mathbf{r}\rho(\mathbf{r})\, dv \tag{24.4}$$

If the total charge of a charge distribution is 0, the electric dipole moment is often the best simple characterization of the distribution, giving a good approximation for the electric field the distribution creates.

Of course, the total charge and the electric dipole moment of a charge distribution could both be 0. In that case, you can define an electric quadrupole moment that characterizes the distribution. In fact, any charge distribution can be viewed as a combination of an electric monopole (point charge), electric dipole, electric quadrupole, electric octupole, and higher terms, called a *multipole expansion*. We will not explore the expansion in this book, except to note that the electric dipole is the second term in the expansion. Like mathematical series expansions such as the Taylor series, the first nonzero term in the multipole expansion often gives a simple approximation for a charge distribution.

The function `electricDipoleMoment` computes the electric dipole moment for any charge distribution.

```
electricDipoleMoment :: ChargeDistribution -> Vec
electricDipoleMoment (PointCharge   q      r)
    = q *^ displacement origin r
electricDipoleMoment (LineCharge    lambda c)
    = vectorLineIntegral    (curveSample  1000) (\r -> lambda r *^ rVF r) c
electricDipoleMoment (SurfaceCharge sigma  s)
    = vectorSurfaceIntegral (surfaceSample 200) (\r -> sigma  r *^ rVF r) s
electricDipoleMoment (VolumeCharge  rho    v)
    = vectorVolumeIntegral  (volumeSample   50) (\r -> rho    r *^ rVF r) v
electricDipoleMoment (MultipleCharges ds    )
    = sumV [electricDipoleMoment d | d <- ds]
```

The function uses pattern matching on the input to split the definition into cases based on the constructor for the distribution. The dipole moment for a point charge is the product of the charge and a displacement vector from the origin to the position of the point charge. For a line charge, we do a vector line integral, similar to Equation 24.4. A surface charge requires a surface integral, and a volume charge uses Equation 24.4 itself. Finally, the dipole moment of a combination distribution is the vector sum of the dipole moments of each component.

Another charge distribution that behaves like an electric dipole is a line charge with a linear charge density that smoothly varies from negative to positive. The function `lineDipole` produces such a distribution, with the linear charge density changing linearly with position. The dipole is centered at the origin, where the linear charge density is 0.

```
lineDipole :: Vec  -- dipole moment
           -> R    -- charge separation
           -> ChargeDistribution
lineDipole p sep
    = let disp = (sep/2) *^ (p ^/ magnitude p)
          curve = straightLine (shiftPosition (negateV disp) origin)
                               (shiftPosition          disp  origin)
          coeff = 12 / sep**3
          lambda r = coeff * (displacement origin r <.> p)
      in LineCharge lambda curve
```

The function takes the same inputs that `simpleDipole` took and determines the linear charge density necessary to produce the desired electric dipole moment.

In the next chapter, we'll see several examples of electric dipoles. We'll look at a simple dipole composed of two point particles, an ideal dipole, and a line dipole. We'll compare their electric fields, noting the commonalities.

Summary

This chapter introduced charge distributions, including line charges, surface charges, and volume charges. We wrote a data type for charge distributions that can handle point charges, line charges, surface charges, volume charges, and combinations of these. We wrote some examples of charge distributions. We wrote functions to calculate the total charge and the electric dipole moment of a distribution. In the next chapter, we will calculate the electric field produced by a charge distribution.

Exercises

Exercise 24.1. Use the functions `chargedLine` and `chargedBall` to create some charge distributions and confirm, using `totalCharge`, that they have the total charge you expect.

Exercise 24.2. Find the total charge and electric dipole moment of the parallel-plate capacitor `diskCap`, choosing some parameters for radius, plate separation, and surface charge density. By varying the parameters, try to determine how the total charge and electric dipole moment depend on radius, plate separation, and surface charge density.

Exercise 24.3. Write a charge distribution for a uniformly charged surface in the shape of a disk, with total charge `q` and radius `radius`.

```
chargedDisk :: Charge -> R -> ChargeDistribution
chargedDisk q radius = undefined q radius
```

Check that your distribution has the total charge you expect by using `totalCharge`.

Exercise 24.4. Write a charge distribution for a uniformly charged circle (one with constant linear charge density), with total charge `q` and radius `radius`.

```
circularLineCharge :: Charge -> R -> ChargeDistribution
circularLineCharge q radius = undefined q radius
```

Check that your distribution has the total charge you expect by using `totalCharge`.

Exercise 24.5. Write a charge distribution for a uniformly charged surface in the shape of a square, with total charge `q` and side length `side`.

```
chargedSquarePlate :: Charge -> R -> ChargeDistribution
chargedSquarePlate q side = undefined q side
```

Check that your distribution has the total charge you expect by using `totalCharge`.

Exercise 24.6. Write a charge distribution for a uniformly charged surface in the shape of a sphere, with total charge `q` and radius `radius`.

```
chargedSphericalShell :: Charge -> R -> ChargeDistribution
chargedSphericalShell q radius = undefined q radius
```

Check that your distribution has the total charge you expect by using `totalCharge`.

Exercise 24.7. Write a charge distribution for a uniformly charged volume in the shape of a cube, with total charge `q` and side length `side`.

```
chargedCube :: Charge -> R -> ChargeDistribution
chargedCube q side = undefined q side
```

Check that your distribution has the total charge you expect by using `totalCharge`.

Exercise 24.8. Use the functions `simpleDipole` and `lineDipole` to create some charge distributions, and confirm, using `electricDipoleMoment`, that they have the electric dipole moments you expect.

Exercise 24.9. Write a charge distribution for a parallel-plate capacitor in which the plates are square surfaces, with side length `side`, separated by a distance `plateSep`. One plate has a uniform surface charge density `sigma`, and the other has a uniform surface charge density `-sigma`.

```
squareCap :: R -> R -> R -> ChargeDistribution
squareCap side plateSep sigma = undefined side plateSep sigma
```

Exercise 24.10. A hydrogen atom in its ground state is made up of a stationary proton at the origin and an electron cloud with volume charge density

$$\rho(r, \theta, \phi) = -\frac{e}{\pi a_0^3} e^{-2r/a_0}$$

where e is the elementary charge of the proton and a_0 is the Bohr radius. Write a charge distribution for this hydrogen atom. Since our volumes are finite, use a spherical ball with a radius of $10a_0$ for the volume that holds the charge density. This will omit a tiny bit of the electron's negative charge. See how close to neutral your hydrogen atom is by using `totalCharge` on the charge distribution you write.

```
hydrogen :: ChargeDistribution
hydrogen = undefined
```

25

ELECTRIC FIELD

In the 1800s, Faraday and Maxwell discovered a new way to think about electric (and magnetic) phenomena that forms the basis for today's electromagnetic theory. In this 19th century view, one particle does not directly apply a force to another particle, as Coulomb's law would imply. Instead, one particle creates an *electric field* that applies a force to the second particle.

We'll start this chapter with a short discussion of exactly what an electric field is. Then, we'll show how to calculate the electric field produced by each of the charge distributions we studied in the previous chapter. We'll begin with the electric field produced by a point charge, which is really the fundamental piece that governs all of the other distributions. After a single point charge, we'll look at the electric field produced by multiple point charges. As an example, we'll compare the electric fields produced by a simple electric dipole composed of two point charges with an ideal electric dipole.

To calculate the electric field, we'll need to introduce a few new mathematical tools. Finding the electric field produced by a line charge requires a vector line integral; finding the electric field produced by a surface charge requires a vector surface integral; and finding the electric field produced by a volume charge requires a vector volume integral. We'll introduce these as needed, and we'll close out the chapter with the details of sampling or discretizing curves, surfaces, and volumes, which will allow us to do numerical line, surface, and volume integrals.

What Is an Electric Field?

An electric field is a *vector field* of the kind we talked about in Chapter 22. The electric field associates a vector $\mathbf{E}(\mathbf{r})$ with each point $\mathbf{r}$ in space; that vector helps determine the force on a particle if there is a particle at point $\mathbf{r}$ in space.

Is the electric field a physical thing, or is it an abstraction we use to think about electricity? I would argue it is both. The electric field is an abstract mathematical construct, as are so many of the ideas of modern theoretical physics. We describe it using mathematical language and posit axioms about it as though it had no more physical reality than a 7×7 matrix of irrational numbers.

In static situations, where charges are not moving or accelerating, we can remain ambivalent about the reality of the electric field, or even deny it. These static situations are the ones where Coulomb's law makes good predictions, and the new electromagnetic theory that speaks of an electric field makes the same predictions. In static situations, then, the electric field can be viewed as merely an abstraction.

In dynamic situations, on the other hand, where electric charge is moving and/or accelerating, it becomes much harder to maintain the view that the electric field is only a calculational tool. The reason is that modern electromagnetic theory is, in addition to being a theory of electricity and magnetism, a theory of light and radiation. Maxwell's insight was that electric and magnetic fields could serve to describe visible light and an entire spectrum of nonvisible light-like waves, including radio waves and microwaves. These waves are now viewed as waves in the electric and magnetic fields. Light is, according to the Faraday-Maxwell theory, an electromagnetic wave. To the extent that light and radiation are physical and real, it seems the electric field is also physical and real.

Introducing the electric field breaks the analysis of electrical situations into two parts. The first part is the creation of the electric field by charge, which we'll discuss in this chapter. The second part is the force that the electric field applies to (a second) charge, which we'll discuss in Chapter 28. Figure 25-1 shows the electric field's role in a situation with two charges.

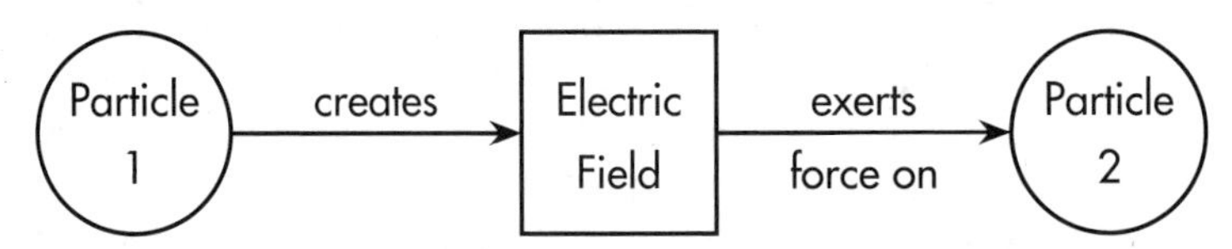

Figure 25-1: Conceptual diagram of the role of the electric field when two charged particles are present. Particle 1 creates the electric field. The electric field exerts a force on particle 2.

We say that the electric field *mediates* the interaction between the two particles. In this chapter, we focus on the electric field created by static, or stationary, charges and charge distributions. The real benefit of the electric field idea appears when charges are moving rapidly or accelerating because Coulomb's law fails to make good predictions in these cases. In general, the Maxwell equations, which we discuss in Chapter 29, describe the creation and evolution of the electric field. The Maxwell equations describe the electric field created by charge in any sort of motion, and they make different predictions from Coulomb's theory when charge is moving or accelerating. Our introduction of the electric field in simpler static situations will put us in a better position to discuss the Maxwell equations later.

Introductory Code

Listing 25-1 shows the first lines of code in the `ElectricField` module we'll write in this chapter.

```
{-# OPTIONS -Wall #-}

module ElectricField where

import SimpleVec
    ( R, Vec, (^+^), (^-^), (*^), (^*), (^/), (<.>), (><)
    , sumV, magnitude, vec, xComp, yComp, zComp, kHat )
import CoordinateSystems
    ( Position, ScalarField, VectorField
    , displacement, shiftPosition, addVectorFields
    , cart, sph, vf3D, vfPNGxy, vfGrad, origin, rVF )
import Geometry ( Curve(..), Surface(..), Volume(..) )
import Charge
    ( Charge, ChargeDistribution(..)
    , diskCap, protonOrigin, simpleDipole, lineDipole )
```

Listing 25-1: Opening lines of code for the `ElectricField` module

We use types and functions from the `SimpleVec` module of Chapter 10, the `CoordinateSystems` module of Chapter 22, the `Geometry` module of Chapter 23, and the `Charge` module of Chapter 24.

Charge Creates an Electric Field

The first part of the modern two-part view of electricity is that electric charge creates the electric field. We want to calculate the electric field created by charges of various sorts. We'll start with a point charge, the simplest charge distribution, and then move to more complex charge distributions.

Electric Field Created by a Point Charge

A particle with charge q_1 located at position $\mathbf{r}_1$ will create an electric field $\mathbf{E}$ given by the following equation:

$$\mathbf{E}(\mathbf{r}) = \frac{1}{4\pi\epsilon_0} q_1 \frac{\mathbf{r} - \mathbf{r}_1}{|\mathbf{r} - \mathbf{r}_1|^3} \tag{25.1}$$

The electric field $\mathbf{E}$ is a function from position to vectors (in other words, a vector field).

A positive point charge produces an electric field that points away from the positive charge. A negative point charge produces an electric field that points toward the negative charge.

We introduced the constant ϵ_0, called the *permittivity of free space*, in Chapter 21; it is defined as

$$\epsilon_0 = \frac{1}{\mu_0 c^2}$$

where c is the speed of light in vacuum and μ_0 is a constant called the *permeability of free space*.

```
epsilon0 :: R
epsilon0 = 1/(mu0 * cSI**2)
```

The speed of light in vacuum is exactly c = 299792458 m/s (the meter is defined to be the length that light travels in a vacuum in 1/299792458 s).

```
cSI :: R
cSI = 299792458  -- m/s
```

Prior to the 2019 SI revision, the constant μ_0 was defined to be exactly $\mu_0 = 4\pi \times 10^{-7}$ N/A^2. The 2019 revision chooses other constants to be exact, leaving μ_0 to be determined by experiment. Nevertheless, it remains very close to this value.

```
mu0 :: R
mu0 = 4e-7 * pi  -- N/A^2
```

We have ϵ_0 and μ_0 because the units in which we measure electric and magnetic fields (or voltage and current) are rooted in experiments measuring the forces between currents and charges. These experiments predate the modern electromagnetic theory based on Maxwell's equations. Keeping the conventional units in the presence of Maxwell's equations requires ϵ_0 and μ_0 to make the units work out. If you're willing to give up the conventional units of current and voltage (amperes and volts), you can write Maxwell's equations without ϵ_0 and μ_0.

The function `eFieldFromPointCharge` encodes Equation 25.1, taking the point particle's charge and position as input and producing the electric field as output.

```
eFieldFromPointCharge
    :: Charge       -- in Coulombs
    -> Position     -- of point charge (in m)
    -> VectorField  -- electric field (in V/m)
eFieldFromPointCharge q1 r1 r
    = let k = 1 / (4 * pi * epsilon0)
          d = displacement r1 r
      in (k * q1) *^ d ^/ magnitude d ** 3
```

The local name `d` stands for the displacement vector $\mathbf{r} - \mathbf{r}_1$ from the charge's position $\mathbf{r}_1$ to the *field point* $\mathbf{r}$. The field point is a position where we are looking for or talking about the electric field. No particle or matter need be present at the field point.

In Chapter 24, we wrote a type for charge distribution. Over the course of the present chapter, we will write functions to calculate the electric field produced by each sort of charge distribution. This allows us to encapsulate the idea that charge creates an electric field in the following function, which produces an electric field given any charge distribution.

```
eField :: ChargeDistribution -> VectorField
eField (PointCharge   q   r) = eFieldFromPointCharge   q   r
eField (LineCharge    lam c) = eFieldFromLineCharge    lam c
eField (SurfaceCharge sig s) = eFieldFromSurfaceCharge sig s
eField (VolumeCharge  rho v) = eFieldFromVolumeCharge  rho v
eField (MultipleCharges cds) = addVectorFields $ map eField cds
```

The function `eField` uses pattern matching on the input to treat each sort of charge distribution separately. For a point charge, it uses the function `eFieldFromPointCharge` we wrote earlier. For line, surface, and volume charges, it uses functions we'll write later in this chapter. For a combination distribution with the constructor `MultipleCharges`, it uses the *principle of superposition*, which says that the electric field produced by multiple charges is the vector sum of the electric fields produced by each individual charge. In this case, we use the function `addVectorFields` from Chapter 22 to combine the electric fields of the component distributions.

The function `eField` is an explicitly recursive function. We see this in the last line of the definition, where `eField` is defined in terms of `eField`. I have tried to avoid writing explicitly recursive functions because they are harder to understand. In this case, the explicit recursion appears only in the `MultipleCharges` clause. It means that when multiple charges are encountered, the appropriate thing to do is, first, find the electric field of each component charge (using the same `eField` function, but probably one of the other clauses), and, second, add these component electric fields together.

The electric field produced by a proton at the origin is given by the vector field `eField protonOrigin`, where `protonOrigin` is the charge distribution we wrote in Chapter 24 for a proton at the origin. Figure 25-2 shows three ways of visualizing the electric field produced by a proton.

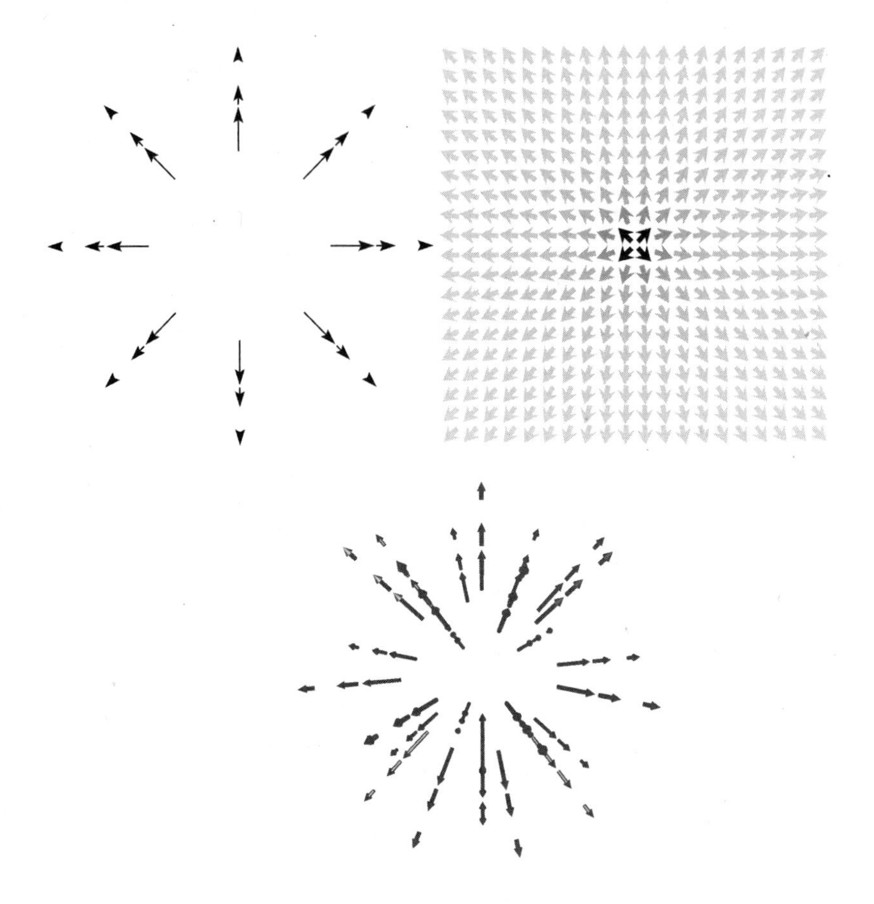

Figure 25-2: Three ways of visualizing the electric field `eField protonOrigin` *produced by a proton. The upper-left image is produced by* `eFieldPicProton2D`*, the upper-right image is produced by* `eFieldPicProtonGrad`*, and the lower picture is a screenshot of the 3D interactive image produced by* `eFieldPicProton3D`*.*

The following code produced the upper-left picture in Figure 25-2.

```
eFieldPicProton2D :: IO ()
eFieldPicProton2D
    = vfPNGxy "eFieldPicProton2D.png" 3e-9 pts (eField protonOrigin)
      where
        pts = [(r * cos th, r * sin th) | r <- [1,1.5,2]
              , th <- [0,pi/4 .. 2*pi]]
```

Sample points lie in the xy-plane, but any plane through the point charge would give the same image. Each arrow represents the electric field at the location of the arrow's tail. In this picture, electric field strength is proportional to the length of arrows displayed. Since the electric field produced by a point charge is inversely proportional to the square of the distance from the point charge, the length of these arrows would get arbitrarily long as we get closer to the point charge. Thus, we chose our sample points and scale factor so that no outrageously long arrows appear on our diagram.

The electric field produced by a proton at a point 1 m away is as follows:

$$\frac{1}{4\pi\epsilon_0}\frac{q_p}{r^2} = \frac{(9 \times 10^9 \text{ N m}^2/\text{C}^2)(1.6 \times 10^{-19} \text{ C})}{(1\text{m})^2} = 1.4 \times 10^{-9} \text{ N/C}$$

The scale factor of `3e-9` means that an electric field of 3×10^{-9} N/C is displayed with an arrow of length 1 m in the scale of the picture. The arrows closest to the proton are 1 m away. Their lengths should be about $(1.4 \times 10^{-9}$ N/C$)/(3 \times 10^{-9}$ N/C$)$, or about half, of the distance from the proton to the tail of the vector, which is true in this picture.

Now consider the following code, which produced the upper-right picture in Figure 25-2:

```
eFieldPicProtonGrad :: IO ()
eFieldPicProtonGrad
    = vfGrad (**0.2) (\(x,y) -> cart x y 0) (\v -> (xComp v, yComp v))
      "eFieldPicProtonGrad.png" 20 (eField protonOrigin)
```

Again, sample points lie in the xy-plane. Each arrow represents the electric field at the location of the arrow's center. In this picture, the electric field is stronger at the darker arrows and weaker at the lighter arrows.

Lastly, we have the following code, which produced the lower picture in Figure 25-2:

```
eFieldPicProton3D :: IO ()
eFieldPicProton3D = vf3D 4e-9
                    [sph r th ph | r  <- [1,1.5,2]
                                 , th <- [0,pi/4..pi]
                                 , ph <- [0,pi/4..2*pi]] (eField protonOrigin)
```

If you run the program `eFieldPicProton3D`, perhaps by making a standalone program and calling it `main`, a 3D vector field will pop up. You can move it around and rotate it with your mouse or pointing device.

This concludes our discussion of the electric field produced by a single point charge. Next we'll look at multiple charges.

Electric Field Created by Multiple Charges

Equation 25.1 is the fundamental equation for the electric field produced by a point charge. The principle of superposition says that the electric field produced by multiple charges is the vector sum of the electric fields produced by each charge alone. For a collection of point charges, with i labeling the particles, q_i the charge of particle i, and $\mathbf{r}_i$ the location of particle i, the electric field produced at field point $\mathbf{r}$ by the collection is

$$\mathbf{E}(\mathbf{r}) = \frac{1}{4\pi\epsilon_0}\sum_i q_i \frac{\mathbf{r} - \mathbf{r}_i}{|\mathbf{r} - \mathbf{r}_i|^3} \tag{25.2}$$

This principle of superposition is already encoded in our `eField` function. Whenever we have multiple charges in a charge distribution, they are tagged with a `MultipleCharges` data constructor, which tells the `eField` function to apply the superposition principle. We used this `MultipleCharges` data constructor in the last chapter when we wrote the `simpleDipole` charge distribution. So, although Equation 25.2 is a useful and important equation, telling us how to find the electric field from multiple point charges, we do not need to write any additional code; the superposition principle is already being applied automatically by the `eField` function.

As a first example of multiple charges, we'll look at a simple electric dipole composed of two particles: one positively charged and the other with an equal but opposite negative charge.

Electric Field of a Simple Electric Dipole

In the previous chapter, we introduced `simpleDipole`, a charge distribution composed of two oppositely charged point particles separated by some distance. The electric dipole moment of sodium chloride is 2.99×10^{-29} C m. The interatomic distance between the sodium and chlorine atoms is 2.36×10^{-10} m. If we think of NaCl as composed of two point particles, the sodium has an effective charge about 0.8 of the proton charge and the chlorine has an effective charge about −0.8 of the proton charge. The effective charge is not an integer multiple of the elementary charge because the electrons are shared between the ions. Here is a charge distribution for NaCl, viewed as a simple electric dipole:

```
simpleDipoleSodiumChloride :: ChargeDistribution
simpleDipoleSodiumChloride = simpleDipole (vec 0 0 2.99e-29) 2.36e-10
```

To find the electric field of this charge distribution, we simply use the `eField` function. Since a `simpleDipole` is a charge distribution constructed with `MultipleCharges` of two `PointCharges`, the `eField` function comes across the `MultipleCharges` constructor first and uses that clause in the definition. The `MultipleCharges` clause says to first find the electric field of each point charge, which it does by using `eField` itself, but this time with the `PointCharge` constructor. So `eField` actually gets used three times. We use it once, but then it uses itself two more times to find the electric fields of each point charge, and then adds them together and sends the result back to us.

```
eFieldSodiumChloride :: VectorField
eFieldSodiumChloride = eField simpleDipoleSodiumChloride
```

The left picture in Figure 25-3 shows the electric field produced by NaCl. The electric field points away from the positively charged sodium atom at the top of the picture and toward the negatively charged chlorine atom at the bottom of the picture. The right picture shows the electric field produced by an ideal dipole, described in the next section. We show them side by side in one figure to see what's common and what's different. The central parts of the pictures are different because charge is located in different places. The outer parts of the pictures are similar; any electric dipole produces this same electric field pattern at places a little bit away from the source.

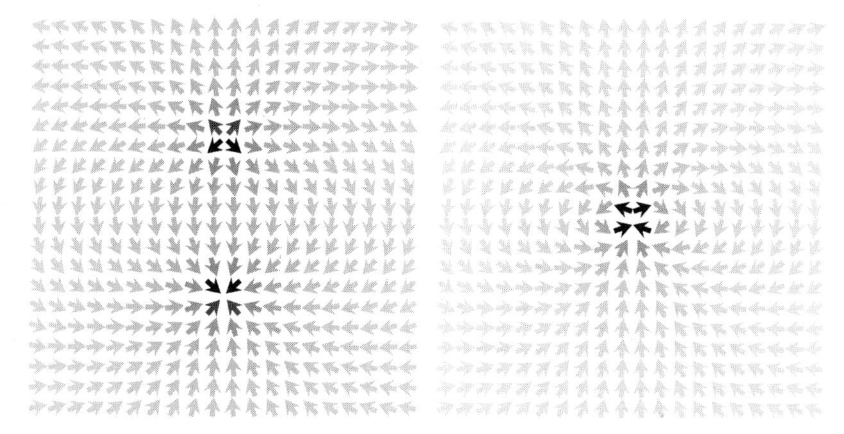

Figure 25-3: Electric field produced by a simple dipole (left) and an ideal dipole (right). The simple dipole consists of one positive point charge and one negative point charge. The left image shows the electric field `eFieldSodiumChloride` and is produced by `eFieldPicSimpleDipole`; the right image shows the electric field `eFieldIdealDipole kHat` and is produced by `eFieldPicIdealDipole`.

Here is the code that produced the picture on the left in Figure 25-3:

```
eFieldPicSimpleDipole :: IO ()
eFieldPicSimpleDipole
    = vfGrad (**0.2) (\(y,z) -> cart 0 (3e-10*y) (3e-10*z))
      (\v -> (yComp v, zComp v)) "eFieldPicSimpleDipole.png" 20
      eFieldSodiumChloride
```

The picture shows a square in the yz-plane, in which y runs from -3×10^{-10} m to 3×10^{-10} m, and z runs across the same range. This is achieved by the mapping in the second input to `vfGrad`, which scales the parameters `y` and `z` by `3e-10`. Recall that the second input of the `vfGrad` function we wrote in Chapter 22 specifies the region of interest by mapping a square from $(-1, -1)$ to $(1, 1)$ into the region we wish to visualize.

Ideal Electric Dipole

Equally but oppositely charged particles are an example of an electric dipole. An *ideal electric dipole* is a source of an electric field formed by letting the distance between these two particles approach zero while the charges increase in magnitude so that the electric dipole moment stays constant. Let's look at the electric field produced by an ideal electric dipole.

The electric field produced by an ideal electric dipole at the origin is

$$\mathbf{E}(\mathbf{r}) = \frac{1}{4\pi\epsilon_0}\frac{1}{r^3}[3(\mathbf{p}\cdot\hat{\mathbf{r}})\hat{\mathbf{r}} - \mathbf{p}] \tag{25.3}$$

where **p** is the electric dipole moment. Here it is in Haskell:

```
eFieldIdealDipole :: Vec          -- electric dipole moment
                  -> VectorField  -- electric field
eFieldIdealDipole p r
    = let k = 1 / (4 * pi * epsilon0)  -- SI units
          rMag = magnitude (rVF r)
          rUnit = rVF r ^/ rMag
      in k *^ (1 / rMag**3) *^ (3 *^ (p <.> rUnit) *^ rUnit ^-^ p)
```

The right side of Figure 25-3 shows the electric field produced by an ideal electric dipole. The magnitude of the electric dipole moment **p** is not so important in this picture because the darkest arrows are those where the electric field is largest in magnitude, regardless of what that magnitude is. Equation 25.3 shows that the electric field increases linearly with the dipole moment, so the picture would be the same for any dipole moment in the z-direction.

Comparing the two pictures in Figure 25-3, we see that the electric field differs in the centers of the pictures, close to the sources of the field. The fields around the edges of the pictures, farther from the sources, are very similar in the two pictures. It is the similarity of the fields a bit farther from the sources that entitles both sources to be called electric dipoles.

Here is the code to produce the picture on the right in Figure 25-3:

```
eFieldPicIdealDipole :: IO ()
eFieldPicIdealDipole
    = vfGrad (**0.2) (\(y,z) -> cart 0 (3e-10*y) (3e-10*z))
      (\v -> (yComp v, zComp v)) "eFieldPicIdealDipole.png" 20
                                 (eFieldIdealDipole kHat)
```

The only differences in this program compared to `eFieldPicSimpleDipole` for NaCl are the filenames and the electric fields. The electric field here is `eFieldIdealDipole kHat`, where we use the unit vector `kHat` for the electric dipole moment because its magnitude doesn't change the picture.

We've seen an example of the electric field produced by multiple point charges and compared it to that of an ideal electric dipole. Continuous charge distributions can be though of as a case of multiple charges, where there are many charges spread over some region. Before we turn to the question of finding the electric field produced by specific continuous charge distributions, there are a few general remarks we can make.

Continuous Distributions

When we move from discrete point charges to a charge that is continuously distributed, we replace the sum in Equation 25.2 with an integral. A general way to write this integral is as follows:

$$\mathbf{E}(\mathbf{r}) = \frac{1}{4\pi\epsilon_0} \int \frac{\mathbf{r} - \mathbf{r}'}{\left|\mathbf{r} - \mathbf{r}'\right|^3} \, dq' \tag{25.4}$$

In Equation 25.4, we've replaced discrete quantities with continuous ones, as indicated in Table 25-1, which shows the correspondence between discrete and continuous quantities.

Table 25-1: Correspondence Between Discrete and Continuous Quantities in the Electric Field Produced by Charge

	Discrete	Continuous
Aggregation method	$\sum_i$	$\int$
Quantity of charge	q_i	dq'
Location of charge	$\mathbf{r}_i$	$\mathbf{r}'$

The form of the integral in Equation 25.4 is so general that we have not even specified whether the charge is distributed along a 1D curve, across a 2D surface, or throughout a 3D volume. In any case, the integral is defined by a limiting process in which the quantities of charge dq' become arbitrarily small and the number of charges becomes arbitrarily large.

Sometimes we can evaluate such an integral exactly; however, more often, we need to compute an approximation to the integral. We do this computationally by turning the continuous integral back into a discrete sum, basically by turning Equation 25.4 back into Equation 25.2. The details depend on whether we are integrating over a 1D curve, a 2D surface, or a 3D volume. We take up each of these cases in turn in the sections that follow.

Electric Field Created by a Line Charge

As we discussed in the previous chapter, a line charge is specified by a curve C and a scalar field λ representing the linear charge density at any point on the curve. When charge is spread along a one-dimensional curve, a little bit of charge dq' at position $\mathbf{r}'$ is given by the product of the linear charge density $\lambda(\mathbf{r}')$ and the length dl' of a small section of the curve near $\mathbf{r}'$.

$$dq' = \lambda(\mathbf{r}')\, dl'$$

We then write the integral of Equation 25.4 as follows:

$$\mathbf{E}(\mathbf{r}) = \frac{1}{4\pi\epsilon_0} \int_C \frac{\mathbf{r} - \mathbf{r}'}{\left|\mathbf{r} - \mathbf{r}'\right|^3} \lambda(\mathbf{r}')\, dl' \qquad (25.5)$$

Such an integral of a vector field over a curve is called a *vector line integral*. Let's explain the vector line integral in more detail.

Vector Line Integral

The vector line integral takes a vector field and a curve as input and returns a vector as output.

```
type VectorLineIntegral = VectorField -> Curve -> Vec
```

The vector line integral of a vector field **F** over a curve C is written

$$\int_C \mathbf{F}(\mathbf{r}')\, dl'$$

What does this integral mean? The integral is defined by dividing the curve C into many small segments. The vector field **F** is evaluated at each point $\mathbf{r}_i$ on or near the segment $\Delta\mathbf{l}_i$ and scaled (multiplied) by the length Δl_i of the segment. We then add up these vectors to form the sum

$$\sum_i \mathbf{F}(\mathbf{r}_i)\Delta l_i$$

The integral is the limit of this vector sum as the lengths of the segments approach 0 and the number of segments becomes arbitrarily large. The definition of the integral involves careful specification of the limiting process, which we'll leave to textbooks on vector calculus.

The integral is not only defined by, but also approximated by, a finite sum.

$$\int_C \mathbf{F}(\mathbf{r}')\, dl' \approx \sum_i \mathbf{F}(\mathbf{r}_i)\Delta l_i \tag{25.6}$$

Our approximate calculation of the integral will use a finite number of segments. We will need a method to approximate a curve by a finite list of segments. We'll represent a segment $\Delta\mathbf{l}_i$ as a short displacement vector lying along the curve. In addition to the displacement vector $\Delta\mathbf{l}_i$, which describes the length and orientation of the segment, we also need a position $\mathbf{r}_i$ indicating where the segment is located on the curve. An approximation to a curve consists of a list of pairs of positions and displacement vectors. Notice that the lengths of the segments need not be the same. A curve approximation method is a function that returns such a list when given a curve.

```
type CurveApprox = Curve -> [(Position,Vec)]
```

A pair looks like $(\mathbf{r}_i, \Delta\mathbf{l}_i)$. We could use many curve approximation methods; we'll delay discussion of a curve approximation method to the final section of this chapter.

Here's the Haskell definition of a vector line integral:

```
vectorLineIntegral :: CurveApprox -> VectorField -> Curve -> Vec
vectorLineIntegral approx vF c
    = sumV [vF r' ^* magnitude dl' | (r',dl') <- approx c]
```

The curve `c` is approximated by the function `approx`, which is provided as an input. The function `approx` will specify the number of segments into which the curve is to be divided, as well as the method for the division and determination of the associated positions of the segments. For each segment `dl'` at position `r'`, the vector field `vF` is evaluated at position `r'` and scaled by the magnitude of `dl'`. We then sum these vectors to give the approximation to the integral. Table 25-2 shows a correspondence between mathematical notation and Haskell notation in the definition of the vector line integral.

Table 25-2: Correspondence Among Continuous Mathematical Notation, Discrete Mathematical Notation, and Haskell Notation for the Vector Line Integral

Continuous math	Discrete math	Haskell
$\int$	$\sum_i$	`sumV` ❶
$\mathbf{r}'$	$\mathbf{r}_i$	`r'` ❷
$\mathbf{F}$	$\mathbf{F}$	`vF`
$\mathbf{F}(\mathbf{r}')$	$\mathbf{F}(\mathbf{r}_i)$	`vF r'`
C		`c` ❸
$d\mathbf{l}'$	$\Delta\mathbf{l}_i$	`dl'` ❹
dl'	Δl_i	`magnitude dl'`
$\mathbf{F}(\mathbf{r}')\,dl'$	$\mathbf{F}(\mathbf{r}_i)\Delta l_i$	`vF r' ^* magnitude dl'`

Surprisingly, Haskell notation is closer to discrete notation only in the summation ❶. The Haskell names we chose for segment location ❷, curve ❸, and segment displacement ❹ are closer to the continuous notation. This is because Haskell does not require the index i used in the discrete notation, so there is no reason to introduce it. Haskell computes the integral in a discrete way, but Haskell's list syntax avoids the need to talk about the index i that would number the list elements.

Back to Electric Field

From Equation 25.5, we see that the vector field **F** we want to integrate to find the electric field produced by a line charge is the function that maps the position $\mathbf{r}'$ of a piece of charge to the vector

$$\lambda(\mathbf{r}')\frac{\mathbf{r}-\mathbf{r}'}{|\mathbf{r}-\mathbf{r}'|^3}$$

in which **r** is the field point, the fixed position where we want to know the electric field. In the function `eFieldFromLineCharge` below, this function is given the local name `integrand`. The local name `d` stands for the displacement $\mathbf{r}-\mathbf{r}'$ from the source point $\mathbf{r}'$ to the field point **r**. Because the source point `r'` is introduced as a name local to the `integrand` function, we must define `d` with a `where` clause, rather than alongside the definitions of `k` and `integrand`.

```
eFieldFromLineCharge
    :: ScalarField  -- linear charge density lambda
    -> Curve        -- geometry of the line charge
    -> VectorField  -- electric field (in V/m)
eFieldFromLineCharge lambda c r
    = let k = 1 / (4 * pi * epsilon0)
          integrand r' = lambda r' *^ d ^/ magnitude d ** 3
              where d = displacement r' r
      in k *^ vectorLineIntegral (curveSample 1000) integrand c
```

We need to give only two items to find the electric field produced by a line charge: the linear charge density, expressed as a scalar field, and the curve that describes the geometry of the line charge. The type signature of the function `eFieldFromLineCharge` makes it clear that the electric field depends only on these two items. We use the `vectorLineIntegral` we defined earlier. The curve approximation method `curveSample 1000` divides the curve into 1,000 segments and is defined later in the chapter.

Now that we've shown how to find the electric field for a line charge, let's look at the electric field produced by the line dipole `lineDipole` we discussed in the previous chapter.

Example of a Line Dipole

In the previous chapter, we introduced `lineDipole`, a line charge with a linearly varying charge density. Suppose we had reason to believe that the charge distribution of NaCl looked more like that of a line dipole than a simple dipole. I have no such evidence, and the charge distribution of NaCl is probably complicated, but it's certainly reasonable to imagine that the charge density varies smoothly, if not linearly, from sodium to chlorine. To model NaCl as a line dipole, we can use the same electric dipole moment and interatomic separation we used earlier for the simple dipole. Here is a charge distribution for NaCl, viewed as a line dipole:

```
lineDipoleSodiumChloride :: ChargeDistribution
lineDipoleSodiumChloride = lineDipole (vec 0 0 2.99e-29) 2.36e-10
```

We can find the electric field of this charge distribution with the `eField` function, which in this case will use the `eFieldFromLineCharge` function we wrote earlier.

```
eFieldLineDipole :: VectorField
eFieldLineDipole = eField lineDipoleSodiumChloride
```

Exercise 25.11 asks you to make a vector field picture for this electric field, similar to the ones we made earlier for the simple dipole and the ideal dipole.

Now that we've discussed how to find the electric field from our first continuous charge distribution, that of a line charge, and shown an example, let's turn to our second continuous charge distribution, that of a surface charge.

Electric Field Created by a Surface Charge

As we discussed in the previous chapter, a surface charge is specified by a surface S and a scalar field σ representing the surface charge density at any point on the surface. When charge is spread along a 2D surface, a little bit of charge dq' at position $\mathbf{r}'$ is given by the product of the surface charge density $\sigma(\mathbf{r}')$ and the area da' of a small patch of the surface near $\mathbf{r}'$.

$$dq' = \sigma(\mathbf{r}')\, da'$$

We then write the integral of Equation 25.4 as follows:

$$\mathbf{E}(\mathbf{r}) = \frac{1}{4\pi\epsilon_0} \int_S \frac{\mathbf{r} - \mathbf{r}'}{\left|\mathbf{r} - \mathbf{r}'\right|^3} \sigma(\mathbf{r}')\, da' \tag{25.7}$$

Such an integral of a vector field over a surface is called a *vector surface integral*. Let's explain the vector surface integral in more detail.

Vector Surface Integral

The vector surface integral takes a vector field and a surface as input and returns a vector as output.

```
type VectorSurfaceIntegral = VectorField -> Surface -> Vec
```

The vector surface integral of a vector field $\mathbf{F}$ over a surface S is written as follows:

$$\int_S \mathbf{F}(\mathbf{r}')\, da'$$

We define the integral by dividing the surface S into many small patches. You can think of a patch as a quadrilateral, but what's really important is the location, area, and orientation of the patch rather than its shape. Each patch $\Delta\mathbf{a}_i$ is a vector area whose magnitude gives the area of the patch and whose direction points perpendicular to the patch. We assume each patch is small enough to be considered flat. Because our surfaces are oriented, as we discussed in Chapter 23, the direction for the patch is unambiguous. The vector field $\mathbf{F}$ is evaluated at a point $\mathbf{r}_i$ on or near the patch $\Delta\mathbf{a}_i$ and scaled by the area Δa_i of the patch. We then add up these vectors to form the sum

$$\sum_i \mathbf{F}(\mathbf{r}_i)\Delta a_i$$

The integral is the limit of this vector sum as the areas of the patches approach 0 and the number of patches becomes arbitrarily large.

The integral is not only defined by, but also approximated by, a finite sum.

$$\int_S \mathbf{F}(\mathbf{r}')\, da' \approx \sum_i \mathbf{F}(\mathbf{r}_i)\Delta a_i$$

Our approximate calculation of the integral will use a finite number of patches. We will need a method to divide a surface into a list of patches. We'll represent an oriented patch $\Delta\mathbf{a}_i$ as a vector perpendicular to the surface. In addition to the vector $\Delta\mathbf{a}_i$, which describes the area and orientation of the patch, we also need a position $\mathbf{r}_i$ indicating where the patch is located on the surface. An approximation to a surface consists of a list of pairs of positions and area vectors.

```
type SurfaceApprox = Surface -> [(Position,Vec)]
```

A pair looks like $(\mathbf{r}_i, \Delta\mathbf{a}_i)$. There are many ways the surface approximation could be done, and we'll delay discussion of this point to the last section of this chapter.

Here's the Haskell definition of a vector surface integral:

```
vectorSurfaceIntegral :: SurfaceApprox -> VectorField -> Surface -> Vec
vectorSurfaceIntegral approx vF s
    = sumV [vF r' ^* magnitude da' | (r',da') <- approx s]
```

The surface s is approximated by the function approx, which is provided as an input. The function approx will specify the number of patches into which the surface is to be divided, as well as the method for the division and determination of the associated positions of the patches. For each patch da' at position r', the vector field vF is evaluated at position r' and scaled by the magnitude of da'. We then sum these vectors to give the approximation to the integral. Table 25-3 shows a correspondence between mathematical notation and Haskell notation in the definition of the vector surface integral.

Table 25-3: Correspondence Among Continuous Mathematical Notation, Discrete Mathematical Notation, and Haskell Notation for the Vector Surface Integral

Continuous math	Discrete math	Haskell
$\int$	$\sum_i$	`sumV`
$\mathbf{r}'$	$\mathbf{r}_i$	`r'`
$\mathbf{F}$	$\mathbf{F}$	`vF`
$\mathbf{F}(\mathbf{r}')$	$\mathbf{F}(\mathbf{r}_i)$	`vF r'`
S		`s`
$d\mathbf{a}'$	$\Delta\mathbf{a}_i$	`da'`
da'	Δa_i	`magnitude da'`
$\mathbf{F}(\mathbf{r}')\,da'$	$\mathbf{F}(\mathbf{r}_i)\Delta a_i$	`vF r' ^* magnitude da'`

The first four rows of Table 25-3 are identical to those of Table 25-2 for the vector line integral because they refer to the vector field and the integration. The last four rows of this table are analogous to those of the previous table, with surface patches substituting for line segments.

Back to Electric Field

From Equation 25.7, we see that the vector field $\mathbf{F}$ we want to integrate to find the electric field produced by a line charge is the function that maps the position $\mathbf{r}'$ of a piece of charge to the vector

$$\sigma(\mathbf{r}')\frac{\mathbf{r}-\mathbf{r}'}{\left|\mathbf{r}-\mathbf{r}'\right|^3}$$

Here we consider $\mathbf{r}$ the fixed position where we want to know the electric field. In the function eFieldFromSurfaceCharge next, which is a Haskell translation of Equation 25.7, we give this function the local name integrand. The local name d stands for the displacement $\mathbf{r} - \mathbf{r}'$ from the charge to the field point. Because we introduce the source point `r'` as a name local to the `integrand` function, we must define `d` with a `where` clause rather than alongside the definitions of `k` and `integrand`.

```
eFieldFromSurfaceCharge
    :: ScalarField  -- surface charge density sigma
    -> Surface      -- geometry of the surface charge
    -> VectorField  -- electric field (in V/m)
eFieldFromSurfaceCharge sigma s r
    = let k = 1 / (4 * pi * epsilon0)
          integrand r' = sigma r' *^ d ^/ magnitude d ** 3
              where d = displacement r' r
      in k *^ vectorSurfaceIntegral (surfaceSample 200) integrand s
```

We need to give only two items to find the electric field produced by a surface charge: the surface charge density σ, expressed as a scalar field, and the surface that describes the geometry of the surface charge. The type signature of `eFieldFromSurfaceCharge` makes it clear that the electric field depends only on these two items. We use the `vectorSurfaceIntegral` we defined earlier. The function `surfaceSample 200` divides the curve into $2(200)^2 = 80{,}000$ patches and is defined later in the chapter.

Now that we've shown how to find the electric field for a surface charge, let's look at the electric field produced by the capacitor `diskCap` we discussed in the previous chapter.

Example of a Capacitor

Let's find the electric field produced by a parallel-plate capacitor whose plates are uniformly charged. The assumption of uniform charge is good when the plates are close together and worsens as the plates are moved farther apart. Suppose we have a capacitor with a plate separation of 4 cm in which the plates are disks with radius 5 cm. The positive plate has a surface charge density of 20 nC/m^2, and the negative plate has a surface charge density of −20 nC/m^2. The expression

```
diskCap 0.05 0.04 2e-8 :: ChargeDistribution
```

represents this charge distribution.

We can find the electric field with the `eField` function.

```
eFieldDiskCap :: VectorField
eFieldDiskCap = eField $ diskCap 0.05 0.04 2e-8
```

In this case, `eField` uses the function `eFieldFromSurfaceCharge` we defined earlier. To start, let's look at the electric field at the center of the capacitor.

```
Prelude> :l ElectricField
[ 1 of 12] Compiling Newton2          ( Newton2.hs, interpreted )
[ 2 of 12] Compiling Mechanics1D      ( Mechanics1D.hs, interpreted )
[ 3 of 12] Compiling SimpleVec        ( SimpleVec.hs, interpreted )
```

```
[ 4 of 12] Compiling Mechanics3D       ( Mechanics3D.hs, interpreted )
[ 5 of 12] Compiling MultipleObjects   ( MultipleObjects.hs, interpreted )
[ 6 of 12] Compiling MOExamples        ( MOExamples.hs, interpreted )
[ 7 of 12] Compiling Electricity       ( Electricity.hs, interpreted )
[ 8 of 12] Compiling CoordinateSystems ( CoordinateSystems.hs, interpreted )
[ 9 of 12] Compiling Geometry          ( Geometry.hs, interpreted )
[10 of 12] Compiling VectorIntegrals   ( VectorIntegrals.hs, interpreted )
[11 of 12] Compiling Charge            ( Charge.hs, interpreted )
[12 of 12] Compiling ElectricField     ( ElectricField.hs, interpreted )
Ok, 12 modules loaded.
*ElectricField> eFieldDiskCap (cart 0 0 0)
vec 0.0 0.0 (-1419.9046806406095)
```

The electric field is in the negative z-direction, pointing from the positive plate above toward the negative plate below.

For comparison, a parallel-plate capacitor with infinitely wide plates and surface charge density σ has electric field σ/ϵ_0. Physicists like the ideal parallel-plate capacitor with infinitely wide plates because there is a simple expression for the electric field it produces. The electric field outside the ideal capacitor is 0. The electric field inside the ideal capacitor (between the two plates) has a uniform value of σ/ϵ_0 and points from the positive plate toward the negative plate. The electric field produced is independent of the plate separation. For a surface charge density of $\sigma = 20 \text{ nC/m}^2$, the electric field magnitude would be

$$\frac{\sigma}{\epsilon_0} = \frac{2 \times 10^{-8} \text{ C/m}^2}{8.85 \times 10^{-12} \text{ C}^2/\text{N m}^2} = 2259 \text{ N/C}$$

The value we found earlier is less than this ideal value (in magnitude) because the radii of our plates are rather modest.

Figure 25-4 shows the electric field produced by our disk capacitor.

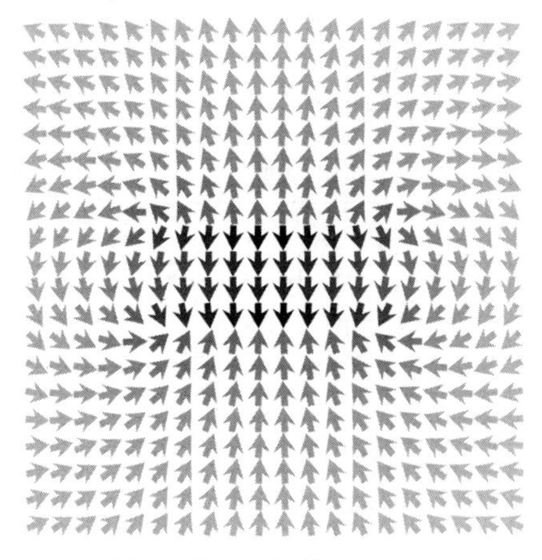

Figure 25-4: Electric field `eFieldDiskCap` produced by a parallel-plate capacitor. The plates have uniform surface charge density. (Image produced by `eFieldPicDiskCap`.)

Figure 25-4 shows the region from −10 cm to 10 cm in the x-direction, and from −10 cm to 10 cm in the z-direction. The radii of the disks are 5 cm, so they extend horizontally over half the width of the figure. The electric field is largest in magnitude where the arrows are darkest, between the plates. The field is smaller outside the plates, but it's not 0. The electric field appears to be fairly uniform between the plates. We see from the shading how the electric field transitions from its maximum between the plates, to moderate values near the edges of the disks, to minimal values farther from the plates. Around the outside of the picture, the electric field looks like that of an electric dipole, which is not surprising since the capacitor is an electric dipole consisting of a positive plate and a negative plate.

Here's the code that produced Figure 25-4:

```
eFieldPicDiskCap :: IO ()
eFieldPicDiskCap = vfGrad (**0.2) (\(x,z) -> cart (0.1*x) 0 (0.1*z))
                   (\v -> (xComp v, zComp v)) "eFieldPicDiskCap.png" 20
                   eFieldDiskCap
```

The program might take several minutes to run. The surface and volume charge integrals involve many computations and can be slow. The methods used here are ones that are conceptually simple rather than numerically efficient. The slowness is mostly a consequence of the simple approach and is not inherent in the use of the Haskell language. There are data structures, such as unboxed vectors, that would speed up many of the operations here, at the expense of making the code less concise.

We've talked about line charges and surface charges. Let's turn to the third and last of our continuous charge distributions: the volume charge.

Electric Field Created by a Volume Charge

As we discussed in the previous chapter, a volume charge is specified by a volume V and a scalar field ρ representing the volume charge density at any point in the volume. When charge is spread throughout a 3D volume, a little bit of charge dq' at position $\mathbf{r}'$ is given by the product of the volume charge density $\rho(\mathbf{r}')$ and the volume dv' of a small portion of the volume near $\mathbf{r}'$.

$$dq' = \rho(\mathbf{r}')\, dv'$$

We then write the integral of Equation 25.4 as follows:

$$\mathbf{E}(\mathbf{r}) = \frac{1}{4\pi\epsilon_0} \int_V \frac{\mathbf{r} - \mathbf{r}'}{|\mathbf{r} - \mathbf{r}'|^3} \rho(\mathbf{r}')\, dv' \qquad (25.8)$$

Such an integral of a vector field over a volume is called a *vector volume integral*. Let's explore the vector volume integral in more detail.

Vector Volume Integral

The vector volume integral takes a vector field and a volume as input and returns a vector as output.

```
type VectorVolumeIntegral = VectorField -> Volume -> Vec
```

The vector volume integral of a vector field **F** over a volume *V* is written as follows:

$$\int_V \mathbf{F}(\mathbf{r}')\, dv'$$

The integral is defined by dividing the volume *V* into many small portions. The vector field **F** is evaluated at each point $\mathbf{r}_i$ on or near the portion and scaled by the volume Δv_i of the portion. We then add up these vectors to form the sum

$$\sum_i \mathbf{F}(\mathbf{r}_i)\Delta v_i.$$

The integral is the limit of this vector sum as the volumes of the portions approach 0 and the number of portions becomes arbitrarily large.

The integral is both defined by and approximated by a finite sum:

$$\int_V \mathbf{F}(\mathbf{r}')\, dv' \approx \sum_i \mathbf{F}(\mathbf{r}_i)\Delta v_i$$

Our approximate calculation of the integral will use a finite number of portions. We'll represent a portion by its location $\mathbf{r}_i$ and its volume Δv_i. An approximation to a volume consists of a list of pairs of positions and portion volumes. A volume approximation method is a function that returns such a list when given a volume.

```
type VolumeApprox = Volume -> [(Position,R)]
```

A pair looks like $(\mathbf{r}_i, \Delta v_i)$. Many volume approximation methods could be used, and, as before, we'll delay discussion of this point until later in the chapter.

Here's the Haskell definition of a vector volume integral:

```
vectorVolumeIntegral :: VolumeApprox -> VectorField -> Volume -> Vec
vectorVolumeIntegral approx vF vol
    = sumV [vF r' ^* dv' | (r',dv') <- approx vol]
```

The volume `vol` is approximated by the function `approx`, giving a list of portion locations and volumes. For each portion `(r',dv')`, the vector field `vF` is evaluated at position `r'` and scaled by `dv'`. We then sum these vectors to give the approximation to the integral.

Table 25-4 shows a correspondence between mathematical notation and Haskell notation in the definition of the vector volume integral.

Table 25-4: Correspondence Among Continuous Mathematical Notation, Discrete Mathematical Notation, and Haskell Notation for the Vector Volume Integral

Continuous math	Discrete math	Haskell
V		`vol`
dv'	Δv_i	`dv'`
$\mathbf{F}(\mathbf{r}')\,dv'$	$\mathbf{F}(\mathbf{r}_i)\Delta v_i$	`vF r' ^* dv'`

This table is analogous to the last few rows of Table 25-3. One important difference between the two is that each patch for a surface integral is a vector, while each bit of volume for a volume integral is a scalar.

Back to Electric Field

We need to give only two items to find the electric field produced by a volume charge: the volume charge density ρ, expressed as a scalar field, and the volume that describes the geometry of the charge. The type signature of the following function, which is a Haskell translation of Equation 25.8, makes it clear that the electric field depends only on these two items:

```
eFieldFromVolumeCharge
    :: ScalarField  -- volume charge density rho
    -> Volume       -- geometry of the volume charge
    -> VectorField  -- electric field (in V/m)
eFieldFromVolumeCharge rho v r
    = let k = 1 / (4 * pi * epsilon0)
          integrand r' = rho r' *^ d ^/ magnitude d ** 3
              where d = displacement r' r
      in k *^ vectorVolumeIntegral (volumeSample 50) integrand v
```

Now that we have functions to calculate the electric field produced by a line charge, a surface charge, and a volume charge, we have completed the definition of `eField` we began earlier in the chapter. We now have a method to find the electric field produced by any charge distribution. In writing functions to calculate the electric field, we spent some time discussing three types of vector integrals: the vector line integral for line charges, the vector surface integral for surface charges, and the vector volume integral for volume charges. With these integrals fresh in our minds, now is a good time to extend our integral methods to scalar integrals, which we used in the last chapter to calculate total charge.

Scalar Integrals

In the process of calculating the electric field produced by a line charge, a surface charge, or a volume charge, we introduced the vector line integral, the vector surface integral, and the vector volume integral. We used these integrals to add up the vector contributions to the electric field from each piece of charge. There are situations in which we need to add up scalar contributions to some quantity that come from sources on a curve, a surface, or a volume. This is the purpose of the scalar line integral, scalar surface integral, and scalar volume integral. Now that we have gone through the details for the vector integrals, it will be relatively easy to understand the scalar integrals.

Scalar Line Integral

The scalar line integral takes a scalar field f and a curve C as input and returns a scalar as output.

```
type ScalarLineIntegral = ScalarField -> Curve -> R
```

The integral is defined, and also approximated, by dividing the curve C into many small segments, exactly as we did for the vector line integral. The scalar field f is evaluated at each point $\mathbf{r}_i$ at the segment $\Delta\mathbf{l}_i$, multiplied by the length Δl_i of the segment, and then added.

$$\int_C f(\mathbf{r}')\, dl' \approx \sum_i f(\mathbf{r}_i)\Delta l_i \tag{25.9}$$

Here is the Haskell definition of a scalar line integral:

```
scalarLineIntegral :: CurveApprox -> ScalarField -> Curve -> R
scalarLineIntegral approx f c
    = sum [f r' * magnitude dl' | (r',dl') <- approx c]
```

The curve `c` is approximated by the function `approx`, which is provided as an input. For each segment `dl'` at position `r'`, the scalar field `f` is evaluated at position `r'` and multiplied by the magnitude of `dl'`. We then sum these numbers to give the approximation to the integral.

Scalar Surface Integral

The scalar surface integral takes a scalar field f and a surface S as input and returns a scalar as output.

```
type ScalarSurfaceIntegral = ScalarField -> Surface -> R
```

The integral is defined, and also approximated, by dividing the surface S into many small patches, exactly as we did for the vector surface integral. The scalar field f is evaluated at each point $\mathbf{r}_i$ at the patch $\Delta\mathbf{a}_i$, multiplied by the magnitude Δa_i of the patch, and then added.

$$\int_S f(\mathbf{r}')\,da' \approx \sum_i f(\mathbf{r}_i)\Delta a_i \tag{25.10}$$

Here is the Haskell definition of a scalar surface integral:

```
scalarSurfaceIntegral :: SurfaceApprox -> ScalarField -> Surface -> R
scalarSurfaceIntegral approx f s
    = sum [f r' * magnitude da' | (r',da') <- approx s]
```

The surface `s` is approximated by the function `approx`. For each patch `da'` at position `r'`, the scalar field `f` is evaluated at position `r'` and multiplied by the magnitude of `da'`. We then sum these numbers to give the approximation to the integral.

Scalar Volume Integral

The scalar volume integral takes a scalar field f and a volume V as input and returns a scalar as output.

```
type ScalarVolumeIntegral = ScalarField -> Volume -> R
```

The integral is defined, and also approximated, by dividing the volume V into many small portions, exactly as we did for the vector volume integral. The scalar field f is evaluated at each portion, multiplied by the volume Δv_i of the portion, and then added.

$$\int_V f(\mathbf{r}')\,dv' \approx \sum_i f(\mathbf{r}_i)\Delta v_i \tag{25.11}$$

Here is the Haskell definition of a scalar volume integral:

```
scalarVolumeIntegral :: VolumeApprox -> ScalarField -> Volume -> R
scalarVolumeIntegral approx f vol
    = sum [f r' * dv' | (r',dv') <- approx vol]
```

The volume `vol` is approximated by the function `approx`. For each portion `dv'` at position `r'`, the scalar field `f` is evaluated at position `r'` and multiplied by `dv'`. We then sum numbers to give the approximation to the integral.

Before we leave this chapter, there is one detail left to discuss, and that's the method we'll use to approximate curves, surfaces, and volumes when we do integrals over these shapes. Let's turn to that detail now.

Approximating Curves, Surfaces, and Volumes

We've seen multiple situations in which we want to add things up over a curve, surface, or volume by doing a line integral, surface integral, or volume integral. When the things we want to add up are vectors, we use a vector line integral, vector surface integral, or vector volume integral. Similarly, when the things we want to add up are scalars, we use a scalar line integral, scalar surface integral, or scalar volume integral. Whether we are adding scalars or vectors, our methods require us to approximate curves, surfaces, and volumes as finite lists of data. This approximation is the subject of this section; although there are many ways this approximation can be done, we'll give only one method for approximating each geometrical object. The subject of numerical analysis is concerned with exploring different ways to make these approximations, examining the trade-offs involved, and doing it cleverly and efficiently. In our case, we're concerned with doing it simply and understandably.

Approximating a Curve

Recall that a curve approximation is a way of turning a curve into a list of locations and displacement vectors.

```
-- introduced earlier in the Chapter
type CurveApprox = Curve -> [(Position,Vec)]
```

Our function curveSample approximates a curve as a number of segments, returning a list of segment positions and displacement vectors.

```
curveSample :: Int -> Curve -> [(Position,Vec)]
curveSample n c
    = let segCent :: Segment -> Position
          segCent (p1,p2) = shiftPosition ((rVF p1 ^+^ rVF p2) ^/ 2) origin
          segDisp :: Segment -> Vec
          segDisp = uncurry displacement
      in [(segCent seg, segDisp seg) | seg <- segments n c]
```

The function takes an integer n that controls the number of segments generated, each segment being a pair of starting position and ending position.

```
type Segment = (Position,Position)
```

Most of the work is done by the function segments, defined next, which returns a list of segments when given the integer n and the curve c. The local function segCent finds the center of each segment. The rVF vector field, introduced in Chapter 22, converts positions to displacement vectors, which the shiftPosition function then averages and converts back to positions.

The local function segDisp then computes the displacement vector of each segment. The displacement vector points from the beginning position of a segment to the ending position of the segment. The function segDisp is the uncurried version of displacement from Chapter 22, taking the two positions as a pair rather than a curried function.

We defined segCent and segDisp locally because they're not used in any other functions. Note that we can still give them type signatures if we wish, although this is not required. Special-purpose functions like these are nice to define locally because it keeps fewer items in the global namespace and helps the reader of the code understand the relationship between the local function segCent and its parent function curveSample. The fact that segCent is local informs the reader that this function is not used elsewhere.

The function segments returns a list of segments when given a curve.

```
segments :: Int -> Curve -> [Segment]
segments n (Curve g a b)
    = let ps = map g $ linSpaced n a b
      in zip ps (tail ps)
```

Note the first argument to segments is an integer n that controls how many segments will be produced. We pass in the curve Curve g a b using pattern matching on the input because this function needs to refer to the curve's parameter limits a and b, as well as to the parameterizing function g. The function begins by breaking the curve's parameter interval from a to b into n equal subintervals using the function linSpaced defined below. At each of the n+1 endpoints of these subintervals, we apply the function g to form a list ps of n+1 positions along the curve. We then zip the list ps with its tail to produce the desired list of n segments. Zipping a list with its tail pairs the first and second items, the second and third items, the third and fourth items, and so on.

The function linSpaced returns a list of numbers that are linearly spaced.

```
linSpaced :: Int -> R -> R -> [R]
linSpaced n x0 x1 = take (n+1) [x0, x0+dx .. x1]
    where dx = (x1 - x0) / fromIntegral n
```

The input n is the number of intervals, so the function returns a list of n+1 numbers, starting at x0 and going up to and including x1. Since n has type Int and x1 - x0 has type R, we need the fromIntegral function to convert n to type R before we do the division. We use the take function to treat the case in which the initial value x0 and the final value x1 are the same, in which case dx is 0 and the arithmetic sequence is an infinite list of the same number. The take function returns only the first n+1 items of the infinite list.

Here are two examples of the use of linSpaced:

```
*ElectricField> :l ElectricField
[ 1 of 12] Compiling Newton2            ( Newton2.hs, interpreted )
[ 2 of 12] Compiling Mechanics1D        ( Mechanics1D.hs, interpreted )
[ 3 of 12] Compiling SimpleVec          ( SimpleVec.hs, interpreted )
[ 4 of 12] Compiling Mechanics3D        ( Mechanics3D.hs, interpreted )
[ 5 of 12] Compiling MultipleObjects    ( MultipleObjects.hs, interpreted )
[ 6 of 12] Compiling MOExamples         ( MOExamples.hs, interpreted )
[ 7 of 12] Compiling Electricity        ( Electricity.hs, interpreted )
[ 8 of 12] Compiling CoordinateSystems ( CoordinateSystems.hs, interpreted )
[ 9 of 12] Compiling Geometry           ( Geometry.hs, interpreted )
[10 of 12] Compiling VectorIntegrals    ( VectorIntegrals.hs, interpreted )
[11 of 12] Compiling Charge             ( Charge.hs, interpreted )
[12 of 12] Compiling ElectricField      ( ElectricField.hs, interpreted )
Ok, 12 modules loaded.
```

```
*ElectricField> linSpaced 4 0 2
[0.0,0.5,1.0,1.5,2.0]
*ElectricField> linSpaced 4 3 3
[3.0,3.0,3.0,3.0,3.0]
```

Now that we've explored one way to approximate a curve, let's do the same for a surface.

Approximating a Surface

Recall that a surface approximation is a way of turning a surface into a list of locations and vector areas.

```
-- introduced earlier in the Chapter
type SurfaceApprox = Surface -> [(Position,Vec)]
```

Our function `surfaceSample` approximates a surface as a number of triangles, returning a list of triangle positions and vector areas.

```
surfaceSample :: Int -> Surface -> [(Position,Vec)]
surfaceSample n s = [(triCenter tri, triArea tri) | tri <- triangles n s]
```

The function takes an integer `n` that controls the number of triangles generated. Most of the work is done by the function `triangles`, which returns a list of triangles when given the integer `n` and the surface `s`. The function `triCenter` finds the center of each triangle, and the function `triArea` computes the vector area of each triangle. These two functions are defined below.

A triangle is described by specifying the positions of its three vertices.

```
data Triangle = Tri Position Position Position
```

The triangles have an orientation, so the order in which we specify the vertices is important. If we are looking at a triangle from a place where the positions `p1`, `p2`, and `p3` of the vertices occur in a counterclockwise order, the orientation points from the triangle toward our viewing location, perpendicular to the triangle surface. The triangles `Tri p1 p2 p3`, `Tri p2 p3 p1`, and `Tri p3 p1 p2` all represent the same triangle with the same orientation, but the triangles `Tri p1 p3 p2`, `Tri p2 p1 p3`, and `Tri p3 p2 p1` represent triangles with the same vertices but an opposite orientation.

We find the center of a triangle by averaging the displacement vectors for the three vertices.

```
triCenter :: Triangle -> Position
triCenter (Tri p1 p2 p3)
    = shiftPosition ((rVF p1 ^+^ rVF p2 ^+^ rVF p3) ^/ 3) origin
```

We convert positions to displacement vectors with the `rVF` vector field, introduced in Chapter 22. We then average them and convert them back to a position with the `shiftPosition` function.

The vector area of a triangle is half the cross product of two of its vector edges. Since we care about the orientation of these triangles, we need to be careful about which way we do the cross product.

```
triArea :: Triangle -> Vec  -- vector area
triArea (Tri p1 p2 p3) = 0.5 *^ (displacement p1 p2 >< displacement p2 p3)
```

The function triangles returns a list of triangles when given a surface.

```
triangles :: Int -> Surface -> [Triangle]
triangles n (Surface g sl su tl tu)
    = let sts = [[(s,t) | t <- linSpaced n (tl s) (tu s)]
                        | s <- linSpaced n sl su]
          stSquares = [( sts !! j     !! k
                       , sts !! (j+1) !! k
                       , sts !! (j+1) !! (k+1)
                       , sts !! j     !! (k+1))
                      | j <- [0..n-1], k <- [0..n-1]]
          twoTriangles (pp1,pp2,pp3,pp4)
              = [Tri (g pp1) (g pp2) (g pp3),Tri (g pp1) (g pp3) (g pp4)]
      in concatMap twoTriangles stSquares
```

Note the first argument to triangles is an integer n that controls how many triangles will be produced. We pass in the surface Surface g sl su tl tu using pattern matching on the input because this function needs to refer to the surface's parameter limits sl and su, as well as all of the other attributes of the surface.

The function begins by breaking the surface's parameter interval from sl to su into n equal subintervals. At each of the n+1 endpoints of these subintervals, we break the parameter interval from tl s to tu s into n equal subintervals, where s is the *s* parameter value at each subinterval endpoint. The local variable sts :: [[(R,R)]] is a list of lists, which can be thought of as an n+1-by-n+1 matrix of parameter pairs corresponding to points on the surface. The local variable stSquares :: [((R,R),(R,R),(R,R),(R,R))] is a list of n^2 "squares" of parameter pairs. Each of these squares is turned into two triangles by the local function twoTriangles. The function triangles returns a list of 2*n^2 triangles approximating the surface.

Now that we've shown one way to approximate a surface, let's turn to the question of approximating a volume.

Approximating a Volume

A volume approximation is a way of turning a volume into a list of locations and numerical volumes.

```
-- introduced earlier in the Chapter
type VolumeApprox = Volume -> [(Position,R)]
```

Our function volumeSample approximates a volume by a number of tetrahedrons. A tetrahedron is a four-sided solid where each side is a triangle. The function returns a list of tetrahedron positions and numerical volumes.

```
volumeSample :: Int -> Volume -> [(Position,R)]
volumeSample n v = [(tetCenter tet, tetVolume tet) | tet <- tetrahedrons n v]
```

The function takes an integer n that controls the number of tetrahedrons used. Most of the work is done by the function tetrahedrons, defined below, which returns a list of tetrahedrons when given the integer n and the volume v. The function tetCenter finds the center of each tetrahedron, and the function tetVolume computes the numerical volume of each tetrahedron. These two functions are also defined below.

We can describe a tetrahedron by specifying the positions of its four vertices.

```
data Tet = Tet Position Position Position Position
```

We can find the center of a tetrahedron by averaging displacement vectors for the four vertices.

```
tetCenter :: Tet -> Position
tetCenter (Tet p1 p2 p3 p4)
    = shiftPosition ((rVF p1 ^+^ rVF p2 ^+^ rVF p3 ^+^ rVF p4) ^/ 4) origin
```

This function is the natural extension of the triCenter function for triangles to tetrahedrons.

The volume of a tetrahedron is 1/6 of the *scalar triple product*, defined as $\mathbf{a} \cdot (\mathbf{b} \times \mathbf{c})$, of three vector edges originating at one vertex or terminating at one vertex. The scalar triple product is also the determinant of the matrix whose columns are the three vector edges.

```
tetVolume :: Tet -> R
tetVolume (Tet p1 p2 p3 p4)
    = abs $ (d1 <.> (d2 >< d3)) / 6
      where
        d1 = displacement p1 p4
        d2 = displacement p2 p4
        d3 = displacement p3 p4
```

We use the abs function to guarantee a positive numerical volume.

Just as we used "parameter squares" to cover the parameter space of a surface, now we use "parameter cubes" to cover the parameter space of a volume. Let's define a data type for parameter cube.

```
data ParamCube
    = PC { v000 :: (R,R,R)
         , v001 :: (R,R,R)
         , v010 :: (R,R,R)
         , v011 :: (R,R,R)
```

```
    , v100 :: (R,R,R)
    , v101 :: (R,R,R)
    , v110 :: (R,R,R)
    , v111 :: (R,R,R)
    }
```

The function tetrahedrons returns a list of tetrahedrons when given a volume.

```
tetrahedrons :: Int -> Volume -> [Tet]
tetrahedrons n (Volume g sl su tl tu ul uu)
    = let stus = [[[(s,t,u) | u <- linSpaced n (ul s t) (uu s t)]
                            | t <- linSpaced n (tl s) (tu s)]
                            | s <- linSpaced n sl su]
          stCubes = [PC (stus !! j     !! k     !! l    )
                        (stus !! j     !! k     !! (l+1))
                        (stus !! j     !! (k+1) !! l    )
                        (stus !! j     !! (k+1) !! (l+1))
                        (stus !! (j+1) !! k     !! l    )
                        (stus !! (j+1) !! k     !! (l+1))
                        (stus !! (j+1) !! (k+1) !! l    )
                        (stus !! (j+1) !! (k+1) !! (l+1))
                    | j <- [0..n-1], k <- [0..n-1], l <- [0..n-1]]
          tets (PC c000 c001 c010 c011 c100 c101 c110 c111)
              = [Tet (g c000) (g c100) (g c010) (g c001)
                ,Tet (g c011) (g c111) (g c001) (g c010)
                ,Tet (g c110) (g c010) (g c100) (g c111)
                ,Tet (g c101) (g c001) (g c111) (g c100)
                ,Tet (g c111) (g c100) (g c010) (g c001)
                ]
      in concatMap tets stCubes
```

Note the first argument to tetrahedrons is an integer n that controls how many tetrahedrons will be produced. We use pattern matching on the input to pass in the volume Volume g sl su tl tu ul uu because this function needs to refer to the volume's parameter limits sl and su as well as all of the other attributes of the volume.

The function begins by breaking the volume's parameter interval from sl to su into n equal subintervals. At each of the n+1 endpoints of these subintervals, we break the parameter interval from tl s to tu s into n equal subintervals, where s is the *s* parameter value at each subinterval endpoint. Finally, for the third dimension, we break each parameter interval from ul s t to uu s t into n equal subintervals, where t is the *t* parameter value at each subinterval endpoint. The local variable stus :: [[[(R,R,R)]]] is a list of lists of lists, which can be thought of as an n+1-by-n+1-by-n+1 array of parameter triples corresponding to points in the volume. The local variable stCubes :: [ParamCube] is a list of n^3 parameter cubes. The local function tets turns each of these cubes into five tetrahedrons. The function `tetrahedrons` returns a list of `5*n^3` tetrahedrons approximating the volume.

Summary

This chapter showed how to calculate the electric field produced by a charge distribution. We wrote functions to find the electric field produced by a point charge, a line charge, a surface charge, and a volume charge. On the way to that goal, we introduced the vector line integral, the vector surface integral, and the vector volume integral. We wrote a function

```
eField :: ChargeDistribution -> VectorField
```

that calculates the electric field of any charge distribution by combining the functions we wrote for each charge distribution.

After introducing the three sorts of vector integrals (line, surface, and volume), we took the opportunity to define three sorts of scalar integrals, in which we were adding up numbers rather than vectors. The chapter ended with a section on approximation of curves, surfaces, and volumes. Performing numerical integrals over these geometric objects requires that we have some way to carve the object up into a finite number of pieces. We showed one way to do this for each object. The next chapter on current distributions parallels the previous chapter on charge distributions. Just as charge is the source of electric fields, current is the source of magnetic fields.

Exercises

Exercise 25.1. Consider a line segment of charge, with uniform linear charge density λ_0. Let us place this line segment on the x-axis from $x = -L/2$ to $x = L/2$. We want to find the electric field produced by this line segment at some point in the xy-plane. Write code to make a picture of the electric field produced by this line segment of charge. You may restrict your attention to the xy-plane. The electric field produced by a line segment of charge is an exactly solvable problem. Find or calculate the exact solution. Produce a picture of the exact electric field for comparison.

Exercise 25.2. Make a picture of the electric field produced by a uniformly charged disk. Put the disk in the xy-plane and show the electric field in the xz-plane.

Exercise 25.3. Make a graph of the magnitude of the electric field produced by a uniformly charged ball versus the distance from the ball's center. Compare our numeric method to the exact solution.

Exercise 25.4. Consider a ring of charge in the xy-plane with radius R and linear charge density $\lambda(\mathbf{r}) = \lambda_0 \cos\phi$, where ϕ is the cylindrical coordinate and R and λ_0 are constants you can choose. Create a 3D visualization of the electric field produced by this charge distribution.

Exercise 25.5. An exercise in the previous chapter asked you to write a charge distribution for the hydrogen atom in its ground state. Find the electric field produced by the hydrogen atom at an arbitrary point in space. Make a graph of the magnitude of the electric field as a function of the distance from the proton.

Exercise 25.6. Produce an image of the electric field created by four equal positive point charges located at the corners of a square.

Exercise 25.7. Produce an image of the electric field in the xz-plane created by a square plate with uniform positive surface charge density in the xy-plane.

Exercise 25.8. Produce images of the electric field in the xy-plane and the yz-plane created by a uniformly charged circular ring in the xy-plane centered on the origin.

Exercise 25.9. If the functions `scalarLineIntegral` and `vectorLineIntegral` seem like they are basically doing the same thing, and that we ought to be able to exploit some sort of commonality to combine them into one function that can do both, you are right.

First, we define a general field that could be a scalar field, a vector field, or something else.

```
type Field a = Position -> a
```

The type `ScalarField` is the same as `Field R` and the type `VectorField` is the same as `Field Vec`.

Next, we make a type class for abstract vectors, which are types that have a zero vector, can be added, and can be scaled by a real number.

```
class AbstractVector a where
    zeroVector :: a
    add   :: a -> a -> a
    scale :: R -> a -> a
```

We write a function `sumG` to add a list of abstract vectors. This function is modeled after the `sumV` function from Chapter 10.

```
sumG :: AbstractVector a => [a] -> a
sumG = foldr add zeroVector
```

With these tools, we can write a general line integral function that performs the role of both the scalar line integral and the vector line integral.

```
generalLineIntegral
    :: AbstractVector a => CurveApprox -> Field a -> Curve -> a
generalLineIntegral approx f c
    = sumG [scale (magnitude dl') (f r') | (r',dl') <- approx c]
```

Write instance declarations that will make the types `R` and `Vec` instances of type class `AbstractVector`.

Exercise 25.10. *Gauss's law* asserts that the *electric flux* through a closed surface is proportional to the charge enclosed by the surface. The *flux* of a vector field describes the total flow through a surface if we imagine that the vectors are the velocity of some fluid. The electric flux Φ_E through a surface S is defined to be the dotted surface integral of the electric field.

$$\Phi_E = \int_S \mathbf{E} \cdot d\mathbf{a}$$

The definition for the dotted surface integral is a lot like that for the vector surface integral we defined in this chapter. The integral is defined by dividing the surface S into many small patches. Each patch $\Delta\mathbf{a}_i$ is a vector whose magnitude gives the area of the patch and whose direction points perpendicular to the patch. The electric field $\mathbf{E}$ is evaluated at a point $\mathbf{r}_i$ on or near the patch $\Delta\mathbf{a}_i$ and dotted with the vector area $\Delta\mathbf{a}_i$ of the patch. We then add up these numbers to form the sum:

$$\sum_i \mathbf{E}(\mathbf{r}_i) \cdot \Delta\mathbf{a}_i$$

The integral is the limit of this vector sum as the areas of the patches approach 0 and the number of patches becomes arbitrarily large.

Here is Haskell code for the dotted surface integral, also known as a *flux integral*:

```
dottedSurfaceIntegral :: SurfaceApprox -> VectorField -> Surface -> R
dottedSurfaceIntegral approx vF s
    = sum [vF r' <.> da' | (r',da') <- approx s]
```

(a) Write a function

```
electricFluxFromField :: VectorField -> Surface -> R
electricFluxFromField = undefined
```

that accepts an electric field and a surface as input and returns electric flux as output.

(b) Write a function

```
electricFluxFromCharge :: ChargeDistribution -> Surface -> R
electricFluxFromCharge dist = undefined dist
```

that returns the electric flux through a given surface produced by the electric field of a given charge distribution.

Exercise 25.11. Compare the electric field `eFieldLineDipole` produced by NaCl, viewed as a line dipole, to the electric fields of the simple dipole and the ideal dipole. Make a vector field picture similar to those we made for the simple dipole and the ideal dipole.

Exercise 25.12. The point charge is the fundamental source of electric field. Given a surface approximation, we can find the electric field of a surface charge by viewing it as a collection of point charges. The surface approximation tells us where to place the point charges and what values they should have. In this way, we skip over the vector surface integral that was our primary method for calculating the field.

```
eFieldFromSurfaceChargeP :: SurfaceApprox -> ScalarField -> Surface
                         -> VectorField
eFieldFromSurfaceChargeP approx sigma s r
    = sumV [eFieldFromPointCharge (sigma r' * magnitude da') r' r
              | (r',da') <- approx s]
```

Write similar functions to calculate the electric field from a line charge and a volume charge.

Exercise 25.13. Write a function

```
surfaceArea :: Surface -> R
surfaceArea = undefined
```

that uses `scalarSurfaceIntegral` to calculate the surface area of a surface.

Exercise 25.14. The electric potential is a scalar field that can be defined in terms of the electric field as follows:

$$V(\mathbf{r}) = -\int_0^{\mathbf{r}} \mathbf{E} \cdot d\mathbf{l}' \tag{25.12}$$

The integral is a dotted line integral over any curve C that begins at the origin and ends at the field point $\mathbf{r}$. The conservative nature of the electrostatic field $\mathbf{E}$ guarantees the result is independent of the chosen curve C.

The dotted line integral is defined and approximated by dividing the curve C into many small segments, exactly as we did for the vector line integral.

$$\int_C \mathbf{F}(\mathbf{r}') \cdot d\mathbf{l}' \approx \sum_i \mathbf{F}(\mathbf{r}_i) \cdot \Delta\mathbf{l}_i \tag{25.13}$$

The vector field $\mathbf{F}$ is evaluated at each point $\mathbf{r}_i$ of the approximation, dotted with the displacement $\Delta\mathbf{l}_i$ of the segment, and then added.

Here is the Haskell definition of the dotted line integral:

```
dottedLineIntegral :: CurveApprox -> VectorField -> Curve -> R
dottedLineIntegral approx f c = sum [f r' <.> dl' | (r',dl') <- approx c]
```

Write a function

```
electricPotentialFromField :: VectorField  -- electric field
                           -> ScalarField  -- electric potential
electricPotentialFromField ef r = undefined ef r
```

that takes the electric field as input and returns the electric potential as output. To write this function, you will need to construct a curve that begins at the origin and ends at the field point where we wish to find the electric potential. This curve can then be passed to `dottedLineIntegral`.

26

ELECTRIC CURRENT

Electric current is electric charge in motion. The current in electric circuits, such as those in our homes and offices, consists of charge flowing along a wire, but it's also useful to consider the possibility of charge flowing across a surface or throughout a volume. These three current distributions—line, surface, and volume—are the subject of this chapter.

This chapter parallels Chapter 24 on electric charge. We'll introduce the ideas of current, surface current density, and volume current density. We'll then define a data type for current distribution capable of representing a line current, a surface current, a volume current, or any combination of these. As electric charge is the source of electric fields, so electric current is the source of magnetic fields. We'll show how to calculate the magnetic dipole moment of any current distribution, and we'll discuss similarities and differences with the electric dipole moment of a charge distribution. Having a good language for current distributions prepares us for the next chapter, in which we'll find the magnetic field produced by a current distribution.

Current Distributions

Current is clearly an electrical phenomenon, being the flow of electric charge. But in 1820, Hans Christian Oersted showed that current was also a magnetic phenomenon, establishing the first connection between electricity and magnetism. Modern electromagnetic theory views current as the fundamental source of magnetic field.

In other words, electric current is the fundamental quantity responsible for magnetic effects (although this took thousands of years to discover after magnetic phenomena were first observed) and plays a key role in electromagnetic theory. We use three types of current distributions. First, there is current flowing along a one-dimensional path such as a line or a curve, which we often refer to simply as *current*. The SI unit for current is the ampere, or amp (A). An ampere of current in a wire means that 1 Coulomb of charge is passing a fixed point on the wire in each second. We typically use the symbol I for current. By convention, current is the flow of positive charge. The convention was established before people knew that it's the negatively charged electrons in metals that are free to move and conduct current. Electrons that are flowing to the left produce a current to the right in our convention.

A second current distribution is current flowing across a two-dimensional surface. In this case, we speak of the *surface current density* **K**, meaning the current per unit of cross-sectional length. The SI unit for surface current density is the ampere per meter (A/m).

Lastly, there is current flowing throughout a three-dimensional volume. In this case, we speak of the *volume current density* **J**, meaning the current per unit of cross-sectional area. The SI unit for volume current density is the ampere per square meter (A/m^2). Table 26-1 summarizes these current distributions.

Table 26-1: Current Distributions

Current distribution	Dimensionality	Symbol	SI unit
Point current	0	Not possible	Not possible
Current	1	I	A
Surface current density	2	**K**	A/m
Volume current density	3	**J**	A/m^2

Let's now turn to our Haskell code.

Introductory Code

Listing 26-1 shows the first lines of code in the `Current` module we'll write in this chapter.

```
{-# OPTIONS -Wall #-}

module Current where
```

```
import SimpleVec
    ( R, Vec, sumV, (><), (*^) )
import CoordinateSystems
    ( VectorField, rVF, cyl, phiHat )
import Geometry
    ( Curve(..), Surface(..), Volume(..) )
import ElectricField
    ( CurveApprox, curveSample, surfaceSample, volumeSample
    , vectorSurfaceIntegral, vectorVolumeIntegral )
```

Listing 26-1: Opening lines of code for the `Current` module

Here we've used types and functions from the `SimpleVec` module of Chapter 10, the `CoordinateSystems` module of Chapter 22, the `Geometry` module of Chapter 23, and the `ElectricField` module of Chapter 25.

Let's define a type synonym for current.

```
type Current = R
```

This is analogous to the type synonym we made for `Charge`. It's a simple way to create a type for current, but because `Current`, `Charge`, and `R` are all the same type, the compiler will not be able to help us from mistakenly using a `Charge` where a `Current` should go, or vice versa.

Now that we've specified a type for current, let's look at a type for current distribution, which will be a little more involved.

A Type for Current Distribution

Just as we did with `ChargeDistribution` in Chapter 24, here we want a new data type, `CurrentDistribution`, that can hold a line current, a surface current, a volume current, or a combination of these. What information do we need to specify each of these? For a line current, we need to specify a curve along which the current flows and the numerical value of the current. A line current requires that we give a `Current` and a `Curve`.

To specify a surface current, we need to give a vector field for the surface current density, which may vary from place to place, as well as a surface across which the current flows. A surface current is specified by giving a `VectorField` and a `Surface`. Similarly, a volume current is specified by giving a `VectorField` and a `Volume`. Finally, a combination of current distributions is specified by giving a list of current distributions.

Let's take a look at the code defining the data type `CurrentDistribution`.

```
data CurrentDistribution
  = LineCurrent     Current     Curve
  | SurfaceCurrent VectorField Surface
  | VolumeCurrent  VectorField Volume
  | MultipleCurrents [CurrentDistribution]
```

The type CurrentDistribution has four data constructors, one for each situation we described earlier. To construct a CurrentDistribution, we use one of the four data constructors along with the relevant information for that sort of current distribution.

Examples of Current Distributions

Let's write some examples of current distributions. The current distribution for current flowing around a circular loop in the xy-plane centered at the origin is called circularCurrentLoop.

```
circularCurrentLoop :: R  -- radius
                    -> R  -- current
                    -> CurrentDistribution
circularCurrentLoop radius i
    = LineCurrent i (Curve (\phi -> cyl radius phi 0) 0 (2*pi))
```

This is one of the simplest current distributions. The function circular CurrentLoop takes a radius and a current as input, and it returns a current distribution. In the following chapter, we'll calculate the magnetic field produced by this current distribution. A circular current loop is also an example of a magnetic dipole, which we'll discuss later in this chapter.

A *solenoid* consists of many turns of wire around a cylindrical frame. The function wireSolenoid returns a current distribution when provided with a radius for the solenoid, a length for the solenoid, the number of turns of wire per unit length, and the current in the wire.

```
wireSolenoid :: R  -- radius
             -> R  -- length
             -> R  -- turns/length
             -> R  -- current
             -> CurrentDistribution
wireSolenoid radius len n i
    = LineCurrent i (Curve (\phi -> cyl radius phi (phi/(2*pi*n)))
                                (-pi*n*len) (pi*n*len))
```

The curve for the wire is a helix. We use the cylindrical coordinate ϕ to parameterize the curve. The z-coordinate increases as ϕ increases to make the helix. If n is the number of turns of wire per unit length, and L is the length, then there are nL turns of wire on the solenoid. To produce this number of turns, the parameter ϕ must go through $2\pi nL$ radians from start to finish. We set the limits for ϕ to be from $-\pi nL$ to πnL so that the solenoid will be centered at the origin. We want the limits for z to be $-L/2$ to $L/2$; if we divide ϕ by $2\pi n$, we will achieve this, so we use $\phi/2\pi n$ for the z-coordinate in the cyl function.

In a *sheet solenoid*, we imagine the turns of wire are so close together that the current is effectively a surface current. We use the same inputs to sheetSolenoid as we did for wireSolenoid.

```
sheetSolenoid :: R  -- radius
              -> R  -- length
              -> R  -- turns/length
              -> R  -- current
              -> CurrentDistribution
sheetSolenoid radius len n i
    = SurfaceCurrent (\r -> (n*i) *^ phiHat r)
      (Surface (\(phi,z) -> cyl radius phi z)
       0 (2*pi) (const $ -len/2) (const $ len/2))
```

Since the sheet solenoid is a surface current, it requires a surface current density **K**. Surface current density is current per unit of cross-sectional length, so we have $K = nI$; the magnitude of the surface current density is the current in one wire times the number of turns per unit length. The direction of the surface current density is $\hat{\phi}$, so the surface current density is $\mathbf{K} = nI\hat{\phi}$, which is given in the code as `\r -> (n*i) *^ phiHat r`. The surface is a cylinder, parameterized by the cylindrical coordinates ϕ and z. The limits for ϕ are 0 to 2π, although we could have chosen $-\pi$ to π and achieved the same result. The limits for z are $-L/2$ to $L/2$. We need to use the function `const` because a surface requires limits on the second parameter that are functions of the first parameter. If the turns are close together, the wire solenoid will produce a very similar magnetic field to the sheet solenoid.

A *toroid* is formed by wrapping wire around a torus, as shown in Figure 26-1.

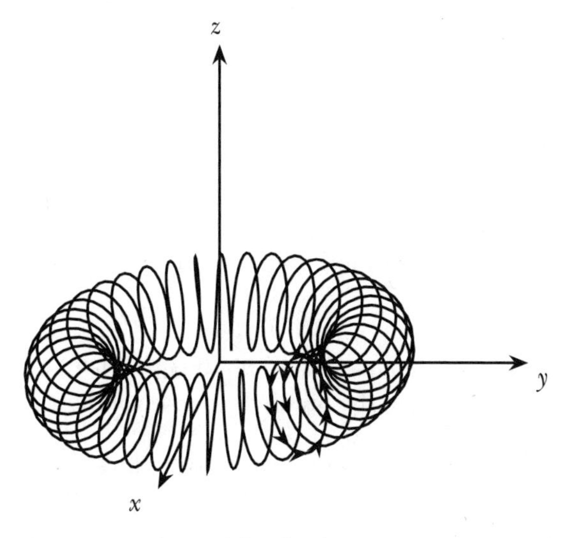

Figure 26-1: A toroidal coil with 40 turns. Arrows on the wire show the direction of the current.

NOTE *Solenoids and toroids are both used in circuits as inductors, which are circuit elements that can help to smooth out rapid changes in voltage. For something like a light dimmer, toroids are often better because most of the magnetic field resides in the torus, which means less electromagnetic noise is spewed into the room as the magnetic field changes. The frequency of the noise is the frequency of the generated electricity (60 Hz in the US), and multiples thereof, and can produce an irritating buzzing sound, so it is good to minimize the noise.*

The function `wireToroid` returns a current distribution when provided with a small radius for the torus, a big radius for the torus, the total number of turns, and the current in the wire.

```
wireToroid :: R  -- small radius
           -> R  -- big radius
           -> R  -- number of turns
           -> R  -- current
           -> CurrentDistribution
wireToroid smallR bigR n i
    = let alpha phi = n * phi
          curve phi = cyl (bigR + smallR * cos (alpha phi)) phi
                      (smallR * sin (alpha phi))
      in LineCurrent i (Curve curve 0 (2*pi))
```

The toroidal curve is based on the parametrization of the torus we use in Exercise 26.3. The two parameters for the toroidal surface are the cylindrical coordinate ϕ and an angle α that advances around the small cross-sectional circle of the torus. The cylindrical coordinates of points on the torus are given in terms of the two parameters ϕ and α, as follows:

$$s = R + r\cos\alpha$$
$$\phi = \phi$$
$$z = r\sin\alpha$$

To make the curve for the wire toroid, we choose the cylindrical coordinate ϕ to be our single parameter, and we let α now depend on ϕ. We choose

$$\alpha(\phi) = n\phi$$

so that α advances $2\pi n$ radians around the small circle (n revolutions) as ϕ advances 2π radians around the big circle (one revolution). The limits on the parameter ϕ are simply 0 to 2π.

We'll see the magnetic field produced by the wire toroid in the next chapter.

Conservation of Charge and Constraints on Steady Current Distributions

Electric charge is conserved. This means that the total amount of charge in any isolated region stays the same over time. Actually, an even stronger statement about charge conservation is true. The amount of charge in any region of space will change precisely to the extent that current flows across the boundary of the region. Current flowing into the region will increase the charge in the region while current flowing out of the region will decrease the charge.

If $Q(t)$ is the charge in some region at time t, and $I(t)$ is the current flowing out of the region at time t, then

$$\frac{dQ(t)}{dt} + I(t) = 0 \tag{26.1}$$

In other words, the rate at which charge increases in the region is the negative of the net current flowing out of the region, which is to say the net current flowing into the region.

The current flowing through any (closed or open) surface S by a volume current density $\mathbf{J}(t, \mathbf{r})$ is given by

$$I(t) = \int_S \mathbf{J}(t, \mathbf{r}) \cdot d\mathbf{a} \tag{26.2}$$

This is the same dotted surface integral, or flux integral, that we used to calculate electric flux in Exercise 25.10. Returning to our region of space with charge $Q(t)$ and net out-flowing current $I(t)$, we can use a time-dependent version of Equation 24.3,

$$Q(t) = \int_V \rho(t, \mathbf{r})\, dv \tag{26.3}$$

to rewrite Equation 26.1 in terms of charge density and current density:

$$\frac{d}{dt} \int_V \rho(t, \mathbf{r})\, dv + \int_{\partial V} \mathbf{J}(t, \mathbf{r}) \cdot d\mathbf{a} = 0 \tag{26.4}$$

Here, V is the region of space we are concerned with and ∂V is the closed surface that constitutes the boundary of V. If we allow the region V to become very small, we can divide both sides of Equation 26.4 by the volume of V to obtain an equation known as the *continuity equation*. See [19] for the mathematical details.

$$\frac{\partial \rho(t, \mathbf{r})}{\partial t} + \nabla \cdot \mathbf{J}(t, \mathbf{r}) = 0 \tag{26.5}$$

The quantity $\nabla \cdot \mathbf{J}(t, \mathbf{r})$ is called the *divergence* of the current density. The divergence of a vector field is the flux of the vector field per unit volume, where the flux is calculated over the closed boundary surface of the volume, in the limit where the volume is allowed to become very small. Since divergence is flux per unit volume, a place where a vector field has positive divergence is a place that the vectors point away from. Similarly, a place where a vector field has negative divergence is a place that the vectors point toward.

The symbol ∇ is called the *del operator*, and in Cartesian coordinates it is given by

$$\nabla = \hat{\mathbf{i}}\frac{\partial}{\partial x} + \hat{\mathbf{j}}\frac{\partial}{\partial y} + \hat{\mathbf{k}}\frac{\partial}{\partial z} \tag{26.6}$$

The word *operator* here is used as physicists use it, meaning something that takes a function as input and produces a function as output. Functional programmers call such a thing a *higher-order function.* The combination of the del operator followed by a dot product symbol forms the divergence. In Cartesian coordinates, the divergence of a vector field looks like the following:

$$\nabla \cdot \mathbf{J} = \frac{\partial J_x}{\partial x} + \frac{\partial J_y}{\partial y} + \frac{\partial J_z}{\partial z} \tag{26.7}$$

A consequence of the continuity equation is dramatically illustrated when you toss a crumpled sheet of aluminum foil into a microwave oven. The microwaves induce large currents in the foil, causing large amounts of charge to pile up in some parts of the foil. This gives rise to intense electric fields, and, finally *SPAAAKKKK!* as the electric field becomes strong enough to ionize the air and allow a spark.

Not every vector field can serve as a steady current density. In this chapter, we are interested in steady current distributions that do not change in time. If the charge density $\rho(t, \mathbf{r})$ and the current density $\mathbf{J}(t, \mathbf{r})$ are independent of time t, the continuity equation demands that $\nabla \cdot \mathbf{J} = 0$ (that is, that the current density be divergenceless). A vector field that is divergenceless is also known as a *solenoidal* vector field, from the shape of the solenoid, or pipe.

The computer will not check to make sure the vector field you use for current density is divergenceless. In this case, as in so many others when we model systems on a computer, the programmer is responsible for ensuring that the modeled system makes sense.

Magnetic Dipole Moment

Just as an electric dipole moment can be associated with any charge distribution, a magnetic dipole moment can be associated with any current distribution. In fact, the analogy extends to a multipole expansion. Just as a charge distribution can be thought of as a combination of monopole, dipole, quadrupole, and higher electric multipoles, so a current distribution can be thought of as a combination of magnetic multipoles, *except* that there is never any magnetic monopole in this expansion.

NOTE *One of the four Maxwell equations we study in Chapter 29 enforces this "no magnetic monopoles" rule. A current distribution has a magnetic dipole moment in close analogy with the electric dipole moment that a charge distribution has. A current distribution also has magnetic multipole moments in analogy with the electric multipole moments that a charge distribution has. But the analogy does not extend to monopole moment. Our universe contains electric charge (electric monopole moment), but so far no one has found any magnetic charge (magnetic monopole moment).*

A magnetic dipole creates a magnetic field and also responds to a magnetic field by feeling a force and/or a torque, so it can be thought of as a magnetically active entity similar to electric current.

We characterize a magnetic dipole by its *magnetic dipole moment* **m**. A magnetic dipole moment can be associated with any current distribution. Unfortunately, the simple picture of the electric dipole moment as a vector from negative charge to positive charge does not extend to the magnetic dipole moment. A current I has a magnetic dipole moment given by

$$\mathbf{m} = \frac{1}{2} I \int_C \mathbf{r}' \times d\mathbf{l}' \tag{26.8}$$

The magnetic dipole moment of a current loop carrying current I is $\mathbf{m} = I\mathbf{a}$, where **a** is the vector area of the loop, a vector whose magnitude gives the area and whose direction is perpendicular to the area.

A surface current density **K** has a magnetic dipole moment given by

$$\mathbf{m} = \frac{1}{2} \int_S \mathbf{r}' \times \mathbf{K}(\mathbf{r}')\, da' \tag{26.9}$$

A volume current density **J** has a magnetic dipole moment given by

$$\mathbf{m} = \frac{1}{2} \int_V \mathbf{r}' \times \mathbf{J}(\mathbf{r}')\, dv'. \tag{26.10}$$

The magnetic dipole moment is often a good simple characterization of a localized current distribution, such as a current loop, giving a good approximation for the magnetic field that the distribution creates.

The magnetic dipole moment for a line current is defined in terms of a crossed line integral, defined as follows:

```
crossedLineIntegral :: CurveApprox -> VectorField -> Curve -> Vec
crossedLineIntegral approx vF c
    = sumV [vF r' >< dl' | (r',dl') <- approx c]
```

This is similar to the vector line integral we defined earlier, except it involves a cross product.

Here is the definition for the magnetic dipole moment of a current distribution:

```
magneticDipoleMoment :: CurrentDistribution -> Vec
magneticDipoleMoment (LineCurrent    i c)
    = crossedLineIntegral   (curveSample  1000) (\r -> 0.5 *^ i *^ rVF r) c
magneticDipoleMoment (SurfaceCurrent k s)
    = vectorSurfaceIntegral (surfaceSample 200) (\r -> 0.5 *^ (rVF r >< k r)) s
magneticDipoleMoment (VolumeCurrent  j v)
    = vectorVolumeIntegral  (volumeSample   50) (\r -> 0.5 *^ (rVF r >< j r)) v
magneticDipoleMoment (MultipleCurrents ds    )
    = sumV [magneticDipoleMoment d | d <- ds]
```

The expression `rVF r` is the displacement vector pointing from the origin to the position `r`. The function `magneticDipoleMoment` encodes Equations 26.8, 26.9, and 26.10.

Summary

This chapter introduced current distributions, which are the fundamental source of magnetic field. We defined a type `CurrentDistribution`, which is capable of holding a line current, a surface current, a volume current, or a combination of these. Our simplest example of a current distribution is a wire loop. We also wrote solenoids and a toroid as examples of current distributions.

There is a multipole expansion for a current distribution that views the current as composed of a magnetic dipole, a magnetic quadrupole, and higher terms. However, there is no magnetic monopole term in this expansion. From far away, a current distribution often looks like a magnetic dipole, so we sometimes think of the magnetic dipole as a source of a magnetic field, similar to how electric charge (monopole) and the electric dipole can be thought of as sources of electric fields.

With the code we wrote in this chapter, we can now calculate the magnetic dipole moment associated with any current distribution. In the next chapter, we'll show how to compute the magnetic field produced by a current distribution.

Exercises

Exercise 26.1. A *Helmholtz coil* consists of two circular loops of wire, parallel to each other and sharing the same central axis, that each carry a current I in the same direction. The loops each have radius R and are separated by a distance equal to the radius R. This particular value of separation allows for a rather uniform magnetic field at the center of the Helmholtz coil. Write a current distribution for the Helmholtz coil.

```
helmholtzCoil :: R  -- radius
              -> R  -- current
              -> CurrentDistribution
helmholtzCoil radius i = undefined radius i
```

In practice, many loops of wire are coiled at the location of each of the two circles so that a moderate current through the wire will produce the effect of a very large current around each single loop.

Exercise 26.2. A simple and popular current distribution is an infinitely long straight wire carrying current I. It is not convenient for us to write a current distribution for an infinitely long wire, so let's give the wire length as a parameter. Write a current distribution for a long straight wire.

```
longStraightWire :: R  -- wire length
                 -> R  -- current
                 -> CurrentDistribution
longStraightWire len i = undefined len i
```

Exercise 26.3. If the turns of wire in a toroid are very close together, we can well approximate the current distribution by a surface current. Write a current distribution for a sheet toroid, similar to our sheet solenoid from earlier. Here is a torus to get you started. The function `torus` takes a small radius and a big radius as input.

```
torus :: R -> R -> Surface
torus smallR bigR
    = Surface (\(phi,alpha) -> cyl (bigR + smallR * cos alpha) phi
                                    (smallR * sin alpha))
      0 (2*pi) (const 0) (const $ 2*pi)
```

Exercise 26.4. Consider a solenoid that has so much wire wrapped around it that it has become fat. The wire on the inside is a distance a from the central axis, and the wire on the outside is a distance b from the central axis ($a < b$). We model this with a volume current density $\mathbf{J} = J_0\hat{\phi}$ in the region $a < s < b$, where J_0 is a constant. Outside of this region, there is no current. The length of the solenoid is L. Write a current distribution for the fat solenoid.

Exercise 26.5. For a steady current density that does not change with time, we can write Equation 26.2 without the time dependence as follows:

$$I = \int_S \mathbf{J}(\mathbf{r}) \cdot d\mathbf{a} \tag{26.11}$$

Write a function that calculates the total current flowing through a surface.

```
totalCurrent :: VectorField  -- volume current density
             -> Surface
             -> Current      -- total current through surface
totalCurrent j s = undefined j s
```

27

MAGNETIC FIELD

Magnets produce magnetic fields, but so do electric currents. Since magnets are made of materials that have circular currents at a microscopic level, physicists consider electric current to be the fundamental source of magnetic fields. In this chapter, we'll explore how magnetic fields are produced by currents. We'll write functions to find the magnetic field for all the current distributions we looked at in the last chapter, and we'll make pictures of the magnetic fields produced by a wire loop, an ideal magnetic dipole, a solenoid, and a toroid. But first, let's begin with a simple example of magnetism.

A Simple Magnetic Effect

Two parallel wires carrying current in the same direction will attract each other. In the modern view of magnetism, one current does not directly apply a force to another current. Instead, one current creates a magnetic field, and that magnetic field applies a force to the second current, as shown in Figure 27-1.

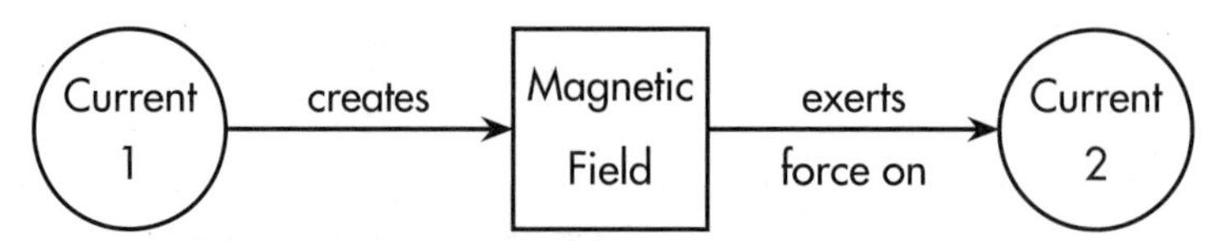

Figure 27-1: Conceptual diagram of the role of the magnetic field when two currents are present

The magnetic field, like the electric field, is a vector field of the kind we talked about in Chapter 22. The magnetic field associates a vector $\mathbf{B}(\mathbf{r})$ with each point $\mathbf{r}$ in space; that vector helps determine the force on a particle if there is a particle at point $\mathbf{r}$ in space. Figure 27-1 shows for the magnetic field what Figure 25-1 showed for the electric field. As the electric field is a mediator of electrical forces between charges, the magnetic field is a mediator of magnetic forces between currents.

Introducing the magnetic field breaks the analysis of magnetic situations into two parts. The first part is current creating the magnetic field, which we'll discuss in this chapter. The second part is the force the magnetic field applies to (a second) current, which we'll discuss in Chapter 28.

Let's now turn to some introductory code.

Introductory Code

Listing 27-1 shows the first lines of code in the `MagneticField` module we'll write in this chapter.

```
{-# OPTIONS -Wall #-}

module MagneticField where

import SimpleVec ( Vec(..), R
                 , (^-^), (*^), (^/), (<.>), (><)
                 , magnitude, kHat, zComp )
import CoordinateSystems
    ( VectorField
    , rVF, displacement, addVectorFields, cart, vfGrad )
import Geometry ( Curve(..), Surface(..), Volume(..) )
import ElectricField
    ( curveSample, surfaceSample, volumeSample
    , vectorSurfaceIntegral, vectorVolumeIntegral, mu0 )
import Current
    ( Current, CurrentDistribution(..)
    , wireSolenoid, wireToroid, crossedLineIntegral, circularCurrentLoop )
```

Listing 27-1: Opening lines of code for the `MagneticField` module

We use types and functions from the `SimpleVec` module of Chapter 10, the `CoordinateSystems` module of Chapter 22, the `Geometry` module of Chapter 23, the `ElectricField` module of Chapter 25, and the `Current` module of Chapter 26.

Current Creates Magnetic Field

The first part of the modern two-part view of magnetism is that current creates magnetic field. We'll start with the magnetic field created by a line current, the simplest current distribution, and then move to more complex current distributions.

Magnetic Field Created by a Line Current

Unlike the situation with charge, in which a point charge serves as the simplest form of charge, there is no such thing as a point current. Current, by definition, must flow, and the simplest way it can flow is along a curve or wire. We'll assume that charge does not pile up anywhere along our wires, a very reasonable assumption since you must go to some trouble to get charge to pile up. Consequently, the current at any point in a wire is the same.

The *Biot-Savart law* provides a way to calculate the magnetic field produced by a current-carrying wire. The wire can be of any shape, so this is a perfect opportunity to use the `Curve` data type we defined in Chapter 23.

Although there is no such thing as a point current, the Biot-Savart law nevertheless claims that the magnetic field of a current-carrying wire can be found as the superposition (that is, sum) of small magnetic field contributions from small segments of the wire. Each small current segment could not exist on its own since the current needs to keep flowing, but we can still compute the magnetic field contribution from a small current segment.

Consider a small segment of wire carrying current I. The segment is characterized by a displacement vector $d\mathbf{l}'$, whose length dl' is short enough that the segment can be considered straight, and whose direction is tangent to the wire. The contribution $\mathbf{dB}(\mathbf{r})$ to the magnetic field at location $\mathbf{r}$ by the small current segment is

$$\mathbf{dB}(\mathbf{r}) = \frac{\mu_0 I}{4\pi} d\mathbf{l}' \times \frac{\mathbf{r} - \mathbf{r}'}{|\mathbf{r} - \mathbf{r}'|^3} \tag{27.1}$$

This says the contribution is proportional to the current I, proportional to the segment length dl', inversely proportional to the square of the displacement $\mathbf{r} - \mathbf{r}'$ from the source point $\mathbf{r}'$ to the field point $\mathbf{r}$, and points in a direction perpendicular to both the current and the displacement.

We compute the magnetic field produced by the entire wire by adding together the contributions of all the small segments.

$$\mathbf{B}(\mathbf{r}) = -\frac{\mu_0 I}{4\pi} \int_C \frac{\mathbf{r} - \mathbf{r}'}{|\mathbf{r} - \mathbf{r}'|^3} \times d\mathbf{l}' \tag{27.2}$$

This integral is the crossed line integral we defined in Chapter 26. The minus sign enters because the cross product is anti-commutative.

The integrand of the crossed line integral is the function that maps a source point $\mathbf{r}'$ to the vector

$$\frac{\mathbf{r} - \mathbf{r}'}{|\mathbf{r} - \mathbf{r}'|^3}$$

Note that $\mathbf{r}$, the field point where we wish to know the magnetic field, is simply a parameter in this integrand; it is *not* the variable of integration. We should think of $\mathbf{r}'$ as the variable of integration since it is the locations $\mathbf{r}'$ of the curve at which we must evaluate the integrand and sum the results.

We'll give the integrand the local name `integrand` in the `bFieldFromLineCurrent` function we'll write next. The type signature of `bFieldFromLineCurrent` makes clear, in a computer-checked way, the two inputs required to find the magnetic field: the `Current` and the `Curve` along which the current flows. For a reader of Haskell, the function `bFieldFromLineCurrent` is a clearer description of what is going on than Equation 27.2 since the latter does not make it terribly clear that the magnetic field depends only on the curve and the current.

```
bFieldFromLineCurrent
    :: Current       -- current (in Amps)
    -> Curve
    -> VectorField  -- magnetic field (in Tesla)
bFieldFromLineCurrent i c r
    = let coeff = -mu0 * i / (4 * pi)  -- SI units
          integrand r' = d ^/ magnitude d ** 3
              where d = displacement r' r
      in coeff *^ crossedLineIntegral (curveSample 1000) integrand c
```

We define a local constant `coeff` to hold the numerical value of $-\mu_0 I/4\pi$ in SI units, and we define a local function `integrand` to hold the integrand. We want to define a local variable `d` for the displacement from `r'` to `r`, but because `r'` exists locally to the function `integrand`, the definition for `d` must occur within the definition for `integrand` and cannot be placed parallel to the definitions of `coeff` and `integrand`.

In Chapter 26, we wrote a type for current distribution. Over the course of the present chapter, we will write functions to calculate the magnetic field produced by each sort of current distribution. This allows us to encapsulate the idea that current creates magnetic field in the following function, which produces a magnetic field given any current distribution:

```
bField :: CurrentDistribution -> VectorField
bField (LineCurrent     i  c) = bFieldFromLineCurrent     i  c
bField (SurfaceCurrent kC s) = bFieldFromSurfaceCurrent kC s
bField (VolumeCurrent   j  v) = bFieldFromVolumeCurrent   j  v
bField (MultipleCurrents cds) = addVectorFields $ map bField cds
```

The function `bField` uses pattern matching on the input to treat each sort of current distribution separately. For a line current, it uses the function `bFieldFromLineCurrent` we wrote earlier. For surface and volume currents, it uses functions we'll write later in this chapter. For a combination distribution with the constructor `MultipleCurrents`, it uses the principle of superposition to find the magnetic field by summing the magnetic fields each individual current produced. We use the function `addVectorFields` from Chapter 22 to combine the magnetic fields of the component distributions.

Magnetic Field of a Circular Current Loop

One of the simplest and most natural ways to produce a magnetic field is with a circular loop of current. A circular loop of current is also a nice model of a magnetic dipole, as we discussed in Chapter 26. Surprisingly, there is no analytical solution for the magnetic field produced by a circular current loop. However, we can get a good approximate solution using the numerical integration embedded in our crossed line integral.

Consider a circular loop in the xy-plane, centered at the origin, with radius 0.25 m. This loop carries a current of 10 A in a counterclockwise direction when viewed from the positive z-axis. We can use the function `circularCurrentLoop` from Chapter 26 to make this current distribution, and we can use the function `bField`, which uses `bFieldFromLineCurrent` we wrote earlier, to find the magnetic field produced by this circular loop.

```
circleB :: VectorField  -- magnetic field
circleB = bField $ circularCurrentLoop 0.25 10
```

The expression `circularCurrentLoop 0.25 10` has type `CurrentDistribution`; we could have given it a name, either at the top level, as we do for `circleB`, or using a `let` or `where` construction. Deciding what to name something is part of the creative process of writing in a functional language. Would it help you or a reader of the code to see a good name for something, or would the name just get in the way, diverting our attention from more important ideas? This is a decision you get to make over and over again. In this case, I thought the current distribution was well-enough named by the function `circularCurrentLoop` and its parameters that another name was unnecessary.

The left side of Figure 27-2 shows the magnetic field in the yz-plane for this current loop. The x-direction comes out of the page in this figure, so the loop in the xy-plane appears as a horizontal line in the center of the picture where the field strength is largest. The magnetic field points through the current loop and back around.

The right side of Figure 27-2 shows the magnetic field of an ideal dipole, which we'll explain next.

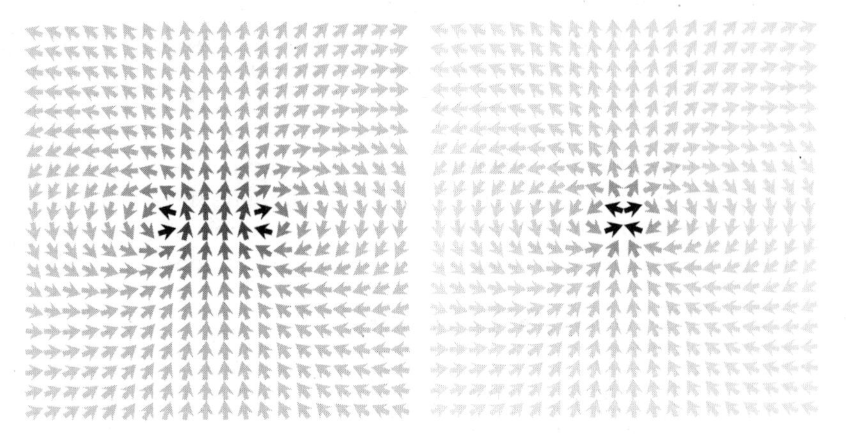

Figure 27-2: Magnetic field produced by a current loop (left) and an ideal magnetic dipole (right). The left image is produced by `bFieldPicLoop`; the right image is produced by `bFieldPicIdealDipole`. The magnetic field around the edges of the figures is very similar, indicating that, away from the source, a current loop looks like a magnetic dipole.

Here is the code that produced the left picture:

```
bFieldPicLoop :: IO ()
bFieldPicLoop
    = vfGrad (**0.2) (\(y,z) -> cart 0 y z) (\v -> (yComp v, zComp v))
      "bFieldPicLoop.png" 20 circleB
```

We use `vfGrad` from Chapter 22 to make a gradient vector field picture. The expression `(**0.2)` is a Haskell section denoting the function `\x -> x**0.2`, a scaling function we use because the magnitude of the field decreases rapidly as we move away from the loop. The other inputs to `vfGrad` declare that we want to look at the field in the yz-plane, assign a name to the output file, specify the number of arrows in each direction, and provide the name of the vector field to be pictured.

As mentioned earlier, a current loop is an example of a magnetic dipole. In the next section, we'll take a closer look at magnetic dipoles.

Ideal Magnetic Dipole

An *ideal magnetic dipole* is a source of magnetic field formed by letting the radius R of the loop approach 0 while the current in the loop gets larger so that the magnetic dipole moment $I\pi R^2$ stays constant. Let's look at the magnetic field produced by an ideal magnetic dipole.

The magnetic field produced by an ideal magnetic dipole at the origin is

$$\mathbf{B}(\mathbf{r}) = \frac{\mu_0}{4\pi}\frac{1}{r^3}[3(\mathbf{m}\cdot\hat{\mathbf{r}})\hat{\mathbf{r}} - \mathbf{m}] \tag{27.3}$$

where $\mathbf{m}$ is magnetic dipole moment. Here's the same equation translated into Haskell:

```
bFieldIdealDipole :: Vec          -- magnetic dipole moment
                  -> VectorField  -- magnetic field
```

```
bFieldIdealDipole m r
    = let coeff = mu0 / (4 * pi)      -- SI units
          rMag = magnitude (rVF r)
          rUnit = rVF r ^/ rMag
      in coeff *^ (1 / rMag**3) *^ (3 *^ (m <.> rUnit) *^ rUnit ^-^ m)
```

With the exception of the coefficient in front, Equation 27.3 is the same as Equation 25.3 for the electric field produced by an ideal electric dipole. This similarity is reflected in the functions `bFieldIdealDipole` and `eFieldIdealDipole`, which differ only in local variable names and the coefficient in front.

The right side of Figure 27-2 shows the magnetic field produced by an ideal magnetic dipole. The magnitude of the magnetic dipole moment **m** is not so important in this picture because the darkest arrows are those where the magnetic field is largest in magnitude, regardless of what that magnitude is. Equation 27.3 shows that the magnetic field increases linearly with the dipole moment, so the picture would be the same for any dipole moment in the z-direction.

Comparing the two pictures in Figure 27-2, we see that the magnetic field differs in the centers of the pictures, close to the sources of the field. The fields around the edges of the pictures, farther from the sources, are very similar in the two pictures. The similarity of the fields a bit farther from the sources entitles both sources to be called magnetic dipoles.

Here's the code to produce the picture on the right in Figure 27-2:

```
bFieldPicIdealDipole :: IO ()
bFieldPicIdealDipole
    = vfGrad (**0.2) (\(y,z) -> cart 0 y z) (\v -> (yComp v, zComp v))
      "bFieldPicIdealDipole.png" 20 (bFieldIdealDipole kHat)
```

The only differences in this program compared to `bFieldPicLoop` for the circular loop are the filenames and the magnetic fields. The magnetic field here is `bFieldIdealDipole kHat`, where we use the unit vector `kHat` for the magnetic dipole moment because its magnitude doesn't change the picture.

We've seen one example of the magnetic field produced by a line current, the circular loop, and compared it to that of an ideal magnetic dipole. Let's look at a second example of magnetic field from a line current, a solenoid.

Wire Solenoid

We defined the wire solenoid as a current distribution in the last chapter. Now let's calculate its magnetic field. We'll look at two wire solenoids. Each has a radius of 1 cm, a length of 10 cm, and a current of 10 A. The first solenoid has 100 turns per meter, for a total of 10 turns. The second solenoid has 1,000 turns per meter, for a total of 100 turns.

To start, let's look at the magnetic field at the center of the two solenoids.

```
Prelude> :l MagneticField
[ 1 of 14] Compiling Newton2          ( Newton2.hs, interpreted )
[ 2 of 14] Compiling Mechanics1D      ( Mechanics1D.hs, interpreted )
```

```
[ 3 of 14] Compiling SimpleVec        ( SimpleVec.hs, interpreted )
[ 4 of 14] Compiling Mechanics3D      ( Mechanics3D.hs, interpreted )
[ 5 of 14] Compiling MultipleObjects  ( MultipleObjects.hs, interpreted )
[ 6 of 14] Compiling MOExamples       ( MOExamples.hs, interpreted )
[ 7 of 14] Compiling Electricity      ( Electricity.hs, interpreted )
[ 8 of 14] Compiling CoordinateSystems ( CoordinateSystems.hs, interpreted )
[ 9 of 14] Compiling Geometry         ( Geometry.hs, interpreted )
[10 of 14] Compiling Integrals        ( Integrals.lhs, interpreted )
[11 of 14] Compiling Charge           ( Charge.hs, interpreted )
[12 of 14] Compiling ElectricField    ( ElectricField.hs, interpreted )
[13 of 14] Compiling Current          ( Current.hs, interpreted )
[14 of 14] Compiling MagneticField    ( MagneticField.hs, interpreted )
Ok, 14 modules loaded.
*MagneticField> bField (wireSolenoid 0.01 0.1 100 10) (cart 0 0 0)
vec 1.3405110355080298e-18 (-9.787828127867364e-7) 1.2326629789010703e-3
*MagneticField> bField (wireSolenoid 0.01 0.1 1000 10) (cart 0 0 0)
vec 9.429923508719186e-17 7.58448310225564e-6 1.2767867386980748e-2
```

We see that in both cases, the magnetic field is primarily in the z-direction, which is along the central axis, as expected. The x-components of the magnetic field are essentially 0. The y-components are small but not 0. Because the wire is in the shape of a helix, there is a small transverse component of the magnetic field. For our particular helix, this small transverse component shows up in the y-direction. Longer solenoids have smaller transverse components.

For comparison, an ideal solenoid has radius R, infinite length, n turns per unit length, and carries current I. Physicists like the ideal solenoid because, as long as n is reasonably large so that current is flowing at essentially all places on the surface of the cylinder, there is a simple expression for the magnetic field it produces. The magnetic field outside the ideal solenoid (at field points farther than R from the central axis) is 0. The magnetic field inside the ideal solenoid has a uniform value of $\mu_0 n I$ and points along the central axis of the solenoid. The magnetic field produced is independent of the radius.

An ideal solenoid with the same number of turns per unit length and same current as our first solenoid would have $n = 100/\text{m}$ and $I = 10$ A, so its magnetic field at the center (and at any other point inside) is

$$\mu_0 n I = (4\pi \times 10^{-7}\ \text{N/A}^2)(100/\text{m})(10\ \text{A}) = 1.26 \times 10^{-3}\ \text{T}$$

The z-component of our first solenoid is pretty close this value, even though our first solenoid is non-ideal in two ways: its length is not infinite and its winding has a spacing of one turn per centimeter (that is, not terribly close together).

An ideal solenoid with the same parameters as our second solenoid would have the same current but with 10 times the winding density n, so its magnetic field would be 10 times what we just calculated. Our second solenoid also has a magnetic field that is pretty close to the value for an ideal solenoid.

Figure 27-3 shows the magnetic fields produced by our two solenoids.

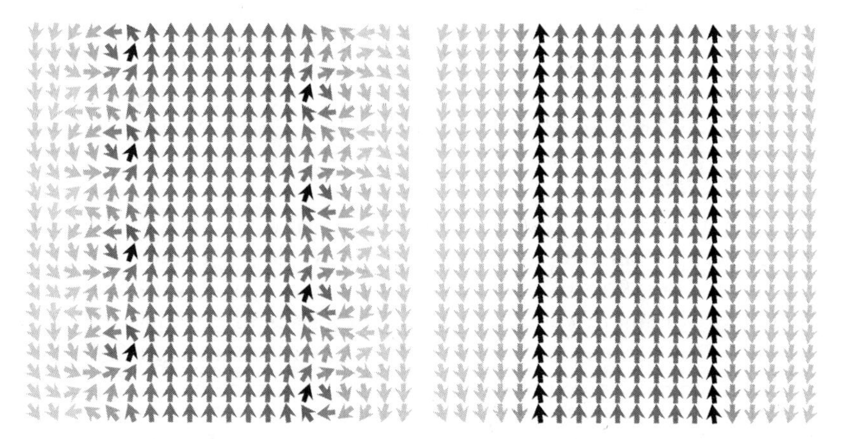

Figure 27-3: Magnetic fields produced by two solenoids. Both solenoids have a 1-cm radius, a 10-cm length, and a 10-A current. The solenoid on the left has 100 turns per meter, for a total of 10 turns. The solenoid on the right has 1,000 turns per meter, for a total of 100 turns. The pictures focus on a 4-cm × 4-cm region at the center of the solenoid. The left image is produced by `bFieldPicSolenoid10`*; the right image is produced by* `bFieldPicSolenoid100`*.*

The pictures in Figure 27-3 show the magnetic field in the yz-plane. The magnetic field produced by the first solenoid is shown on the left. The picture is a 4-cm × 4-cm area in the yz-plane at the center of the solenoid. Since this solenoid has one turn per centimeter, we see four turns of the solenoid. The places where the wire cuts through the yz-plane, where the magnetic field circles around the wire, are noticeable. The magnetic field is highest near the wires. The magnetic field inside the solenoid is clearly larger than that outside the solenoid, and in the opposite direction.

The magnetic field produced by the second solenoid is shown on the right. The picture is also a 4-cm × 4-cm area in the yz-plane at the center of the solenoid. This solenoid has 10 turns per centimeter, so 40 turns over the height of the picture. This is too many for us to see the individual wires. As with the first solenoid, the magnetic field is highest near the wires, larger on the inside than the outside, and has the opposite direction on the outside compared to the inside.

Here's the code that made the images in Figure 27-3. The program `bFieldPicSolenoid10`, named for 10 total turns, creates the picture on the left of the figure, while `bFieldPicSolenoid100`, named for 100 total turns, creates the picture on the right.

```
bFieldPicSolenoid10 :: IO ()
bFieldPicSolenoid10 = vfGrad (**0.2) (\(y,z) -> cart 0 (0.02*y) (0.02*z))
                      (\v -> (yComp v, zComp v)) "bFieldPicSolenoid10.png" 20
                      (bField $ wireSolenoid 0.01 0.1 100 10)

bFieldPicSolenoid100 :: IO ()
bFieldPicSolenoid100 = vfGrad (**0.2) (\(y,z) -> cart 0 (0.02*y) (0.02*z))
```

```
            (\v -> (yComp v, zComp v)) "bFieldPicSolenoid100.png" 20
            (bField $ wireSolenoid 0.01 0.1 1000 10)
```

We get pictures that are 4 cm wide by 4 cm high by mapping the visible square corners at $(-1, -1)$ and $(1, 1)$ to Cartesian coordinates $(0, -0.02, -0.02)$ and $(0, 0.02, 0.02)$ with the function `\(y,z) -> cart 0 (0.02*y) (0.02*z)`.

Wire Toroid

We defined the wire toroid as a current distribution in the last chapter. Now let's calculate its magnetic field. We'll look at a wire toroid with a small radius of 0.3 m, a big radius of 1 m, 50 turns of wire, and 10 A of current. This current distribution has the expression `wireToroid 0.3 1 50 10`.

The program `bFieldWireToroid` gives the magnetic field of this wire toroid.

```
bFieldWireToroid :: VectorField
bFieldWireToroid = bField (wireToroid 0.3 1 50 10)
```

Figure 27-4 shows the magnetic field produced by our toroid.

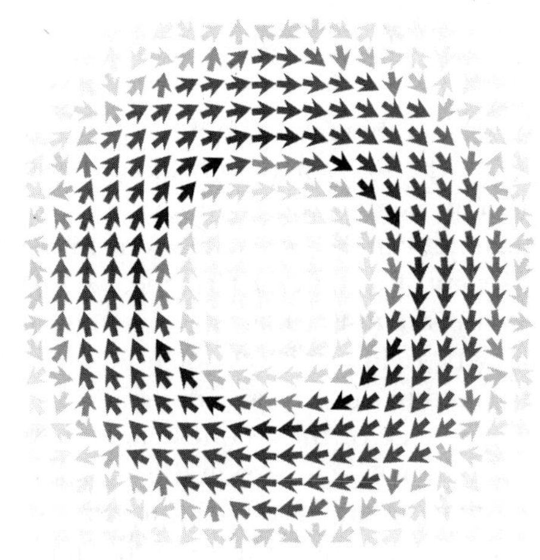

Figure 27-4: Magnetic field `bFieldWireToroid` produced by a toroidal coil of wire. The image is produced by `bFieldPicWireToroid`.

The picture in Figure 27-4 shows the magnetic field in the xy-plane. The picture is a 1.5-m × 1.5-m area in the xy-plane at the center of the toroid. The picture shows how the magnetic field is confined inside the torus.

Here is the code that produced the image:

```
bFieldPicWireToroid :: IO ()
bFieldPicWireToroid
    = vfGrad (**0.2) (\(x,y) -> cart (1.5*x) (1.5*y) 0)
      (\v -> (xComp v, yComp v)) "bFieldPicWireToroid.png" 20 bFieldWireToroid
```

The wire loop, wire solenoid, and wire toroid are examples of line currents. Let's move up a dimension and look at how to find the magnetic field produced by a surface current.

Magnetic Field Created by a Surface Current

Current flowing across a surface produces a magnetic field. To calculate this magnetic field, we must supply two pieces of information: the `Surface` across which the current flows and the surface current density **K**, represented as a `VectorField`.

Here's the version of the Biot-Savart law that gives the magnetic field produced by a surface current:

$$\mathbf{B}(\mathbf{r}) = \frac{\mu_0}{4\pi} \int_S \mathbf{K}(\mathbf{r}') \times \frac{\mathbf{r} - \mathbf{r}'}{|\mathbf{r} - \mathbf{r}'|^3} da' \quad (27.4)$$

This is a vector surface integral of the same type we dealt with in Chapter 25. The integrand of the vector surface integral is the function that maps a source point $\mathbf{r}'$ to the vector

$$\mathbf{K}(\mathbf{r}') \times \frac{\mathbf{r} - \mathbf{r}'}{|\mathbf{r} - \mathbf{r}'|^3}$$

Note that $\mathbf{r}$, the field point where we wish to know the magnetic field, is simply a parameter in this integrand; it is *not* the variable of integration. We should think of $\mathbf{r}'$ as the variable of integration since it is the locations $\mathbf{r}'$ that lie on the surface at which we must evaluate the integrand and sum the results. In the `bFieldFromSurfaceCurrent` function we'll write next, we give the integrand the local name `integrand`. The type signature of `bFieldFromSurfaceCurrent` makes clear, in a computer-checked way, the two inputs that must be provided and the return type of a vector field. For a reader of Haskell, the function `bFieldFromSurfaceCurrent` is a clearer description of what is going on than Equation 27.4.

```
bFieldFromSurfaceCurrent
    :: VectorField  -- surface current density
    -> Surface      -- surface across which current flows
    -> VectorField  -- magnetic field (in T)
bFieldFromSurfaceCurrent kCurrent s r
    = let coeff = mu0 / (4 * pi)  -- SI units
          integrand r' = (kCurrent r' >< d) ^/ magnitude d ** 3
              where d = displacement r' r
      in coeff *^ vectorSurfaceIntegral (surfaceSample 200) integrand s
```

We define a local constant `coeff` to hold the numerical value of $\mu_0/4\pi$ in SI units, and we define a local function `integrand` to hold the integrand. As before, we want to define a local variable `d` for the displacement from `r'` to `r`, but because `r'` exists locally to the function `integrand`, the definition for `d` must occur within the definition for `integrand` and cannot be placed parallel to the definitions of `coeff` and `integrand`.

As with `bFieldFromLineCurrent`, this function is used by the function `bField`, which computes the magnetic field due to any current distribution. The final function used by `bField` finds the magnetic field due to a volume current, and we turn to that next.

Magnetic Field Created by a Volume Current

Current flowing though a volume produces a magnetic field. To calculate this magnetic field, we must supply two pieces of information: the `Volume` through which the current flows and the volume current density $\mathbf{J}$, represented as a `VectorField`.

Here's the version of the Biot-Savart law that gives the magnetic field produced by a volume current:

$$\mathbf{B}(\mathbf{r}) = \frac{\mu_0}{4\pi} \int_V \mathbf{J}(\mathbf{r}') \times \frac{\mathbf{r} - \mathbf{r}'}{\left|\mathbf{r} - \mathbf{r}'\right|^3} dv' \tag{27.5}$$

This is a vector volume integral of the same type we dealt with in Chapter 25. The integrand of the vector volume integral is the function that maps a source point $\mathbf{r}'$ to the vector

$$\mathbf{J}(\mathbf{r}') \times \frac{\mathbf{r} - \mathbf{r}'}{\left|\mathbf{r} - \mathbf{r}'\right|^3}$$

The function `bFieldFromVolumeCurrent` calculates the magnetic field produced by a volume current density.

```
bFieldFromVolumeCurrent
    :: VectorField  -- volume current density
    -> Volume       -- volume throughout which current flows
    -> VectorField  -- magnetic field (in T)
bFieldFromVolumeCurrent j vol r
    = let coeff = mu0 / (4 * pi)  -- SI units
          integrand r' = (j r' >< d) ^/ magnitude d ** 3
              where d = displacement r' r
      in coeff *^ vectorVolumeIntegral (volumeSample 50) integrand vol
```

The only difference between this function and `bFieldFromSurfaceCurrent` is that this function uses `vectorVolumeIntegral` to perform a volume integral rather than a surface integral. As with the functions `bFieldFromLineCurrent` and `bFieldFromSurfaceCurrent`, this function is used by the function `bField`.

Summary

This chapter showed how to calculate the magnetic field produced by a current distribution. We wrote functions to find the magnetic field produced by a line current, a surface current, and a volume current. We wrote a function

```
bField :: CurrentDistribution -> VectorField
```

that calculates the magnetic field of any current distribution by combining the functions we wrote for each current distribution. We looked at the magnetic fields produced by a wire loop, an ideal magnetic dipole, a wire solenoid, and a wire toroid. Over the last four chapters, we have focused on the part of electromagnetic theory in which charge (and current, which is moving charge) produces fields. In the next chapter, we'll turn to the other side of electromagnetic theory in which fields apply force to charge.

Exercises

Exercise 27.1. *Faraday's law* asserts a relationship between the *magnetic flux* through a surface and the inclination for a current to flow around the boundary of the surface.

The magnetic flux Φ_B through a surface S is defined to be the dotted surface integral of the magnetic field:

$$\Phi_B = \int_S \mathbf{B} \cdot d\mathbf{a}$$

We defined the dotted surface integral in Chapter 25.

(a) Write a function

```
magneticFluxFromField :: VectorField -> Surface -> R
magneticFluxFromField = undefined
```

that accepts a magnetic field and a surface as input and returns magnetic flux as output.

(b) Write a function

```
magneticFluxFromCurrent :: CurrentDistribution -> Surface -> R
magneticFluxFromCurrent = undefined
```

that returns the magnetic flux through a given surface produced by the magnetic field of a given current distribution.

Exercise 27.2. Use `vf3D` from Chapter 22 to show the magnetic field `circleB` of a wire loop. You will need to search for an appropriate scale factor. The magnitude of the magnetic field at the center of a circular loop is $\mu_0 I/2R$, where I is the current in the loop and R is the radius. You can use this expression to take a good first guess at a scale factor.

```
visLoop :: IO ()
visLoop = undefined
```

Exercise 27.3. Here is a program that computes and displays the magnetic field of a wire solenoid with 1,000 turns of wire. It's not so different from the programs we wrote in this chapter for solenoids with 10 and 100 turns.

```
bFieldPicSolenoid1000 :: IO ()
bFieldPicSolenoid1000
    = vfGrad (**0.2) (\(y,z) -> cart 0 (0.02*y) (0.02*z))
             (\v -> (yComp v, zComp v)) "bFieldPicSolenoid1000.png" 20
             (bField $ wireSolenoid 0.01 0.1 10000 10)
```

The resulting picture, shown here, looks nothing like the magnetic field of a solenoid.

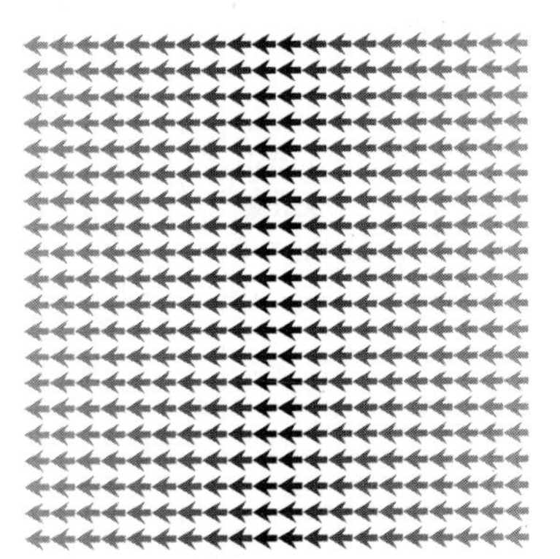

Identify the cause of this strange behavior and fix it.

Exercise 27.4. Make a picture of the magnetic field produced by a sheet solenoid. Pick values for the radius, length, number of turns per unit length, and current. Decide which of the visualization methods you will use.

Exercise 27.5. Consider the magnetic field `circleB` produced by a circular wire in the xy-plane, centered at the origin, with radius 0.25 m carrying 10 A of current. Make a graph of the z-component of the magnetic field as a function of location along the x-axis. It should look something like Figure 27-5.

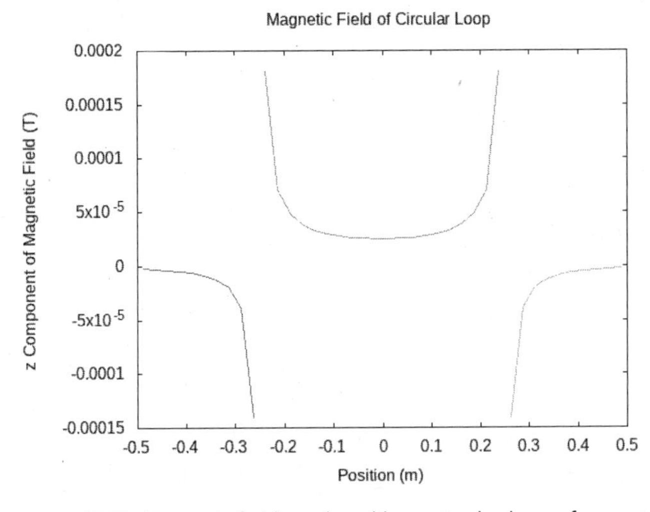

Figure 27-5: Magnetic field produced by a circular loop of current

Exercise 27.6. Consider the magnetic field produced by a circular loop carrying current. Make a graph of the z-component of the magnetic field as a function of location along the z-axis. Include both positive and negative values of z. The maximum magnetic field should occur at $z = 0$.

Exercise 27.7. In one of the exercises in the last chapter, we introduced the Helmholtz coil as a current distribution. Make a graph of the z-component of the magnetic field as a function of location along the z-axis. Include both positive and negative values of z. You should find that the magnetic field is much more uniform near the center than that produced by a single loop of wire.

28

THE LORENTZ FORCE LAW

We've seen that modern electromagnetic theory is a field theory. The existence of electric and magnetic fields explains electric and magnetic phenomena. Electric charge (and its moving form, current) is both the source of these fields and the recipient of force from these fields. Electromagnetic theory, then, has two aspects: charge creates electric and magnetic fields, and electric and magnetic fields apply force to charge. In the last four chapters, we dealt with the first of these aspects in static and steady situations. We showed how charge creates electric fields and how current creates magnetic fields.

In this chapter, we'll consider the *Lorentz force law*, which addresses the second aspect of electromagnetic theory by describing how electric and magnetic fields apply force to charge. Then, in the next and final chapter of this book, we'll return to the first aspect of electromagnetic theory with the *Maxwell equations*, which describe how electric and magnetic fields are created and evolve in dynamic situations.

Our goal in this chapter is to describe the motion of a charged particle in the presence of an electric field and a magnetic field. After a short discussion about statics and dynamics in electromagnetic theory, we'll turn to the question of an appropriate state for a particle in the presence of electric and magnetic fields. We'll then introduce the Lorentz force law, which describes the force on such a particle. We'll explain the purpose of the electric field and then address the question of how to perform state-update functions on a particle experiencing forces from electric and magnetic fields. We'll describe a particle's response to an imposed electric and magnetic field, but we will save consideration of the electromagnetic field radiated by moving charges for the final chapter of the book. We'll finish the chapter by making some animations of a particle in electric and magnetic fields.

Let's begin with some introductory code.

Introductory Code

Listing 28-1 shows the first lines of code in the `Lorentz` module we will write in this chapter.

```
{-# OPTIONS -Wall #-}
{-# LANGUAGE MultiParamTypeClasses #-}

module Lorentz where

import SimpleVec ( R, Vec, (^+^), (*^), (^*), (^/), (><), zeroV, magnitude )
import Mechanics1D ( RealVectorSpace(..), Diff(..), rungeKutta4 )
import Mechanics3D ( HasTime(..), simulateVis )
import CoordinateSystems ( Position(..), VectorField, cart, v3FromPos, origin
                         , shiftPosition, addVectorFields, visVec )
import qualified Vis as V
```

Listing 28-1: Opening lines of code for the `Lorentz` module

We use types and functions from the `SimpleVec` module of Chapter 10, the `Mechanics1D` module of Chapter 15, the `Mechanics3D` module of Chapter 16, and the `CoordinateSystems` module of Chapter 22.

This is the first time we're dealing with the dynamics of electromagnetic theory, so a short discussion of statics and dynamics will help set the stage for this chapter.

Statics and Dynamics

Part II of this book dealt with mechanics, which is a *dynamic* subject, in that the quantities we're interested in, including position, velocity, acceleration, force, momentum, and energy, are all changing in time. Newton's second law, in particular, is a dynamic equation because it tells us how the velocity of a particle changes in the presence of forces on the particle. So far in Part III of this book, electromagnetic theory has appeared to be a *static* subject, in that electric and magnetic fields were not changing in time. In the last four chapters, we spent our time exploring how static charges produce electric fields and how steady currents produce magnetic fields. But in fact, electromagnetic theory is just as much a dynamic subject as mechanics.

Each of the two aspects of electromagnetic theory has dynamics associated with it. Because charges move and accelerate, electric and magnetic fields change in time. The Maxwell equations we'll explore in the last chapter of this book describe how these fields change in time. Because a particle experiences forces, electromagnetic or otherwise, its velocity changes in time. This is the dynamics of mechanics we discussed in Part II. Once we know the forces that act on a particle, we use Newton's second law and the ideas in Part II to find the motion of the particle.

A good way to deal with dynamics is to ask how the state of a physical system changes with time. The practice of focusing on how the state of a physical system evolves in time will be just as useful in electromagnetic theory as it was in mechanics. Deciding what state variables to include in the state is just as important a question now as it was then. The strategy of choosing state variables and finding a differential equation that expresses how those variables change in time is one that transcends many theories in physics and beyond. It is not the only way to view physics, but it is a very important and useful way.

In preparation to do electrodynamics, then, we need to give some thought to an appropriate state. What do we put in a state? We can start with quantities we care about. In mechanics, we certainly care about the positions of particles, so we put them in the state. If the rate of change of a quantity we care about depends on other quantities, we may need to put them in the state as well. In mechanics, the rate of change of position is velocity, so we put velocity in the state. The rate of change of velocity depends on the forces that act on the particle, so we might choose to put quantities on which the forces depend in the state.

Thus, we're going to include the electric field and the magnetic field in the state. Conceptually and computationally, this is a pretty big step upward in complexity. Conceptually, this is the first time we are including functions in the state. Computationally, the sheer amount of information we need to keep track of in the state gets a major bump upward in going from mechanics to electromagnetic theory. In the mechanics of point particles, we need six numbers in the state for each particle in our system. We live in three-dimensional space, so we need three numbers to record the position of each particle, and we need three numbers to record the velocity of each particle. The electric field, a function from space to vectors, is closer to an infinite collection of numbers since the electric field contains a vector for every point in space.

There are two reasons to include electric and magnetic fields in the state of a system. One reason is that we care about the electric and magnetic fields, and we want to know how they change in time, although we won't address the changes in electric and magnetic fields until the next chapter. A second reason to include the fields in the state is that they are needed to determine the electric force and the magnetic force on particles. We'll deal with this issue in the present chapter.

Besides the electric and magnetic fields, what else should we include in the state? One option for state is to keep track only of the electric field and the magnetic field, excluding particle information from the state. This choice for state is useful for electromagnetic waves, or radiation from a known source, but it is not useful if we care about the motion of particles, as we do in this chapter. Hence, we will not pursue this option in this chapter.

A second option for state is to include the position and velocity of each particle we care about in addition to the electric and magnetic fields. The two-body force we introduced in Chapter 21 to describe the Coulomb force between charged particles would no longer be needed because the electric and magnetic fields are now the entities that produce force on particles. The Maxwell equations would describe how the electric and magnetic fields change based on the charged particles present in the system. Newton's second law would describe how the velocities of the particles change from electromagnetic and possibly other forces. In fact, removing two-body forces by introducing fields has been a theme in physics since 1865. Newton's law of universal gravity, a two-body force, can be removed once the field theory of general relativity is incorporated.

However, there is a technical problem with this combination of multiple particles and fields. The problem is that each particle contributes to the electric field and the magnetic field. The contribution that a charged point particle makes to the electric field increases without bound as the field point approaches the location of the particle. Applying the equations in the most obvious, naive way provides an infinite, or undefined, force on every particle, produced by itself. This is both conceptually and computationally troubling. The best conceptual resolution is to argue that point particles had to be an idealization anyway and to treat charge as spread throughout some volume, abandoning the particle theory of charge for a field theory of charge. One computational "quick fix" is, for each particle, to keep track of the fields produced only by the other particles. We don't want to get involved in any of these complexities, interesting as they may be, so we will not consider a state of multiple particles and fields.

Instead, we will choose a third option: a state that includes the position and velocity for a single particle, along with the electric and magnetic fields. This will allow us to focus on the motion of a single particle in an electric and a magnetic field. Let's see what this state looks like.

State of One Particle and Fields

By adding vector fields for the electric field and the magnetic field, we extend the type `ParticleState` we used in Chapter 16 for one particle to the type `ParticleFieldState`, which includes the fields. Here's the definition of the data type using record syntax:

```
data ParticleFieldState = ParticleFieldState { mass          :: R
                                             , charge        :: R
                                             , time          :: R
                                             , position      :: Position
                                             , velocity      :: Vec
                                             , electricField :: VectorField
                                             , magneticField :: VectorField }
```

As you can see, we have mass, charge, time, position, and velocity in the state, the same five state variables in `ParticleState` from Chapter 16. Now, we also include the electric and magnetic fields. Besides the two new slots for the fields, we are making one small change from the `ParticleState` type we used in Chapter 16. We are now using the `Position` data type we defined in Chapter 22 for position, rather than the `Vec` we used in Part II.

As we have done in past chapters, when we write a new data type for state, we also write a new data type for state derivative. In other words, we write a data structure to hold the time derivatives of the state variables. Following the pattern we've used for naming this state derivative type, we call the new type `DParticleFieldState`.

```
data DParticleFieldState = DParticleFieldState { dmdt :: R
                                               , dqdt :: R
                                               , dtdt :: R
                                               , drdt :: Vec
                                               , dvdt :: Vec
                                               , dEdt :: VectorField
                                               , dBdt :: VectorField }
```

We want to be able to use the Euler and fourth-order Runge-Kutta methods for solving differential equations in this setting, and that requires us to make the new data type an instance of the `RealVectorSpace` type class. This amounts to defining what it means to add `ParticleFieldStates` and to scale such expressions by real numbers. Here is the instance declaration:

```
instance RealVectorSpace DParticleFieldState where
    dst1 +++ dst2
        = DParticleFieldState { dmdt = dmdt dst1  +  dmdt dst2
                              , dqdt = dqdt dst1  +  dqdt dst2
                              , dtdt = dtdt dst1  +  dtdt dst2
                              , drdt = drdt dst1 ^+^ drdt dst2
                              , dvdt = dvdt dst1 ^+^ dvdt dst2
                              , dEdt = addVectorFields [dEdt dst1, dEdt dst2]
                              , dBdt = addVectorFields [dBdt dst1, dBdt dst2]
                              }
    scale w dst
        = DParticleFieldState { dmdt = w *  dmdt dst
                              , dqdt = w *  dqdt dst
```

```
                            , dtdt = w *  dtdt dst
                            , drdt = w *^ drdt dst
                            , dvdt = w *^ dvdt dst
                            , dEdt = (w *^) . (dEdt dst)
                            , dBdt = (w *^) . (dBdt dst)
                            }
```

NOTE *Unfortunately, this code is some of the most repetitive and uninformative code in the book. It just recounts the obvious thing we must mean by adding or scaling each of the derivatives of state variables. It gives me pain to write such repetitive, boilerplate code. It's a very interesting question to ask how we might avoid writing such code. I would love to give you the answer, but we must stay focused on the task at hand, which is defining a new data type and making sure it can be used with code we've already written.*

We need a `Diff` instance to describe how the types `ParticleFieldState` and `DParticleFieldState` are related. Recall that this involves defining the function `shift` to show how state variables are shifted by derivatives over a small time interval.

```
instance Diff ParticleFieldState DParticleFieldState where
    shift dt dst st
        = ParticleFieldState
          { mass          = mass     st  +  dmdt dst  * dt
          , charge        = charge   st  +  dqdt dst  * dt
          , time          = time     st  +  dtdt dst  * dt
          , position      = shiftPosition (drdt dst ^* dt) (position st)
          , velocity      = velocity st ^+^ dvdt dst ^* dt
          , electricField = \r -> electricField st r ^+^ dEdt dst r ^* dt
          , magneticField = \r -> magneticField st r ^+^ dBdt dst r ^* dt
          }
```

As usual, we shift each state variable by the product of the corresponding derivative and the time step.

There is one more type class instance declaration we need. When we wrote the `simulateVis` function in Chapter 16 to make 3D animations, we wanted to use it with any of the state spaces we had defined, or any we might define in the future. The one requirement that `simulateVis` made on a state space was that it include a notion of time. The instance declaration below simply gives the function that returns the time of a state.

```
instance HasTime ParticleFieldState where
    timeOf = time
```

Now that we've defined a new data type for the state of one particle with the electric and magnetic fields, let's turn to a discussion of the force that electric and magnetic fields apply to a particle.

Lorentz Force Law

Electric and magnetic fields produce forces on a charged particle. The force on a particle with charge q, position $\mathbf{r}(t)$, and velocity $\mathbf{v}(t)$ in an electric field $\mathbf{E}$ and a magnetic field $\mathbf{B}$ is given by the Lorentz force law.

$$\mathbf{F}_{\text{Lorentz}} = q[\mathbf{E}(\mathbf{r}(t)) + \mathbf{v}(t) \times \mathbf{B}(\mathbf{r}(t))] \tag{28.1}$$

The electric force,

$$\mathbf{F}_{\text{electric}} = q\mathbf{E}(\mathbf{r}(t))$$

gives a sense of the meaning of the electric field. The electric field represents a force per unit charge at a position in space. There may or may not be any charge at a point in space, but if there is, the product of the charge and the electric field vector at that point gives the force on the charge. Positive charge feels a force in the same direction as the electric field vector; negative charge feels a force in the opposite direction.

The magnetic force

$$\mathbf{F}_{\text{magnetic}} = q\mathbf{v}(t) \times \mathbf{B}(\mathbf{r}(t))$$

is harder to interpret because of the cross product, which indicates that the magnetic force on a charged particle is perpendicular to the magnetic field vector at the location of the particle and to the velocity of the particle. There is some symmetry between the creation of a magnetic field by a moving charge and the force from a magnetic field on a moving charge in that both processes are governed by equations containing a cross product. The appearance of charge and velocity in the magnetic force equation means that to feel a magnetic force, a particle must have charge and be moving. This is another symmetry between the creation and effect of magnetic field. Just as moving charge, or current, creates a magnetic field, it's moving charge that feels a force from the field.

The Lorentz force law, Equation 28.1, is simply the sum of the electric and magnetic forces. The function `lorentzForce` expresses the Lorentz force law in Haskell.

```
lorentzForce :: ParticleFieldState -> Vec
lorentzForce (ParticleFieldState _m q _t r v eF bF)
    = q *^ (eF r ^+^ v >< bF r)
```

Two of the terms, `eF` and `bF`, belong to the fields that produce the force. Three of the terms, `q`, `r`, and `v`, belong to the particle that feels the force. Because our state contains state variables for both the particle and the fields, the Lorentz force depends only on the state of the system.

Do We Really Need an Electric Field?

If we use the two-part modern view of electricity to first calculate the electric field $\mathbf{E}$ produced by a particle with charge q_1 at position $\mathbf{r}_1$

$$\mathbf{E}(\mathbf{r}) = kq_1 \frac{\mathbf{r} - \mathbf{r}_1}{|\mathbf{r} - \mathbf{r}_1|^3}$$

and then apply the Lorentz force law to find the force on a particle with charge q_2 at position $\mathbf{r}_2$

$$\mathbf{F}_2 = q_2\mathbf{E}(\mathbf{r}_2) = kq_1q_2\frac{\mathbf{r}_2 - \mathbf{r}_1}{\left|\mathbf{r}_2 - \mathbf{r}_1\right|^3} = kq_1q_2\frac{\mathbf{r}_{21}}{\left|\mathbf{r}_{21}\right|^3}$$

we recover Equation 21.2, Coulomb's 18th century electricity law. This is somewhat comforting since it gives us an opportunity to see how the electric field relates to Coulomb's law. But what have we gained by introducing the electric field? It appears to be nothing but a large piece of ontological and mathematical baggage! Why introduce the electric field if we just get back Coulomb's result?

The answer is that the electric field offers no new predictions over Coulomb's theory in static situations, which is when charges are not moving or accelerating. In the world of statics, the electric field is at best a convenience and at worst an irritation. However, when charges are moving or accelerating, Coulomb's law, as embodied in Equations 21.1, 21.2, and 21.3, no longer holds. The methods we developed in Chapter 25 also cease to hold. If charged particles move slowly, Coulomb's law and the equations of Chapter 25 are good approximations, but as charges approach the speed of light, those theories fail completely.

What's more, when charged particles accelerate, they radiate. In other words, they produce electric and magnetic fields that carry energy and momentum far away from the accelerating particle. The electric and magnetic fields take on a life of their own, and their description becomes an important part of the state of the system. Radiating electromagnetic fields are not treated in this chapter, but we do cover them in the final chapter of the book.

The electric and magnetic fields help enforce a principle of locality, the idea that interactions between entities (particles or fields) happen close to each other, not at a distance. Newton's law of universal gravitation and Coulomb's law are examples of *action at a distance*. They suggest that one object has a direct and instantaneous effect on another object arbitrarily far away. People have discussed the philosophical implications of this for centuries. Mathematically, it is no problem to have action at a distance in a framework that accommodates the idea of universal time. If a particle far away wiggles now, I feel the gravitational differences now. But since the acceptance of Einstein's relativity, we have given up the idea of a universal time. What's more, relativity tells us that the notion of simultaneity is observer dependent (or at least reference-frame dependent). In a framework of relativity, action-at-a-distance laws like Newton's universal gravity and Coulomb's law are quite problematic. Einstein realized that his 1905 special relativity was incompatible with Newton's universal gravity, and by 1915 he had developed a new theory of gravity called general relativity.

By 1865, Faraday and Maxwell had done for electricity what Einstein did for gravity in 1915. If a charged particle wiggles in one place, that modifies the electric field produced by the particle. Changes ripple through the electric field at the speed of light, and only later affect a second particle. Faraday and Maxwell made electromagnetic theory into a field theory that removed the need for action at a distance, explained the relationships between electricity and magnetism, predicted radiation, and gave a theory of light. The electric field is now viewed as a small price to pay for all these benefits.

Now that we've introduced the Lorentz force law, discussed it a bit, and seen how it governs the aspect of electromagnetic theory in which fields apply force to charge, let's combine it with the state-based approach to dynamics that will allow us to predict the motion of a charged particle in electric and magnetic fields.

State Update

Unlike our previous chapters on mechanics in Part II, in which we considered a list of forces from any source that act on a particle, in this chapter we'll assume that the electric and magnetic forces that comprise the Lorentz force are the only forces acting on our particle. The reason for this is simply to keep our focus on electromagnetic theory. It is not too difficult to extend the code we write in this chapter to include arbitrary forces beyond electromagnetism that act on a particle.

The function `newtonSecondPFS` (`PFS` for `ParticleFieldState`) expresses the differential equation that gives the rates of change of the state variables. The only interesting rate of change is that of the velocity, which is based on the net force on the particle divided by its mass, according to Newton's second law.

```
newtonSecondPFS :: ParticleFieldState -> DParticleFieldState
newtonSecondPFS st
    = let v = velocity st
          a = lorentzForce st ^/ mass st
      in DParticleFieldState { dmdt = 0              -- dm/dt
                              , dqdt = 0              -- dq/dt
                              , dtdt = 1              -- dt/dt
                              , drdt = v              -- dr/dt
                              , dvdt = a              -- dv/dt
                              , dEdt = const zeroV    -- dE/dt
                              , dBdt = const zeroV    -- dB/dt
                              }
```

The mass and charge of the particle do not change, so their rates are 0. Time changes at a rate of 1 second per second, so its rate is 1. Position changes at a rate given by velocity; that's just the definition of velocity. Velocity changes at a rate given by acceleration, which by Newton's second law is net force divided by mass. Net force here is just the Lorentz force since we decided to limit our attention to electromagnetic forces. In this chapter, we are not allowing the electric and magnetic fields to change, so their rates are 0. Since they are vector functions, their rates of change are the constant function that returns the zero vector, `const zeroV`.

Table 28-1 compares the function `newtonSecondPFS` to the other functions that expressed Newton's second law in Part II when we worked with other state spaces. Because we are confining our attention to electromagnetic forces, and because these forces can be determined solely by information contained in the state of the particle-field system, the function newtonSecondPFS does not require a list of forces as input, as all of the other functions in the table do.

Table 28-1: Functions for Newton's Second Law

Function	Type
newtonSecondV	Mass -> [Velocity -> Force] -> Velocity -> R
newtonSecondTV	Mass -> [(Time, Velocity) -> Force] -> (Time, Velocity) -> (R, R)
newtonSecond1D	Mass -> [State1D -> Force] -> State1D -> (R, R, R)
newtonSecondPS	[OneBodyForce] -> ParticleState -> DParticleState
newtonSecondMPS	[Force] -> MultiParticleState -> DMultiParticleState
newtonSecondPFS	ParticleFieldState -> DParticleFieldState

Recall that a numerical method allows us to transform a differential equation into a state-update function. State-update functions are important for animation and also for solving problems by obtaining a list of states.

The function pfsUpdate (pfs for ParticleFieldState, but lowercase since functions must begin with a lowercase letter) will serve as a state-update function for animation, or for producing a list of states.

```
pfsUpdate :: R  -- time step
          -> ParticleFieldState -> ParticleFieldState
pfsUpdate dt = rungeKutta4 dt newtonSecondPFS
```

The function uses fourth-order Runge-Kutta, simply because it tends to give the best results, but we could use any numerical method.

With a state-update function in hand, we are ready to animate the motion of a charged particle in electric and magnetic fields.

Animating a Particle in Electric and Magnetic Fields

We want to use the simulateVis function to do 3D animation. This requires that we supply the five inputs to that function, which are a time-scale factor, an animation rate, an initial state, a display function, and a state-update function. The function pfsUpdate will serve as our state-update function.

Providing an initial state is made easier by defining a default state. With a default state in hand, we can specify an initial state by listing the items that differ from the default state. The default state has every state variable set to 0.

```
defaultPFS :: ParticleFieldState
defaultPFS = ParticleFieldState { mass          = 0
                                , charge        = 0
                                , time          = 0
                                , position      = origin
                                , velocity      = zeroV
```

```
                                    , electricField = const zeroV
                                    , magneticField = const zeroV }
```

The following function, `pfsVisObject`, is a display function that displays the particle as a green sphere, the electric field as a collection of blue vectors, and the magnetic field as a collection of red vectors.

```
pfsVisObject :: R  -- cube width
             -> ParticleFieldState -> V.VisObject R
pfsVisObject width st
    = let r = position st
          xs = [-width/2, width/2]
          es :: [(Position,Vec)]
          es = [(cart x y z, electricField st (cart x y z))
                    | x <- xs, y <- xs, z <- xs]
          maxE = maximum $ map (magnitude . snd) es
          bs :: [(Position,Vec)]
          bs = [(cart x y z, magneticField st (cart x y z))
                    | x <- xs, y <- xs, z <- xs]
          maxB = maximum $ map (magnitude . snd) bs
          metersPerVis = width/2
      in V.VisObjects [ vectorsVisObject metersPerVis (2*maxE) es V.blue
                      , vectorsVisObject metersPerVis (2*maxB) bs V.red
                      , V.Trans (v3FromPos (scalePos metersPerVis r))
                            (V.Sphere 0.1 V.Solid V.green)
                      ]
```

This function takes a real number as input that specifies the width of the cube of space we want to display. It calculates the electric field at a collection of eight locations and then finds the maximum electric field magnitude so that it can scale the displayed electric field vectors. It then calculates the magnetic field at the same eight locations and finds the maximum magnetic field magnitude so that it can scale the displayed magnetic field vectors.

The function uses another function, called `vectorsVisObject`, to make the pictures of electric field and magnetic field.

```
vectorsVisObject :: R  -- scale factor, meters per Vis unit
                 -> R  -- scale factor, vector field units per Vis unit
                 -> [(Position,Vec)]  -- positions to show the field
                 -> V.Color
                 -> V.VisObject R
vectorsVisObject metersPerVis unitsPerVis pvs color
    = V.VisObjects [V.Trans (v3FromPos (scalePos metersPerVis r)) $
                      visVec color (v ^/ unitsPerVis) | (r,v) <- pvs]
```

This function takes two scale factors as input: one specifying the number of meters per Vis unit, and another giving the number of vector field units per Vis unit. It then takes a list of pairs of positions and vectors to be displayed and a color for the vectors. The function pfsVisObject uses vectorsVisObject twice: once for the electric field and once for the magnetic field.

Both pfsVisObject and vectorsVisObject use another helping function, called scalePos, that scales positions.

```
scalePos :: R -> Position -> Position
scalePos metersPerVis (Cart x y z)
    = Cart (x/metersPerVis) (y/metersPerVis) (z/metersPerVis)
```

This function works by scaling each position coordinate.

Our main program, called animatePFS, is a fun toy to play with. We can set the electric and magnetic fields to be whatever we wish and the initial conditions of the particle to be whatever we wish, and we can see what happens.

```
animatePFS :: R                      -- time scale factor
           -> Int                    -- animation rate
           -> R                      -- display width
           -> ParticleFieldState     -- initial state
           -> IO ()
animatePFS tsf ar width st
    = simulateVis tsf ar st (pfsVisObject width) pfsUpdate
```

The function takes a time-scale factor and an animation rate as inputs, along with a display width and an initial state. It calls simulateVis to do the animation.

The next two subsections show specific animations for a proton in uniform electric and magnetic fields and for a classical electron orbiting a proton.

Uniform Fields

Listing 28-1 displays the motion of a proton in uniform electric and magnetic fields. The initial state of the system gives the proton mass, the proton charge, the initial proton velocity, the electric field, and the magnetic field. The electric field is a uniform field in the y-direction, and the magnetic field is a uniform field in the z-direction. The proton makes a curious hopping motion in the presence of these fields.

```
{-# OPTIONS -Wall #-}

import SimpleVec ( vec )
import Electricity ( elementaryCharge )
import Lorentz ( ParticleFieldState(..), animatePFS, defaultPFS )

main :: IO ()
main = animatePFS 1e-5 30 0.05
       ( defaultPFS { mass          = 1.673e-27  -- proton in kg
                    , charge        = elementaryCharge
```

```
       , velocity      = vec 0 2000 0
       , electricField = \_ -> vec 0 20 0
       , magneticField = \_ -> vec 0  0 0.01 } )
```

Listing 28-2: Stand-alone program animating a proton in uniform electric and magnetic fields

By changing the electric field, the magnetic field, or the initial velocity of the proton, you can see many different sorts of motion. Figure 28-1 shows a snapshot of the animation.

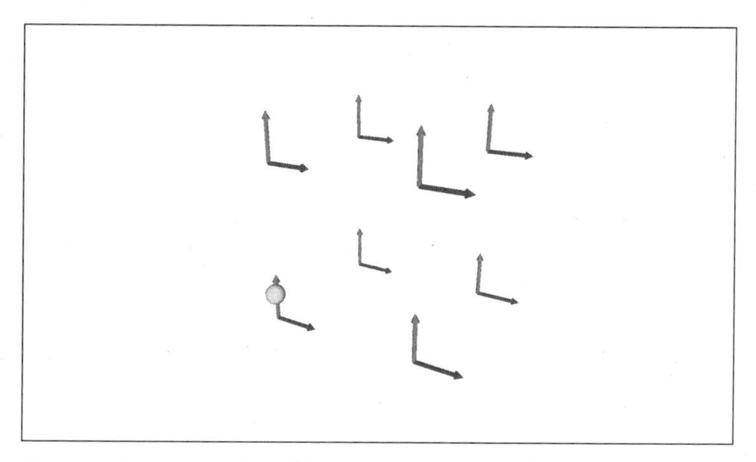

Figure 28-1: A screenshot of the animation showing the motion of a proton in particular electric and magnetic fields

Classical Hydrogen

Our second specific animation is a classical hydrogen atom. Hydrogen is the simplest atom, composed only of a proton and an electron. We need quantum mechanics to describe the properties of hydrogen correctly, but we will explore the Newtonian motion of an electron in the electric field created by a proton. This electric field provides an example of a non-uniform field, showing that our code can handle arbitrary electric and magnetic fields. It is important to note that our classical hydrogen atom uses the classical theory of Newtonian mechanics, but it does not use the full Faraday-Maxwell theory of electrodynamics, which is often regarded as "classical" because it is non-quantum. The Faraday-Maxwell theory, which we explore in the final chapter of this book, predicts that the electron radiates electromagnetic energy, causing it to spiral inward. Here we treat a simpler version of classical hydrogen in which the electron does not radiate.

Listing 28-3 displays the motion of an electron in the presence of the electric field produced by a proton at the origin.

```
{-# OPTIONS -Wall #-}

import SimpleVec ( vec )
import Electricity ( elementaryCharge )
import CoordinateSystems ( cart )
```

```
import Charge ( protonOrigin )
import ElectricField ( eField, epsilon0 )
import Lorentz ( ParticleFieldState(..), animatePFS, defaultPFS )

main :: IO ()
main = animatePFS period 30 (4*bohrRadius)
       ( defaultPFS { mass          = electronMass
                    , charge        = -elementaryCharge  -- electron charge
                    , position      = cart bohrRadius 0 0
                    , velocity      = vec 0 v0 0
                    , electricField = eField protonOrigin } )
           where electronMass = 9.109e-31  -- kg
                 bohrRadius   = 0.529e-10  -- meters
                 v0 = elementaryCharge
                      / sqrt (4 * pi * epsilon0 * electronMass * bohrRadius)
                 period = 2 * pi * bohrRadius / v0
```

Listing 28-3: Stand-alone program animating the electron in classical hydrogen

In this case, we use several local variables to determine the initial velocity for circular motion, which is needed in the initial state, and the period of the motion, which is used in the time-scale factor. The electron executes circular motion, much like a satellite orbiting a planet. Figure 28-2 shows a snapshot of the animation of the electron in the electric field produced by a proton.

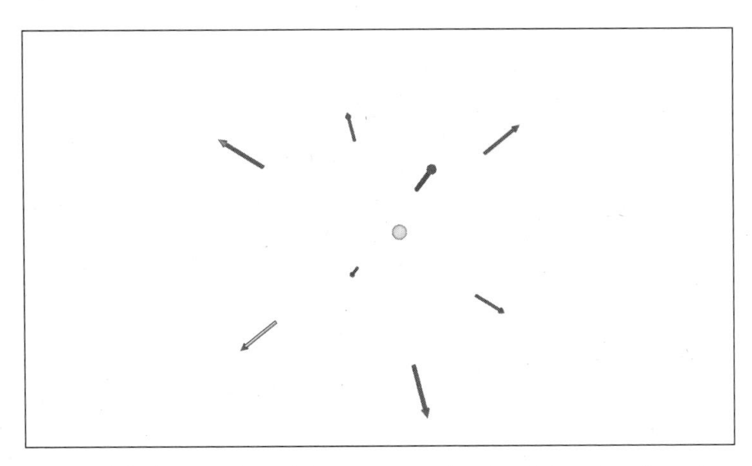

Figure 28-2: A screenshot of the animation showing the motion of an electron in the electric field produced by a proton

Summary

This chapter dealt with the aspect of electromagnetic theory in which fields apply force to charge. The Lorentz force law describes the force that electric and magnetic fields exert on charge. We then found the motion of a charged particle in electric and magnetic fields by using Newton's second law, as we did in Part II of the book. We defined a new state space for one particle with electric and magnetic fields. Our differential equation and state-update rule modified only the particle state, leaving the state of the electric and magnetic fields unchanged. We complete the state-update project in the next chapter, where the Maxwell equations describe how the electric and magnetic fields change in time.

Exercises

Exercise 28.1. Write a function called `eulerCromerPFS`, analogous to the function `eulerCromerPS` from Chapter 16, that implements the Euler-Cromer method with the `ParticleFieldState` data type. Test it out by rewriting `pfsUpdate` with `eulerCromerPFS` in place of `rungeKutta4`. Recompile the animation code and see if there are any noticeable differences.

Exercise 28.2. Suppose a particle with charge 9 nC is fixed at the origin. (9 nC is about the charge your sock might have coming out of the dryer.) A proton released from rest a distance 1 mm away will accelerate away from the origin. Make graphs of the proton position and proton velocity as functions of time.

Exercise 28.3. Suppose a 1-m $\times$ 1-m plate has a uniform surface charge density of 9 nC/m^2. (9 nC of charge would raise its potential to a few tens of volts.) A proton is released from rest a distance 1 mm from the center of the plate and accelerates away from the plate. Make graphs of the proton position, velocity, and acceleration as functions of time.

Exercise 28.4. A current loop with a radius of 10 cm carries a current of 100 A. (A 100-A current is usually the maximum a small house or large apartment will draw.) Let's fix the loop in the xy-plane, center it at the origin, and have the current flow counterclockwise as viewed from the positive z-axis. We wish to see the effect on a proton traveling near the loop. Let's start the proton at position

$$(x, y, z) = (11 \text{ cm}, -1 \text{ m}, 0)$$

and give it an initial velocity in the positive y-direction. The proton should deflect to the left when it gets near the current loop since currents flowing in the same direction attract. Make pictures of the proton's trajectory in the xy-plane. Explore how the deflection depends on the initial proton velocity.

Exercise 28.5. Explore the motion of a proton in a uniform magnetic field with no electric field. By changing the initial velocity, you should be able to get the proton to move in a circle, a helix, and a line. Which initial velocities produce each motion?

Exercise 28.6. Rewrite the function `newtonSecondPFS` so that it takes a list of non-electromagnetic forces as input.

```
newtonSecondPFS' :: [ParticleFieldState -> Vec]
                 -> ParticleFieldState -> DParticleFieldState
newtonSecondPFS' fs st = undefined fs st
```

29

THE MAXWELL EQUATIONS

In the past several chapters, we've introduced electric and magnetic fields and suggested that these fields are properly part of the description of the state of a physical system, at least one in which electric charge plays a role. What we have not yet done is shown how the electric and magnetic fields evolve in time. This chapter addresses that issue by introducing the Maxwell equations.

The Maxwell equations describe how electric and magnetic fields are created by charge and current, how the fields are related to each other, and how the fields evolve in time. Together with the Lorentz force law, the Maxwell equations express modern electromagnetic theory.

We'll begin the chapter with some introductory code, after which we'll present the Maxwell equations. We'll then discuss four relationships between electricity and magnetism implied by these equations and how the equations relate to our treatment of electric and magnetic fields in previous chapters. We'll also show how the state-update technique can be applied to the Maxwell equations. Finally, we'll present the finite difference time domain (FDTD) method for solving the Maxwell equations and use it to animate the electric field produced by oscillating charge.

Introductory Code

Listing 29-1 shows the first lines of the code in the Maxwell module we'll write in this chapter.

```
{-# OPTIONS -Wall #-}

module Maxwell where

import SimpleVec
    ( R, Vec(..), (^/), (^+^), (^-^), (*^)
    , vec, negateV, magnitude, xComp, yComp, zComp, iHat, jHat, kHat )
import CoordinateSystems
    ( ScalarField, VectorField
    , cart, shiftPosition, rVF, magRad )
import ElectricField ( cSI, mu0 )
import qualified Data.Map.Strict as M
import qualified Diagrams.Prelude as D
import Diagrams.Prelude
    ( Diagram, Colour
    , PolyType(..), PolyOrientation(..), PolygonOpts(..), V2(..)
    , (#), rotate, deg, rad, polygon, sinA, dims, p2
    , fc, none, lw, blend )
import Diagrams.Backend.Cairo ( B, renderCairo )
```

Listing 29-1: Opening lines of code for the Maxwell module

We use types and functions from the SimpleVec module of Chapter 10, the CoordinateSystems module of Chapter 22, and the ElectricField module of Chapter 25. We also do a qualified import of Data.Map.Strict, giving it the short name M, for use in the section on the FDTD method. We import several types and functions from the Diagrams package to do asynchronous animation at the end of the chapter.

The Maxwell Equations

In SI units, the Maxwell equations consist of the following four equations:

$$\nabla \cdot \mathbf{E} = \frac{1}{\epsilon_0}\rho \tag{29.1}$$

$$\nabla \cdot \mathbf{B} = 0 \tag{29.2}$$

$$\nabla \times \mathbf{E} + \frac{\partial \mathbf{B}}{\partial t} = 0 \tag{29.3}$$

$$\nabla \times \mathbf{B} - \frac{1}{c^2}\frac{\partial \mathbf{E}}{\partial t} = \mu_0 \mathbf{J} \tag{29.4}$$

The electric field is denoted $\mathbf{E}$, the magnetic field $\mathbf{B}$, the current density $\mathbf{J}$, and the charge density ρ. Remember that ϵ_0 is the permittivity of free space, first introduced in Chapter 21. The symbol ∇ is called the *del operator*, and in Cartesian coordinates it is given by:

$$\nabla = \hat{\mathbf{i}}\frac{\partial}{\partial x} + \hat{\mathbf{j}}\frac{\partial}{\partial y} + \hat{\mathbf{k}}\frac{\partial}{\partial z} \tag{29.5}$$

The word *operator* here is used as physicists use it, meaning something that takes a function as input and produces a function as output. Functional programmers call such a thing a higher-order function.

The combination of the del operator followed by a dot product symbol, as in Equations 29.1 and 29.2, is called the *divergence*, a higher-order function that takes a vector field as input and produces a scalar field as output. The definition of divergence is flux per unit volume, so a place where a vector field has positive divergence is a place that the vectors point away from. Similarly, a place where a vector field has negative divergence is a place that the vectors point toward. In Cartesian coordinates, the divergence of a vector field looks like the following:

$$\nabla \cdot \mathbf{E} = \frac{\partial E_x}{\partial x} + \frac{\partial E_y}{\partial y} + \frac{\partial E_z}{\partial z} \tag{29.6}$$

The combination of the del operator followed by a cross product symbol, as in Equations 29.3 and 29.4, is called the *curl*, a higher-order function that takes a vector field as input and produces a vector field as output. The definition of curl is circulation per unit area, so it describes how the vectors form a pattern of circulation. A place where a vector field has curl in the z-direction is a place where the vectors point in a counterclockwise sense parallel to the xy-plane. In Cartesian coordinates, the curl of a vector field looks like the following:

$$\nabla \times \mathbf{E} = \left(\frac{\partial E_z}{\partial y} - \frac{\partial E_y}{\partial z}\right)\hat{\mathbf{i}} + \left(\frac{\partial E_x}{\partial z} - \frac{\partial E_z}{\partial x}\right)\hat{\mathbf{j}} + \left(\frac{\partial E_y}{\partial x} - \frac{\partial E_x}{\partial y}\right)\hat{\mathbf{k}} \tag{29.7}$$

Equation 29.1 is called Gauss's law (you may have encountered Gauss's law in the exercises in Chapter 25). Gauss's law says that charge dictates the divergence of the electric field. Since vectors point away from a place of positive divergence and toward a place of negative divergence, Gauss's law says that the electric field points away from positive charge and toward negative charge.

Equation 29.2 is called Gauss's law for magnetism, or "no magnetic monopoles." Since the divergence of the magnetic field must be 0 at all points in space and time, there is no magnetic charge for the magnetic field to point toward or away from.

Equation 29.3 is called Faraday's law (you may have encountered Faraday's law in the exercises in Chapter 27). It asserts a link between the curl of the electric field and the time rate of change of the magnetic field. Faraday's law explains electric generators and transformers.

Equation 29.4 is the *Ampere-Maxwell law*. It asserts a relationship among the curl of the magnetic field, the time rate of change of the electric field, and the current density.

There are four independent variables in the Maxwell equations: three space coordinates and one time coordinate. There are six dependent variables: three electric field components and three magnetic field components. We can think of the charge and current densities as source terms. They are inputs to the Maxwell equations that determine the fields and how they change.

With the Maxwell equations, we have electric field and magnetic field appearing in the same equations for the first time. The Maxwell equations describe four relationships between electricity and magnetism, which we describe next.

Relationships Between Electricity and Magnetism

Equation 29.1 is purely electric, and Equation 29.2 is purely magnetic. The remaining two Maxwell equations express four relationships between electricity and magnetism.

First, electric charge produces a magnetic field when it moves. This was Hans Christian Oersted's 1820 discovery. He saw that electric current could deflect a compass needle. This relationship is described by the Ampere-Maxwell law, Equation 29.4, in that the current density $\mathbf{J}$ is related to the curl of the magnetic field, $\nabla \times \mathbf{B}$. An earlier version of this law, called Ampere's law, omitted the time derivative of the electric field and thus expressed in a simpler, if less comprehensive, way the dependence of magnetic field on current.

Second, a changing magnetic field produces an electric field. This was Faraday's discovery and goes by the name of Faraday's law, Equation 29.3. So, there are, in a sense, two sources of electric fields: one is charge and the other is a changing magnetic field. Coulomb's law is missing this changing magnetic field contribution to the electric field and is incompatible with relativity. Most electric generation plants today use Faraday's law to produce alternating electric current. The rotation of a turbine produces a changing magnetic field, which produces an electric field that drives the current.

Third, a changing electric field produces a magnetic field. In 1865, Maxwell added the term with the time derivative of the electric field to Ampere's law, creating the Ampere-Maxwell law, Equation 29.4. The added term is known as *displacement current* because, although it is not an electrical current, it serves a similar role in the creation of a magnetic field.

Finally, electric and magnetic fields constitute light. The modern theory of optics asserts that light is an electromagnetic wave. There is no such thing as an electric wave or a magnetic wave by itself. Wavelike electric fields are always accompanied by wavelike magnetic fields.

Connection to Coulomb's Law and Biot-Savart Law

If the Maxwell equations describe how electric and magnetic fields are created and how they evolve, why did we not use them in Chapters 25 and 27?

Those chapters gave methods for how an electric field is created by charge and how a magnetic field is created by current, respectively. How do the methods given in those chapters relate to the Maxwell equations?

Remember that Chapter 25 gave methods for calculating the electric field in static situations (that is, in situations where charge is not moving or accelerating). In practice, the methods of that chapter, which are equivalent to Coulomb's law, work reasonably well for charges moving slowly compared with the speed of light. In static situations, we can remove the two time derivative terms from the Maxwell equations, causing the equations for electricity to decouple from those for magnetism. Thus, in static situations, Equations 29.1 and 29.3 become

$$\nabla \cdot \mathbf{E} = \frac{1}{\epsilon_0}\rho \tag{29.8}$$

$$\nabla \times \mathbf{E} = 0 \tag{29.9}$$

and describe static electricity. Coulomb's law is the solution to Equations 29.8 and 29.9.

Similarly, in static situations, Equations 29.2 and 29.4 become

$$\nabla \cdot \mathbf{B} = 0 \tag{29.10}$$

$$\nabla \times \mathbf{B} = \mu_0 \mathbf{J} \tag{29.11}$$

and describe the magnetic field created by steady currents. The Biot-Savart law of Chapter 27 is the solution to Equations 29.10 and 29.11.

The static methods we introduced in Chapters 25 and 27 are useful and substantially simpler than the Maxwell equations, but they do not account for dynamic situations where charge is moving quickly or accelerating. We turn now to the task of solving the Maxwell equations, employing a state-update similar to what we used to solve Newton's second law in Part II of the book.

State Update

To understand how electric and magnetic fields change in time, it is helpful to rearrange Maxwell Equations 29.3 and 29.4 so that they give the rates of change of the fields in terms of the fields and the current density.

$$\frac{\partial \mathbf{E}}{\partial t} = c^2 \left[\nabla \times \mathbf{B} - \mu_0 \mathbf{J}\right] \tag{29.12}$$

$$\frac{\partial \mathbf{B}}{\partial t} = -\nabla \times \mathbf{E} \tag{29.13}$$

Equations 29.1 and 29.2 serve as constraints; as long as the electric and magnetic fields satisfy them at one point in time, they will continue to satisfy them as they change according to Equations 29.12 and 29.13.

Figure 29-1 shows a schematic diagram for the Maxwell equations.

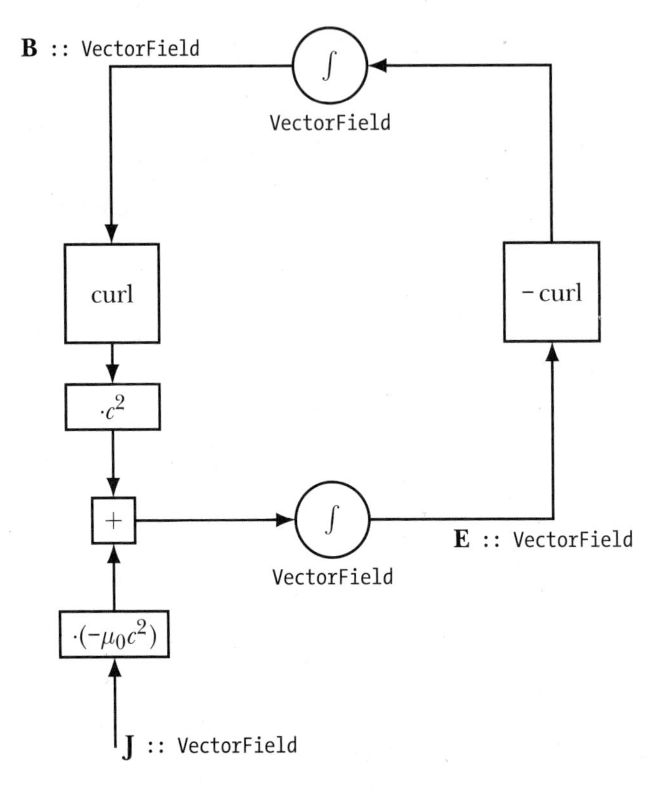

Figure 29-1: A schematic diagram representing the Maxwell equations

This diagram is similar to the diagrams we made in Part II of the book for Newton's second law. The most striking difference is that the values carried by the wires here are vector fields, whereas in mechanics they were numbers or vectors. As we integrated acceleration to obtain velocity in mechanics, here we integrate the curl of the electric field to obtain the magnetic field. As a loop in this kind of schematic diagram signaled a differential equation in mechanics, the loop here also signals that the Maxwell equations are differential equations.

As you can see in Figure 29-1, magnetic field change is controlled by the negative curl of the electric field. Electric field change is controlled by both current density and the curl of the magnetic field. This schematic diagram represents Equations 29.3 and 29.4. The integrators are with respect to time, as they have been in all of the schematic diagrams like this. The type `VectorField` below each integrator indicates the nature of the state contained in the integrator. Each integrator here contains a full vector field as state that gets updated by the vector field acting as input to the integrator. Equations 29.1 and 29.2 place constraints on the vector fields that can be held as state by the integrators. The multiplications by c^2 and $-\mu_0 c^2$ are appropriate for SI units.

As in Part II of the book, our methods for solving the Maxwell equations involve treating time discretely, updating the quantities we care about over a time step that is small compared with time scales on which significant change occurs, and then iterating this update procedure over many small time steps. For the Maxwell equations, the quantities we care about are the electric and magnetic fields. We update the electric and magnetic fields using Equations 29.12 and 29.13 to provide the rate at which electric and magnetic fields change. The updated fields change by the product of the rate with the time step.

$$\mathbf{E}(t + \Delta t, \mathbf{r}) \approx \mathbf{E}(t, \mathbf{r}) + c^2[\nabla \times \mathbf{B}(t, \mathbf{r}) - \mu_0 \mathbf{J}(t, \mathbf{r})]\Delta t \qquad (29.14)$$

$$\mathbf{B}(t + \Delta t, \mathbf{r}) \approx \mathbf{B}(t, \mathbf{r}) - [\nabla \times \mathbf{E}(t, \mathbf{r})]\Delta t \qquad (29.15)$$

To encode Equations 29.14 and 29.15 in Haskell, we need to take the curl of a vector field. We'll turn next to the question of how to write the curl in Haskell.

Spatial Derivatives and the Curl

The divergence and curl that appear in the Maxwell equations are types of spatial derivatives. Conceptually, the simplest spatial derivative is the *directional derivative*, defined to be the rate at which a field changes as we move in a specified direction. If f is a scalar field and $\hat{\mathbf{n}}$ is a unit vector, the directional derivative of f in the direction $\hat{\mathbf{n}}$ is defined to be the limit of the ratio of the difference of the values of the scalar field at two points separated by distance ϵ in the direction $\hat{\mathbf{n}}$, divided by ϵ.

$$D_{\hat{\mathbf{n}}} f(\mathbf{r}) = \lim_{\epsilon \to 0} \frac{f(\mathbf{r} + \epsilon\hat{\mathbf{n}}/2) - f(\mathbf{r} - \epsilon\hat{\mathbf{n}}/2)}{\epsilon} \qquad (29.16)$$

Our computational directional derivative will not take a limit but instead will simply calculate the ratio using a small displacement $\epsilon\hat{\mathbf{n}}$, which we call `d` in the following code:

```
directionalDerivative :: Vec -> ScalarField -> ScalarField
directionalDerivative d f r
    = (f (shiftPosition (d ^/ 2) r) - f (shiftPosition (negateV d ^/ 2) r))
      / magnitude d
```

The displacement vector `d` that serves as the first input to `directionalDerivative` has two roles. Its direction specifies the direction in which we want the derivative. In calculus, where we take limits, this is its only role. But in computation, where our derivatives involve small but finite steps, the second role of this input is for its magnitude to specify the step size for the derivative. We evaluate the field at two points: one shifted half the displacement vector from `r` and the other shifted minus half the displacement vector from `r`. We find the difference of these two field values and divide it by the magnitude of the displacement vector.

Recall from Equation 29.7 that we can find the curl from partial derivatives along the three coordinate directions. The partial derivative with respect to x is the directional derivative in the $\hat{\mathbf{i}}$ direction. The local functions `derivX`, `derivY`, and `derivZ` in the code for `curl` are partial derivatives.

We write the curl of a vector field in terms of Cartesian coordinates and partial derivatives, as in Equation 29.7.

```
curl :: R -> VectorField -> VectorField
curl a vf r
    = let vx = xComp . vf
          vy = yComp . vf
          vz = zComp . vf
          derivX = directionalDerivative (a *^ iHat)
          derivY = directionalDerivative (a *^ jHat)
          derivZ = directionalDerivative (a *^ kHat)
      in      (derivY vz r - derivZ vy r) *^ iHat
          ^+^ (derivZ vx r - derivX vz r) *^ jHat
          ^+^ (derivX vy r - derivY vx r) *^ kHat
```

The input `a` is a real number specifying the spatial step size to use for the curl. The input `vf` is the vector field for which we want the curl. The local variables `vx`, `vy`, and `vz` have type `ScalarField` and are the components of the vector field `vf`. The partial derivatives `derivX`, `derivY`, and `derivZ` have type `ScalarField -> ScalarField`. Finally, we use Equation 29.7 to find the curl.

Now that we can take the curl of a vector field, we are ready to try encoding the Maxwell equations in Haskell.

A Naive Method

The simplest encoding of the Maxwell equations uses a state space consisting of the current time, the electric field, and the magnetic field. We use the type synonym `FieldState` to describe a triple consisting of a real number for time, a vector field for electric field, and a vector field for magnetic field.

```
type FieldState = (R            -- time t
                  ,VectorField  -- electric field E
                  ,VectorField  -- magnetic field B
                  )
```

The function `maxwellUpdate` encodes Equations 29.14 and 29.15, which describe how the electric field and the magnetic field can be updated in time.

```
maxwellUpdate :: R                    -- dx
              -> R                    -- dt
              -> (R -> VectorField)  -- J
              -> FieldState -> FieldState
maxwellUpdate dx dt j (t,eF,bF)
    = let t'    = t + dt
          eF' r = eF r ^+^ cSI**2 *^ dt *^ (curl dx bF r ^-^ mu0 *^ j t r)
          bF' r = bF r ^-^             dt *^  curl dx eF r
      in (t',eF',bF')
```

The input `dx` to `maxwellUpdate` is a real number describing the spatial step size to use for the curl present in Equations 29.14 and 29.15. The input `dt` is a real number describing the time step. The input `j` is a time-dependent vector field describing the current density $\mathbf{J}$. Table 29-1 gives a correspondence between the mathematical notation of Equations 29.14 and 29.15 with the Haskell notation in `maxwellUpdate`.

Table 29-1: Correspondence Between Mathematical Notation and Haskell Notation for the Maxwell Equations

	Mathematics	Haskell
Time	t	`t`
Position	$\mathbf{r}$	`r`
Time step	Δt	`dt`
Speed of light	c	`cSI`
Permeability of free space	μ_0	`mu0`
Current density	$\mathbf{J}$	`j`
Current density	$\mathbf{J}(t, \mathbf{r})$	`j t r`
Electric field	$\mathbf{E}(t, \mathbf{r})$	`eF r`
Magnetic field	$\mathbf{B}(t, \mathbf{r})$	`bF r`
Updated electric field	$\mathbf{E}(t + \Delta t, \mathbf{r})$	`eF' r`
Updated magnetic field	$\mathbf{B}(t + \Delta t, \mathbf{r})$	`bF' r`
Curl	$\nabla\times$	`curl dx`
Curl of electric field	$\nabla \times \mathbf{E}(t, \mathbf{r})$	`curl dx eF r`
Curl of magnetic field	$\nabla \times \mathbf{B}(t, \mathbf{r})$	`curl dx bF r`
Vector addition	$+$	`^+^`
Vector subtraction	$-$	`^-^`
Scalar multiplication	Juxtaposition	`*^`

We update the time by adding the time step `dt` to the current time `t` to form the updated time `t'`. We update the electric field by adding $c^2[\nabla \times \mathbf{B}(t, \mathbf{r}) - \mu_0 \mathbf{J}(t, \mathbf{r})]\Delta t$ to the current electric field to form the updated electric field. We update the magnetic field by subtracting $[\nabla \times \mathbf{E}(t, \mathbf{r})]\Delta t$ from the current magnetic field to form the updated magnetic field.

To find the electric and magnetic fields as functions of time, we could iterate the `maxwellUpdate` function to produce a long list of states. The `maxwellEvolve` function does this.

```
maxwellEvolve :: R                      -- dx
              -> R                      -- dt
              -> (R -> VectorField)     -- J
              -> FieldState -> [FieldState]
maxwellEvolve dx dt j st0 = iterate (maxwellUpdate dx dt j) st0
```

Sadly, there is problem. While the code we have written compiles and in principle can be run, it is hopelessly inefficient. The trouble is that the computer does not automatically remember the function values it has already calculated, and it recalculates the same things over and over again. A function, to the Haskell compiler, is a rule for calculating outputs from inputs. If we know that we will need the output of a function in the future, it is up to us as Haskell programmers to see that it is available, usually by giving it a name. The values of the electric field at various places are just such function outputs in this naive method. They are not stored anywhere and must be recalculated each time they are needed. By the time we get to the eighth time step, for example, the computer needs to know the values of the electric and magnetic fields at the seventh time step, but these were not stored, so they must be recalculated. But the values for the seventh time step depend on those for the sixth, which were not stored and therefore must be recalculated.

The state `FieldState` and update method `maxwellUpdate` we have written in this section, while elegant and illustrative of what we want the computer to do, are not usable in practice, which is why we call them "naive" methods. Nevertheless, I would argue that there is value in this code. It type checks, indicating that the compiler agrees we are asking for something that makes sense. It is written in a style that is readable and can help us to understand what the Maxwell equations are about. Perhaps some day compilers will be smart enough to be able to plan what values should be remembered because they will be used again.

However, for today, we want to write code that runs and produces results. To do that, we'll turn to a new method.

The FDTD Method

We saw in the naive method that using functions to describe the state of a system, while clear in meaning and elegant in exposition, is not an efficient way to solve the Maxwell equations. To get decent executable code, we want numbers describing the state of our system rather than functions. To achieve this, we will select a large but finite number of positions in space at which to keep track of the electric and magnetic field components. The method we describe in detail is called the *finite difference time domain* (*FDTD*) method for solving the Maxwell equations. It is the simplest method used by people who need to numerically solve the Maxwell equations. The FDTD method is described more fully in [18].

The FDTD method is still based on Equations 29.12 and 29.13. Each of these two equations is a vector equation. It is helpful to write out the Cartesian components of these equations. Using Equation 29.7 to express the curl, Equations 29.17, 29.18, and 29.19 list the x-, y-, and z-components of Equation 29.12.

$$\frac{\partial E_x}{\partial t} = c^2 \left(\frac{\partial B_z}{\partial y} - \frac{\partial B_y}{\partial z} - \mu_0 J_x \right) \tag{29.17}$$

$$\frac{\partial E_y}{\partial t} = c^2 \left(\frac{\partial B_x}{\partial z} - \frac{\partial B_z}{\partial x} - \mu_0 J_y \right) \tag{29.18}$$

$$\frac{\partial E_z}{\partial t} = c^2 \left(\frac{\partial B_y}{\partial x} - \frac{\partial B_x}{\partial y} - \mu_0 J_z \right) \tag{29.19}$$

Similarly, the x-, y-, and z-components of Equation 29.13 are as follows:

$$\frac{\partial B_x}{\partial t} = -\left(\frac{\partial E_z}{\partial y} - \frac{\partial E_y}{\partial z}\right) \tag{29.20}$$

$$\frac{\partial B_y}{\partial t} = -\left(\frac{\partial E_x}{\partial z} - \frac{\partial E_z}{\partial x}\right) \tag{29.21}$$

$$\frac{\partial B_z}{\partial t} = -\left(\frac{\partial E_y}{\partial x} - \frac{\partial E_x}{\partial y}\right) \tag{29.22}$$

The FDTD method consists of approximating each partial derivative with a symmetric finite difference. By symmetric, we mean that we can approximate the partial derivative of a field component (E_x, E_y, E_z, B_x, B_y, or B_z) at a particular point (t, x, y, z) in time and space by sampling the field component at two points equidistant from (t, x, y, z). In the case of a partial derivative with respect to time, the sample points are $(t + \Delta t/2, x, y, z)$ and $(t - \Delta t/2, x, y, z)$. In the case of a partial derivative with respect to space, say in the y-direction, the sample points are $(t, x, y+\Delta y/2, z)$ and $(t, x, y-\Delta y/2, z)$. For example, the partial derivative of E_x with respect to time is approximated as

$$\frac{\partial E_x(t, x, y, z)}{\partial t} \approx \frac{E_x\left(t + \frac{\Delta t}{2}, x, y, z\right) - E_x\left(t - \frac{\Delta t}{2}, x, y, z\right)}{\Delta t} \tag{29.23}$$

and the partial derivative of E_z with respect to y is approximated as

$$\frac{\partial E_z(t, x, y, z)}{\partial y} \approx \frac{E_z\left(t, x, y + \frac{\Delta y}{2}, z\right) - E_z\left(t, x, y - \frac{\Delta y}{2}, z\right)}{\Delta y} \tag{29.24}$$

Applying this finite difference approximation to Equation 29.17 and performing a bit of algebra results in an equation that tells us how to update the value of the x-component of electric field.

$$\begin{aligned} E_x\left(t + \frac{\Delta t}{2}, x, y, z\right) &\approx E_x\left(t - \frac{\Delta t}{2}, x, y, z\right) \\ &+ c^2\left[\frac{B_z\left(t, x, y + \frac{\Delta y}{2}, z\right) - B_z\left(t, x, y - \frac{\Delta y}{2}, z\right)}{\Delta y}\right. \\ &\quad - \frac{B_y\left(t, x, y, z + \frac{\Delta z}{2}\right) - B_y\left(t, x, y, z - \frac{\Delta z}{2}\right)}{\Delta z} \\ &\quad \left. - \mu_0 J_x(t, x, y, z)\right]\Delta t \end{aligned} \tag{29.25}$$

There are five analogous equations for E_y, E_z, B_x, B_y, and B_z. Equation 29.25 and the other five are concisely expressed in vector form as

$$\mathbf{E}\left(t+\frac{\Delta t}{2},\mathbf{r}\right) \approx \mathbf{E}\left(t-\frac{\Delta t}{2},\mathbf{r}\right) + c^2[\nabla \times \mathbf{B}(t,\mathbf{r}) - \mu_0 \mathbf{J}(t,\mathbf{r})]\Delta t \tag{29.26}$$

$$\mathbf{B}\left(t+\frac{\Delta t}{2},\mathbf{r}\right) \approx \mathbf{B}\left(t-\frac{\Delta t}{2},\mathbf{r}\right) - [\nabla \times \mathbf{E}(t,\mathbf{r})]\Delta t \tag{29.27}$$

where components of the curl are approximated as follows:

$$[\nabla \times \mathbf{B}(t,\mathbf{r})]_x \approx \left[\frac{B_z\left(t,x,y+\frac{\Delta y}{2},z\right) - B_z\left(t,x,y-\frac{\Delta y}{2},z\right)}{\Delta y} - \frac{B_y\left(t,x,y,z+\frac{\Delta z}{2}\right) - B_y\left(t,x,y,z-\frac{\Delta z}{2}\right)}{\Delta z}\right]\Delta t \tag{29.28}$$

Notice the similarity between Equations 29.26 and 29.27 on one hand and Equations 29.14 and 29.15 on the other. The only difference is that the curls of the fields and the current density are evaluated at a time midway between the original and updated values of the fields in the FDTD Equations 29.26 and 29.27, while the curls and current density are evaluated at the time of the original values of the fields in Equations 29.14 and 29.15, which is closer to the Euler method.

The curl of Equation 29.28 requires values of B_y and B_z that are half a spatial step away. Updating E_x at a point in time and space depends on E_x at the same place one time step Δt earlier. It also depends on B_y half a spatial step to either side in the z-direction half a time step earlier, and it depends on B_z half a spatial step to either side in the y-direction half a time step earlier.

These half-spatial-step dependencies imply that the locations at which we keep track of the six components should be staggered. The places at which we keep track of E_x will be shifted slightly from the places where we keep track of E_y or B_y. Equation 29.25 and the five analogous equations for E_y, E_z, B_x, B_y, and B_z determine where we should track each component. We turn next to describing the locations at which we will keep track of the electric and magnetic field components.

The Yee Cell

We will use a triple `(nx,ny,nz)` of `Int`s to specify a location where we keep track of the field components. The integer `nx` measures the number of half spatial steps in the x-direction from the origin of our coordinate system. In other words, if `dx` is the spatial step size in the x-direction, equivalent to Δx in mathematical notation, then the x-coordinate of the position associated with `(nx,ny,nz)` is `fromIntegral nx * dx / 2`. Even integers denote whole steps from the origin, while odd integers denote an odd number of half steps. Table 29-2 shows the locations at which each of the six field components is tracked.

Table 29-2: Locations at Which We Calculate Components of the Electric and Magnetic Fields

Component	nx	ny	nz
E_x	odd	even	even
E_y	even	odd	even
E_z	even	even	odd
B_x	even	odd	odd
B_y	odd	even	odd
B_z	odd	odd	even

The locations at which values of E_x are kept are held in a list called `exLocs`, which is formed using a list comprehension to allow the integer `nx` to range over a sequence of consecutive odd integers, `ny` to range over a sequence of consecutive even integers, and `nz` to range over a sequence of consecutive even integers, as specified in Table 29-2. Other lists with similar names hold the locations of other field components.

```
exLocs, eyLocs, ezLocs, bxLocs, byLocs, bzLocs :: [(Int,Int,Int)]
exLocs = [(nx,ny,nz) | nx <- odds , ny <- evens, nz <- evens]
eyLocs = [(nx,ny,nz) | nx <- evens, ny <- odds , nz <- evens]
ezLocs = [(nx,ny,nz) | nx <- evens, ny <- evens, nz <- odds ]
bxLocs = [(nx,ny,nz) | nx <- evens, ny <- odds , nz <- odds ]
byLocs = [(nx,ny,nz) | nx <- odds , ny <- evens, nz <- odds ]
bzLocs = [(nx,ny,nz) | nx <- odds , ny <- odds , nz <- evens]
```

The constant `spaceStepsCE` (CE for center to edge) gives the number of full spatial steps from the center to the edge of our grid.

```
spaceStepsCE :: Int
spaceStepsCE = 40
```

We use integers to specify locations in the grid. The largest even integer, called `hiEven`, is twice the number of full steps from center to edge.

```
hiEven :: Int
hiEven =  2 * spaceStepsCE
```

The even numbers used to specify locations range from `-hiEven` to `hiEven`.

```
evens :: [Int]
evens = [-hiEven, -hiEven + 2 .. hiEven]
```

The odd numbers used to specify locations begin one above the lowest even number and end one below the highest even number.

```
odds :: [Int]
odds = [-hiEven + 1, -hiEven + 3 .. hiEven - 1]
```

The pattern of locations for storing the field components is called a *Yee cell* and is shown in Figure 29-2. The Yee cell is named after Kane S. Yee, who pioneered the FDTD method in the 1960s.

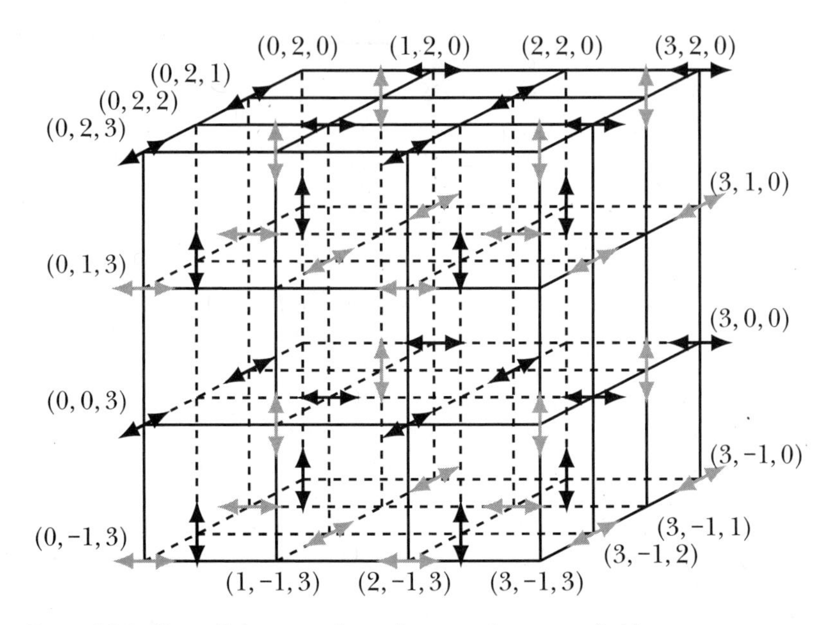

Figure 29-2: Yee cell showing where electric and magnetic field components are calculated

Figure 29-2 shows a patch of three-dimensional space, with a coordinate system in which x increases to the right, y increases up the page, and z increases out of the page. Double-headed arrows are placed at the locations in space where field components are tracked. Black arrows are electric field components, and gray arrows are magnetic field components. The direction in which the arrow points indicates which of the three components is being shown. A left-right arrow, for example, is an x-component. Figure 29-2 is a visual way of expressing the information in Table 29-2. For example, an E_x-component is stored at `(nx,ny,nz) = (1,2,0)` since `nx` is odd, `ny` is even, and `nz` is even.

One of the features of the Yee cell is that each component's nearest neighbors contain the information needed to update that component in time.

Let's talk next about how we represent the state of the electric and magnetic fields in the FDTD method and how we update that state.

A Type for State

Instead of the `FieldState` data type we used in the naive method, which contains functions for the electric and magnetic fields, we wish to have a state that holds numbers for the components of the electric and magnetic fields at the positions identified in the Yee cell of Figure 29-2.

One of the standard Haskell library modules, `Data.Map.Strict`, includes a data structure called a `Map` that is good for this purpose. The type `Map k v` is the type of a lookup table of key-value pairs, with `k` being the type of the key and `v` being the type of the value. In Chapter 9 we showed how a list of pairs with type `[(k,v)]` can be used as a lookup table, but the type `Map k v` is better because it will store the keys in such a way that they can be looked up quickly.

For the key, we will use a triple `(nx,ny,nz)` of `Int`s to describe the location of a field component, and for the value we will use a real number `R`. So, the type we want to hold our field data is `Map (Int,Int,Int) R`.

Since electric field components are stored at different locations from magnetic field components, we could get away with a single lookup table, but we choose to use two tables, one for electric field and one for magnetic field, to make the code easier to read.

Our state space, called `StateFDTD`, consists of the time, three real numbers for the spatial step size in each direction, a `Map (Int,Int,Int) R` for the electric field, and a `Map (Int,Int,Int) R` for the magnetic field. It is not strictly necessary to include the spatial step sizes in the state, but it is convenient because functions that have a state as input often need to know the spatial step sizes to do their jobs. A function that computes the curl of a field, for example, needs the spatial step size.

```
data StateFDTD = StateFDTD {timeFDTD :: R
                           ,stepX    :: R
                           ,stepY    :: R
                           ,stepZ    :: R
                           ,eField   :: M.Map (Int,Int,Int) R
                           ,bField   :: M.Map (Int,Int,Int) R
                           } deriving Show
```

In the introductory code, the line `import qualified Data.Map.Strict as M` allows us to access all of the functions and types defined in `Data.Map.Strict` if we prefix them with a capital `M`. Because of the way we imported the module, we need to refer to the type `Map (Int,Int,Int) R` as `M.Map (Int,Int,Int) R`, which we see above is the type we use to hold the electric field and the magnetic field.

The function `initialStateFDTD` takes a real number that it uses for the spatial step size in all three directions as input, and it returns a state in which the electric field and the magnetic field are 0 at all positions.

```
initialStateFDTD :: R -> StateFDTD
initialStateFDTD spatialStep
    = StateFDTD {timeFDTD  = 0
                ,stepX = spatialStep
                ,stepY = spatialStep
                ,stepZ = spatialStep
                ,eField = M.fromList [(loc,0) | loc <- exLocs++eyLocs++ezLocs]
                ,bField = M.fromList [(loc,0) | loc <- bxLocs++byLocs++bzLocs]
                }
```

The function M.fromList from the Data.Map.Strict module turns a list-of-pairs lookup table into a Map lookup table. We use a list comprehension to form a list of pairs in which the key is one of the locations for a field component and the value is 0.

The Data.Map.Strict module uses strict evaluation instead of Haskell's default lazy evaluation. When doing numerical calculations, we almost never want to use lazy evaluation. Lazy evaluation is good when we might, depending on input data, compute only a subset of the possible outputs of our program. But when we are numerically evaluating a model, we just want to compute the quantities of interest at all specified points. In this case, we don't need to pay the price in memory space (a memory pointer to either the code to evaluate a function or the result of a previous evaluation) of lazy evaluation. A general rule of thumb is that the strict version of a function is usually what you want, unless you really know what you are doing.

Let's look now at how we will compute curls in the FDTD method.

FDTD and the Curl

Equation 29.28 showed how to compute the x-component of the curl of the magnetic field in the FDTD method. There are five other analogous equations: two for the y- and z-components of the curl of the magnetic field and three for the components of the curl of the electric field. The approximation of the curl in Equation 29.28 is based on the approximation of the partial derivative in Equation 29.24, so we'll want to encode the partial derivative first. However, even more basic than computing the partial derivative is simply looking up values from the key-value lookup table, so let's address that now.

Looking Up Values in the Lookup Table

The Data.Map.Strict module provides the function lookup, which we write as M.lookup, to retrieve values from a lookup table. Let's look at the type of this function.

```
Prelude> :l Maxwell
[ 1 of 13] Compiling Newton2            ( Newton2.hs, interpreted )
[ 2 of 13] Compiling SimpleVec          ( SimpleVec.hs, interpreted )
[ 3 of 13] Compiling Mechanics1D        ( Mechanics1D.hs, interpreted )
[ 4 of 13] Compiling Mechanics3D        ( Mechanics3D.hs, interpreted )
[ 5 of 13] Compiling MultipleObjects    ( MultipleObjects.hs, interpreted )
[ 6 of 13] Compiling MOExamples         ( MOExamples.hs, interpreted )
[ 7 of 13] Compiling Electricity        ( Electricity.hs, interpreted )
[ 8 of 13] Compiling CoordinateSystems ( CoordinateSystems.hs, interpreted )
[ 9 of 13] Compiling Geometry           ( Geometry.hs, interpreted )
[10 of 13] Compiling Integrals          ( Integrals.lhs, interpreted )
[11 of 13] Compiling Charge             ( Charge.hs, interpreted )
[12 of 13] Compiling ElectricField      ( ElectricField.hs, interpreted )
[13 of 13] Compiling Maxwell            ( Maxwell.hs, interpreted )
```

```
Ok, 13 modules loaded.
*Maxwell> :t M.lookup
M.lookup :: Ord k => k -> M.Map k a -> Maybe a
```

If we wanted to see the type of lookup from Data.Map.Strict without loading the code in this chapter, we could do the following:

```
*Maxwell> :m Data.Map.Strict
Prelude Data.Map.Strict> :t Data.Map.Strict.lookup
Data.Map.Strict.lookup :: Ord k => k -> Map k a -> Maybe a
```

Here we ask for the type of lookup using the fully qualified name of the function, prefixing the module name before the function name to distinguish the lookup of Data.Map.Strict from the lookup of Prelude.

From the type, we see that M.lookup wants a key and a lookup table and will return something of type Maybe a. If it finds the key in the table, it will return the value associated with it, wrapped in the Just constructor of the Maybe a type. If it doesn't find the key, it will return Nothing.

Our helping function lookupAZ uses M.lookup to do its work.

```
lookupAZ :: Ord k => k -> M.Map k R -> R
lookupAZ key m = case M.lookup key m of
                   Nothing -> 0
                   Just x  -> x
```

The function lookupAZ (AZ for assume zero) has a slightly simpler type than M.lookup. The function serves two purposes. First, it relieves us of the need to case analyze the results each time we do a lookup. Second, when we compute a curl for a location at the edge of our grid, we will be trying to look up values that don't exist because they are just off the grid. For these two reasons, we write a function that treats nonexistent keys as if they had values of 0. This is not the safest procedure, because it will not help us find errors in our code if we are asking for nonexistent keys because of a mistake we made in programming. I'm generally a pretty careful and conservative person, but in this one case, I chose to live on the wild side.

The partial derivative requires that we look up values of the relevant component half a spatial step to either side of where we want to compute the derivative. Half a spatial step means one integer higher and one integer lower in the relevant direction. The functions partialX, partialY, and partialZ all have the same type.

```
partialX,partialY,partialZ :: R -> M.Map (Int,Int,Int) R -> (Int,Int,Int) -> R
partialX dx m (i,j,k) = (lookupAZ (i+1,j,k) m - lookupAZ (i-1,j,k) m) / dx
partialY dy m (i,j,k) = (lookupAZ (i,j+1,k) m - lookupAZ (i,j-1,k) m) / dy
partialZ dz m (i,j,k) = (lookupAZ (i,j,k+1) m - lookupAZ (i,j,k-1) m) / dz
```

Each function takes a spatial step size, a lookup table (called m for map), and a location as input. Each works by using the lookupAZ function to retrieve values on either side of the given location. The difference between these values is divided by the step size to obtain the approximation to the partial derivative.

Computing the Curl

With partial derivative in hand, we now turn to the curl. Here are six functions that compute components of the curl of the electric field and the magnetic field. Equation 29.7 gives the components of the curl.

```
curlEx,curlEy,curlEz,curlBx,curlBy,curlBz :: StateFDTD -> (Int,Int,Int) -> R
curlBx (StateFDTD _ _ dy dz _ b) loc = partialY dy b loc - partialZ dz b loc
curlBy (StateFDTD _ dx _ dz _ b) loc = partialZ dz b loc - partialX dx b loc
curlBz (StateFDTD _ dx dy _ _ b) loc = partialX dx b loc - partialY dy b loc
curlEx (StateFDTD _ _ dy dz e _) loc = partialY dy e loc - partialZ dz e loc
curlEy (StateFDTD _ dx _ dz e _) loc = partialZ dz e loc - partialX dx e loc
curlEz (StateFDTD _ dx dy _ e _) loc = partialX dx e loc - partialY dy e loc
```

Each curl function takes a `StateFDTD` and a location as input. The function `curlBx` computes the x-component of the curl of the magnetic field. According to Equation 29.7, this is the difference between the partial derivative with respect to y of B_z, denoted `partialY dy b loc` in the code above, and the partial derivative with respect to z of B_y, denoted `partialZ dz b loc`, each evaluated at the given location. Why do we not specify in the expression `partialY dy b loc` that it is the z-component we want the derivative of? The answer is that, because of the way the Yee cell is constructed, every location at which we need to compute `curlBx` has B_z living one integer away in the y-direction. The x-component of the curl of the magnetic field is needed only to update the x-component of the electric field. We use `curlBx` only when we update E_x, and B_z is its neighbor in the y-direction, so taking `partialY` at that location automatically takes the partial of B_z.

Let's look now at how we will update the state.

State Update

The function `stateUpdate` takes a time step, a time-dependent current density, and a state as input, and it uses that information to produce an updated state as output. It passes the real work off to the functions `updateE` and `updateB`, which update the electric and magnetic fields, respectively.

```
stateUpdate :: R                     -- dt
            -> (R -> VectorField)   -- current density J
            -> StateFDTD -> StateFDTD
stateUpdate dt j st0@(StateFDTD t _dx _dy _dz _e _b)
    = let st1 = updateE dt (j t) st0
          st2 = updateB dt st1
      in st2
```

Updating the electric field, as Equation 29.25 shows, requires knowledge of the current density, so we include the current density at the present time, `j t`, as an input to `updateE`.

The role of updateE is to carry out Equation 29.25 and the analogous equations for E_y and E_z that come from Equations 29.18 and 29.19.

```
updateE :: R             -- time step dt
        -> VectorField   -- current density J
        -> StateFDTD -> StateFDTD
updateE dt jVF st
    = st { timeFDTD = timeFDTD st + dt / 2
         , eField   = M.mapWithKey (updateEOneLoc dt jVF st) (eField st) }
```

The function updateE uses record syntax to update two of the items in the state: the time and the electric field. The function updates the current time, timeFDTD st, by adding half a time step to it. The function updateB adds the other half time step.

We update the electric field components at every place they are stored using the mapWithKey function from Data.Map.Strict. Let's look at the type of mapWithKey.

```
Prelude Data.Map.Strict> :m Data.Map.Strict
Prelude Data.Map.Strict> :t mapWithKey
mapWithKey :: (k -> a -> b) -> Map k a -> Map k b
```

The mapWithKey function takes a higher-order function k -> a -> b as input. For us, this will be a function (Int,Int,Int) -> R -> R. It describes how to use the key and value of a key-value pair to produce a new value. The function updateEOneLoc dt jVF st, defined later, serves this role for us, describing how to update an electric field component at a particular position in space.

The function updateB updates the magnetic field. It does for the magnetic field what updateE does for the electric field. The only difference is that the current density is not required to update the magnetic field, so it is not an input to updateB.

```
updateB :: R -> StateFDTD -> StateFDTD
updateB dt st
    = st { timeFDTD = timeFDTD st + dt / 2
         , bField   = M.mapWithKey (updateBOneLoc dt st) (bField st) }
```

As promised, updateB increases the time by half a time step, so after we have used both updateE and updateB, the time has increased by one whole time step. As with updateE, updateB uses mapWithKey to carry out the update over all locations we are tracking in the state. For the magnetic field, the function we map across the lookup table is called updateBOneLoc dt st. We define it later, and it describes how to update the magnetic field at one particular location in space.

Now we look at the functions that update the electric and magnetic fields at one point in space. Here we finally see the Maxwell equations. The function updateEOneLoc is responsible for updating electric field components at one location in space.

```
updateEOneLoc :: R -> VectorField -> StateFDTD -> (Int,Int,Int) -> R -> R
updateEOneLoc dt jVF st (nx,ny,nz) ec
    = let r = cart (fromIntegral nx * stepX st / 2)
                   (fromIntegral ny * stepY st / 2)
                   (fromIntegral nz * stepZ st / 2)
          Vec jx jy jz = jVF r
      in case (odd nx, odd ny, odd nz) of
           (True , False, False)
               -> ec + cSI**2 * (curlBx st (nx,ny,nz) - mu0 * jx) * dt  -- Ex
           (False, True , False)
               -> ec + cSI**2 * (curlBy st (nx,ny,nz) - mu0 * jy) * dt  -- Ey
           (False, False, True )
               -> ec + cSI**2 * (curlBz st (nx,ny,nz) - mu0 * jz) * dt  -- Ez
           _ -> error "updateEOneLoc passed bad indices"
```

It requires a time step, a current density, the state, a location, and a present electric field component value as inputs. It uses a `let` construction to define a few local variables. The local variable `r` holds the position described by the integer triple `(nx,ny,nz)`. The current density needs this position, and we compute it by multiplying each integer by half a spatial step in the appropriate direction. The local variables `jx`, `jy`, and `jz` are the components of the current density at the relevant location. Finally, we decide which component is being updated by examining the oddness or evenness of the three integers. As Table 29-2 indicates, an odd-even-even triple of integers means we are updating E_x, an even-odd-even triple of integers means we are updating E_y, and an even-even-odd triple of integers means we are updating E_z. We include a final line to catch a triple that does not fall into one of these three cases, which would indicate an error in our code since `updateEOneLoc` should only ever be used at locations that hold electric field components.

Depending on the case analysis, we update the electric field component, called `ec`, using one of the three Cartesian components of Equation 29.26. The local variable `ec` contains the present value of the electric field component to be updated (that is, one of E_x, E_y, or E_z).

The function `updateBOneLoc` does for the magnetic field what `updateEOneLoc` does for the electric field.

```
updateBOneLoc :: R -> StateFDTD -> (Int,Int,Int) -> R -> R
updateBOneLoc dt st (nx,ny,nz) bc
    = case (odd nx, odd ny, odd nz) of
        (False, True , True ) -> bc - curlEx st (nx,ny,nz) * dt  -- Bx
        (True , False, True ) -> bc - curlEy st (nx,ny,nz) * dt  -- By
        (True , True , False) -> bc - curlEz st (nx,ny,nz) * dt  -- Bz
        _ -> error "updateBOneLoc passed bad indices"
```

This function is simpler because it does not involve the current density. Again, there is a case analysis on the oddness or evenness of the three integers describing the location to determine which magnetic field component we are asking the function to update. As Table 29-2 indicates, an even-odd-odd triple of integers means we are updating B_x, an odd-even-odd triple of integers means we are updating B_y, and an odd-odd-even triple of integers means we are updating B_z.

Depending on the case analysis, we update the magnetic field component, called `bc`, using one of the three Cartesian components of Equation 29.27. The local variable `bc` contains the magnetic field component to be updated, which could be B_x, B_y, or B_z, depending on the oddness or evenness of the integers in the triple.

This completes the description of the FDTD method. The function `stateUpdate` is the entry point for those wishing to use the method. It requires a time step, a time-dependent current density, and an initial state, and it gives back an updated state one time step later. We will likely want to iterate this `stateUpdate` function to see the fields evolve over time.

Let's turn now to a use of the FDTD method by producing an animation of the fields for the radiation produced by an oscillating current density.

Animation

An accelerating charge radiates. In other words, an accelerating charge creates wavelike electric and magnetic fields that emanate away from the source charge. We can track how the electric and magnetic fields evolve in time by solving the Maxwell equations using the FDTD method we developed.

Current Density

In this section, we'll produce an animation of the electric field produced by an oscillating current density. Our current density will be localized in space, and we'll center our coordinate system on the current density. There are several ways we could produce a localized current density. Since we have discretized space in the FDTD method, one way to specify a localized current density is to allow the current density to be nonzero at a single location in the FDTD grid. Slightly more convenient for us is to specify a current density that extends over several grid points but quickly drops off with distance from its center.

One function that drops off in this way depends on the distance r from the origin as e^{-r^2/l^2}. Such a function is called a *Gaussian*. Its largest value is at the origin, and its value decreases with distance from the origin. The parameter l has dimensions of length and gives a sense of the region over which the value is significant. At $r = l$, the Gaussian value is 36.8 percent of its value at the origin. At $r = 2l$, its value is only 1.8 percent of its value at the origin. And by $r = 3l$, its value is only about a hundredth of a percent. Equation 29.29 gives the current density we use for our radiation animation.

$$\mathbf{J}(t, \mathbf{r}) = J_0 e^{-r^2/l^2} \cos(2\pi f t)\hat{\mathbf{k}} \tag{29.29}$$

We need three parameters to fully specify this current density: an amplitude J_0, a localization length l, and a frequency f. We can think of this current density as representing a charge at the origin that is oscillating in the z-direction.

The function jGaussian describes the current density in Equation 29.29.

```
jGaussian :: R -> VectorField
jGaussian t r
    = let wavelength = 1.08                  -- meters
          frequency = cSI / wavelength      -- Hz
          j0 = 77.5                          -- A/m^2
          l = 0.108                          -- meters
          rMag = magnitude (rVF r)           -- meters
      in j0 *^ exp (-rMag**2 / l**2) *^ cos (2*pi*frequency*t) *^ kHat
```

The function jGaussian uses some local variables to specify its behavior. We wish for the oscillation to occur with a frequency that will produce radiation with a wavelength of 1.08 m. The frequency (in Hz) is the speed of light divided by the wavelength. We chose an amplitude of 77.5 A/m^2 because this radiates about 100 W of power. We chose the parameter l to be 0.108 m, which is the same value we will choose later for the spatial step size of the grid. This means that only grid points near the origin will contain any significant current density.

Having decided on a current density to serve as the source of our electric and magnetic fields, we turn to a few comments about the boundary of the grid.

Grid Boundary

The FDTD method uses a finite grid that keeps track of the electric and magnetic fields at a finite number of places. We use neighboring grid points to calculate the curl needed by the Maxwell equations, as explained earlier. What happens at the edges of the grid? The simple choice we made is to assume that electric and magnetic fields beyond the grid are 0. This choice is enforced by the lookupAZ function, which returns 0 for any point off the grid. While this choice is simple and seems reasonable, it has some undesirable properties. An outgoing wave will reflect at the edge of the grid, bounce back, and interfere. However, if the grid is very large, the amplitude of the reflected wave may be very small and its presence may be tolerable. In our case, we display only a portion of the grid on which calculations occur. Our animation terminates before the wave hits the boundary, so we do not see any reflected wave. In general, the results obtained using our simple boundary conditions are only valid until the wave propagates to the edge of the grid. There are more sophisticated methods available for dealing with the boundary conditions at the edge of the grid. One way is to model a material that absorbs all of the radiation incident; this more or less acts like an infinite box, without requiring computing an infinite number of points. The book by Inan and Marshall [18] has a nice discussion of boundary conditions for the FDTD method.

Even without sophisticated boundary conditions, the calculations we are doing are computationally intensive. It could take 20 minutes or more to generate all of the PNG files that will be sewn together to produce the final animation. The files are produced as the information becomes available, so you can see on your own machine how many files are produced in a minute and estimate how long the entire batch will take.

Now we are ready to turn to the question of producing frames for the asynchronous animation.

Display Function

We want a function that will produce a picture given a StateFDTD. The function makeEpng serves this role. It produces a PNG graphics file from a state of the electromagnetic field. We intend to produce one such graphics file at each time step and then sew them together into an animation. The picture we produce in makeEpng is that of the electric field in the xz-plane. We use shading to indicate the strength of the field, transitioning from one color (usually black or white) indicating a zero field to another color indicating some maximum strength.

```
makeEpng :: (Colour R, Colour R) -> (Int,StateFDTD) -> IO ()
makeEpng (scol,zcol) (n,StateFDTD _ _ _ _ em _)
    = let threeDigitString = reverse $ take 3 $ reverse ("00" ++ show n)
          pngFilePath = "MaxVF" ++ threeDigitString ++ ".png"
          strongE = 176  -- V/m
          vs = [((fromIntegral nx, fromIntegral nz),(xComp ev, zComp ev))
                | nx <- evens, nz <- evens, abs nx <= 50, abs nz <= 50
                , let ev = getAverage (nx,0,nz) em ^/ strongE]
      in gradientVectorPNG pngFilePath (scol,zcol) vs
```

The function makeEpng takes a pair of colors as input, as well as a pair containing an integer n and the state of the electromagnetic field. The color pair consists of a strong color scol for the color of the strongest fields and a zero color zcol for the color of a zero field. The integer n that is paired with the state serves as part of the name of the PNG file.

The function makeEpng uses local variables to name the PNG file, the threshold for a strong electric field, and a list of electric field values to be displayed. The local name pngFilePath is a String whose value is the name of the PNG file to be produced. This name is *MaxVF* followed by three digits from the integer n, followed by *.png*. We use the threshold strongE for a strong electric field to choose the display color for each electric field arrow. We color electric field values of 176 V/m or higher with the strong color scol, values of 0 with the zero color zcol, and values in between with a blend of the two colors.

The list vs has type [((R,R),(R,R))] and contains the two-dimensional locations and components of the electric field to be displayed. The function getAverage, defined next, takes a triple of even integers as input and averages the values on either side of the Yee cell to produce a vector at a single point in space. Finally, we use the function gradientVectorPNG, defined below, to make the picture.

Two Helping Functions

The function getAverage, used in makeEpng earlier, produces field vectors at particular locations by averaging the values around the location. Since the Yee cell stores different field components at different locations, we might ask whether there is any natural way to recombine the components into a single vector. The answer is yes, as long as we are willing to use the average of the values at two locations. At any point in the Yee cell labeled by an even-even-even triple, electric field components are stored at each adjacent location. By averaging these, we can produce an electric field vector at any even-even-even location. Similarly, we can produce a magnetic field vector at any odd-odd-odd location.

```
getAverage :: (Int,Int,Int)  -- (even,even,even) or (odd,odd,odd)
           -> M.Map (Int,Int,Int) R
           -> Vec
getAverage (i,j,k) m
    = let vXl = lookupAZ (i-1,j  ,k  ) m
          vYl = lookupAZ (i  ,j-1,k  ) m
          vZl = lookupAZ (i  ,j  ,k-1) m
          vXr = lookupAZ (i+1,j  ,k  ) m
          vYr = lookupAZ (i  ,j+1,k  ) m
          vZr = lookupAZ (i  ,j  ,k+1) m
      in vec ((vXl+vXr)/2) ((vYl+vYr)/2) ((vZl+vZr)/2)
```

The function getAverage takes an integer triple as input, which should be either even-even-even for an electric field or odd-odd-odd for a magnetic field, along with a lookup table, and produces a vector. It does this by sampling the six locations adjacent to the input location, averaging the values in each direction, and putting the averaged components into a vector.

The function gradientVectorPNG, used by makeEpng earlier, is similar to vfGrad from Chapter 22. It produces a gradient vector field picture.

```
gradientVectorPNG :: FilePath
                  -> (Colour R, Colour R)
                  -> [((R,R),(R,R))]
                  -> IO ()
gradientVectorPNG fileName (scol,zcol) vs
    = let maxX = maximum $ map fst $ map fst $ vs
          normalize (x,y) = (x/maxX,y/maxX)
          array = [(normalize (x,y), magRad v) | ((x,y),v) <- vs]
          arrowMagRadColors :: R  -- magnitude
                            -> R  -- angle in radians, ccw from x axis
                            -> Diagram B
```

```
        arrowMagRadColors mag th
            = let r      = sinA (15 D.@@ deg) / sinA (60 D.@@ deg)
                  myType = PolyPolar [120 D.@@ deg,  0 D.@@ deg, 45 D.@@ deg
                                     , 30 D.@@ deg, 45 D.@@ deg,  0 D.@@ deg
                                     ,120 D.@@ deg]
                           [1,1,r,1,1,r,1,1]
                  myOpts = PolygonOpts myType NoOrient (p2 (0,0))
              in D.scale 0.5 $ polygon myOpts # lw none #
                 fc (blend mag scol zcol) # rotate (th D.@@ rad)
        step = 2 / (sqrt $ fromIntegral $ length vs)
        scaledArrow m th = D.scale step $ arrowMagRadColors m th
        pic = D.position [(p2 pt, scaledArrow m th) | (pt,(m,th)) <- array]
    in renderCairo fileName (dims (V2 1024 1024)) pic
```

The function gradientVectorPNG takes three inputs: a name for the PNG file, a pair of colors to use, and a list of two-dimensional vector locations and components. It gives the local name fileName to the string given as a name for the PNG file. It gives the local names scol and zcol to the strong color and zero color to be used in the picture. The list vs :: [((R,R),(R,R))] gives the locations (first pair of real numbers) and components (second pair of real numbers) of the vectors to be displayed. The magnitudes of these two-dimensional vectors are expected to be in the range 0 (which will get the zero color) to 1 (which will get the strong color).

The function gradientVectorPNG assigns the local name maxX to the largest value of x describing the locations of the arrows. The local function normalize takes an (x, y) pair as input and returns a pair in the square from $(-1, -1)$ to $(1, 1)$. The function normalize assumes that the region to be displayed is a square patch in the xy-plane centered at the origin. The local list array contains the normalized locations at which arrows are to be placed, along with the magnitude and orientation of each arrow.

The function arrowMagRadColors is a helping function that produces a diagram of a single arrow. We define it as a local function because gradientVectorPNG is the only function that uses it. Since it is a local function, it can use the local colors scol and zcol without these colors being inputs to arrowMagRadColors. The function arrowMagRadColors expects the magnitude of arrows to be in the range from 0 to 1, assigning the zero color to 0 and the strong color to 1.

We use the local variable step to scale the size of the arrows. It is based on the number of arrows to be displayed on each horizontal row, that number being equal to the square root of the total number of arrows to be displayed over the entire square. The local variable pic holds the entire picture that the final line of the function renders.

Main Program

Listing 29-2 sets the time step size, the number of time steps to be taken, and the spatial step size through its specification of the initial state.

```
{-# OPTIONS -Wall #-}

import Maxwell ( makeEpng, stateUpdate, jGaussian, initialStateFDTD )
import Diagrams.Prelude ( black, yellow )

main :: IO ()
main = let dt = 0.02e-9   -- 0.02 ns time step
           numTimeSteps = 719
       in sequence_ $ map (makeEpng (yellow,black)) $ zip [0..numTimeSteps] $
          iterate (stateUpdate dt jGaussian) (initialStateFDTD 0.108)
```

Listing 29-2: Stand-alone program to produce PNG files for an electric field animation

It uses the `sequence_` function, described in Chapter 20, to turn a list of actions into a single action. Since the function application operator `$` is right associative (recall Table 1-2), it's easiest to read the definition of `mainPNGs` from right to left. The rightmost phrase,

```
iterate (stateUpdate dt jGaussian) (initialStateFDTD 0.108)
```

is an infinite list of states, starting with an initial state in which the electric and magnetic fields are 0 everywhere and the spatial step size in each direction is 0.108 m. Applying `zip [0..numTimeSteps]` to this infinite list produces a finite list, each element being a pair of an integer with a state. Applying `map (makeEpng (yellow,black))` to this list of pairs produces a finite list with type `[IO ()]`. Finally, applying `sequence_` converts the list of actions into a single action. This program will produce 720 files, named *MaxVF000.png* through *MaxVF719.png*, that we can combine into an MP4 movie with an external program such as `ffmpeg`.

The following command asks the external program `ffmpeg` to combine all PNG files named *MaxVFDDD.png*, where the capital Ds are digits. We ask for a frame rate of 25 frames/second. The final movie is called *MaxVF.mp4*.

```
$ ffmpeg -framerate 25 -i MaxVF%03d.png MaxVF.mp4
```

We use a spatial step size of 0.108 m because it is one tenth of the wavelength we expect from our current density. 10 spatial steps per wavelength is about the smallest I'd like to go. More spatial steps per wavelength would produce more accurate results but would take longer to run, assuming we increase the number of grid points to allow the same number of wavelengths to fit across the grid.

The time step needs to be a bit smaller than the time it takes light to travel one spatial step; otherwise, the method becomes unstable. (See [18] for details on the stability criterion.) It takes light about 0.36 ns to travel one spatial step. Our time step of 0.02 ns is plenty small to avoid instability. Of course, a smaller time step produces more accurate results at the cost of longer computation time.

Figure 29-3 shows one of the frames of the animation, except that we used black as the strong color and white as the zero color.

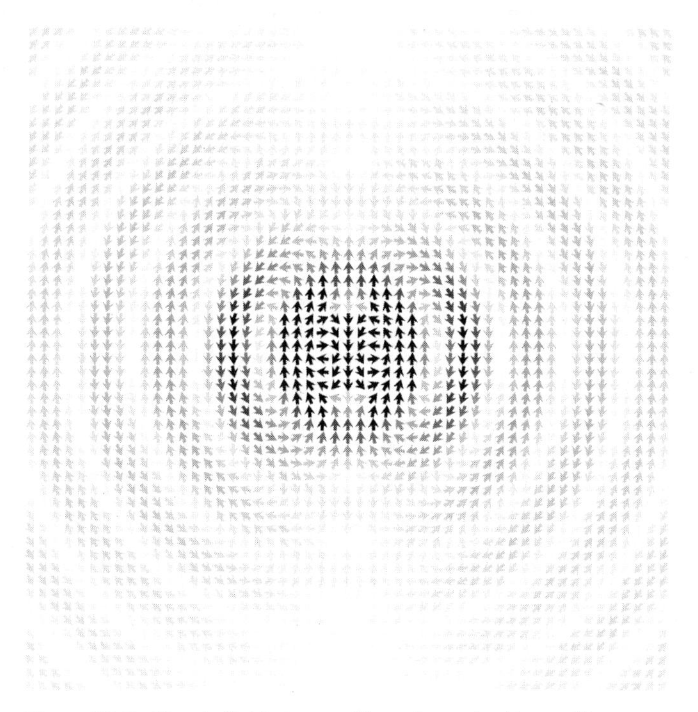

Figure 29-3: Electric field obtained by solving the Maxwell equations using the current density `jGaussian`. The image is one of the frames produced by the main program and shows the xz-plane.

The wavelike nature of the electric field is apparent. The magnetic field, not shown in Figure 29-3, points into or out of the page. The magnitude of the electric field decreases with distance from the source at the center. The radiating electric field is stronger in the $z = 0$ plane and weaker above and below the source in the z-direction.

Summary

In this chapter, we saw how the Maxwell equations describe the evolution of electric and magnetic fields. We identified four relationships between electricity and magnetism, and we explained how the Maxwell equations relate to our description of electric and magnetic fields in previous chapters. We saw how the Maxwell equations, like Newton's second law, can be viewed as rules for a state-update technique. We described the FDTD method for solving the Maxwell equations, and we applied it to the radiation produced by an oscillating charge and current density. We produced an animation of the wavelike electric field generated by an oscillating current density.

We've covered a lot of ground in this book. Many of the ideas are really cool but not so easy to grasp right away. If you're like most people, you've understood some things and gotten stuck on others. Patience and perseverance are my advice when you're stuck. Patience is especially important, and it sometimes means skipping ahead to the beginning of the next section or chapter. I've had books on my shelves that were unreadable for years, but somehow I got the background I needed and then one day I could read them.

I hope you've enjoyed the introduction to computational physics in Haskell that this book provides. You can, of course, do computational physics in any programming language you like. You would learn a lot by translating the work we've done here into another language. Let's take a brief look back at what we've done to recall the benefits of a functional language for physics. A pure functional language allows and encourages us to express what is central and important in a single function. The function `newtonSecondPS` from Chapter 16 expresses Newton's second law. The function `maxwellUpdate` in this chapter expresses the Maxwell equations.

Pure functional programming provides a simpler model of computation than imperative languages like Python since names (variables) refer to quantities that never change. This encourages the naming of verbs (functions) that change nouns, rather than giving names to nouns that change. Physics is a natural candidate to take advantage of pure functional writing because the central ideas of physics, like Newton's second law and the Maxwell equations, are expressible as verbs.

Moreover, a typed functional language allows us to express with precision the nature of the verb a function describes. The type of `newtonSecondPS` expresses that we can produce a differential equation from a list of one-body forces. The type of `bFieldFromLineCurrent` from Chapter 27 expresses that we can calculate a magnetic field from a curve and a current.

The effectful functions we've written, those like `gradientVectorPNG` with `IO ()` in their type that *do* something, are certainly useful for producing graphs and animations, but they are not really part of the "elegant code" promised in this book's subtitle. Haskell is as powerful as any other language for doing these things, but the real strength of a functional language for physics lies in the elegant expression of its central ideas, which are purely functional. Programming in a functional language has allowed me to write code that parallels the organization of the subject in my mind. I find that it helps me to think about the subject.

The practice of writing physics in a functional language is in its infancy. Two advanced books on the subject are [20] and [11]. There is a lot more to discover about what physics and functional programming can offer each other. I hope you continue to explore these areas if they interest you.

Exercises

Exercise 29.1. Use `gnuplot` to graph the Gaussian function

$$G(x) = e^{-x^2/l^2}$$

for several values of l.

Exercise 29.2. Modify the main program and the function `makeEpng` to produce an animation of the magnetic field in the xy-plane produced by the current density `jGaussian`. Try using 10^{-6} T as the threshold for a strong magnetic field.

Exercise 29.3. The current density

$$\mathbf{J}(t, \mathbf{r}) = J_0 e^{-r^2/l^2} \cos(2\pi f t)\hat{\phi} \tag{29.30}$$

has an oscillating magnetic dipole moment, while that of Equation 29.29 has an oscillating electric dipole moment. The radiation produced by the current density of Equation 29.30 is called magnetic dipole radiation. Produce an animation of the magnetic field in the xz-plane. It should look similar to the electric field animation we made for the electric dipole radiation from the current density of Equation 29.29. Try using 2×10^{-7} T as the threshold for a strong magnetic field.

APPENDIX

INSTALLING HASKELL

This appendix explains how to install the Glasgow Haskell Compiler and libraries other people have written.

Installing GHC

The Glasgow Haskell Compiler (GHC) is the Haskell compiler we use in this book. It is free, open source software anyone can download and install.

The installation procedure depends on which operating system you use. For GNU/Linux and macOS users, I recommend going to *https://www.haskell.org* and choosing **Downloads**. Follow the instructions for your operating system. You will know you have succeeded when you can start the GHCi interactive compiler, usually by typing `ghci` at the command prompt. At this point, you are ready to get started with Chapter 1. In addition to GHC itself, the installation method you use will install either Cabal or Stack. Cabal and Stack are the two most common tools for installing additional library packages. I describe their use later in this appendix.

For Microsoft Windows users, I recommend following the instructions at *https://www.fpcomplete.com/haskell/get-started/windows*. FPComplete is a company that provides services for industrial Haskell users. The installer they provide will install both the Glasgow Haskell Compiler and the Stack library package manager. You will know you have succeeded when you can start the GHCi interactive compiler and obtain a GHCi prompt by typing `stack ghci` at the PowerShell prompt. At this point, you are ready to get started with Chapter 1.

Installing a Text Editor

To write source code files, you will need a text editor. You can use a basic text editor like Notes on macOS or gedit on Linux, or you can choose from a number of more sophisticated text editors available. These more sophisticated editors often have helpful features for programmers, like text highlighting, that can often be configured to be sensitive to the language you are programming in.

You can find advice for getting your Haskell environment to work smoothly with your editor on the Haskell wiki at *https://wiki.haskell.org/Haskell*. Good editors for Haskell are Emacs, Vim, Visual Studio Code, and Atom. Simple text editors like Notes are typically included with the operating system. Emacs is available at *https://www.gnu.org/software/emacs*, Vim is available at *https://www.vim.org*, Atom is available at *https://atom.io*, and Visual Studio Code is available at *https://code.visualstudio.com*. Follow the instructions for your operating system. (Users who want to run Emacs on macOS should download it from *https://emacsforosx.com*. This link provides standard Emacs built to run in the macOS environment. Since it is standard Emacs, it is possible to reliably customize it according to the advice you find online. The first stop in customization is *https://www.emacswiki.org*.)

Installing Gnuplot

Beginning in Chapter 7, we use `gnuplot` to make graphs. `Gnuplot` is a standalone graphing program, independent of Haskell, with a web page at *http://gnuplut.info*. Installing `gnuplot` so that it can be used with Haskell is a two-step process. First, you must install the `gnuplot` program so that it works independently from Haskell. Second, you must install the Haskell `gnuplot` package so that Haskell code can access `gnuplot`'s functionality. This section deals with installing the `gnuplot` program, while the following section explains how to install the Haskell `gnuplot` package.

The process for installing the `gnuplot` program depends on your operating system. For GNU/Linux, you can usually use your package manager. For example, on Ubuntu Linux, the command

```
$ sudo apt install gnuplot
```

will install the `gnuplot` program.

On macOS, I recommend the Homebrew package manager at *https://brew.sh*. After you follow the instructions to install Homebrew, you can issue the following command to install the `gnuplot` program:

```
$ brew install gnuplot
```

On Microsoft Windows, follow the instructions at *http://www.gnuplot.info* to download the `gnuplot` installer for Windows. Run the installer, which asks a series of questions about where to install things and other installation details. Make a note of the directory in which `gnuplot` gets installed (perhaps *C:\Program Files\gnuplot\bin*). You can accept the default settings for all of the installer's questions except one: when the installer gives the opportunity to "Add application directory to your PATH environment variable," check that box. After the installer has completed its work, there is one more thing you must do. Using a file browser, navigate to the directory where `gnuplot` is installed and find the file named *wgnuplot_pipes*. Copy this file to a new file in the same directory called *pgnuplot*. If the file is named *wgnuplot_pipes.exe*, copy it to a new file in the same directory called *pgnuplot.exe*. This will allow Haskell to use `gnuplot`.

At this point, whatever your operating system, you should be able to run the `gnuplot` program independently from Haskell. From a command line, you would type the following:

```
$ gnuplot
```

After starting `gnuplot`, you should be able to issue a command at the `gnuplot` prompt, such as

```
gnuplot> plot cos(x)
```

and a window containing a plot should pop open. Once you succeed in installing the `gnuplot` program, you are ready to install the Haskell `gnuplot` package, which lets you control `gnuplot` from Haskell.

Installing Haskell Library Packages

There are functions other people have written that we will want to use that are not included in the Prelude (the standard collection of functions available by default). Such functions exist in library modules that can be imported in our source code file or loaded directly into GHCi. There is a standard set of library modules that comes with GHC, and there are others you can install with Cabal or Stack. Library modules outside of the standard libraries are organized into *packages*, each containing one or more modules.

Suppose we want access to the `plotFunc` function in the `Graphics.Gnuplot.Simple` module provided by the Haskell `gnuplot` package. We must install the `gnuplot` package.

The two major tools for installing Haskell library packages are Cabal and Stack. You need to use only one of these. At least one of them will be available by following the instructions for GHC installation.

Cabal (Common Architecture for Building Applications and Libraries) existed first. At the time it was written (around 2005), it was considered very important to minimize the number of required downloads, so Cabal was designed to install a global set of packages, and all applications were supposed to build against this common set of packages. And again, in the interest of efficiency, Cabal allowed only one version of each package to be installed.

This led to a problem: many libraries were evolving quickly, adding features and changing their interfaces. A common problem encountered was that an application might build against libraries that, in turn, depended on different versions of a common ancestor. This sometimes required uninstalling and reinstalling all of your packages, and occasionally reloading different versions of all your packages to build a new application. The problem was called "dependency hell" or "Cabal hell," and the name tells you all you need to know about how painful it was.

The solution was to allow multiple versions of packages to be installed, and Cabal now allows this.

The Stack system provides many of the same features as Cabal, and in fact can smoothly work alongside it, but its goals are slightly different. Stack is aimed at meeting the requirements of commercial users who need assurance that their applications will build even as the Haskell library infrastructure evolves. Stack calls this goal "reproducible builds." To get reproducible builds, Stack's default mode of operation is to let you specify a compiler version and a set of curated packages known to work properly with that compiler. The curated sets include more than 2,000 packages, so you are likely to find most of what you need there (and if you don't, it's not hard to specify the additional packages you want downloaded and built). The upside of this apparent complexity is that not only does your Haskell program build the same way each time, it runs the same way.

Stack and Cabal are generally able to avoid the problem of inconsistent dependencies breaking the builds of large, complicated projects. There is a price for this, though. They may download many more packages than you expect. Stack, in particular, may download multiple compilers to ensure that the packages and compiler are known to produce consistent results. This may seem unnecessary, but it's required by the way the GHC compiler works. For important, but fussy, technical reasons, the GHC compiler does not have a standardized "application binary interface" (ABI). This means you can't use libraries compiled with one version of GHC with an application compiled by another. This is not a bug—it turns out that to get a pure functional language with lazy evaluation and decent performance, you need to give up something. And one of those somethings is a stable ABI.

Using Cabal

To load a module into GHCi, the working directory must have access to the module. For a module outside the standard modules provided by the GHC installation itself, the package that contains the module must be installed. There are two ways to install a package using Cabal: globally, so that the package can be accessed from any directory, and locally, so that it can be accessed only from the current working directory.

Using Cabal to Install a Package Globally

To install the `gnuplot` package globally, issue the following command:

```
$ cabal install --lib gnuplot
```

On my computer, this command creates or changes the file */home/walck/.ghc/x86_64-linux-8.10.5/environments/default*, which contains the list of globally installed Haskell packages. After you have installed one or more packages globally, a Cabal command such as the one we just issued may fail to install a new package if Cabal cannot find a version of the requested package that is compatible with existing globally installed packages. One way to solve this problem is to rename the file containing the global package list and then try to install all of the packages you want simultaneously. For example, to install the `gnuplot`, `gloss`, and `cyclotomic` packages simultaneously, you would issue the following command:

```
$ cabal install --lib gnuplot gloss cyclotomic
```

Because we renamed the global package list, Cabal will not find a global package list and consequently will make a new one.

Using Cabal to Install a Package Locally

To install the `gnuplot` package locally (in the current working directory), issue the following command:

```
$ cabal install --lib gnuplot --package-env .
```

The dot at the end of the command refers to the current working directory. This command creates or changes a file with a name like *.ghc.environment.x86_64-linux-8.10.5* in the current working directory. This file contains a list of packages installed locally (in the current working directory). After you have installed one or more packages locally in some directory, a Cabal command such as the one we just issued may fail to install a new package if Cabal cannot find a version of the requested package that is compatible with existing locally installed packages. One way to solve this problem is to rename the file containing the local package list and then try to install all of the packages you want simultaneously. For example, to install the `gnuplot`, `gloss`, and `cyclotomic` packages simultaneously, you would issue the following command:

```
$ cabal install --lib gnuplot gloss cyclotomic --package-env .
```

Because we renamed the local package list, Cabal will not find a local package list and consequently will make a new one.

Using Stack

To install the gnuplot package using Stack, issue the command

```
$ stack install gnuplot
```

at the command prompt. Stack keeps track of more things behind the scenes than Cabal does, and global installation with Stack is usually all you need.

After installing the gnuplot package, you can load the Graphics.Gnuplot .Simple module into GHCi. If you are using Stack, you should start GHCi with stack ghci rather than ghci. In this way, Stack will be able to find the modules of the packages you have installed.

```
Prelude> :m Graphics.Gnuplot.Simple
Prelude Graphics.Gnuplot.Simple> :t plotFunc
plotFunc
  :: (Graphics.Gnuplot.Value.Atom.C a,
      Graphics.Gnuplot.Value.Tuple.C a) =>
     [Attribute] -> [a] -> (a -> a) -> IO ()
```

Here we ask for the type of the function plotFunc, simply to show that it is available now that we've loaded the module that defines it.

To use the plotFunc function in a source code file, include the line

```
import Graphics.Gnuplot.Simple
```

at the top of your source code file.

Installing Gloss

Beginning in Chapter 13 we use gloss to make animations. Unlike gnuplot, gloss is not a stand-alone program; it is only a Haskell package. However, gloss uses the freeglut graphics libraries to do its work, and the freeglut functionality is supplied by non-Haskell libraries that must be installed separately from the gloss package itself. So, like installing gnuplot, installing gloss is a two-step process. First, you must install the non-Haskell freeglut libraries. Second, you must install the Haskell gloss package.

The procedure for installing the freeglut libraries depends on your operating system. For a GNU/Linux system, a command such as

```
$ sudo apt install freeglut3
```

should do the trick. On macOS, a command like

```
$ brew install freeglut3
```

is what you want. You will need to install the brew package manager to use this command. On macOS, you may also need the xquartz package to use freeglut, which you can install with

```
$ brew install xquartz
```

For a Microsoft Windows system, search the web for "freeglut windows" and follow the instructions you find.

After you install the freeglut libraries, you can install the `gloss` package with a command like

```
$ cabal install --lib gloss
```

or

```
$ stack install gloss
```

depending on whether you are using Cabal or Stack.

Installing Diagrams

Beginning in Chapter 22 we use the `diagrams` package to visualize vector fields. Actually, the `diagrams` package is just a wrapper around three packages called `diagrams-core`, `diagrams-lib`, and `diagrams-contrib`. The purpose of the wrapper is to make it easier to install because you can issue one command instead of three. We will use two of these three packages plus one other. We will use `diagrams-core`, `diagrams-lib`, and `diagrams-cairo`.

Similar to `gloss`, the `diagrams-cairo` package uses some graphics libraries to do its work, and you must install these non-Haskell libraries separately from the `diagrams-cairo` package itself. So, like installing `gnuplot` and `gloss`, installing `diagrams` is a two-step process. First, you must install the non-Haskell graphics libraries. Second, you must install the Haskell `diagrams` packages.

The graphics libraries needed are called `cairo` and `pango`. The procedure for installing these libraries depends on your operating system. For a GNU/Linux system, a command such as

```
$ sudo apt install libcairo2-dev libpango1.0-dev
```

should do the trick. On macOS, you can use a similar command with the `brew` package manager.

After you've installed the `cairo` and `pango` libraries, you can install the `diagrams` packages with a command like

```
$ cabal install --lib diagrams-core diagrams-lib diagrams-cairo
```

or

```
$ stack install diagrams-core diagrams-lib diagrams-cairo
```

depending on whether you are using Cabal or Stack.

Setting Up Your Coding Environment

As this book progresses, our code gets more complicated because we use modules other people have written and modules that we have written ourselves. We want to load some of our code into GHCi, and we also want to write stand-alone programs. Thus, we need a way to stay organized so that we have access to the modules we need so we are empowered to do the things we want to do. There are two main ways to stay organized:

(1) Keep all of your source code files in a single directory. This includes files intended for loading into GHCi as well as stand-alone programs. Install packages so that this directory has access to them. Arrange for this directory to have access to the modules from the book.

(2) Create a fresh directory for each project you work on. See that this directory has access to whatever modules and packages the project needs. Each directory might have a *.cabal* file and, if you are using `stack`, a *stack.yaml* file. These files describe the requirements for your project.

I recommend method (1), at least until you see some advantage in making a new directory for a new project. For the purposes of this book, the exercises you are asked to do are not so large that each demands its own directory.

What We Want in a Coding Environment

Before I give two specific suggestions for organizing your coding environment, let's lay out what we are trying to achieve. What follows are four desired properties we want our coding environment to have:

(a) We want to be able to load a source code file we have written into GHCi with GHCi's `:l` command. Such a source code file may or may not have a module name. Such a source code file also may or may not import modules using Haskell's `import` keyword.

(b) We want to be able to load a module that someone else has written, such as `Graphics.Gnuplot.Simple`, into GHCi with GHCi's `:m` command.

(c) We want to be able to produce an executable program from a source code file we have written. Such a source code file may or may not import modules using Haskell's `import` keyword.

(d) We want to be able to use the modules defined in this book by loading them into GHCi and by writing source code to `import` them.

To load a source code file into GHCi, as desired in (a), we will need to start GHCi in the directory where our file lives. If our source code file imports modules, it needs access to them. If a module our source code file imports is provided by a package, the current working directory must have access to that package. This can be either local or global access, as defined earlier in this appendix. If the module is defined in a source code file, such as one of the modules written in this book, that file must live in the working directory, or in a place where GHC knows to look for it.

To load a module written by someone else into GHCi, as desired in (b), the working directory needs to have access to the package that provides the module we wish to load. This can be local or global access, as described earlier.

Producing a stand-alone program, as desired in (c), is the subject of Chapter 12. There, we discuss three methods to produce a stand-alone program: one using GHC, one using Cabal, and one using Stack. Using Cabal or Stack as described in that chapter is a form of method (2) since we are allowed to have only one *.cabal* file in each directory. However, that *.cabal* file is allowed to specify multiple stand-alone programs, so it is possible to use Cabal or Stack with method (1).

To achieve (d), the simplest thing is to put all of the module-defining *.hs* files (such as *Mechanics3D.hs*, which defines the `Mechanics3D` module) into your working directory. Since the source code file you write is also in this directory, GHC will look for modules your source code file imports in the working directory when you load your file into GHCi, or when you compile it with GHC.

The following two sections give specific suggestions about where to put the modules defined in this book, which you can download at *https://lpfp.io*. The two suggestions are alternatives, so you need to follow only one of them.

All Code in One Directory

As suggested earlier, the simplest method for staying organized is to put everything in one directory. This includes:

- Source code files you intend to load into GHCi
- Source code files you intend to compile into executable programs
- Source code files, such as *Mechanics3D.hs*, for the modules defined in this book

This one directory will be your working directory for all of your Haskell work. If you continue to program in Haskell, you will outgrow this method. You will want to work on different projects with different purposes and different needs, and you won't want all of your code in one directory. When you get to this point, there are many ways forward. The Cabal and Stack tools offer many ways to organize your work.

For now, we need to make sure our one working directory has access to all of the packages we need for the projects in this book. The following command, to be entered as one long line at a command prompt, will locally install all the packages we need for this book.

```
$ cabal install --lib gnuplot gloss not-gloss spatial-math diagrams-lib
  diagrams-cairo --package-env .
```

One disadvantage of this method is that we can load the book modules with GHCi's `:l` command but not with GHCi's `:m` command, and this means we can load only one book module into GHCi at a time. This could be inconvenient if we want access in GHCi to functions defined in different book modules. One way around this is to make a new source code file that imports all the modules we want, and then load that source code file into GHCi with :l.

Another way around this disadvantage is to use the Stack tool to manage the modules from this book, as explained in the next section.

One Way to Use Stack

The Cabal and Stack tools provide many (maybe too many) ways for you to organize your work in Haskell. Here we will look at one way in detail. In this method, we still have one directory that contains all of our Haskell work, but this directory has two subdirectories: one for the book modules, and one for stand-alone programs. So there are three places where source code files can exist. They can live in the main working directory, they can live in the module subdirectory, or they can live in the stand-alone program subdirectory. A source code file you intend to load into GHCi will probably live in the main working directory.

Stack needs two configuration files to manage things. One is named *LPFP.cabal*, and the other is named *stack.yaml*. These two files will live in the main working directory. The file *LPFP.cabal* describes the modules we want to have access to, as well as the executable programs we want Stack to build for us. Listing A-1 gives this file.

```
cabal-version: 1.12

name:           LPFP
version:        1.0
description:    Code for the book Learn Physics with Functional Programming
homepage:       http://lpfp.io
author:         Scott N. Walck
maintainer:     walck@lvc.edu
copyright:      2022 Scott N. Walck
license:        BSD3
license-file:   LICENSE
build-type:     Simple

library
  exposed-modules:
      Charge, CoordinateSystems, Current, ElectricField, Electricity, Geometry
    , Integrals, Lorentz, MagneticField, Maxwell, Mechanics1D, Mechanics3D
    , MOExamples, MultipleObjects, Newton2, SimpleVec
  hs-source-dirs: src
  build-depends:
      base >=4.7 && <5, gnuplot, spatial-math, gloss, not-gloss, diagrams-lib
    , diagrams-cairo, containers
  default-language: Haskell2010

executable LPFP-VisTwoSprings
  main-is: VisTwoSprings.hs
```

```
  hs-source-dirs: app
  build-depends: LPFP, base >=4.7 && <5, not-gloss
  default-language: Haskell2010

executable LPFP-GlossWave
  main-is: GlossWave.hs
  hs-source-dirs: app
  build-depends: LPFP, base >=4.7 && <5, gloss
  default-language: Haskell2010
```

Listing A-1: The file LPFP.cabal *describing the modules we want access to and the executables we want produced*

After some introductory matter appear one library stanza and two executable stanzas. The library stanza lists all of the modules from this book that we want to have access to. It says that the source code for these modules is in the subdirectory *src* and that these modules depend on several packages, such as `gnuplot` and `gloss`. The `base` module contains most of the essential libraries for the simplest data types. The version specification means "version 4.7 or newer, but the major version must be less than 5." The "default-language" specification tells us that we're using the 2010 version of the Haskell language specification, which is the current version. The previous version was Haskell98, which gives you a hint of how much time elapses between major revisions to the language.

There is one executable stanza for each stand-alone program we want Stack to build for us. Two are listed here, but you can have as many as you like. The first executable stanza describes the stand-alone program for the source code file *VisTwoSprings.hs* in the subdirectory called *app*. The executable program will be called *LPFP-VisTwoSprings* and will be available globally to run from any directory. The packages required by this stand-alone program are listed as well.

At the time of writing, the `diagrams` packages are not included in the curated list of packages Stack uses by default, so we must list some extra packages in a file called *stack.yaml*. Listing A-2 shows this file.

```
resolver: lts-18.21

packages:
- .

extra-deps:
- diagrams-cairo-1.4.1.1
- diagrams-lib-1.4.4
- active-0.2.0.15
- cairo-0.13.8.1
- diagrams-core-1.5.0
- dual-tree-0.2.3.0
- monoid-extras-0.6.1
- pango-0.13.8.1
```

```
- statestack-0.3
- glib-0.13.8.1
- gtk2hs-buildtools-0.13.8.2
```

Listing A-2: The file stack.yaml *describing the extra dependencies the modules in this book need*

For each version of the compiler, Stack supports a collection of curated packages that are known to build with that compiler and to be generally compatible with each other. A compiler and package set is specified by a version number. In Listing A-2, `lts-18.21` in the `resolver` field means "GHC 8.10.7 and packages compatible with it." This particular compiler/package collection has long-term support (the `lts-` prefix). This means you can count on it being around for awhile, typically a few years.

If you need to live on the bleeding edge to get the features you need, there are snapshot collections and, for the very latest, the nightly build.

The next field, `packages` refers to packages that *you* have written, typically libraries useful for your own project. In Listing A-2, the packages are simply files in the current directory, or "`.`" in Unix-speak.

The `extra-deps` are additional packages your application depends on that are not part of the curated set specified by the `resolver` field. (There is in fact not much difference between a `package` and an `extra-dep` except that it is possible to write test and benchmark targets for our own packages—very important parts of a large application—and these are not available for `extra-deps`.)

Your first stop for questions about the *stack.yaml* file should be *https://docs.haskellstack.org/en/stable/README*.

You can see the packages `diagrams-core`, `diagrams-lib`, and `diagrams-cairo` we are interested in. The remaining packages are packages `diagrams` depends on. Specific versions of these packages are listed. By the time this book appears in print, newer versions of these packages may be available.

To build the executable programs, issue the following command in the main working directory (the directory in which the *stack.yaml* and *LPFP.cabal* files live):

```
$ stack install
```

To start a GHCi session in which all of the book modules are automatically loaded, you can issue the following command:

```
$ stack ghci
```

With this method, we can load any or all of the book modules into GHCi. To remove a module, you can use GHCi's `:m` command with the module prefixed by a minus sign. To remove the `Newton2` module, type the following:

```
ghci> :m -Newton2
```

Similarly, to add an additional module, use a plus sign prefix. To add the `Graphics.Gnuplot.Simple` module, type the following:

```
ghci> :m +Graphics.Gnuplot.Simple
```

Issuing `stack ghci` will also give you the option of loading one of the executable programs into GHCi if you wish.

Summary

This appendix described how to install the Haskell compiler and a text editor, and it went over methods for installing additional library packages using Cabal and Stack. It also showed different ways to organize libraries and source code files for building projects in Haskell.

BIBLIOGRAPHY

[1] Miran Lipovaca. *Learn You a Haskell for Great Good!: A Beginner's Guide*. No Starch Press, 2011.

[2] Bryan O'Sullivan, John Goerzen, and Don Stewart. *Real World Haskell: Code You Can Believe In*. O'Reilly Media, 2008.

[3] Harold Abelson, Gerald Jay Sussman, and Julie Sussman. *Structure and Interpretation of Computer Programs*. MIT Press, 2nd edition, 1996.

[4] Daniel P. Friedman, Matthias Felleisen, and Duane Bibby. *The Little Schemer*. MIT Press, 4th edition, 1996.

[5] David Goldberg. What every computer scientist should know about floating-point arithmetic. *ACM Comput. Surv.*, 23(1): 5–48, March 1991.

[6] Kip S. Thorne and Roger D. Blandford. *Modern Classical Physics: Optics, Fluids, Plasmas, Elasticity, Relativity, and Statistical Physics*. Princeton University Press, 2017.

[7] Conal Elliott. The vector-space package. *https://hackage.haskell.org/package/vector-space*, 2008–2019.

[8] Chris Doran and Anthony Lasenby. *Geometric Algebra for Physicists*. Cambridge University Press, 2007.

[9] David Hestenes. *Space-Time Algebra*. Springer International Publishing, 2015.

[10] Charles W. Misner, Kip S. Thorne, John A. Wheeler, and David I. Kaiser. *Gravitation*. Princeton University Press, 2017.

[11] Gerald J. Sussman, Jack Wisdom, and Will Farr. *Functional Differential Geometry*. Functional Differential Geometry. MIT Press, 2013.

[12] Wolfgang Rindler. *Essential Relativity: Special, General, and Cosmological*. Springer, 2013.

[13] Sean M. Carroll. *Spacetime and Geometry: An Introduction to General Relativity*. Cambridge University Press, 2019.

[14] Bernard Schutz. *A First Course in General Relativity*. Cambridge University Press, 2009.

[15] Isaac Newton, I. Bernard Cohen, Anne Whitman, and Julia Budenz. *The Principia: Mathematical Principles of Natural Philosophy*. University of California Press, 1999.

[16] Richard C. Dorf and Robert H. Bishop. *Modern Control Systems*. Pearson, 2017.

[17] Ken Dutton, Steve Thompson, and Bill Barraclough. *The Art of Control Engineering*. Pearson Education. Addison Wesley, 1997.

[18] Umran S. Inan and Robert A. Marshall. *Numerical Electromagnetics: The FDTD Method*. Cambridge University Press, 2011.

[19] David J. Griffiths. *Introduction to Electrodynamics*. Cambridge University Press, 2017.

[20] Gerald J. Sussman and Jack Wisdom. *Structure and Interpretation of Classical Mechanics*. MIT Press, 2015.

INDEX

B

C

D

E

F

N

Q

R

S

T

U

V

RESOURCES

Visit *https://nostarch.com/learn-physics-functional-programming* for errata and more information.

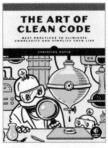

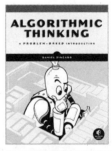